Netherlands-Wetlands

Developments in Hydrobiology 88

Series editor

H. J. Dumont

Netherlands-Wetlands

Proceedings of a Symposium held in Arnhem, The Netherlands, December 1989

Edited by

E.P.H. Best & J.P. Bakker

Reprinted from Hydrobiologia, vol. 265 (1993)

Springer Science+Business Media, B.V.

Library of Congress Cataloging-in-Publication Data

A C.I.P. Catalogue record for this book is available from the Library of Congress.

ISBN 978-94-010-4904-7 ISBN 978-94-011-2042-5 (eBook)
DOI 10.1007/978-94-011-2042-5

Printed on acid-free paper

Contents

ECOLOGY, CONSERVATION, MANAGEMENT, RESEARCH AND POLICY

Hydrobiologia **265**: vii, 1993.
E. P. H. Best & J. P. Bakker (eds), Netherlands-Wetlands.

Preface

The Netherlands is a country rich in water. The variety in water bodies is enormous taking the relatively small size of the country into consideration. Most land is situated below sea level, and would be inundated without the management measures required to keep it dry. Three large rivers, the rivers Rhine, Meuse and Scheldt, cross the country and form extensive estuaries in the coastal areas. Former sea arms have been closed partly or completely, with consequences for water quality and biotic communities. Because the country is so rich in water bodies it also hosts many wetlands.

Scientific and public awareness has been growing during the last decade that the original wetland character of The Netherlands is unique. Therefore, the Netherlands Ecological Society ('Oecologische Kring') and the Netherlands Society for Aquatic Ecology ('Hydrobiologische Vereniging') jointly organized a symposium on 'The Netherlands Wetland', December 1989, in Arnhem (The Netherlands).

The symposium was meant to (1) summarize the existing knowledge on structure and function of the wetland types characteristic for The Netherlands, (2) identify gaps in knowledge, and (3) indicate new avenues for ecological research.

This book is a compilation of most of the seminars presented during the symposium. The introduction treats not only The Netherlands as wetland in a larger landscape but also the various wetland types within the country itself. The subsequent case studies concern representative wetland types. Introduction and case studies, both give an historical, hydrological, physico-chemical and biotic characterization of the wetland concerned, give a description of management measures in the past and in the present, summarizes the threats of the systems and indicate perspectives for the ecosystems given the present environmental conditions and the expectations for short- and long-term changes therein. In the introduction special attention is paid to the border-crossing biotic elements like birds. These sections are followed by a summary of the ecology of the Netherlands wetlands and indications for future ecological research. Finally, environmental policy and its consequences for management on a national, regional and local scale is discussed.

Wageningen/Haren

ELLY P. H. BEST
J. P. BAKKER

Hydrobiologia **265**: 1–14, 1993.
E. P. H. Best & J. P. Bakker (eds), Netherlands-Wetlands.

Netherlands-Wetlands

W.J. Wolff
DLO-Institute for Forestry and Nature Research, P.O. Box 46, 3956 ZR Leersum, The Netherlands and Dept. of Nature Conservation, Agricultural University, Wageningen, The Netherlands

Key words: wetlands, The Netherlands, wetland history, wetland policy

Abstract

The Netherlands occupy 41,864 km^2 of which about 6,600 km^2 (16%) have been classified as internationally important wetlands. About 3000 km^2 (7%) have been included in the Ramsar Convention list of internationally important wetlands.

The situation of the wetlands in The Netherlands has seen large changes during the last 2000 years. Thousands of km^2 of peatlands have disappeared; thousands of km^2 of coastal salt marshes and a similar area of shallow lakes have been reclaimed for agriculture.

The Dutch wetlands occupy a key position in the bird migration of the West-Palaearctic flyway.

The hydrology of The Netherlands and of Dutch wetlands is dominated by the river Rhine. Hence, nutrients and contaminants imported by the Rhine and from other sources have a major impact on Dutch wetlands. Reduction of pollution and eutrophication is a major requirement for conservation of Dutch wetlands. Dutch politics appears to be moving in this direction.

Introduction

Literally translated Nederland ('The Netherlands') means the Low Country. Hence, it is not astonishing that The Netherlands always have been and still are a very wet country. This is illustrated by many topographical names, such as 'Waterland' (water land), 'Zeeland' (sea land), and the many toponyms containing the terms '-dam' or 'polder' (drained land).

However, although every type of wetland in The Netherlands can be named with a specific Dutch word, there is no general term in the Dutch language denoting wetlands. What is wetland elsewhere, is land to the Dutch. The word 'wetland' was introduced into the Dutch language especially because of the participation of The Netherlands in the so-called Ramsar Convention. This 'Convention on Wetlands of International Importance especially as Waterfowl Habitat' was formulated in Ramsar, Iran, in 1971. It entered into force in 1975 when seven countries had acceded the treaty. The Netherlands ratified this convention in 1980 and it is in this period between 1971 and 1980 that the word 'wetlands' became part of the Dutch language, especially the jargon spoken by conservationists and politicians.

The Ramsar Convention aims at the conservation and wise use of wetlands and of the populations of waterbirds occurring in these wetlands. Although in the early years the Convention put much emphasis on waterbirds, the Ramsar Convention is essentially not a bird treaty, but a wetlands treaty. Under this treaty the government of The Netherlands has prepared an official list of wetlands of international importance (Rooth, 1985, 1989). Of this Dutch list enumerating 103 wetlands with a total area of 6,600 km^2 10 wetlands with a total area of about 3,000 km^2 have been designated for inclusion in the international

List of Wetlands of International Importance referred to in Article 2 of the Convention (as per 1 January 1990). This means that about 16% of the territory of The Netherlands is considered to consist of internationally important wetlands and that about 7% of its territory actually has been registrated as an internationally important wetland. However, The Netherlands are wetter than that!

Definitions of wetlands

In general terms, wetlands are areas where saturation with water is the dominant factor determining the nature of soil development and the types of plants and animals living in the soil or on its surface. Wetlands are transitional systems between land and water where the water table is usually at or near the surface or the land is covered by shallow water. According to Cowardin *et al.* (1979) wetlands have one or more of the following attributes: (1) at least periodically, the bottom supports predominantly hydrophytes (water plants); (2) the substrate is predominantly undrained hydric soil and/or (3) the substrate is nonsoil and is saturated with water or covered by shallow water at some time during the growing season of each year.

The 'Convention on Wetlands of International Importance as Waterfowl Habitat' defines wetlands as 'areas of marsh, fen, peatland or water, whether natural or artificial, permanent or temporary, with water that is static or flowing, fresh, brackish or salt, including areas of marine water the depth of which at low tide does not exceed six metres'.

Several authors (e.g. Shaw & Fredine, 1956; Miller, 1976; Cowardin *et al.*, 1979 and Frayer *et al.*, 1983), however, set the lower limit of wetlands at the extreme low water of spring tides or, in non-tidal waters, at a depth of 2 or 3 metres, because this is the maximum depth at which emergent plants normally grow.

According to the Ramsar Convention islands situated within wetlands and channels deeper than six meters situated within coastal wetlands should be considered to form part of these wetlands.

Part of the Dutch wetlands qualify as such under all definitions. However, about half of The Netherlands consists of agricultural land lying below sea level. Originally this land used to be very wet, and hence wetland, but due to centuries of civil engineering (Van Veen, 1962) the water table is now more or less fixed below the surface of the soil. In clayey and sandy arable areas this water table may occur down to 1–2 metres, but in peaty grassland areas the water table may be only 10–20 cm below the surface. The latter type of agricultural areas still may be considered as wetlands, because of the high water table and the organisms occurring. Figure 1 shows the distribution of the different types of wetlands in The Netherlands at present.

Some Dutch words denoting specific types of wetlands have become incorporated in the international scientific literature. Such words are:

- *wad* (singular, plural: wadden) denoting unvegetated tidal flats in the northern (Friesland, Groningen) part of The Netherlands; the plural form is well known in the international scientific literature on coastal areas;
- *slik* denoting unvegetated (muddy) tidal flats in the southern (Zeeland) part of The Netherlands and the NW (Flanders) part of Belgium; due to Belgian scientific studies 'slikke' is now widely used in the French scientific literature;
- *schor* meaning salt marsh in the southern part of the country; again due to Belgian studies 'schorre' is often used in French scientific publications.

Another Dutch word which has acquired some status outside The Netherlands is *polder*, originating from the SW-Netherlands and NW-Belgium (Dekker, 1971) and originally denoting an embanked and reclaimed salt marsh. Along the Wadden Sea, especially in Germany and Denmark, the Frisian word *koog* has obtained the same meaning; it is still used in Germany ('Koog') and Denmark ('kog'). Nowadays 'polder' in The Netherlands also denotes an area with an independent hydrology regulated by man.

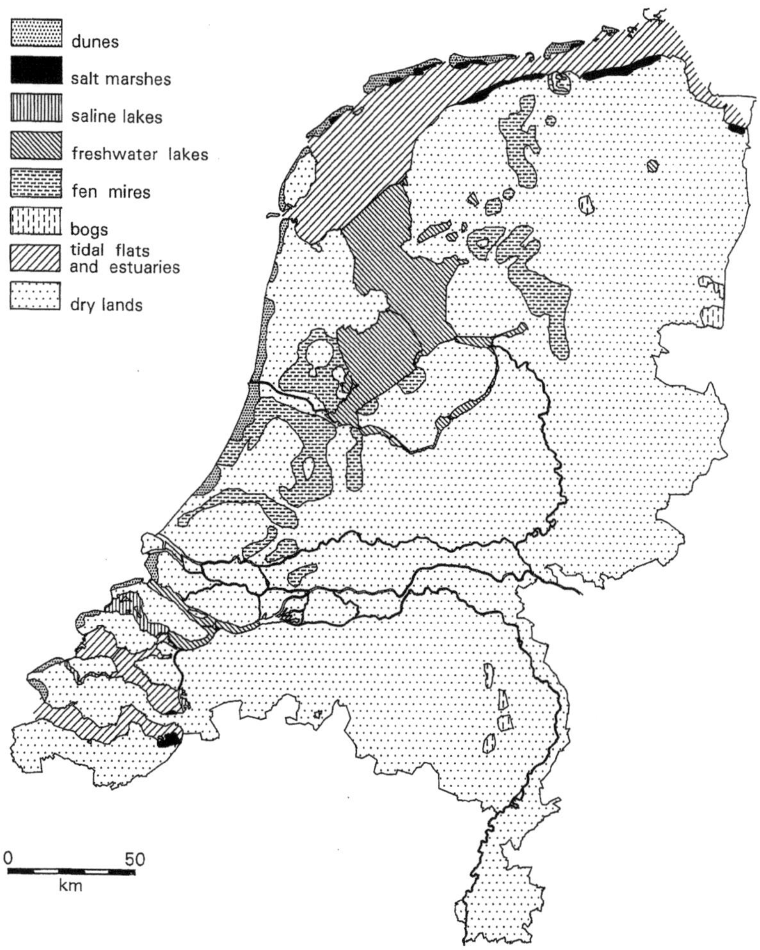

Fig. 1. Present distribution of wetlands in The Netherlands.

Geological and historical development of wetlands in The Netherlands

Geology

Quaternary geology has prospered in The Netherlands during the last decades and has resulted in several palaeographic reconstructions of the topography of the country (e.g. Pons & Wiggers, 1959, 1960; Pons & Van Oosten, 1974; Van Staalduinen *et al.*, 1979). Zagwijn (1986) summarizes all available evidence in an interesting series of maps. Figures 2 and 3 are derived from his maps and show the distribution of wetlands in The Netherlands at about 100 A.D., i.e. during the Roman occupation, and around 1200.

The geological development may be summarized as follows (Zagwijn, 1986). After the last Ice Age sea level in the North Sea rose rapidly. When the rate of sea level rise declined at about 6000 B.P. a system of barrier beaches and islands developed, situated approximately along the present western and northern coastlines. With a sufficient supply of sediment the tidal basins behind the barrier coast filled in until the tidal inlets could be closed by wave action, especially in the areas with smaller tidal amplitudes. In the former tidal basins marsh vegetation developed which gradually changed into ombrotrophic raised bogs. At larger tidal amplitudes and/or in areas with a shorter supply of sediment, viz. the Wadden Sea area and the south-western part of The Netherlands, the

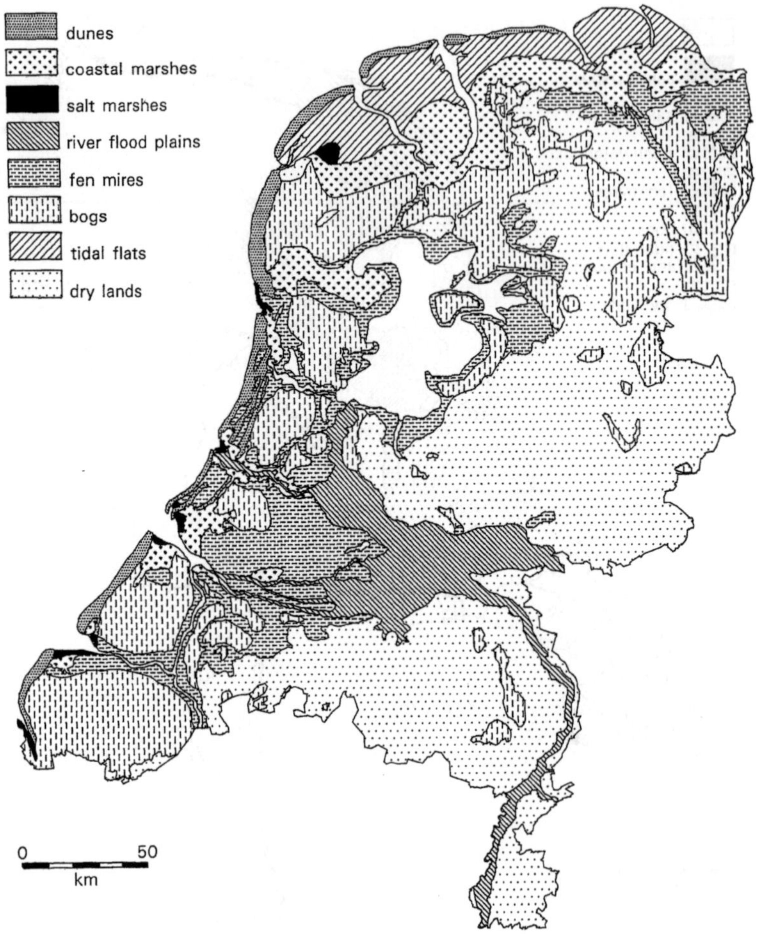

Fig. 2. Distribution of major wetland types in The Netherlands in 100 A.D. After Zagwijn (1986).

coastline did not close and estuarine conditions persisted. Here extensive areas of salt and brackish marshes developed which merged into a belt of fresh marshes and peat bogs at the landward side. In the same period ombrotrophic bogs also developed in depressions inland. Along the distributaries of the rivers Rhine and Meuse extensive floodplains with swamp forests and freshwater tidal areas developed. Later the North Sea started to erode the coastal area again and thus caused increased drainage of the coastal moors. In large areas peat growth came to a standstill. Figure 2 shows a stage in this development (100 A.D.), which has not yet or virtually not been influenced by human activities.

Middle ages

The end of the Roman period in The Netherlands coincided more or less with or even was partly caused by an increased influence of the sea in the coastal area. Human habitation disappeared from large parts of the coastal zone and seawater apparently penetrated much further inland. Only in the northern part of the country habitation of the coastal marshes continued on artificial dwelling mounds ('wierden' or 'terpen'). Elsewhere man reappeared along the coast several centuries later and settled first in the extensive areas of salt and brackish marshes, often on levees and on dune ridges and other sandy outcrops. The new colonists practised grazing and agriculture on alluvial

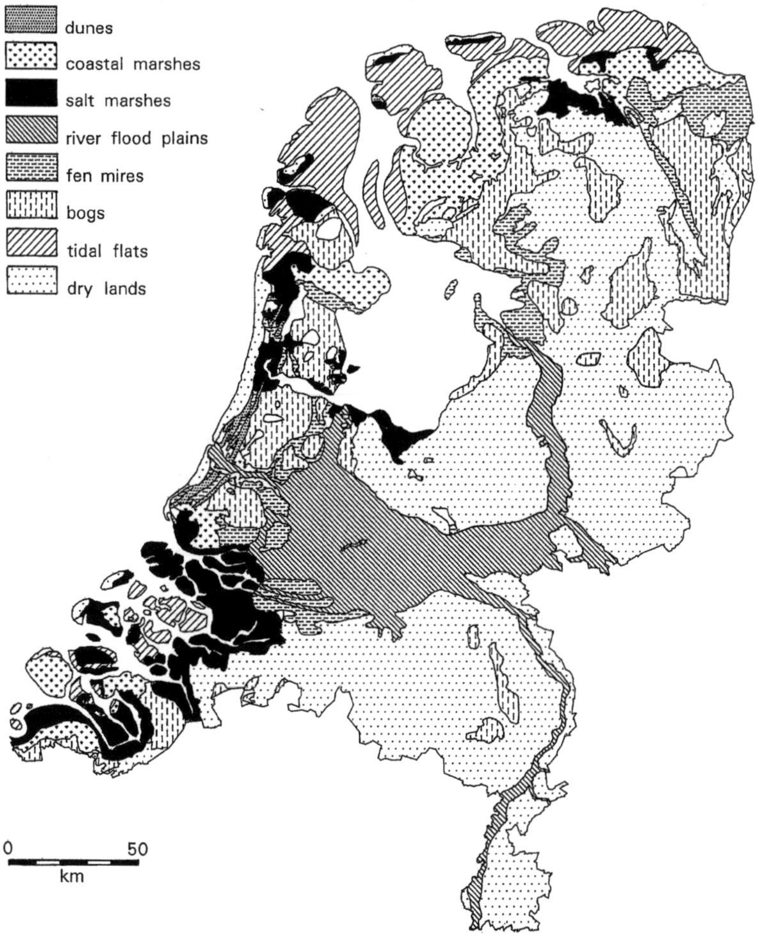

Fig. 3. Distribution of major wetland types in The Netherlands in 1200 A.D. After Zagwijn (1986).

soils. With an increasing population peat and other wetland soils also were cultivated.

Two different types of interventions in the landscape which have their roots in that period, proved to be of major importance for the development of The Netherlands: drainage and embankment.

Drainage by digging ditches was necessary to create sufficiently dry conditions for agriculture, especially for cultivation of cereals. Thus large areas of peat and other wetlands were reclaimed (Edelman, 1974). Hutchinson (1980) documents the effects on peat areas in a recent example. Within 40 years of the start of drainage an area of peatland in East Anglia showed a subsidence of about 3 m due to shrinkage and peat wastage. These processes no doubt will have occurred in the Dutch coastal area in the Middle Ages as well. Most probably the process was speeded up by deliberate burning of the top layer of the peat to provide extra nutrients for agriculture. In combination with a mediaeval population explosion the peat agriculture led to the complete disappearance of the coastal ombrotrophic peat ecosystems.

As soon as subsidence of the land surface started the land became more accessible for sea and river water. In response the population started to construct dams and dikes.

Thus in the 10th–12th centuries the larger part of the inhabited Dutch territory became protected by dikes against extreme high tides and river floods. In connection with embankment gravity drainage had to be maintained or installed at the same time to remove the excess of precipitation. Thus in the Middle Ages the Dutch population had mastered two technological innovations: drainage and embankment, which in the short term were beneficial to agriculture, but which after some centuries proved to be disastrous for the landscape. At that time another intervention in the landscape, viz. the digging of peat for fuel, still was relatively unimportant in view of the vast areas of peat and the small population (Borger, 1975). More important at that time may have been the extraction of peat from areas inundated by seawater to produce salt, a process which may have been practised from Roman times up to about the 16th century. This type of salt production, however, was feasible only around the estuaries. It ended when cheaper salt could be imported from Mediterranean countries.

Within the dikes the land surface subsided often by several metres. Outside the dikes the rivers continued to deposit sand and mud in their winter bed and thus raised its level. Along the rivers and estuaries the vast areas which could be flooded during storm surges and very high river discharges disappeared due to embankments, causing higher water levels during such events. Also eustatic rise of sea level played a part. Hence, both along the rivers and the coast the level of the land sank relative to that of the water (Fig. 4). Problems proved to be unavoidable, especially along the coast. During high water levels dike bursts became a frequently occurring event; their traces

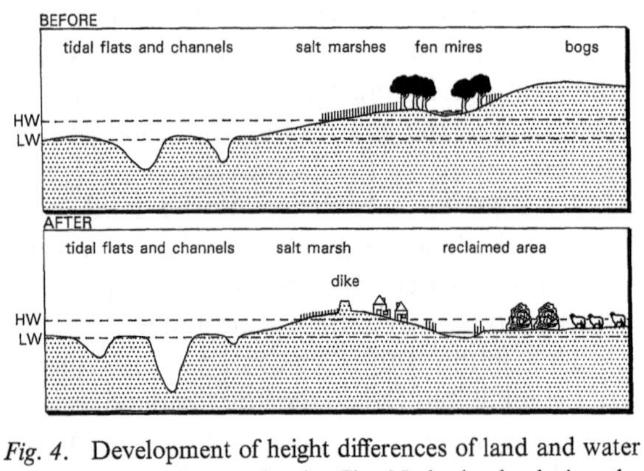

Fig. 4. Development of height differences of land and water along rivers and estuaries in The Netherlands during the Middle Ages.

can still be observed at hundreds of places. As a results of dike bursts often large areas were flooded. Along the rivers the damage normally could be repaired in periods of low river discharge. But along the coast, because of the size and low level of the basins created by embankment, drainage and subsequent subsidence, the tidal volumes of such basins, once inundated, were too large to be mastered by mediaeval technology. Thus vast areas of land became permanently inundated during the late Middle Ages. After inundation, however, accretion and salt marsh development started along the margins of the inundated areas, thus enabling a gradual reclamation of at least part of the land lost. Nowadays the remnants of several of these drowned areas form highly valued wetland areas. Examples are the present Dollard estuary (14th–15th century; at that time about 150 km^2 inundated), the Biesbosch freshwater tidal area (1421; about 300 km^2), the Reimerswaal tidal flats in the Oosterschelde estuary (1530; about 100 km^2) and the saltmarsh area of Saaftinge (1584–85 through military inundations; at that time about 150 km^2). Elsewhere water courses and small lakes developed in the peat landscape through long-lasting inundations and afterwards increased in size through wave erosion of the shores. Thus lakes like the Beemster, the Purmer and the Haarlemmermeer developed. As a result, for instance, over 50% of the land area in the present Dutch province of Noord-Holland changed into lakes or disappeared into the sea between the 10th (Fig. 5) and the 14th century (Fig. 6). The situation did not improve until the 17th century.

The 17th century

In the 17th century the newly gained independence of The Netherlands stimulated the economic and technological development of the country. Capital earned by trade all over the world was used for embankment and land reclamation in The Netherlands. New inventions, such as wind-mills, provided the technology. Thus most of the large lakes north of Amsterdam were

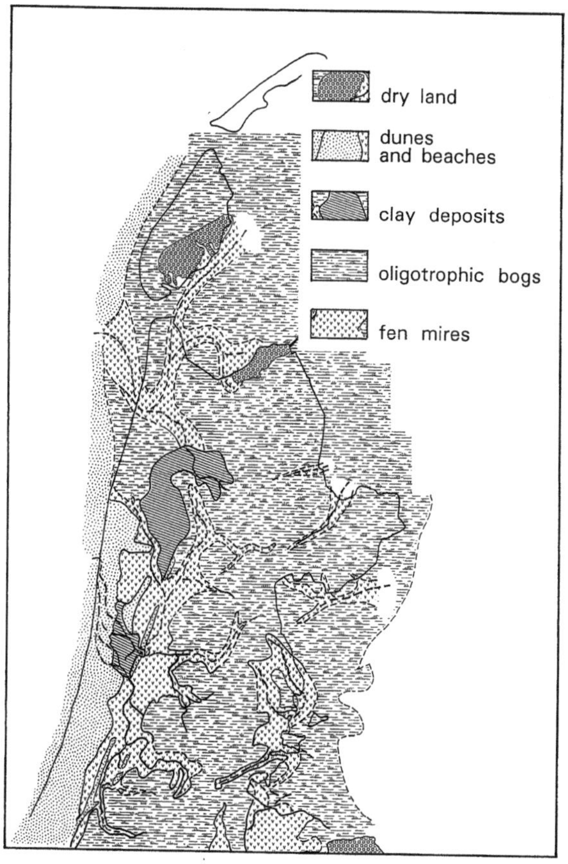

Fig. 5. Distribution of land and water in the Dutch province of Noord-Holland in the 10th century. After Eisma & Wolff (1980).

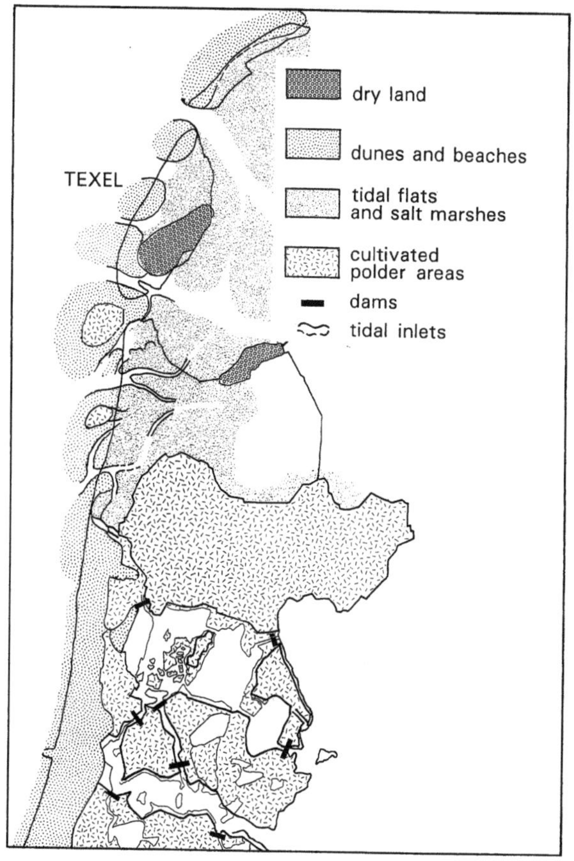

Fig. 6. Distribution of land and water in the Dutch province of Noord-Holland in the 14th century. After Eisma & Wolff (1980).

pumped dry between 1612 and 1635. Also elsewhere many shallow fresh and brackish-water lakes were reclaimed. Altogether nearly 300 km^2 of lakes were pumped dry in the first half of the 17th century. In the same period about 400 km^2 of salt marshes were embanked and reclaimed for agriculture (Van Veen, 1962).

At the same time however the booming Netherlands economy of the 17th century needed energy. Next to windmills peat proved to be a very important source. In this period the earlier small-scale peat extraction for fuel, which had been slowly expanding since the 12th century, reached industrial size. From the second half of the 16th century large areas of high-lying ombrotophic bogs, situated especially in the northern and eastern part of The Netherlands, were dug away until the mineral subsoil was reached. This process peaked in the 19th century and it has continued until well into the 20th century (Stol, 1986). The result is that of the 1800 km^2 of ombrotrophic bogs still occurring at the start of the 17th century only about 36 km^2 still occurs in The Netherlands today. Of this only a small part consists of growing *Sphagnum* bogs (Neijenhuis, 1973). Thus peat extraction led to the virtual disappearance of a characteristic type of wetland in The Netherlands. Peat was extracted also in areas where it occurred below the ground water table. Originally many of these areas were used for agriculture, particularly as grazing land. However, the invention of a special dredge ('baggerbeugel') in the 16th century made exploitation of this peat feasible (Stol, 1986). Here the result of peat extraction has been the creation of large areas of fens and broads as well as several shallow lakes. Altogether at least

600 km^2 of wetlands have been created in this way, although mainly in the province of Zuid-Holland since the 19th century some 300 km^2 of these shallow lakes have been pumped dry again and reclaimed for agriculture. In the 20th century this form of peat extraction has stopped because coal, oil, and gas provide cheaper alternatives as energy sources. There exist, however, still large areas where wetlands could be created by peat extraction.

The 20th century

In the 20th century new technologies based on coal and oil as energy sources made wetland transformation at a very large scale feasible. The scene was set by the damming of the Zuiderzee. This brackish and shallow inlet of the North Sea, covering about 3700 km^2, was dammed in 1932 and changed to a stagnant freshwater lake. Since then about 1680 km^2 of this lake have been pumped dry and reclaimed for agriculture. New wetlands have been developed in these new polders. At present some 65 km^2 of freshwater marshes occur about 4 m below mean sea (and lake) level. Also other estuarine areas have been reclaimed and changed into stagnant lakes: 1950 – Brielse-Maas estuary – 12 km^2; 1952 – Braakman estuary – 15 km^2; 1961 – Veerse-Gat estuary – 40 km^2; 1969 – Lauwerszee estuary – 90 km^2; 1970 – Haringvliet-Hollands-Diep-Biesbosch estuary – 235 km^2; 1971 – Grevelingen estuary – 140 km^2. The closure of the Oosterschelde estuary (482 km^2) was due for 1978, but because of wide-spread concern the plans have been changed. Only about 100 km^2 have been changed into freshwater lakes and the remainder of the estuary has been provided with a storm surge barrier which slightly reduced the tidal amplitude. Another recent development is the disappearance of winter inundations. Until the first half of the 20th century thousands of km^2 of low-lying meadows were inundated during many winters due to insufficient functioning of the drainage system. Nowadays a well-designed drainage system and a large pumping capacity keep these meadows dry all around the year. But new changes are looming: do the economic returns from these areas warrant the costs of the infrastructure and the pumping? If not, other uses of these areas have to be looked for and the creation of new wetlands is one of the alternatives considered seriously (Nature Policy Plan, 1989). The picture shows an ever increasing impact of man on the Dutch landscape. Van Veen (1962) gives some interesting figures about this impact (Fig. 7). Al-

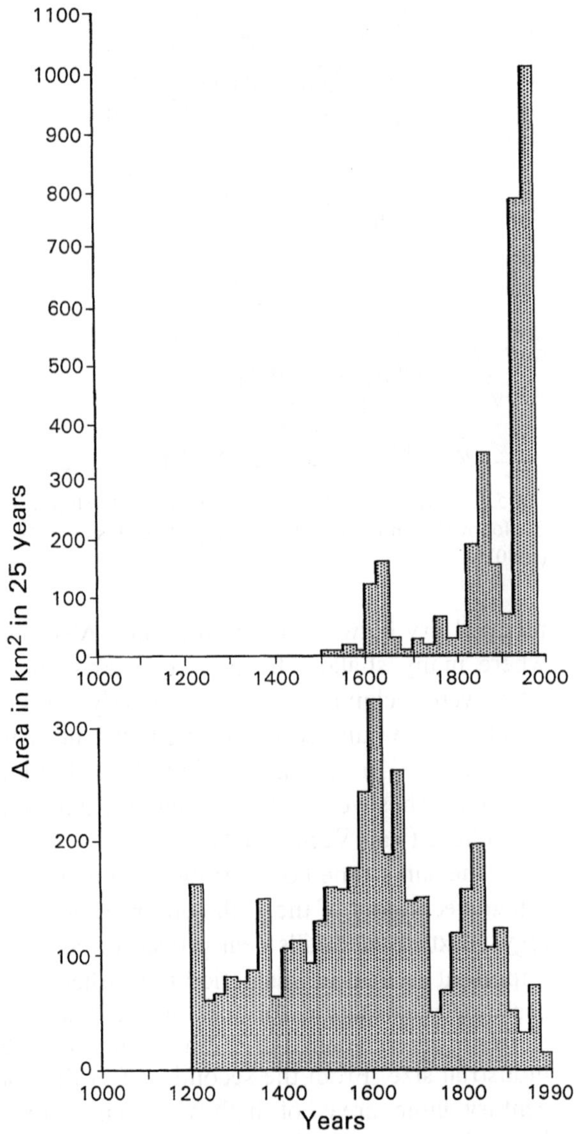

Fig. 7. Area of reclaimed coastal areas (bottom) and freshwater lakes (top) since 1200 A.D. Modified after Van Veen (1962).

together some 4000 km^2 of saltmarshes have been reclaimed on the shore after the original mediaeval large-scale embankments, whereas about 1400 km^2 of lakes have been pumped dry before another 1680 km^2 were pumped dry in the former Zuiderzee in this century. To achieve this some 10,000,000,000 m^3 of peat, sand, and clay were dug by spade and transported by simple means as wheelbarrows, horse carts, and sailing vessels before more modern equipment moved even larger quantities from the 19th century on.

The hydrology of The Netherlands

Table 1 shows the freshwater balance of The Netherlands for an average year and a very dry year as 1976 (Colenbrander, 1986). The paramount importance of the rivers Rhine and Meuse for the freshwater budget of the country is very clear. Originally the rivers influenced only a limited part of the country since they ran largely directly to the sea. Nowadays, however, the river water is extracted from the rivers to supply a very large part of the country with fresh water. This water is used for irrigation purposes, for countering intrusion of salt and brackish water and for dilution of pollution. It serves also to replenish freshwater in those parts of the country where the present very efficient drainage system causes water shortages in the growing season of crops but also of forests and nature areas.

The major part (80–90%) of the fresh water reaches the North Sea west and south-west of Rotterdam through the Rotterdam Waterway and the Haringvliet sluices. Another 10% is discharged into the Wadden Sea through the freshwater lake IJsselmeer, whereas small quantities reach the North Sea all along the Dutch coast. Two other rivers, viz. the Scheldt and the Ems, discharge also in Dutch coastal waters (about 5000 × 10^6 m^3 per year), but do not contribute to the fresh water budget of The Netherlands since the mixing with sea water starts already at the other side of the Dutch border.

Dutch coastal waters and especially the tidal wetland of the Wadden Sea are heavily influenced by the discharge of the Rhine (De Groot, 1963; Van Bennekom *et al.*, 1975; Wolff, 1983). This causes a lowered salinity in these waters and affects the sediment and nutrient budgets. Nowadays, however, the major influence is through pollution of the river water.

Table 1. The water balance of The Netherlands for an average year and a very dry year like 1976 (in mm and in 10^6 m^3). After Colenbrander (1986).

	Average year		Dry year 1976	
	mm	10^6 m^3	mm	10^6 m^3
In				
Precipitation	775	30100	535	20800
Rhine (at the border)	1775	69000	1065	41500
Meuse (at the border)	215	8400	90	3500
Other rivers	75	3000	40	1500
	2840	110500	1730	67300
Out				
Evapotranpiration	501	19500	528	20500
Different uses	129	5000	154	6000
To North Sea	2210	86000	1048	40800
	2840	110500	1730	67300

Sedimentation characteristics of Dutch wetlands

The major rivers Rhine and Meuse transport coarse sand and pebbles into The Netherlands from the German and Belgium parts of their catchments, respectively. Within The Netherlands the transporting power of the rivers decreases due to the decrease of the fall and the influence of the tides (Terwindt *et al.*, 1963). This means that only the finer fractions are still transported seaward. Finally, all of the sand and large quantities of silt and clay settle in the lakes which have been created by damming the Zuiderzee and the Haringvliet estuary, respectively. This leads to huge accumulations of heavily polluted sediment in the freshwater lakes Ketelmeer and Haringvliet and adjacent waters. The Third Water Policy Plan (1989) estimates the amount of heavily polluted sediment thus deposited at 22 $\times$ 10^6 m^3 so far. As a result only the finest fractions of silt and clay, which also escape from the estuarine turbidity zone, reach the North Sea suspended in the river water. This suspended material is transported further by the residual current system in the North Sea which runs from south to north along the Dutch coast and thus arrives in front of the tidal inlets of the Wadden Sea and eventually in the Wadden Sea itself (De Groot, 1963; Van Bennekom *et al.*, 1975). Turbulent mixing processes induced by the tidal currents cause also a transport of suspended matter, part of which is directed southward. Thus river mud is distributed along a large part of the Dutch coast.

Van Straaten & Kuenen (1957), Postma (1961), Groen (1967) and Dronkers (1984) have shown that suspended material occurring in the coastal zone of the North Sea is subject to an inshore transport due to an interaction of tidal transport, gravity, sedimentation, and erosion. This results in sedimentation of clay and silt in quiet places of the Wadden Sea and the estuaries. The harbour basins, notably those of Rotterdam, also happen to be such quiet places, necessitating the annual removal of some 20 * 106 m3 of dredge spoil.

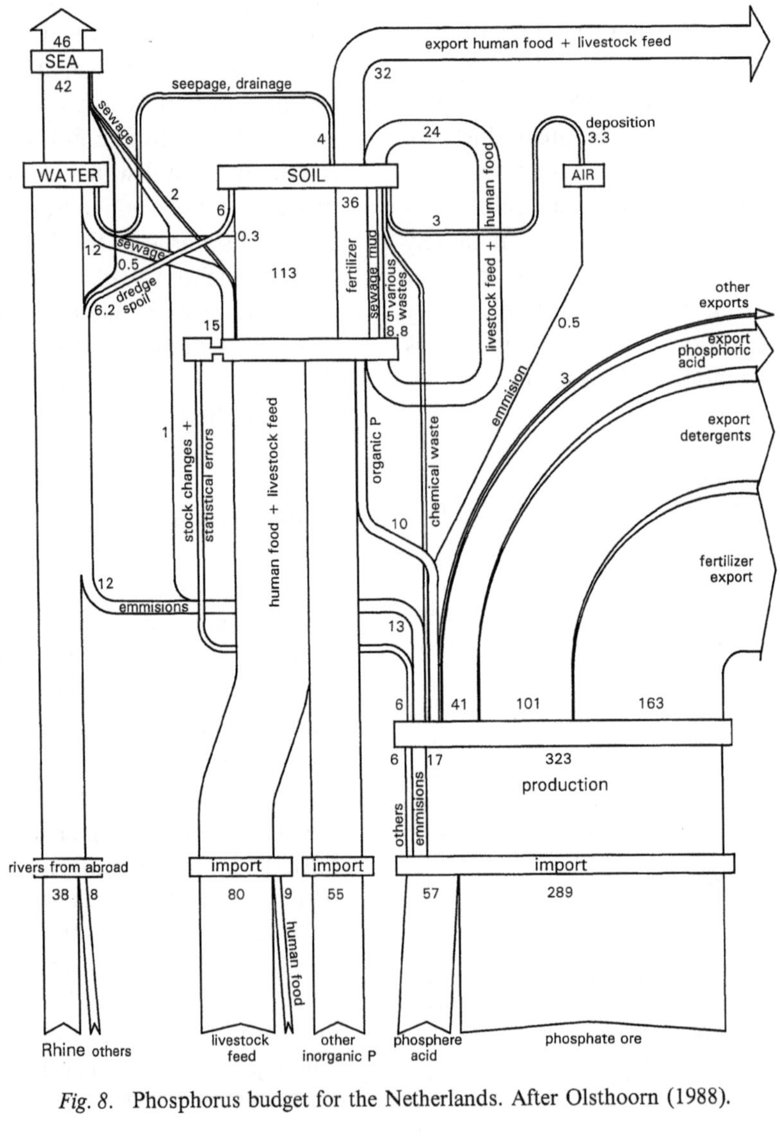

Fig. 8. Phosphorus budget for the Netherlands. After Olsthoorn (1988).

Most of this is dumped offshore in the North Sea from where it probably is transported northward with the residual current and thus arrives in the Wadden Sea as well.

Sand behaves differently since it is transported mainly along the bottom. River sand does not reach the North Sea, nor does the sand at the North Sea floor show much movement. However, an extensive transport of sand occurs in the coastal zone, usually with a net transport into the estuaries and the tidal basins of the Wadden Sea (De Jonge *et al.*, 1993).

Nutrient budgets of The Netherlands

Olsthoorn (1988, 1989) presents comprehensive budgets for phosphorus and nitrogen for The Netherlands as a whole (Figs 8 and 9). Tables 2 and 3, based on these comprehensive budgets, show the budgets for the Dutch fresh surface waters, including most of the wetlands. Important features are the very high loads of phosphorus and nitrogen to Dutch fresh surface waters and especially the high rate of accumulation of phosphorus. Whether nitrogen is accumulating at the

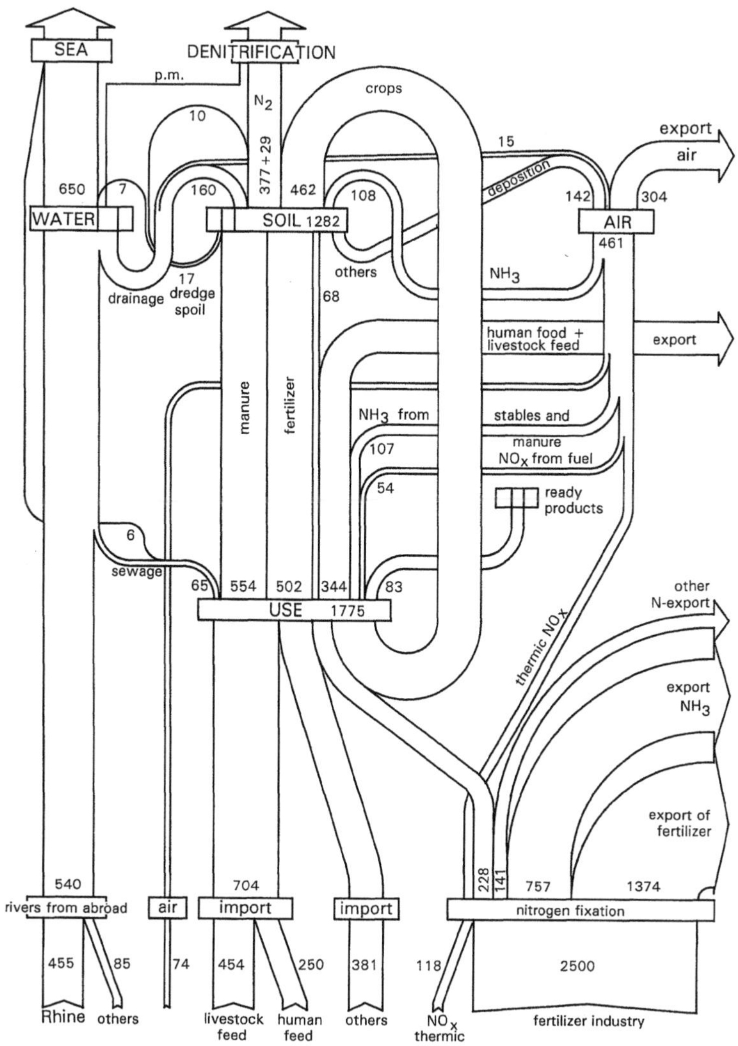

Fig. 9. Nitrogen budget for the Netherlands. After Olsthoorn (1989).

Table 2. Phosphorus budget of Dutch surface fresh waters in 10^6 kg P per year. After Olsthoorn (1988).

In	
Households	19.8
Industry	18.2
Agriculture	2.2–5.2
Precipitation	0.3
Imported by rivers	46
Various	1.5
	88–91
Out	
To North Sea	45
Removal from waste water	9.7
Dredge spoil stored on land	6.2
Various	1.1
Accumulation in aquatic sediments	26–29
	88–91

Table 3. Nitrogen budget of Dutch fresh surface waters in 10^6 kg per year. After Olsthoorn (1989).

In	
From agriculture	125
Waste water	59
Atmospheric deposition	15
Imported by rivers	540
Various	36
	775
Out	
To North Sea	550
Dredge spoil stored on land	8
Various	1
Accumulation or denitrification in sediment	216
	775

same high rate or is mainly denitrified in the aquatic sediments is not completely clear yet.

Also the coastal waters show very high loads of nutrients. Billen *et al.* (1985) present a nitrogen budget for the Westerschelde estuary (Fig. 10) from which may be derived that the present nitrogen load of this estuary is many times the natural load, defined as the load unaffected by human activities. Similarly De Jonge *et al.* (1993) show that the nitrogen and phosphorus loads of the Dutch Wadden Sea have increased many times since about 1950. When we compare the data on

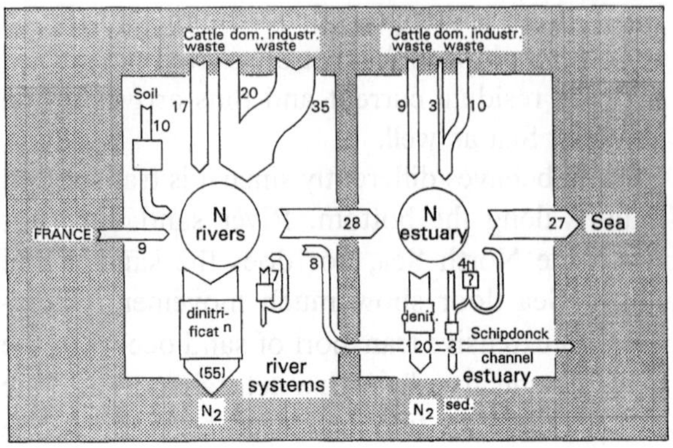

Fig. 10. Nitrogen budget for the Westerschelde. After Billen *et al.* (1985).

the present situation presented by Olsthoorn with the data on the original, natural levels of nutrients in Dutch rivers (Van Bennekom & Salomons, 1980), we are forced to conclude that the present nutrient loading of the Dutch territory is at least tenfold and perhaps even twentyfold the original natural values.

Dutch wetlands in an international context

The Netherlands have an area of about 41,864 km^2 and thus are a very small country. It comprises only 0.3% of the total area of Europe and a minute 0.02% of the land area of the world. Hence, the Dutch share of the wetlands of the world is negligible. Nevertheless, for some types of wetlands The Netherlands have a higher share than follows from the size of the country. For example, 7% of all tidal saltmarshes in Europe occur in The Netherlands (Dijkema, 1989) and even 15% of all European tidal flats are found in this country (Wolff, 1989).

The Netherlands also play a key role for wetland birds (Rooth, 1989a). Major parts of the NW-European populations of Cormorant (50%), Grey heron (30%), Spoonbill (58%), Shoveler (40%), Oystercatcher (41%), Avocet (41%), Black-tailed godwit (64%), Montagu's harrier (77%), Bluethroat (45%), and Sandwich tern (25%) breed in The Netherlands. Over half a million geese (6 species) visit The Netherlands in winter, making this country the principal wintering area along the East Atlantic seaboard. Ducks number well over a million in mild winters. Shorebirds, mainly waders, also number well over a million in autumn, i.e. the peak of the migration season (Smit, 1981).

Within the framework of the Ramsar Convention criteria have been established to determine the international conservation value of wetlands. For example, the occurrence of 1% of the population of a wetland bird species in any time of the year makes a wetland internationally important. According to these Ramsar bird criteria for the definition of international importance of wetlands Dutch wetlands occupy the top rank in Europe on the international lists (Smit, 1981) with the Wadden Sea, the IJsselmeer lake and the Delta estuaries occupying the first three places. This means that no other European wetlands receive the same numbers of birds.

These Dutch wetlands thus play a key part in a bird migration system extending from the arctic tundras of NE-Canada, Greenland, Iceland, Scandinavia, Spitsbergen, Russia, and NW-Siberia to the wetlands south of the Sahara in Mauritania, Senegal, Mali, Guinea-Bissau and even further south (Smit & Piersma, 1989). The tidal wetlands of Wadden Sea and Delta area are also important nurseries for fish and shrimps living in the North Sea (Zijlstra, 1972).

Wetlands and policy in The Netherlands

As stated above The Netherlands are signatory to the Ramsar Convention and, hence, have a wetland policy. Under this policy about 7% of the Dutch territory have been placed on the official list of internationally important wetlands kept by UNESCO and more wetlands may be expected to follow. An obligation imposed by the Convention is 'wise use' of all of the wetlands within Dutch territory. 'Wise use' has been defined formally at the Regina conference of the parties to the Ramsar Convention in 1987, so some remarks seem appropriate.

In the first place wetland reclamations in The Netherlands seem to have become a thing of the past (Wolff, 1992). In 1988 a final decision was reached on the embankment and subsequent reclamation of a salt marsh area in the north of the province of Friesland. After a debate lasting over 20 years it was eventually decided at the highest political level that no embankment will be carried out. With this decision a tradition of 1000 years of coastal reclamations has come to its end, since any further coastal reclamations seem highly unlikely.

Similarly the reclamation of inland wetlands has ended. Recently there was only one plan left: the reclamation of the Markerwaard polder (600 km^2) in the IJsselmeer lake. This project had to be the final part of the reclamation of the former Zuiderzee according to a law passed in 1918. However, in view of the agricultural overproduction in the European Common Market and the increased awareness of wetland values in The Netherlands the costs and benefits of this project are perceived in a different way today. After considerable public and political discussion the project has been abandoned in 1991.

For all other inland wetlands as well as the coastal wetlands the national Nature Policy Plan (1989) aims at preservation of the wetland character. However, this cannot be achieved without a drastic reduction of the nutrient load and the load of contaminants of wetlands (Best *et al.*, 1993). Therefore the policy goals set in the national Environmental Policy Plan and the Third Water Policy Plan are very important for the continued existence of wetland values in The Netherlands. These Policy Plans aim at a very drastic reduction (50–90%) of nutrient and contaminant loads in the territory of The Netherlands within the next 10 years. Whether these goals are politically realistic remains to be seen, but they are certainly urgently required ecologically given the present state of Dutch wetlands.

References

Best, E.P.H., J.T.A. Verhoeven & W.J. Wolff, 1993. The ecology of The Netherlands wetlands: characteristics, threats, prospects and perspectives for ecological research. Hydrobiologia 265: 305–319.

Billen, G., M. Somville, E. de Becker & P. Servais, 1985. Nitrogen budget of the Scheldt hydrographical basin. Neth. J. Sea Res. 19: 223–230.

Borger, G.J., 1975. De Veenhoop. Een historisch-geografisch onderzoek naar het verdwijnen van het veendek in een deel van West-Friesland. Thesis, Amsterdam, 242 pp.

Colenbrander, H.J., 1986. Water in The Netherlands. TNO Committee on Hydrological research, The Hague, 70 pp.
Convention on Wetlands of International Importance especially as Waterfowl Habitat, 1971. Ramsar.
Cowardin, L.M., V. Carter, F.C. Golet & E.T. LaRoe, 1979. Classification of wetlands and deepwater habitats of the United States. U.S. Fish and Wildlife service, Biol. serv. Progr. FWS/OBS-79/31, 103 pp.
De Groot, A.J., 1963. Mangaantoestand van Nederlandse en Duitse holocene sedimenten. Versl. landbouwk. Onderz. 69, 7: 1–164.
De Jonge, V.N., K. Essink & R. Boddeke, 1993. The Dutch Wadden Sea: a changed ecosystem. Hydrobiologia 265: 45–71.
Dekker, C., 1971. Zuid-Beveland. De historische geografie en de instellingen van een Zeeuws eiland in de middeleeuwen. Van Gorcum, Assen, 684 pp.
Dijkema, K.S., 1989. Kwelders en schorren. In: W.J. Wolff (ed.) – De internationale betekenis van de Nederlandse natuur. SDU, 's-Gravenhage: 41.
Dronkers, J., 1984. Import of fine marine sediment in tidal basins. In: Neth. Inst. Sea Res. – Publ. Ser. 10–1984: 83–105.
Edelman, T., 1974. Bijdrage tot de historische geografie van de Nederlandse kuststrook. Rijkswaterstaat Publikatie 14, 82 pp.
Environmental Policy Plan, 1989. Nationaal Milieubeleidsplan. SDU, 's-Gravenhage, 258 pp.
Frayer, W.E., T.J. Monahan, D.C. Bowden & F.A. Graybill, 1983. Status and trends of wetlands and deepwater habitats in the conterminous United States, 1950's to 1970's. U.S. Fish and Wildlife Service, Nat. wetland Inventory, 32 pp.
Groen, P., 1967. On the residual transport of suspended matter by an alternating tidal current. Neth. J. Sea Res. 3: 564–574.
Hutchinson, J.N., 1980. The record of peat wastage in the East Anglian fenlands at Holme Post, 1848–1978 A.D. J. Ecol. 68: 229–250.
Millar, J.B., 1976. Wetland classification in western Canada: a guide to marshes and shallow open water wetlands in the grasslands and parklands of the Prairie Provinces. Can. Wildlife Serv., Spec. Sci. Rep. – Wildlife 20: 14 pp.
Neijenhuis, F., 1973. Hoogveen in Nederland: een verdwijnend landschapstype? Natuur en Landschap 27: 98–126.
Nature Policy Plan, 1989. Natuurbeleidsplan. SDU, 's-Gravenhage, 179 pp.
Olsthoorn, C.S.M., 1988. Fosforbalans van Nederland, 1986. Kwartaalbericht Milieu (CBS) 88/4: 4–9.
Olsthoorn, C.S.M., 1989. Stikstofbalans van Nederland, 1986. Kwartaalbericht Milieu (CBS) 89/4: 35–39.
Pons, L.J. & M.F. van Oosten, 1974. De bodem van Noord-Holland. Stiboka, Wageningen, 193 pp.
Pons, L.J. & A.J. Wiggers, 1959. De holocene wordingsgeschiedenis van Noord-Holland en het Zuiderzeegebied. Tijdschr. Kon. Ned. aardrijksk. Genootschap 76: 104–152.
Pons, L.J. & A.J. Wiggers, 1960. Idem. Idem 77: 1–57.
Postma, H., 1961. Transport and accumulation of suspended matter in the Dutch Wadden Sea. Neth. J. Sea Res. 1: 148–190.
Rooth, J., 1985. Lijst van internationaal belangrijke wetlands in Nederland. Staatscourant.
Rooth, J., 1989. Lijst van internationaal belangrijke wetlands in Nederland. In: A.L. Spaans (ed.) Wetlands en watervogels. Pudoc, Wageningen: 95–110.
Rooth, J., 1989a. Vogels. In: W.J. Wolff (ed.) – De internationale betekenis van de Nederlandse natuur. SDU, 's-Gravenhage: 98–106.
Shaw, S.P. & C.G. Fredine, 1956. Wetlands of the United States. U.S. Fish and Wildlife Service, Circ. 39, 67 pp.
Smit, C.J., 1981. The importance of the Wadden Sea for estuarine birds. In: C.J. Smit & W.J. Wolff (eds) – Birds of the Wadden Sea. Balkema, Rotterdam: 280–289.
Smit, C.J. & T. Piersma 1989. Numbers, mid-winter distribution and migrations of wader populations using the East Atlantic flyway. In: H. Boyd (ed.) – Proc. IWRB/RAMSAR Flyway Workshop. Canadian Wildlife Service, Ottawa.
Staalduinen, C.J. van, H.A. van Adrichem Bogaert, M.J.M. Bless, J.W.C. Doppert, H.M. Harsveldt, H.M. van Montfrans, E.Oele, R.A. Wermuth & W.H. Zagwijn, 1979. The geology of The Netherlands. Meded. Rijks Geol. Dienst 31–2.
Stol, T., 1986. De turfwinningslandschappen. In: S. Barends, J. Renes, T. Stol, J.C. van Triest, R.J. de Vries & F.J. van Woudenberg, 1986. Het Nederlandse landschap: een historisch-geografische benadering. Stichting Matrijs, Utrecht: 47–56.
Terwindt, J.H.J., J.H. de Jong & E. van der Wilk, 1963. Sediment movement and sediment properties in the tidal area of the lower Rhine (Rotterdam Waterway). Verh. Kon. Ned. Geol. Mijnb. Gen. (Geol.) 21: 243–258.
Third Water Policy Plan, 1989. Derde Nota Waterhuishouding. SDU, 's-Gravenhage, 297 pp.
Van Bennekom, A.J., W.W.C. Gieskes & S.B. Tijssen, 1975. Eutrophication of Dutch coastal waters. Proc. Roy. Soc. Lond. B 189: 359–374.
Van Bennekom, A.J. & W. Salomons, 1980. Pathways of nutrients and organic matter from land to ocean through rivers. In: River Inputs to Ocean Systems. Proc. Review Workshop, Rome, 26–30 March 1979. UNEP/Unesco.
Van Straaten, L.M.J.U. & P.H. Kuenen, 1957. Accumulation of fine-grained sediments in the Dutch Wadden Sea. Geol. Mijnb. (N.S.) 19: 329–354.
Van Veen, J., 1962. Dredge, drain, reclaim! The art of a nation. 5th edition. Martinus Nijhoff, The Hague, 200 pp.
Wolff, W.J., 1983. De invloed van de Rijn op de levensgemeenschappen van de Nederlandse kustwateren. In: G.P. Hekstra & W. Joenje (eds) – Rijnwater in Nederland. Rijnwater-symposium Oecol. Kring, 30 September 1983, Rotterdam: 101–124.
Wolff, W.J., 1989. Getijdewateren. In: W.J. Wolff (ed.) – De internationale betekenis van de Nederlandse natuur. SDU, 's-Gravenhage: 92–96.
Wolff, W.J., 1992. The end of a tradition: 1000 years of embankment and reclamation of wetlands in the Netherlands. Ambio, in press.
Zagwijn, W.H., 1986. Nederland in het Holoceen. Geologie van Nederland, deel 1. Rijks Geologische Dienst, Haarlem / Staatsuitgeverij, 's-Gravenhage, 46 pp.
Zijlstra, J.J., 1972. On the importance of the Wadden Sea as a nursery area in relation to the conservation of the southern North Sea fishery resources. Proc. Symp. Roy. Soc. London 29: 233–258.

Hydrobiologia **265**: 15–44, 1993.
E. P. H. Best & J. P. Bakker (eds), Netherlands-Wetlands.

Nutrient cycling and foodwebs in Dutch estuaries

P.H. Nienhuis
Netherlands Institute of Ecology, Vierstraat 28, 4401 EA Yerseke, The Netherlands

Key words: nutrient cycling, eutrophication, water management, long-term changes, estuarine food-web, estuary, coastal lagoon, The Netherlands, man-induced disturbance

Abstract

In this review several aspects of the functioning of the Dutch estuaries (Ems-Dollard, Wadden Sea, Oosterschelde, Westerschelde, Grevelingen and Veerse Meer) have been compared. A number of large European rivers (especially Rhine) have a prevailing influence on the nutrient cycling of most Dutch estuaries. Owing to the increased loading of the estuaries with nitrogen and phosphorus compounds, effects of eutrophication on the biological communities could be demonstrated, mainly in the western Wadden Sea. The causality, however, of the relation between increased nutrient loading and increased biomass and production of primary producers in the turbid tidal Dutch ecosystems is questioned. The most obvious biological effects of eutrophication have been observed in a non-tidal brackish lagoon, Veerse Meer.

The estuarine food web received major attention. Budget studies of the main primary producers revealed a dominance of phytoplankton in all Dutch estuaries, followed by microphytobenthos in the tidal systems and macrophytes in the lagoons. The quantitative distribution of primary producers and primary and secondary consumers shows remarkable similarities along the physical and chemical estuarine gradients, notwithstanding the large variability in space and the considerable inconstancy over time. Among the secondary consumers (waterfowl, marine fish, larger invertebrates) the levels of organic carbon consumption – expressed in g C m^{-2} y^{-1} – are almost the same, when tidal estuaries are compared with non-tidal lagoons, notwithstanding the fact that the consumer populations show large qualitative differences. The transfer from primary consumers to secondary consumers reveals a bottle neck: especially during late winter, when macrozoobenthos reaches its lowest biomass, food may be a serious limiting resource for large numbers of migratory waders foraging on the intertidal flats. The consequences of the Deltaplan , the closure of several estuaries in the southwest of the Netherlands and their subsequent transfer into non-tidal lagoons, offer complicated case studies of ecosystem changes. Several examples of long-term trends in ecosystem development in Grevelingen lagoon have been discussed.

Introduction

For the purpose of this review estuaries are defined as semi-enclosed and coastal bodies of water that have (free) connections with the open sea and within which sea-water is measurably diluted with fresh water derived from land drainage (Cameron & Pritchard, 1963). According to this definition six coastal waters in The Netherlands should be considered as estuaries, viz. Ems-

Dollard, Wadden Sea, Grevelingen, Oosterschelde, Veerse Meer and Westerschelde (Fig. 1). Among the Dutch estuaries several geomorphological types, sensu Fairbridge (1980), can be distinguished: (a) funnel-shaped, coastal plain estuaries (Ems-Dollard, Westerschelde); (b) complicated bar-built estuary (Wadden Sea); (c) blind estuaries, closed off by sea-walls or other civil engineering constructions and transformed into lagoons with a semi-permanent connection with the open sea (Grevelingen, Veerse Meer). A specific type of blind estuary is the Oosterschelde estuary, a man-manipulated system, closed off by a storm surge barrier, a system of flood gates which allows the tides to enter the estuary freely, but which will be closed during extreme storm floods.

In this review the prevailing influence of several major European rivers (Rhine, Meuse and Scheldt) on estuarine functioning will be outlined. Physical characteristics together with the chemical cycling of a number of nutrients will be treated. Eutrophication appears to be one of the most conspicuous trends affecting coastal ecosystem functioning. The estuarine foodweb will receive major attention, exemplified in budget studies of primary producers in all Dutch estuaries. The quantitative role of primary and secondary consumers is only little understood, but the data available offer similarities and contrasts between the Dutch estuaries. The consequences of the Deltaplan, the closure of several estuaries in the southwest of The Netherlands and their subsequent transfer into non-tidal lagoons, offer complicated case studies of ecosystem changes. Finally the carrying capacity of tidal and non-tidal estuaries for migratory waterfowl and fish populations will be discussed. The ecology of the Wadden Sea has been described in more detail by De Jonge *et al.* (1993).

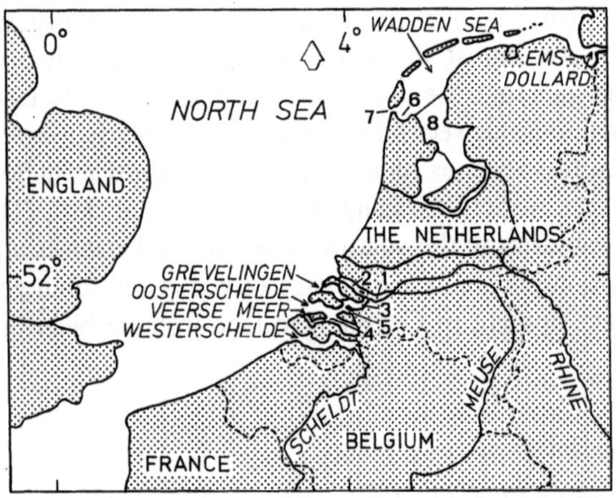

Fig. 1. The geographic position of the Dutch estuaries. Additional geographic names mentioned in the text: 1 = Hollands Diep; 2 = Haringvliet; 3 = Volkerak; 4 = Zandkreek; 5 = Krabbenkreek; 6 = Balgzand; 7 = Marsdiep; 8 = Lake IJssel.

Dutch estuaries and large European rivers: discharge characteristics

All Dutch estuaries share a common feature: they are influenced by the run-off of the large European rivers Rhine, Meuse and Scheldt. The average discharge of Rhine and Meuse together is 2578 m^3 s^{-1}. The Scheldt has an average discharge of only 112 m^3 s^{-1} (De Ruyter *et al.*, 1987). More than 50% of the water of the river Rhine runs via Nieuwe Waterweg directly into the North Sea. About 30% flows into the Haringvliet, and finally also enters the North Sea, depending on the management regime of the Haringvliet-sluices. Owing to the residual currents along the Dutch coast the Rhinewater is mainly transported in a 50 to 70 km wide zone into northern directions (Fig. 2). Roughly 6% of the original Rhinewater finally enters the Wadden Sea (Ridderinkhof, 1990). Via the northern branch of the Rhine, IJssel, 8–15% of the Rhine run-off flows into the IJsselmeer, and finally enters the western Wadden Sea (Van der Veer *et al.*, 1988).

The load of dissolved nitrogen (ammonium and nitrate) of Rhine and Meuse has increased over the period 1950–1985 by a factor 2 to 4, whereas the phosphorus load has increased by a factor 5 to 7 (Fig. 3; Van der Veer *et al.*, 1988). Contrary to the nitrogen load, the phosphorus load decreased slowly after 1982. The concentrations of nutrients in the river increased even more markedly than the load: a 5-fold increase for N and a 10-fold increase for P, with recently a slight de-

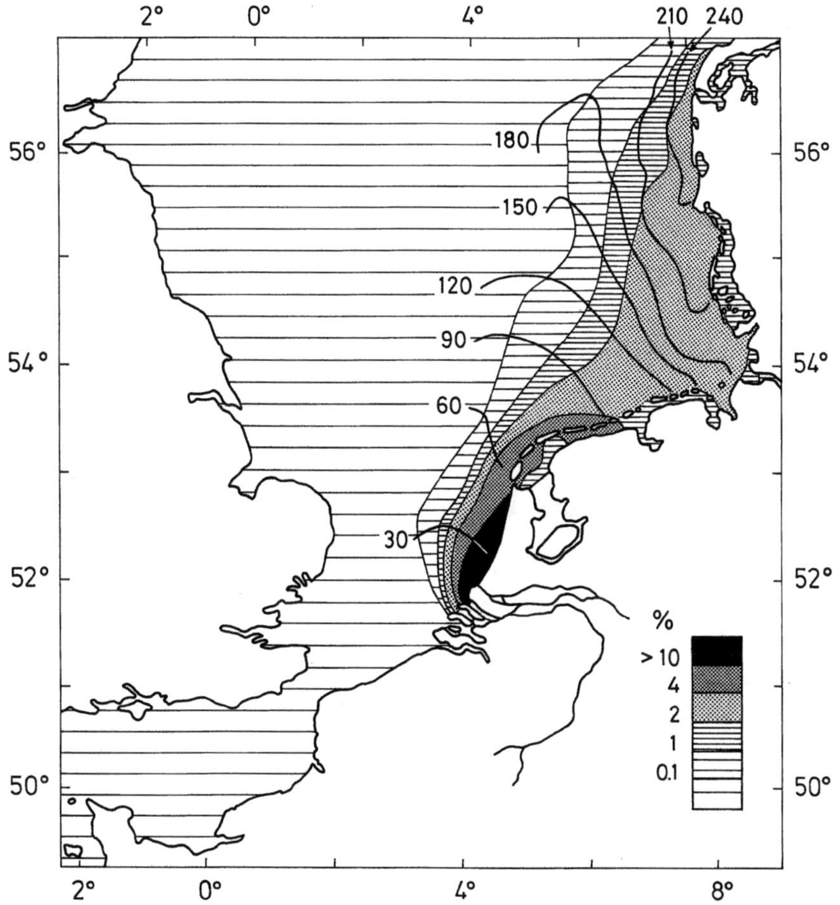

Fig. 2. Flow of Rhine and Meuse water along the coasts of The Netherlands, Germany and Denmark; southwest wind 4.5m s^{-1}; water mass fractions in %, age distribution indicated as solid lines in days (derived from de Ruyter *et al.*, 1987).

crease. The increased concentrations of nutrients in the Rhine-Meuse water resulted in a 3- to 5-fold rise of N and P concentrations in the Dutch coastal waters. The increase of nutrient concentrations and loads in the Westerschelde estuary showed a similar dramatic trend over the past 30 to 40 years (Van Buuren, 1988). The average nutrient loadings on the Dutch coastal waters reflect the discharges of marine and riverine water: Rhine and Meuse – including the Noordzeekanaal – contribute roughly 60% to the total N- and P-load; the residual current of oceanic Channel water adds 30% and only 1 to 2% comes from the river Scheldt (Van Buuren, 1988).

Less than 5% of the original Rhine-Meuse water reaches directly from Hollands Diep-Volkerak or indirectly via the North Sea, the estuaries and brackish lagoons in the southwest of The Netherlands. Particularly the construction of the Volkerakdam in 1969 deprived the southern estuaries of their direct influx of river-water. In the present situation the Oosterschelde, Grevelingen and Veerse Meer are mainly loaded with nutrients from agricultural run-off, treated waste water and drainage canals. The saline water bodies in the southwest of The Netherlands were spatially separated (compartmentalization) owing to the Deltaworks. Consequently each of these estuaries has its own eutrophication history and its own specific water regime. These characteristics mean that each unit has to be managed according to its particular conditions (Nienhuis, 1989a).

Tidal estuaries like the Wadden Sea, Ems-Dollard, Oosterschelde and Westerschelde are

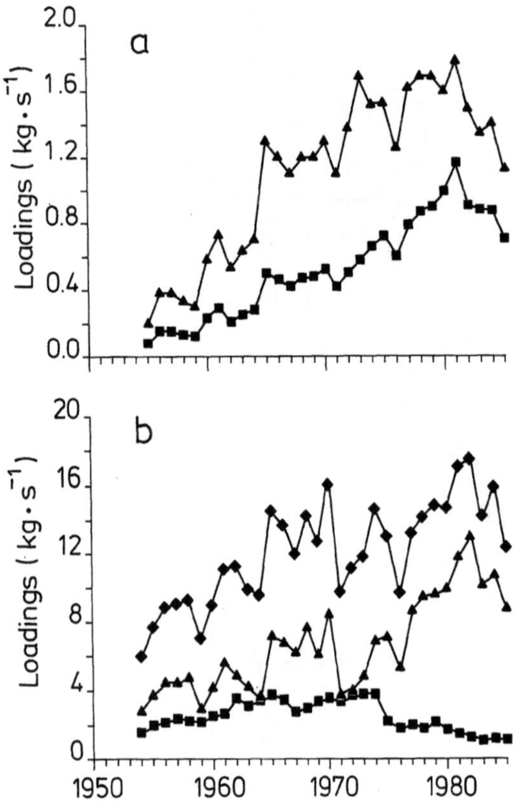

Fig. 3. (A) Loadings of total phosphate (triangles), orthophosphate (squares); (B) total nitrogen (diamonds), nitrate (triangles) and ammonium (squares) of the river Rhine near the German-Dutch border (van der Veer *et al.*,1988).

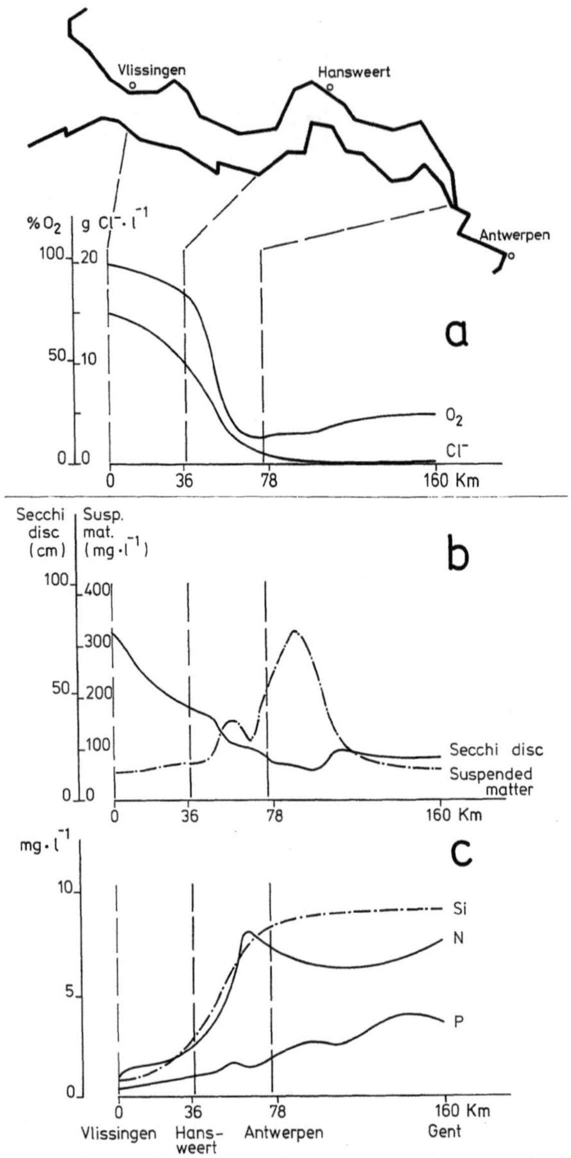

Fig. 4. Gradients in (a) chlorinity, oxygen, (b) suspended matter, Secchi disc visibility, and (c) total Si, N and P in Westerschelde estuary from Gent to Vlissingen (Hummel *et al.*, 1988).

physically controlled, highly dynamic systems. Proper estuaries such as Ems-Dollard and Westerschelde, show a complicated latitudinal and longitudinal gradient in physico-chemical and biological parameters.The abiotic gradient of Westerschelde is shown in Fig. 4 (Hummel *et al.*, 1988). Chlorinity of the watermass in the estuary increases downstream, as a consequence of the increasing mixing of sea-water with fresh water. Similarly, the oxygen concentration increases from 20% south of Antwerp to saturation levels near Vlissingen. The low oxygen concentrations upstream are caused by the heavy load of suspended organic matter, originating from untreated waste water produced by densely populated urban areas in Belgium. As a result of mineralization of these organic compounds high concentrations of inorganic nutrients (N, P, Si) are to be found upstream Antwerp (Fig. 4). Moreover, the high load of suspended matter results in low transparency of the water mass around and upstream of Antwerp.

The upper reaches of many funnel-shaped, coastal plain estuaries are characterized by very turbid water with poor light penetration. The sudden increase in suspended matter concentration, the turbidity maximum (Fig. 4), is caused by the inward flow of marine water and its suspended matter along the bottom of the estuary, until it reaches a stagnation point where inward flow ceases. At this stagnation point the sea-water and its particles rise to mix with the fresh surface water of lower density, but also loaded with suspended matter. The position of the turbidity maximum varies with the strength of the river flow and with the timing of the tidal excursion. Not only the Westerschelde estuary shows a turbidity maximum, but also Ems-Dollard (Baretta & Ruardy, 1988) and several funnel-shaped estuaries abroad, like Weser (Wellershaus, 1984) and Gironde (Laane *et al.*, 1987).

In Table 1 a number of characteristics of Dutch estuaries are summarized. The residence time of the water masses in non-tidal Grevelingen and Veerse Meer is long, compared to the residence time in the tidal estuaries. The net fresh-water load directly derived from the rivers Rhine and Meuse in the Oosterschelde, Veerse Meer and Grevelingen is very small: 1 to 2% of the discharge of the rivers. The Wadden Sea receives approximately 400 $m^3 s^{-1}$ Rhinewater; Westerschelde and Ems-Dollard receive 100 to 150 $m^3 s^{-1}$ of fresh water each from their tributaries. Veerse Meer experiences almost permanent salinity stratification, whereas Grevelingen and Ems-Dollard have only a few deeper channels stratified during summer. The Westerschelde (marine compartment), Oosterschelde and Wadden Sea estuaries are completely mixed tidal systems. The Westerschelde and Ems-Dollard are extremely turbid – extinction coefficient 0.5 to 7 m^{-1}; Grevelingen contains very clear water-extinction coefficient 0.2 to 0.5 m^{-1}, with Oosterschelde, Veerse Meer and Wadden Sea in between.

Cycling of nutrients in estuaries

The cycling of carbon, nitrogen, phosphorus, and sulfur is of importance in estuaries. The estuarine foodweb – often expressed in terms of the cycling of organic carbon – is dependent on the in situ input of energy from sunlight and on the transportation of allochthonous organic matter into the estuary from the rivers and from the sea. Within the estuary the primary producers, convert the energy from the sun into living material. Several categories of plants are distributed over the estuary: phytoplankton, microphytobenthos and macrophytobenthos; macrophytes are subdivided into rooting phanerogams (seagrasses) and macro-algae, both on hard and on soft substrates.

Although estuarine food chains are normally thought as starting with phytoplankton or seagrasses, and ending with larger carnivores such as fish and birds, heterotrophic microbes may represent a major portion of both the first and the last trophic levels within an estuary, and they may dominate the flow in between. When primary producers senesce and die, microbes invade the plant tissues and begin to transform the organic matter. The result is that some of the energy originally

Table 1. Properties of the Dutch estuaries, derived from Wollast (1988) and Bokhorst (1988) for Westerschelde, from Projectgroep Balans (1988) and Wetsteyn & Peperzak (1988) for Oosterschelde, from Nienhuis (1985) and de Vries *et al.* (1988a) for Grevelingen, from Daemen(1985) and Stronkhorst *et al.*, (1985) for Veerse Meer, from EON-projectgroep (1988) for Wadden Sea and from Rijkswaterstaat (1985) and Baretta & Ruardy (1988) for Ems-Dollard. Load = fresh-water load from rivers.

	Westerschelde	Oosterschelde	Grevelingen	Veerse Meer	Ems-Dollard	Wadden Sea
Area (km^2)	300	350	108	22	460	1200
Residence time (d)	30–90	5–40	180–360	± 180	14–70	8–15
Load ($m^3 s^{-1}$)	100	20	5	3	150	400
Tides	+	+	–	–	+	+
Stratification	–	–	±	+	±	–
Extinction coef.(m^{-1})	0.5–7	0.4–1.5	0.2–0.5	0.3–1.4	1–7	0.5–3
Chlorinity (‰)	0–17	15–17	14–16	8–12	0–17	10–17

fixed by the plants is lost by respiration of the microbes, some is leached as dissolved organic matter into the water, some is incorporated into microbial biomass, and some may be transformed to other organic compounds not incorporated into microbial cells (Day *et al.*, 1989).

Dead animal tissues, faeces and pseudofaeces function also as substrates for microbial transformations. Bacteria often use dissolved organic matter unavailable to the rest of the community; *e.g.* amino acids dissolved in estuarine waters are rapidly consumed by bacteria. Thus bacteria prevent the escape of some energy that would otherwise be lost as dissolved organic matter from the estuary and, since the bacteria are eaten by other organisms, redirect it through the foodweb.

The top layer of intertidal sediments is mostly aerobic over several millimeters, as a result of the exposure to air at low tide and oxygen production by benthic microalgae. At deeper levels in the sediment oxygen consuming processes may exhaust the O_2-supply, so that these sediment layers usually become anoxic. The four anaerobic mineralization processes at work are fermentation, dissimilatory nitrogenous oxide reduction, dissimilatory sulphate reduction and methanogenesis. Along the pathways of anaerobic decomposition the complex array of organic compounds available can, in general, be used only by the fermentors and dissimilatory nitrogenous oxide reducers. The metabolism of sulphate reducers and methane producers probably depends largely on the activities of the former groups of decomposers, and primarily on the fermenters. The limitation to growth of any of these bacterial populations, however, may not be carbon but rather insufficient electron donors or receptors; *e.g.* dissimilatory nitrogenous oxide reducers require NO_3 or NO_2, and dissimilatory sulphate reducers require SO_4, for terminal electron acceptors (Day *et al.*, 1989).

Besides live phytoplankton suspended detritus is the main source of food in the water column of estuaries. In the Ems-Dollard estuary, 2 to 4% of the total amount of suspended matter, including inorganic silt and sand, is detritus (Baretta & Ruardy, 1988).

It is well established that the nutritive value of living phytoplankton is higher than that of detritus. It is frequently suggested that the bacterial coating of refractory detritus acts as a food source for detritus feeders (Fenchel & Jorgensen, 1977). Recently there are doubts about the nutritive role of detritus-adhered bacteria. Newell & Field (1983) estimated that bacteria could constitute only 9% of the carbon requirements of the coastal consumer community. Carbohydrates and proteins have the highest nutritional value in estuarine detritus, expressed in calorific content. According to Laane *et al.* (1987) particulate organic carbon calorific content (and not POC concentration) is a good indicator of differences in biomass and production rates of certain bivalves in Ems-Dollard estuary.

Although low concentrations of nutrients, combined with data of nutrient-uptake kinetics experiments, indicate that both silica, nitrogen and phosphorus may act as growth rate limiting nutrients for phytoplankton in Dutch coastal waters (Riegman *et al.*, 1990), nitrogen is supposed to be a major limiting nutrient for primary production in estuaries. All biological transformations of nitrogen are performed by bacteria: nitrogen fixation, nitrification and dissimilatory nitrogeneous oxide reduction. Microbes, *e.g.* blue-green bacteria, are responsible for introducing usable forms of nitrogen into the estuarine community via nitrogen fixation, assimilation of nitrogen into microbial biomass. Nitrification is the biological oxidation of ammonia to nitrite and then to nitrate. In the process of nitrogeneous oxide reduction, nitrate or nitrite are used as the terminal electron acceptors instead of oxygen. It is an anaerobic process, in contrast to the obligately aerobic nitrification. Specific bacteria may reduce these nitrogenous oxides, *e.g.* in estuarine bottom sediments, where aerobic and anaerobic 'patches' alternate. When a gas is produced the process is called denitrification. These gaseous endproducts represent a potential loss of nitrogen from the estuary, and as a result there has been considerable interest in the magnitude of denitrification in estuaries (Kaplan *et al.*, 1979; Sorensen, 1984; Blackburn & Henriksen, 1983).

However, little is known from the Dutch estuaries (Law & Owens, 1990); most data used are model calculations and extrapolations (De Vries *et al.*, 1988a; Rijkswaterstaat, 1990b).

Phosphorus compounds are biologically very active in estuaries, which means that the residence time of dissolved phosphate may be very short in these environments. De Jonge & Villerius (1989) described two antagonistic removal mechanisms in the Ems estuary. First, calcite formed at sea and in the estuary is transported coastward and to the upper reaches of the estuary. During this transport phosphate adsorbs onto this calcite source. In the upper reaches of the estuary, part of the calcite dissolves because of a decrease in pH and high carbon dioxide production owing to remineralization of organic matter; consequently, part of the phosphate will desorb. Second, in the upper reaches of the estuary, where high concentrations of suspended material are present, the conditions are again favorable for adsorption of phosphate, and this time onto the clay particles, the non-calcite mineral fraction of the suspended matter. Apart from biological production and decomposition processes, the non-conservative concentration gradient in estuarine phosphate could also be caused by the process of calcite dissolution. If the processes described by De Jonge & Villerius (1989) have significance under field conditions, than the processes of sediment transport and phosphate adsorption and desorption favors accumulation of marine phosphates together with retention of riverine phosphates in the estuary.

Phosphate is taken up and released by both aerobic and anaerobic sediments, although most exchange occurs between water and anaerobic sediments. Also, more exchange occurs with disturbed than undisturbed sediments (Kelderman, 1985). As oxygen slowly diffuses into sediments it is rapidly consumed by the benthic community. As the oxygen concentration is depleted with depth in the underwater sediment, dissimilatory sulphate reduction dominates, producing sulfide, which decreases the redox potential, causing a solubilization of inorganic phosphate (Fenchel & Riedl, 1970). If the surface of the sediment remains aerobic, much of this phosphate is again made insoluble as it diffuses, or as it is bioturbated, into the aerobic zone. If the overlying waters are anoxic, as is the case in stratified estuaries, the phosphate may enter the water column, stimulating photosynthesis. Besides oxygen availability a temperature-mediated process of mobilization and accumulation of sediment-bound phosphate makes the picture more complicated, as was found by Kelderman (1980) in the Grevelingen lagoon. Thus the availability of inorganic phosphate to estuarine phytoplankton depends on direct and indirect microbial and physical processes both in the water and in the sediment (Day *et al.*, 1989). According to De Jonge (1990) phosphorus and not nitrogen is the main limiting factor for phytoplankton growth in Dutch coastal waters.

Transformations of sulphur compounds in estuaries may be carried out by both macro- and microorganisms. Assimilatory sulphate reduction, the reduction and uptake of sulphate for later incorporation into biomass, can be performed by both plants and a variety of microbes. The reduction of sulphate as a terminal electron acceptor in respiration (dissimilatory sulphate reduction), however, can be done only by a small group of anaerobic bacteria (*e.g. Desulfovibrio*).

In the largely anaerobic sediments of estuaries, dissimilatory sulphate reduction is an extremely important process that passes energy from autotrophs to food chains based on reduced sulphur. Much of the anaerobic carbon and energy flow in estuaries, where it has been studied, is accounted for by this process (Fenchel & Riedl, 1970; Postgate, 1984). Also, the presence of reduced sulphur lowers the Eh in sediments, thus regulating the solubility of various ions and hence the distribution of plants and animals.

Volatile sulphur emitted from estuarine waters, mud flats and salt marshes, which gives them their special smell, includes many low-molecular weight organic compounds containing reduced sulphur as well as hydrogen sulphide. The origins of these compounds have not been investigated fully but may well be microbial (Day *et al.*, 1989).

Anaerobic mineralization by sulphate-reducing bacteria has been investigated intensively in the

Ems-Dollard estuary (Baretta & Ruardy, 1988) and appears to differ widely between separate parts of the estuary. In the central area the anaerobic degradation of organic matter constitutes approximately 50% of total mineralization. In the most seaward part of Ems-Dollard, where aerobic mineralization prevails, this percentage is distinctly lower (11 to 34%), but in the inner part of Dollard it may rise to more than 85% of total mineralization, owing to the extreme high load of organic material.

The range of nutrient concentrations in Dutch estuaries differs greatly. Oosterschelde and Grevelingen reach only seldom values above 1 mg l^{-1} for N, P and Si, and nutrient concentrations frequently approach zero during heavy blooms of phytoplankton in these ecosystems. Veerse Meer has higher maximum values for N and Si, but depletion occurs during the growing season (De Vries *et al.*, 1988b; Daemen, 1985). Westerschelde and Ems-Dollard have by far the highest trophic potential: nutrient concentrations never approach zero during spring and summer and reach high values during winter (8 mg l^{-1} N; 4 mg l^{-1} P and 9 mg l^{-1} Si); (Rijkswaterstaat, 1985; Bokhorst, 1988).

The question arises which factor acts as the limiting constituent for the productivity of algae in the Dutch estuaries. In the turbid Westerschelde and Ems estuaries it seems obvious that the availability of light is limiting phytoplankton production (Colijn, 1983). In parts of the western Wadden Sea phosphate may be limiting (Veldhuis, 1987; De Jonge, 1990). It is assumed that nitrogen is limiting in the Oosterschelde and Grevelingen (De Vries *et al.*, 1988a; Wetsteijn *et al.*, 1990) and occasionally also in the Veerse Meer.

Table 2 shows a large variation in total nitrogen loadings (4–235 g N m^{-2} y^{-1}) in Dutch coastal waters, estuaries and brackish lagoons, whereas N concentrations only show a range of 0.5–4.6 mg l^{-1}. Obviously a nitrogen load of 40 to 200 g N m^{-2} y^{-1} does not give rise to extremely high chlorophyll concentrations during summer in the turbid Dutch estuaries and in the coastal zone. It is assumed that light availability is the limiting factor in these waters, and not nitrogen, explaining the relatively low phytoplankton (chlorophyll) biomass in summer. On the contrary, in the clear water of Veerse Meer a load of 34 g N m^{-2} y^{-1} results in high chlorophyll concentrations (100 mg chl a m^{-3}, or even higher; De Vries *et al.*, 1988b). This large production of phytoplankton biomass in Veerse Meer, and the consequent deposition of particulate organic carbon on the bottom sediments, resulted during the period 1980–1983 in an increase of the anaerobic sediment surface area from 4 to 25% of the bottom surface (Stronkhorst *et al.*, 1985). Obviously, Veerse Meer is vulnerable to eutrophication, owing to the long residence time of the water mass, the low extinction coefficient and the almost permanent stratification.

The Oosterschelde has a very low N loading,

Table 2. Nitrogen and chlorophyll in Dutch estuaries and coastal waters. Data derived from de Vries *et al.*, 1988b; SMOES-model calculations (H. Scholten, pers. comm.); Rijkswaterstaat, 1985).

	Load (gN $m^{-2}yr^{-1}$)	Mean	
		Winter conc. $N\text{-}NH_4 + N\text{-}NO_3$ (mg l^{-1})	Chlorophyll conc. summer (mg Chl a m^{-3})
North Sea coastal	40	0.5	15
Dollard	61	3.5	40
Wadden Sea	50	1.2	30
Westerschelde	235	4.6	30
Oosterschelde	5	1.0	5
Grevelingenmeer	4	0.7	5
Veerse Meer	34	3.0	100

resulting in low chlorophyll concentrations (Table 2). Owing to the execution of the Deltaplan, Oosterschelde has mainly been deprived of its loading with Rhine-water. Model calculations revealed that a reduction of 50% of the nitrogen load of the river Rhine would only lead to a reduction of less than 6% for N-concentrations in the Oosterschelde estuary (SMOES-model; H. Scholten, pers. comm.). Oosterschelde is mainly influenced by marine coastal water, containing low concentrations of total nitrogen (less than 0.5 mg N l^{-1}; Brockmann *et al.*, 1988).

Grevelingen has an extremely low N-loading (4 g N m^{-2} y^{-1}; Table 2), in contrast with Veerse Meer. GREWAQ-model calculations (De Vries *et al.*, 1988a) revealed that the production of phytoplankton in Grevelingen is limited by nitrogen availability and not by light. The N:P ratio of the dissolved nutrients in the water column of Grevelingen lagoon during winter is 2 to 4, pointing in the direction of a surplus of phosphate, notwithstanding the dominance of nitrogen in the discharge water, loading the lagoon (N:P = 22:1). Model calculations showed a high turnover of nitrogen in the water column. The chain of processes: nutrient uptake by phytoplankton – formation of organic matter in algal cells – mineralization of dead algae – regeneration of nutrients – etc. comes about in Grevelingen 8 times a year. This implies that about 90% of the annual primary production of phytoplankton occurs on the basis of regenerated nutrients, especially NH4 (De Vries *et al.*, 1988a).

In many shallow estuarine waters the main food chain is dominated by phytoplankton and benthic filter feeders (mussels, cockles), just as is the case in Grevelingen, Oosterschelde and Wadden Sea. The turnover rate of nutrients in these ecosystems is determined by the filtering capacity of benthic filter feeders. Theoretically every 5 to 10 days the entire volume of water of the estuaries mentioned, circulates through the filtering apparatus of the suspension feeders. Filter feeders act as natural controllers of eutrophication processes (Officer *et al.*, 1982): they deposit organic material from the water column onto the bottom sediments. Moreover, they accelerate the regeneration of nutrients from the deposited particulate organic matter, thereby enhancing the primary production of phytoplankton, as was assumed for Grevelingen (De Vries *et al.*, 1988a).

The chain of processes – partly measured, partly theoretical – from biodeposition to regeneration of nutrients and the coupling between nitrification and denitrification, is triggered by the load of organic material to the sediment. In this context denitrification is undoubtedly a significant process in estuarine ecosystems. Rates of gaseous losses of nitrogen in the range of 5 to 70 mg N m^{-2} d^{-1} were determined in Belgian, Dutch and Danish coastal sediments, representing 8 to 23 % of the amount of nitrogen mineralization in the benthic subsystem (Billen & Lancelot, 1987). However, when the load of organic material on the bottom increases faster than the process of aerobic mineralization, anaerobic conditions will prevail in the sediment, leading to death of bottom fauna and a disconnection of nitrification and denitrification. GREWAQ-model calculations simulate that a loading of approximately 10 g N m^{-2} y^{-1} or more uncouples the eutrophication controlling processes (De Vries *et al.*, 1988a).

Veerse Meer experiences a N-load of 34 g m^{-2} y^{-1} in combination with a long residence time of the water, permanent stratification mainly in the eastern section, and clear water during a considerable part of the year. Obviously, the chlorophyll concentrations in this lagoon cannot be controlled by the bottom fauna, although a substantial benthic biomass is available (Coosen *et al.*, 1990).

The estuarine foodweb – primary producers

Estuarine food webs are characterized by two main food chains, viz. the grazing food chain starting with plant material (phytoplankton, benthic micro- and macrophytes) consumed by herbivorous zooplankton, filterfeeding bivalves, grazing gastropods and many other herbivores, which are in turn consumed by smaller and larger carnivores, such as marine fish, waders and seals. In temperate estuaries much plant material remains ungrazed when still alive. Decomposing organic material together with adhering micro-organisms fuels the detritus food chain (many polychaetes, some bivalves and other invertebrates), finally also ending into an array of carnivores.

The bulk of the food in estuaries is composed of plant material. In Fig. 5 a tentative carbon budget of the main categories of primary producers in all Dutch estuaries is given. Phytoplankton is the dominant primary producer contributing 45 to 71% to the overall annual production of organic material. Notwithstanding the large differences in nutrient loadings of the separate waters (Table 2), primary production of phytoplankton differs by a factor of only 2.4 between the light-limited, turbid inner Ems-Dollard (100 gC m^{-2} y^{-1}) and Westerschelde (125 gC m^{-2} y^{-1}), and the clear, presumably not nutrient-limited Veerse Meer (240 gC m^{-2} y^{-1}). The Wadden Sea, Oosterschelde and Grevelingen hold intermediate positions.

We have to realize that the 'pies' in Fig. 5 depict average, annual, integrated data; for reasons of comparison these idealized pictures are useful. However, phytoplankton primary production is

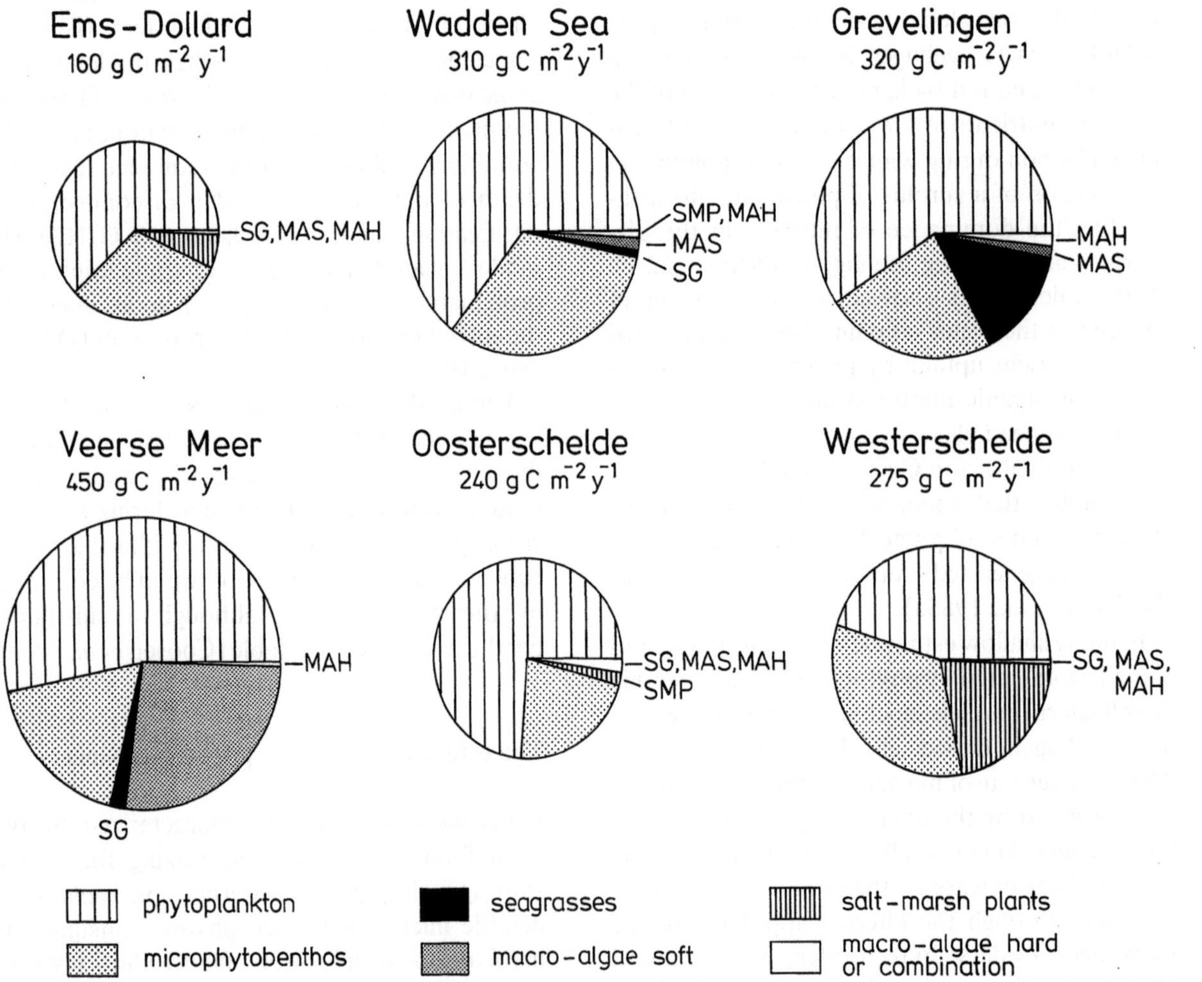

Fig. 5. Preliminary carbon budget of primary producers in Dutch estuaries based on averaged annual data (Nienhuis, 1989a and additional data from Rijkswaterstaat, 1985; Colijn, 1983 [inner part Ems-Dollard], de Wilde & Beukema, 1984; Vosjan, 1987; Baretta & Ruardy, 1988 [Wadden Sea], Wetsteyn *et al.*,1990 [Oosterschelde] and Coosen *et al.*, 1990 [Veerse Meer]. SG = seagrass, MAS = macro-algae on soft substrates; MAH = macro-algae on hard substrates; SMP = salt-marsh plants.

not a continuous process over the course of the year. It usually starts with the spring bloom in March (Grevelingen) or April (turbid estuaries), showing peaks and drops during spring, summer and autumn (Fig. 6), depending on water temperature, insolation, nutrient availability, competition between plankton species and other factors. This implies that year to year changes in phytoplankton production may be rather substantial.

Another factor blurring the average annual picture is the fact that phytoplankton primary production is dependent on the locality in the estuary where the measurements have been performed. Figure 7 shows the spatial variation in production data: in Oosterschelde and Ems-Dollard where the environment changes from the dominance of deep tidal channels near the sea to

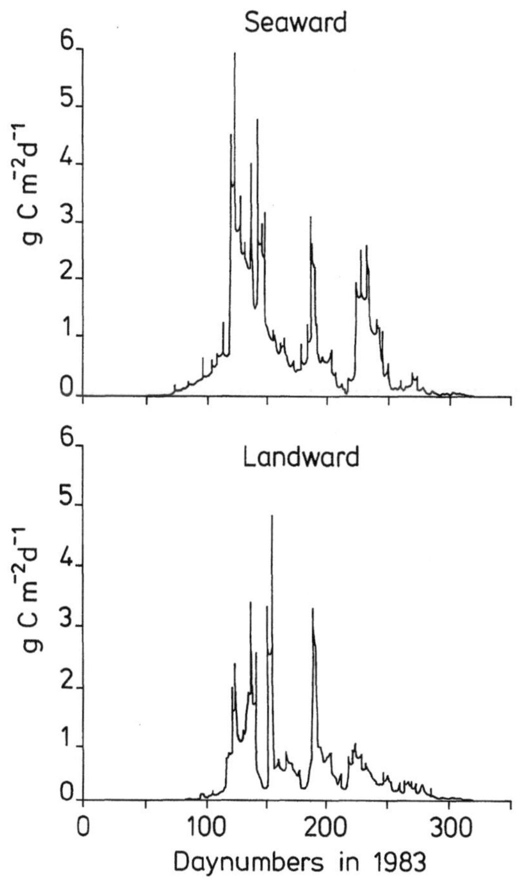

Fig. 6. Gross primary production of phytoplankton, measured daily with the C14 method on two locations in Oosterschelde estuary (Projectgroep Balans, 1988).

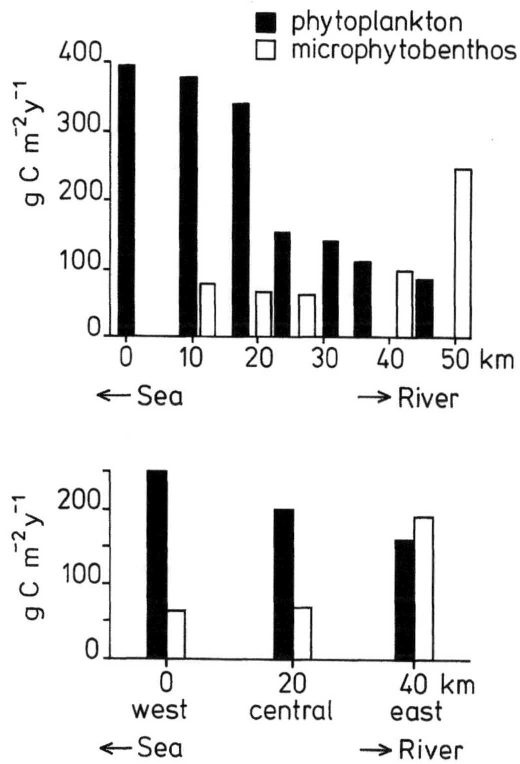

Fig. 7. Annual primary production of phytoplankton and microphytobenthos in Ems-Dollard estuary (Rijkswaterstaat, 1985; Baretta & Ruardy, 1988) [upper panel] and Oosterschelde estuary (Scholten *et al.*, 1990) [lower panel].

shallow intertidal flats near the river, phytoplankton production decreases landwards, whereas microphytobenthos production increases in the same direction.

Production of microphytobenthos (benthic diatoms, green algae etc.) is roughly 30 to 70% of the production of phytoplankton. In Westerschelde estuary the relative share of benthic microphytes is assumed to be rather high, based on a P/B ratio derived from very high biomass data, sampled on intertidal flats (D.J. de Jong, pers. comm., DGW).

Cadée (1984) published data from the western Wadden Sea on microphytobenthos primary production and biomass, over the period 1968–1981 (Fig. 8). Both production and biomass show an increasing trend. Cadée (1984) suggested a relation between the increasing microphytobenthos parameters and the increasing eutrophication of

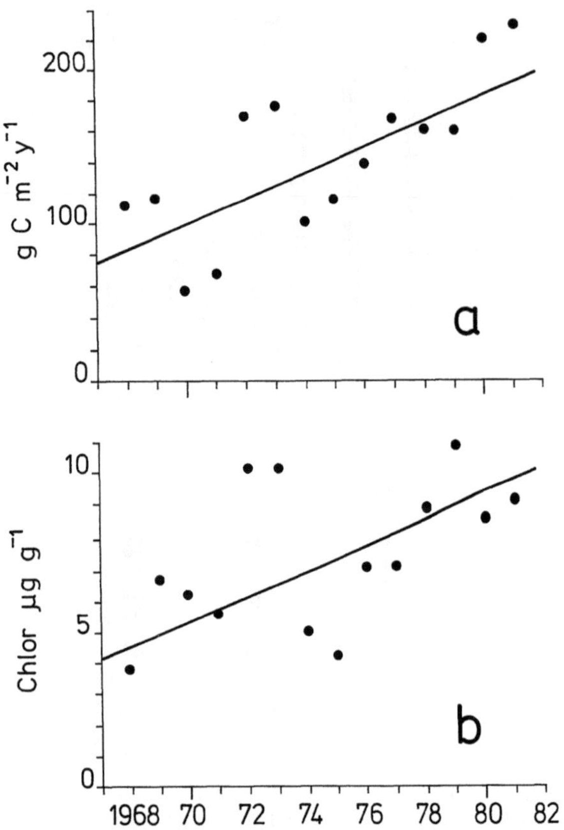

Fig. 8. (A) Microphytobenthos primary production (g C m^{-2} y^{-1}) and (B) biomass (functional chlorophyll a $\mu g\ g^{-1}$ sediment) at one station in the western Wadden Sea (from Cadée, 1984); regression coefficients are not given.

the western Wadden Sea. His statements concerning this relation have been formulated as follows: the large tidal, seasonal and year to year variability in the open Wadden Sea ecosystem, together with geomorphological changes of the permanent sample plot owing to dredging operations, make an assessment of long-term trends of most parameters very difficult. Moreover, changes and improvements in the measuring methods of organic matter, chlorophyll and primary production interfere seriously with the assessment of any long-term trend in input of nutrients in the western Wadden Sea.

It may be questioned whether increasing nutrient loading of the water mass of the western Wadden Sea should necessarily give an increase in production and biomass of microphytobenthos. Stated otherwise: a relation between an increasing nutrient load of the water mass and increasing microphytobenthos activity suggests a limiting role for these nutrients before the process of eutrophication started. This is unlikely in the western Wadden Sea. Concentrations of nutrients in interstitial water of subtidal and intertidal flats are always higher than in the water column, because of the accumulation of decomposing organic matter in shallow-water sediments. In the Oosterschelde estuary, far less eutrophicated than the Wadden Sea (see Table 2), De Jong *et al.* (1990) found out that the nutrient supply from the interstitial water in the sediment was not limiting the growth rate of intertidal microphytobenthos. Phytoplankton productivity should reflect more directly a possible effect of an increasing trend in nutrient loadings. Phytoplankton productivity, measured at two locations in the western Wadden Sea, indeed showed an increase over the period 1965 to 1986 (Fig. 9 data derived from Van der Veer *et al.*, 1988). However, owing to the scanty data, the large variation in the data set and the methodological changes in the course of time (Cadée 1980, 1984, 1986; Cadée & Hegeman 1974a, 1974b, 1977), the significance of the increase is unclear.

The Oosterschelde estuary changed dramatically owing to the construction of a so-called storm surge barrier in the mouth of the estuary and the building of two auxiliary dams, affecting current velocities, transparency of the water mass, nutrient load, seston quantities etc. (Wetsteyn *et al.*, 1990). Figure 9 shows the annual phytoplankton primary production at one location in Oosterschelde estuary over the period 1982 to 1988: note the large fluctuations over the years and a variation between 200 and 340 g C m^{-2} y^{-1} (Vegter & De Visscher, 1987; Wetsteyn *et al.*, 1990). Neither an increasing nor a decreasing trend is visible. The construction of the storm surge barrier was finished in 1986, but no break in the production data could be observed (cf. Fig. 9). Wetsteyn *et al.*, (1990) and Scholten *et al.*, (1990) formulated the hypothesis that the decreased loading of Oosterschelde estuary with nutrients has been compensated by an increase in

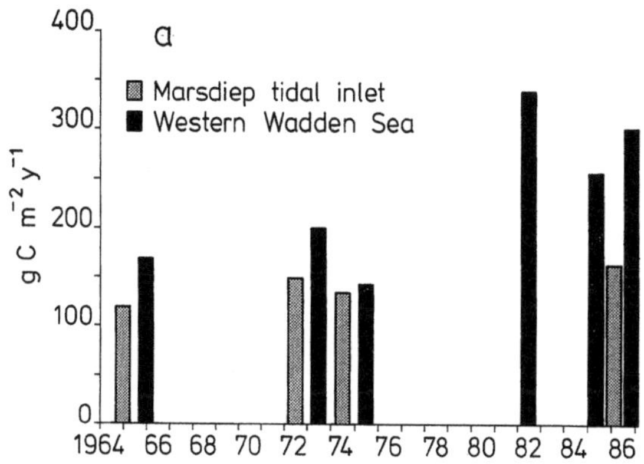

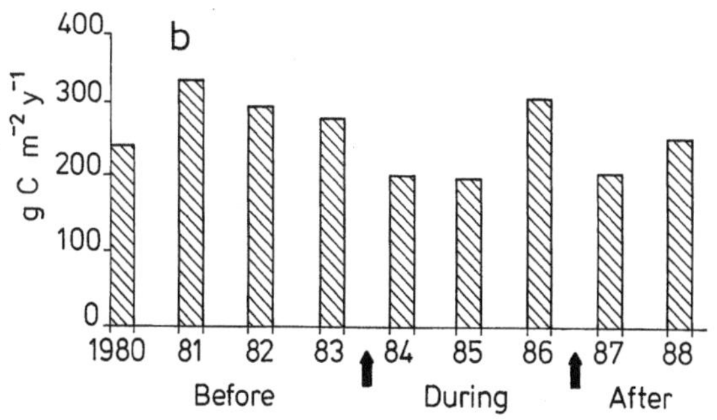

Fig. 9. Annual C14 phytoplankton primary production, integrated over the water column, (A) at two stations in western Wadden Sea, Marsdiep tidal inlet and inner western Wadden Sea, (compilation derived from van der Veer *et al.*, 1988) and (B) at one station in eastern Oosterschelde before, during and after the completion of the storm surge barrier (data derived from Wetsteyn *et al.*, 1990; Nienhuis, 1989b; Vegter & de Visscher, 1987 [data for 1985].

light penetration through the water column, resulting in almost the same integrated level of primary production of phytoplankton on an annual basis.

Data from Oosterschelde show how carefully 'correlations' should be drawn between environmental parameters and phytoplankton productivity. The integrated parameter primary production is a robust characteristic of the pelagic estuarine ecosystem, showing large resilience against changes in the environment (Herman & Scholten, 1990).

Supratidal wetlands overgrown with higher plants, salt marshes, form one of the natural habitats, fringing estuaries in the temperate climatic zones. Salt marshes have an extremely high net primary production, roughly 400 to 1200 g C m^{-2} y^{-1}, above- and below-ground production taken together (Huiskes, 1988; Groenendijk, 1984), strongly dependent on the vegetation type under study (Bakker *et al.*, 1993). The larger part of the organic matter decomposes in the salt marsh itself, and only a small part of the detritus, especially coarse particles during storms, is transported over large stretches of the estuary (Dankers *et al.*, 1984).

Owing to the impact of man, many of the original salt marshes along the Dutch estuaries have been transformed in the past centuries into agricultural land and surrounded by sea-walls. That is the reason why salt-marsh plants contribute only very little to the total organic carbon budget of Ems-Dollard, Wadden Sea and Oosterschelde estuary (Fig. 5). Grevelingen and Veerse Meer, former estuaries and now non-tidal lagoons, lack a fringe of salt marshes. Only Westerschelde meets the standards set for a 'natural' estuary, i.e. that about 5 to 10% of the surface area of the estuary should be covered by salt marshes (Dijkema, 1987).

The contribution of above-ground salt-marsh primary production to the carbon budget of Westerschelde estuary (60 g C m^{-2} y^{-1}) is roughly 20% (Fig. 5). The larger part of the salt marsh production originates from the eastern section of the estuary, the Verdronken Land of Saeftinghe, and is not transported over extensive stretches of the estuary.

Macrophytes growing on hard substrates, i.e. seaweeds attached to sea-walls and stone-clad dikes, make only a minor contribution to the carbon budgets of the entire water bodies (Fig. 5), notwithstanding their high production per m^{-2} habitat (several 100's g C m^{-2} y^{-1}; Nienhuis & Daemen, 1985).

High turbidity and exposure to waves and tides prevent the potential sediment habitats in the Westerschelde and Ems-Dollard from being invaded by macrophytes. The Oosterschelde and western Wadden Sea have only local growth of macrophytes on sediment substrates in sheltered regions, such as Balgzand, Zandkreek and Krabbenkreek embayments (Fig. 1). In the brackish lagoons, like Grevelingen and Veerse Meer, macrophytes living on or rooting in sediment, respectively macro-algae (mainly green algae) and seagrasses, offer a significant share to the carbon budget. In the mesotrophic Grevelingen the rooting seagrass *Zostera marina* dominates, covering 20% of the surface area of the lagoon and showing a within-habitat production of 150 g C m^{-2} y^{-1}, but contributing only 14% to the annual carbon budget, owing to a lower annual average P/B ratio (3) compared to phytoplankton (Fig. 5).

Veerse Meer offers a different picture: Seagrasses cover only 3% of the surface area of the lagoon (Hannewijk, 1988) and have to compete with green algae – mainly *Ulva* – for space and light. The lagoon is dominated by *Ulva* spp. during summer, showing a roughly estimated production of 500 g C m^{-2} y^{-1} in shallow areas. The contribution of *Ulva* to the annual carbon budget of Veerse Meer is roughly 120 g C m^{-2} y^{-1}, which is 27% of the lagoon's budget (Fig. 5). The high nitrogen load of Veerse Meer not only results in a relatively high production of phytoplankton, but also in mass growth of *Ulva* in shallow areas.

Figure 5 shows that eutrophicated Veerse Meer has the highest total primary production of all estuaries in The Netherlands – roughly 450 g C m^{-2} y^{-1} – and that the light limited Ems-Dollard estuary has the lowest values – roughly 160 g C m^{-2} y^{-1}.

The estuarine foodweb – primary and secondary consumers

Phytoplankton is the dominant primary producer in all Dutch estuaries. Although there are large differences from place to place and over time, it can be stated that roughly 10 to 20% of the phytoplankton net production is grazed by herbivorous zooplankton in shallow estuaries; in deeper coastal waters a far larger part is consumed by zooplankton (Valiela, 1984; C. Bakker, pers. comm. for estuaries in the southwest of The Netherlands).

The remainder of the phytoplankton, roughly

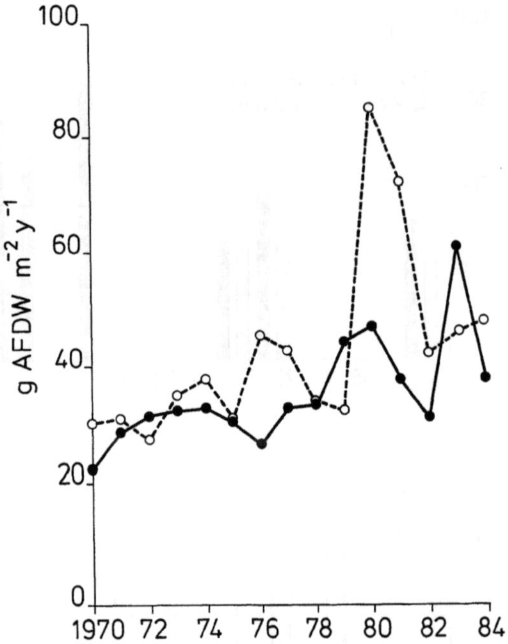

Fig. 10. Changes in total macrozoobenthos production (in g AFDW m^{-2} y^{-1}) during a 15-year period in a tidal flat in the western Wadden Sea; estimates based on samples taken along 12 transects (interrupted lines) and in 3 permanent plots (uninterrupted lines); (from Beukema & Cadée, 1986).

80 to 90%, arrives after one or more tidal excursions and turbulent mixing, dead or alive, on the bottom of the estuary and serves mainly as food for benthic filter-feeding (macro)fauna. One of the most striking facts in estuaries in the temperate climate area, and also in Dutch estuaries, is the tremendous richness in bottom infauna and epifauna of intertidal and shallow subtidal sediments, in an environment that is only superficially aerated –several millimeters to several centimeters– and that deeper down is almost completely anaerobic.

There is still some debate on the quantitative role of the benthic meiofauna. Although the biomass of meiofauna (0.5 g C m^{-2} in the Wadden Sea) is only 5% of the biomass of the macrofauna (10–20 g C m^{-2} in the Wadden Sea), the meiofauna P/B ratio is much higher than the turnover of the macrofauna. Hence the meiofaunal contribution to the overall benthic metabolism might be considerable (Kuipers *et al.*, 1981; Witte & Zijlstra, 1984).

Beukema & Cadée (1986) and Beukema (1989) gave data on changes in biomass and secondary production of macrozoobenthos on Balgzand, a tidal flat area in the westernmost part of the Wadden Sea. Macrozoobenthic production increased significantly to almost double values ($P < 0.01$) over a period of 15 years (Fig. 10). An increasing trend is also observed for macrozoobenthos biomass (Fig. 11). When we focus, however, on the period 1970 to 1980 no increasing trend can be detected; a 'normal' year to year variation in biomass is depicted, between 15 and 26 g ash-free dry weight (AFDW) m^{-2}.

Beukema (1982) clearly showed how large the variation in macrozoobenthos biomass can be in space and time. Figure 12 gives the relation between the average biomass of macrozoobenthos on slightly sloping intertidal flats, and the distance from the coastline at high water spring. Close to the coast, around high water level, biomass is low (less than 10 g AFDW m^{-2}). Maximum biomass of 40 g AFDW m^{-2} is present 1.5 to 3 km from the coastline, rapidly decreasing into the direction of deeper localities. Figure 12 shows also seasonal variations in total biomass of permanent sample plots on Balgzand tidal flats. The lowest biomass values (30% lower than the annual average) are reached in late winter/early spring and the highest values (30% higher than the annual average) show up in July/August.

Balgzand is a rather homogeneous 50 km^2 tidal flat area. Even larger differences in space might be

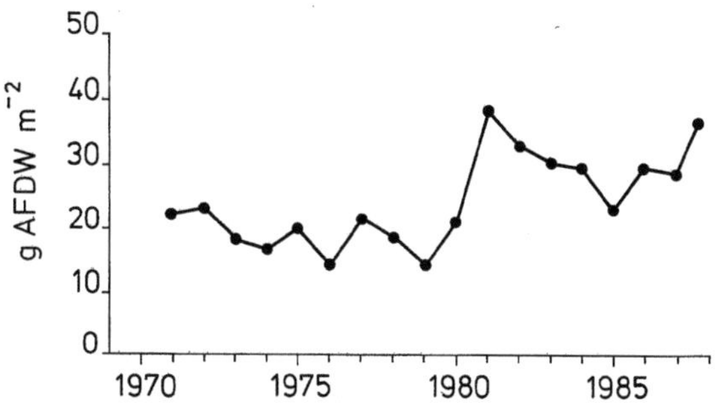

Fig. 11. Significant ($p < 0.01$) increase in biomass (ash-free dry weight m^{-2}) of macrozoobenthos on tidal flats in the western part of the Dutch Wadden Sea. Annual data from March sampling at Balgzand (means of 15 stations) (Beukema, 1989).

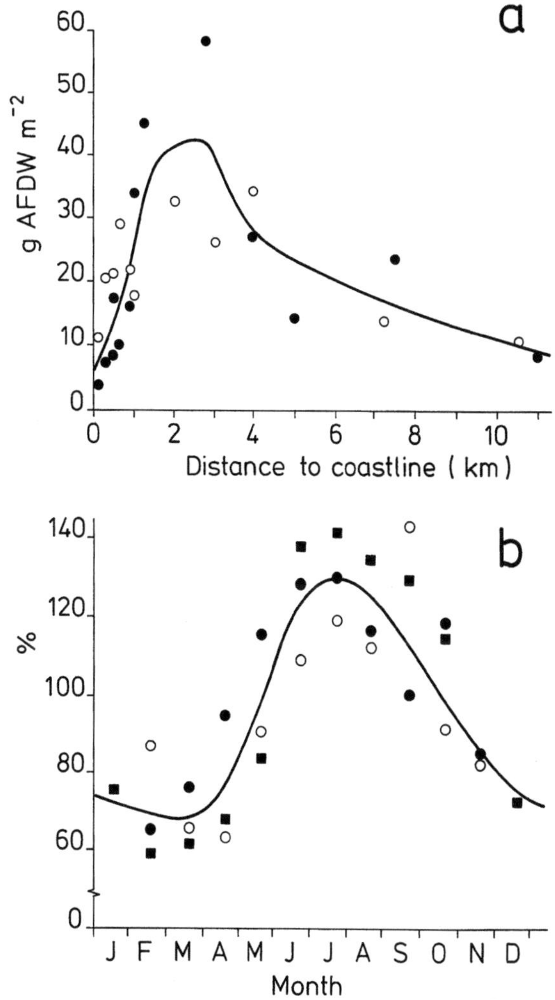

Fig. 12. (A) The relation between average macrozoobenthos biomass (in g AFDW m^{-2}) at intertidal flats in the Wadden Sea and the distance to the coastline of the mainland (closed circles) or a major island (open circles); (from Beukema, 1982). (B) Seasonal changes in macrozoobenthos biomass of 3 sample plots (squares, open circles, closed circles) at Balgzand intertidal flat, expressed as percentage of the integrated average annual biomass values at those sample plots (from Beukema, 1974, 1982).

expected along an estuarine gradient from the sea to the river, taking into account that the level of sampling in the intertidal zone and the time in the year may be excluded as variables. Figure 13 shows a compilation of data from Coosen & Smaal (1985) and Baretta & Ruardy (1988) from Oosterschelde and Ems-Dollard, respectively. There is a conspicuous difference in overall mac-

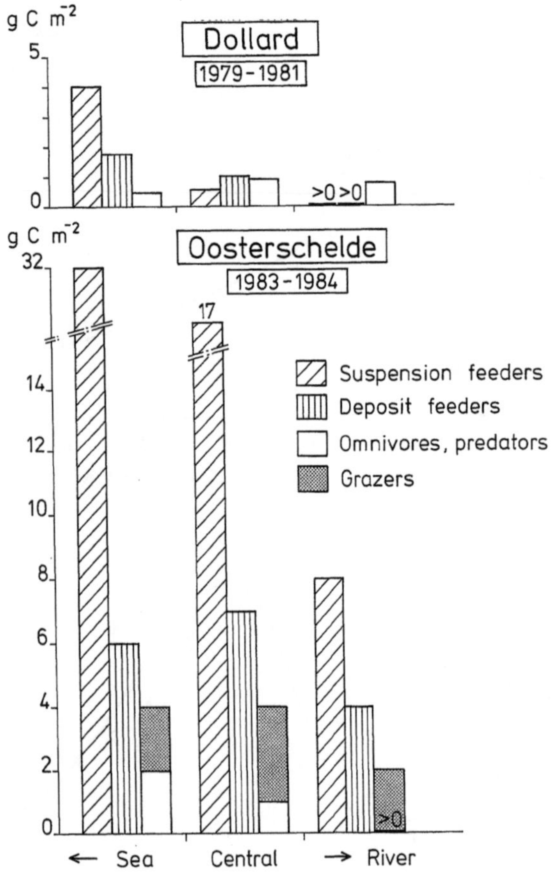

Fig. 13. Intertidal macrozoobenthos biomass along two estuarine gradients in Dollard (Rijkswaterstaat, 1985; Baretta & Ruardy, 1988) and Oosterschelde (Coosen & Smaal, 1985).

rozoobenthos biomass between the turbid, light-limited Ems-Dollard and the clear Oosterschelde: the latter estuary contains roughly 8 times more macrozoobenthos m^{-2} than the former. Besides the better sustainment on the primary trophic level (cf. Fig. 5) of Oosterschelde estuary, the most plausible explanation is the lack of cultivated mussels in Ems-Dollard.

Figure 13 reveals a dominance of suspension feeders in both estuaries, except in the muddy inner part of the Dollard; suspension feeders biomass decreases from the sea towards the river and deposit feeder biomass shows an opposite trend, although not clearly delineated. Grazers (like *Hydrobia ulvae*) have a considerable biomass in Oosterschelde, but are almost completely lacking in Dollard.

Figure 14 shows annual biomass data from a few stations in the Oosterschelde estuary over the period 1983 to 1988. Several samplings per year were organized, and the data show minimum values in late winter and maximum biomass in fall (cf. Fig. 12). The large fluctuations in the seaward station are mainly the result of the fluctuating biomass of *Cerastoderma edule*. The major decrease in cockle biomass in spring 1985 may be related to the severe winter of 1984–1985. The decrease in winter 1987–1988 reflects the extremely large harvest of cockles by man (Smaal *et al.*, 1990). The total macrozoobenthos biomass is ten times lower in the landward station – with-

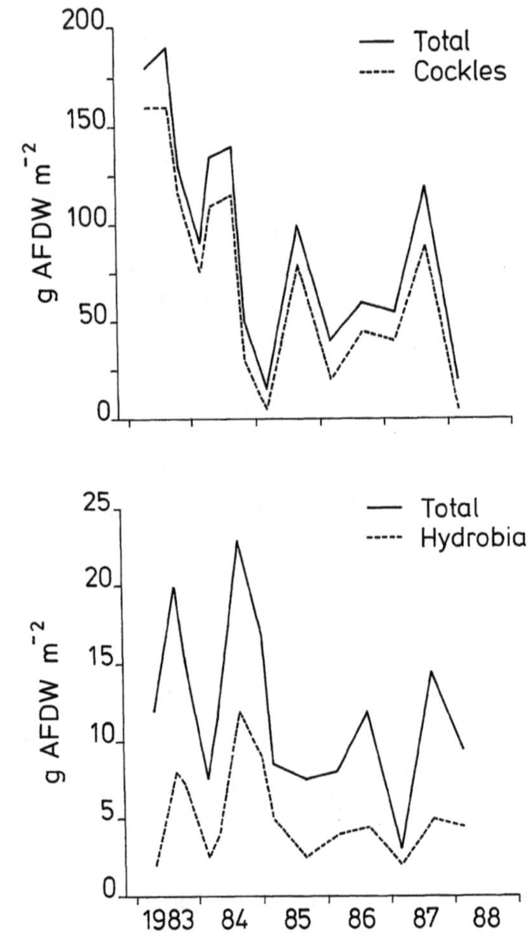

Fig. 14. Intertidal macrofauna biomass in Oosterschelde estuary over the period 1982–1988, at 1 seaward locality (upper panel) and 1 landward locality (lower panel) (Smaal *et al.*, 1990).

out a substantial contribution of cockles – as in the seaward station. Biomass shows seasonal fluctuations but is unusually low in early 1988. This may be related to the slightly reduced tidal range during winter 1986–1987, which caused intertidal benthos to die as a result of extended periods of drought (Hummel *et al.*, 1986).

Data both from the Wadden Sea and from the Oosterschelde reveal how strongly macrozoobenthos biomass and production values vary over time and in space. Many biological and abiological factors modify these patterns: density dependent and density independent population changes, lethal effects of severe winters, impact of the building of large civil engineering constructions and manipulations with the tidal factor (Oosterschelde), fisheries activities (Oosterschelde) and presumably eutrophication (Wadden Sea). We know far too little to apply causal explanations to these variations.

What is the relation between the food – the biomass of microalgae – and the consumers – the macrozoobenthos biomass? From budget calculations, mainly based on estimates of community metabolism, De Wilde & Beukema (1984) concluded that a shortage rather than an abundance of organic matter as food existed for the benthic fauna in the western Wadden Sea. The main assumption on which their statement is founded, is that 90% of the primary food available will be mineralized in the bottom sediments, either aerobically (68%) or anaerobically (21%). In a discussion on the budget of organic material in the Wadden Sea, Beukema & Nienhuis (1985) came to a diverging conclusion: rather than a shortage in food supply, the availability of suitable habitats, predation pressure and the quality of the food (organic material) would decide to what extent macrozoobenthos biomass expansion is limited.

Going higher in the estuarine foodweb, at the level of the secondary consumers, the original energy from the sun, substantiated in primary organic matter, is so far fractioned and dissipated that in terms of consumption and production only a few g C m^{-2} y^{-1} is left. Many carnivores occur in the estuarine ecosystem, and all of them have their own specialized ways of catching their prey, consuming their food and assimilating what is necessary for growth of their body tissue, their respiration and reproduction.

At low water tidal flats are a very well laid table for a wealth of waterfowl (waders, ducks, geese), each having their own feeding strategy. The classic diagram of Green (1968) shows the relation between the depth of the infauna prey in the sediment and the length of the bill of the waders (Fig. 15). This way of exploitation suggests an optimal resource partitioning between the birds.

The Oosterschelde estuary can be taken as an example. Over 200,000 migratory birds, mainly carnivorous waders (max. 150,000) and herbivores ranking second (max. 40,000), use the estuary as wintering grounds. Oystercatcher (*Haematopus ostralegus*) and dunlin (*Calidris alpina*), with maximum numbers in winter of 87,000 and 53,000 respectively, are the dominant waders (data Rijkswaterstaat, Middelburg). Meire & Coosen (1985) calculated that roughly 20% of the macrozoobenthos biomass is annually consumed by birds.

Cerastoderma edule, a suspension feeder living just below the sediment surface, is one of the most important secondary producers in Dutch estuaries (Beukema, 1976). Following severe winters, which kill almost all adult cockles living in a tidal flat, successful spatfalls regularly occur. It is assumed that severe winters not only kill adult cockles, but also decimate or retard the development of the potential predators of cockle spat. Once the small cockles have managed to escape from their predators in the course of their first summer, they are likely to establish an offspring population, lasting about 5 years, unless they are wiped out by the next severe winter (Reise, 1985).

Small cockles in the range of 0.5 to 5 mm length have many predators on tidal flats. The larger the cockles grow, the fewer predators are able to consume them. Cockles of about 2 years of age and 1 to 3 cm in shell length are an important prey to oystercatchers. Cockles larger than 3 cm are almost without natural predators (Fig. 16; Reise, 1985). Only siphon-nipping by fish is a continuous threat for larger cockles (De Vlas, 1979), and

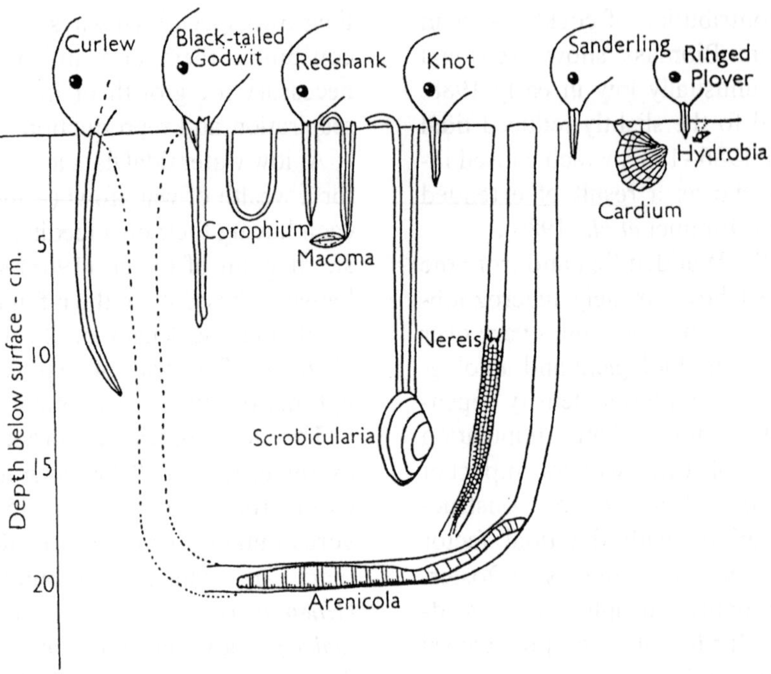

Fig. 15. Relation between the length of the bill of intertidal waders and the depth in the sediment of their macrofauna preys (Green, 1968).

of course cockle fisheries by man may have a tremendous impact on the populations.

Effects of eutrophication in Dutch estuaries

Eutrophication is the process of enrichment of the environment with nutrients, until a level has been reached where unwanted, negative effects on ecosystem functioning become manifest. So far a number of biological effects of increasing eutrophication have been mentioned for the Dutch coastal waters and the western Wadden Sea (see for summary Nelissen & Stefels, 1988): increasing biomass and production of microalgae and an increasing dominance of flagellates over diatoms.

Correlations between increased nutrient loadings and increased primary production as demonstrated for Dutch coastal and Wadden Sea waters, can hardly be achieved for the estuaries and lagoons in the southwest of The Netherlands. Long term series, such as published for the Wadden Sea by Beukema & Cadée (1987) are not available for the Oosterschelde estuary. For the Westerschelde Bokhorst (1988) summarized a series of physico-chemical characteristics over the period 1982 to 1988. The time series of chlorophyll concentrations in the water column show a rising trend. This trend can be explained partly by variations in the spring light extinction. The remaining part to be interpreted is unclear and can hardly be due to an increase in nutrient concentrations (Herman & Hummel, 1989). Besides that, any possible trend in eutrophication of the Delta waters has been obscured (except for Westerschelde) by the execution of the Deltaplan, drastically changing the hydrography of the area. Veerse Meer and Grevelingen have their own specific eutrophication stories, starting respectively in 1961 and 1971, when the estuaries were closed off from the sea.

In Fig. 17 data on the occurrence of the flagellate *Phaeocyctis pouchetii* in the western part of Oosterschelde estuary are compared with data from the western Wadden Sea. The Wadden Sea data, from Cadée & Hegeman (1986), reveal a relation between increasing eutrophication and increasing incidence of *Phaeocyctis* blooms. In a

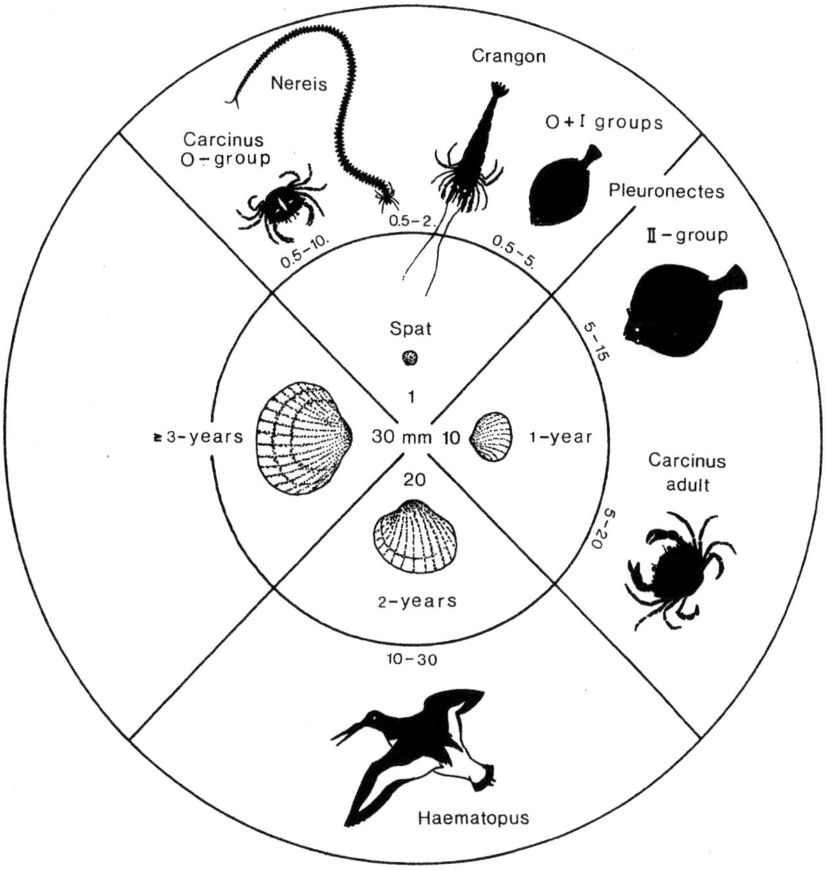

Fig. 16. Predator spectrum of *Cerastoderma edule* in the German Wadden Sea; lengths of the cockle shells (mm) are depicted along the inner circle (Reise, 1985).

recent paper Cadée & Hegeman (1991) presented data on *Phaeocystis* blooms in the Marsdiep (W. Wadden Sea), as early as 1897 en 1899. The long-term trend (1897–1990) shows an increase in the durations of the *Phaeocystis* blooms, surpassing the normal yearly variation, but this increase has probably not been as significant as suggested earlier by Cadée & Hegeman (1986). The Oosterschelde data show a significant decrease in the length of the period of the *Phaeocyctis* blooms, but an increase in bloom intensity. These partly contradictory results question the usefulness of *Phaeocystis pouchetii* as an indicator for increasing eutrophication of Dutch estuaries in general. Older records from before the second World War also reveal incidental strong *Phaeocyctis* blooms in the southern North Sea (C.H.R. Heip, pers. comm.), which questions the use of the flagellate as an indicator even more strongly.

In their analyses of published data on nutrients and primary production of the Western Wadden Sea De Jonge (1990) and De Jonge & Essink (1991) are pertinent with regard to the relation between phosphate loading and primary production of phytoplankton. Figure 18 shows a significant correlation between both parameters over the period 1950–1986. The annual primary production in the Marsdiep tidal basin (W. Wadden Sea) is mainly determined by the phosphate discharge from Lake IJssel, which directly reaches that part of the Wadden Sea.

The outflows of the major mainland rivers (Rhine, Meuse) stay generally close to the Dutch coast and join the Elbe, Ems and Weser inputs,

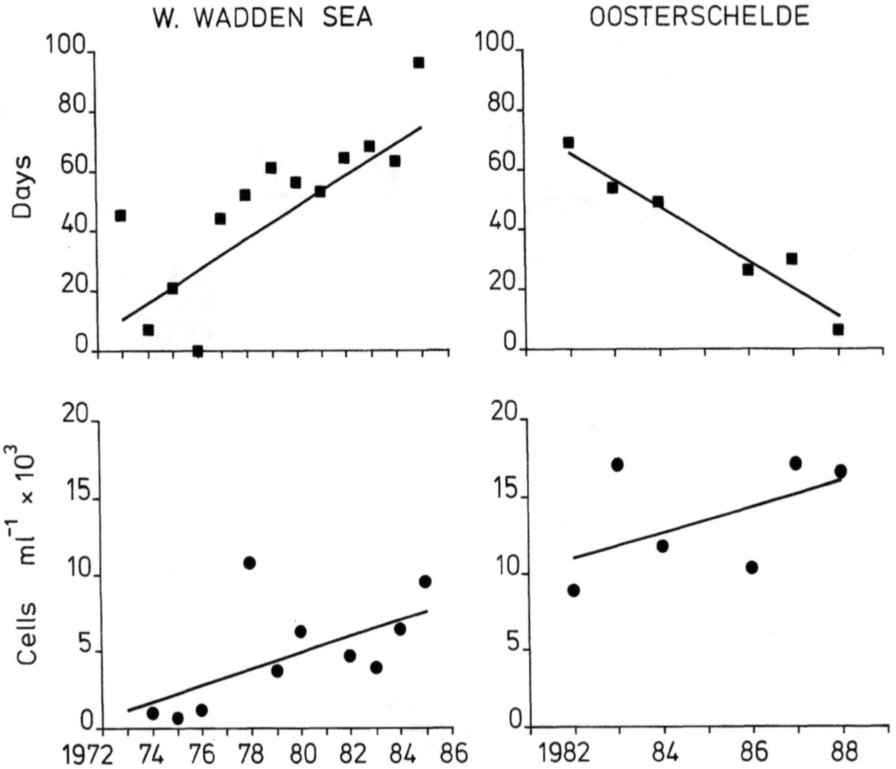

Fig. 17. Duration of the springbloom of *Phaeocystis pouchetii*, expressed as number of days with concentrations of cells above 1,000 ml^{-1}, and average concentration of cells, expressed as number of cells ml^{-1}, for the western Wadden Sea (Cadée & Hegeman, 1986) and the western Oosterschelde (data from C. Bakker, DIHO). For Oosterschelde: number of days: $y = 819–9.178x$, R2 = 0.94399, $P < 0.01$; cells ml^{-1}: $P > 0.05$.

to circulate around in the German Bight before moving north up to the Danish coast and farther north. In the period between 1962 and 1984 in the vicinity of the island of Helgoland surface temperatures increased, salinity decreased, concentrations of phosphate, nitrate and nitrite increased, while those of silicate and ammonium decreased (Radach & Berg, 1986). Compared to data from 1936, nutrient concentrations in the German Bight in 1978 increased by a factor of 2 or 3 (Weichart, 1986). The phytoplankton biomass close to Helgoland has increased significantly, especially flagellates increased by a factor of 16 (Dethlefsen, 1989). Bätje & Michaelis (1986) report on increasing intensities of blooms of *Phaeocyctis pouchetii* in the past decade. In the summers of 1981 and 1983 extensive areas in the German Bight and off the Danish coast were found with low dissolved oxygen concentrations in water very close to the bottom, and consequent mortalities of fish and benthic organisms (Rachor & Albrecht, 1983; Dethlefsen & Von Westernhagen, 1983). Hickel *et al.* (1989) described how low dissolved oxygen occurred as a combination of hydrological and meteorological conditions which lead to stable water stratification and high biological oxygen demand in the German Bight areas studied. Low dissolved oxygen concentrations, a.o., intensifies the toxicity of specific substances as well as their accumulation. Low dissolved oxygen represents a general stress with consequences for the immune system of marine animals (Dethlefsen, 1989).

In addition to an increase in the intensity and frequency of algal blooms and oxygen depletion, it has been suggested that eutrophication will cause a shift in the pelagic algal communities, viz. a decrease in species diversity and an increase of

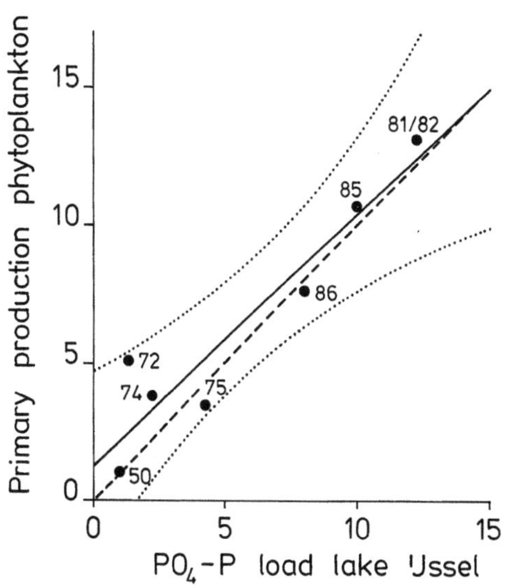

Fig. 18. Plot of the relative change in mean phosphate loads from Lake IJssel and the relative change in primary production in the Marsdiep tidal channel. The reference year is 1950. Dashed line represents the expected curve when primary production is directly dependent on the phosphate discharges. Given is also the regression function (closed line). $y = 0.91x + 1.24$ and confidence limits, $r = 0.94$ ($p < 0.01$); (De Jonge, 1990).

the importance of flagellates (Brockmann *et al.*, 1988; Richardson, 1989).

According to Richardson (1989) blooms of phytoplankton, both of toxic and non-toxic species, are a natural phenomenon in the coastal and central parts of the North Sea. However, eutrophication from anthropogenic sources gives rise to an increase in the intensity and frequency of such algal blooms in the coastal areas, possibly stimulating the frequency of oxygen depletion, and thus seriously affecting ecosystem functioning.

Beukema & Cadée (1986, 1987) postulated the hypothesis that increased nutrient concentrations in Dutch coastal waters and in the western Wadden Sea induced increased primary production, which caused, in turn, increased secondary production. The authors are careful enough not to exclude several alternative factors that may have contributed to the observed trends, such as the decrease during the last decades of various toxic substances, like heavy metals and insecticides in the river Rhine and in the Dutch Wadden Sea (Dijkzeul, 1982; Essink, 1985), an increase in the concentration of suspended organic material and an increase of the sedimentation rate at many of the tidal flats in the western Wadden Sea. According to Beukema & Cadée (1986) it would be virtually impossible to estimate the contribution of these factors to the observed increases in productivity.

De Jonge & Essink (1991) are far more pertinent in their statements. According to these authors the increase in nutrient loadings on the western Wadden Sea over the period 1970 to 1988 correlates significantly with the increase in intertidal macrozoobenthos biomass. As can be derived from the data of De Jonge & Essink (1991), the eastern Wadden Sea, however, shows a 2–3 times higher – fluctuating and not increasing – level of macrozoobenthos biomass (40–90 g AFDW m^{-2}) than the western Wadden Sea (20–35 g AFDW m^{-2}) over the period 1970–1988. It remains fully unexplained why the eastern Wadden Sea, which shows hardly any signs of eutrophication compared to the western Wadden Sea, reveals much higher levels of macrozoobenthos biomass, than the latter area.

In the western Wadden Sea a relation between increasing eutrophication causing increasing algal stocks and primary production and increasing biomass and production of macrozoobenthos cannot be excluded, but interpretation should be done carefully, taking into account all other factors mentioned. Moreover, caution is needed in demonstrating a positive relation between benthos production and biomass and eutrophication, because this relation presupposes the existence of food limitation for the benthos. It is unlikely that in a eutrophicated system further addition of nutrients might enhance the productivity of filter feeders. Model calculations for Oosterschelde estuary showed that a doubling of the nutrient load on the ecosystem gave rise to an increase of only 2.5% in benthic filter-feeder biomass (Herman & Scholten, 1990).

Decomposing algal biomass may easily lead to depletion of oxygen. Low oxygen concentrations have been observed in the 1970's during summer in the western Wadden Sea (Tijssen & Van Bennekom, 1976), and in the 1930's in seagrass beds during the night (Broekhuysen, 1935). No clear signs of negative effects on the zoobenthos were observed in the area during the past 20 years. The few examples of local mass mortalities of zoobenthos may be related to forced mineralization of deposited algal blooms (Beukema & Cadée, 1986). Accumulation of decomposing macro-algal material may cause local death of macrozoobenthos, *e.g.* in places where much *Ulva* and *Enteromorpha* is drifted ashore (cf. Reise, 1984). The same phenomenon has been observed by me during summer in isolated embayments of Oosterschelde estuary. It is impossible to say whether the incidence of these local accumulations of macroalgae has increased over the last 10 to 20 years in the Oosterschelde estuary.

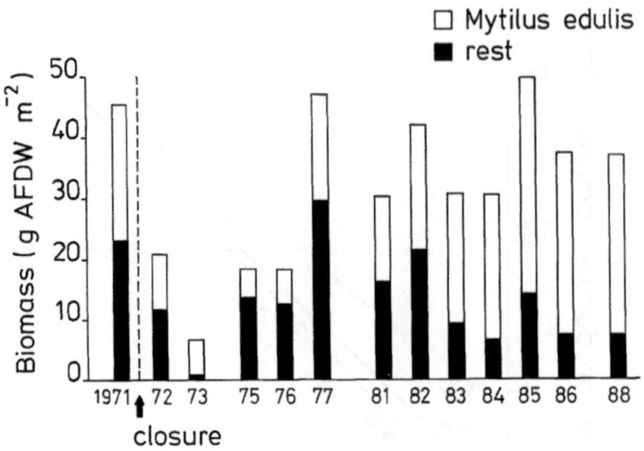

Fig. 19. Long-term changes (1971–1988) in macrozoobenthos biomass, averaged over a number of sampling plots, > 1.5 m waterdepth, in Grevelingen lagoon (Lambeck *et al.*, 1989).

Grevelingen, from an estuary to a saline lagoon

Grevelingen estuary was closed in 1971, and gradually changed into an artificial brackish lagoon (Nienhuis, 1985). Primary production is dominated by phytoplankton (Fig. 5), as is the case in all the other tidal and non-tidal estuaries and lagoons. A specific characteristic of Grevelingen is the extensive perennial population of eelgrass, *Zostera marina*, covering large areas, but declining recently.

Lambeck & Brummelhuis (1985), Lambeck & Pouwer (1986), Lambeck & De Smet (1987) and Lambeck, Wessel & Hannewijk (1989) described the changes in macrozoobenthos biomass and population dynamics after the closure of Grevelingen estuary from 1971 until 1988. In 1971 the intertidal and subtidal populations of macrozoobenthos became isolated from the tidal movements, and consequently large changes set in. Figure 19 shows an overall picture of macrozoobenthos biomass, averaged over a number of sampling spots. In 1971, before the closure, macrozoobenthos biomass was dominated (80%) by *Mytilus edulis* and *Cerastoderma edule*. A strong decrease set in and revealed a lowest point in 1973 (only 5 g AFDW m^{-2}). Thereafter biomass recovered and showed a varying pattern between approximately 30 and 50 g AFDW m^{-2} over the period 1977 to 1988. The spatial and temporal changes in the populations of the individual macrozoobenthos species showed a dynamic pattern, of which most wax and wane is up to now only poorly understood.

Mytilus edulis, although dominant only during the period 1971 to 1977, is since then steadily on the decline (biomass in 1986–1988 only 4 g AFDW m^{-2}). *Cerastoderma edule* decreased in biomass directly after the closure in 1971, but is still present. Its sibling species, *Cerastoderma lamarcki*, characteristic for stagnant, brackish habitats, emerged already in 1973 and since then to 2–5 g AFDW m^{-2}. Any data about competition between those two species are lacking. *Nassarius reticulatus*, a carnivorous gastropod, was first discovered in 1975; after an original success (up to 7 g AFDW m^{-2}) it recently showed a failing reproduction. Since 1983 the colonizer *Crepidula fornicata*, a filter-feeding mollusc, showed a rapid increase to a biomass in 1988 of 10 to 15 g AFDW m^{-2}. Although the overall level of macrozoobenthos biomass shows since 1981 a 'normal', stable pattern, the population structure is highly dynamic, and new colonizers – opportunists – may still get their chance.

Grevelingen offers an example of long-term changes in the numbers and biomass of secondary consumers after a sudden intervention. Figure 20 (derived from Nienhuis, 1985) summarizes a number of qualitative trends, reproduced in smooth lines; annual fluctuations have been left out.

Before May 1971, when Grevelingen was still a tidal estuary, a considerable population of migratory marine flat fish (plaice, dab, sole) lived in the estuary. The sudden closure of the estuary ment a blockade for the remaining population of flat fish. Their migratory behavior was counteracted and hence their usual way of reproduction. The seawalls surrounding the lagoon prevented them from returning to their spawning grounds in the North Sea. Figure 20 shows a gradually declining number of flat fish to a very low level at the end of the eighties.

Migratory birds, carnivorous waders, used to exploit the intertidal flats in Grevelingen estuary before May 1971. Owing to the closure of the estuary the water level was fixed, and consequently all intertidal flats changed into either permanently submerged sediments or permanently dry terrestrial shore areas. Figure 20 shows the dramatic fall of the numbers of waders directly after 1971.

The sudden change in May 1971 induced several other changes on the tertiary trophic level. The numbers of small gobiid fish (*Pomatoschistus minutus*, *P.microps*, *Gobius niger*) and small pelagic fish (*Gasterosteus aculeatus*, *Sprattus sprattus*, *Atherina presbyter*) increased gradually after the closure (Fig. 20). All these fish species took advantage of the decreased exposure of the lagoon, the availability of suitable breeding habitats and proper food (Doornbos, 1987).

The closure of Grevelingen estuary resulted also in a major increase in the numbers of piscivorous birds (great crested grebe – *Podiceps cristatus*; cormorant – *Phalacrocorax carbo*; red-breasted merganser – *Mergus serrator*). This

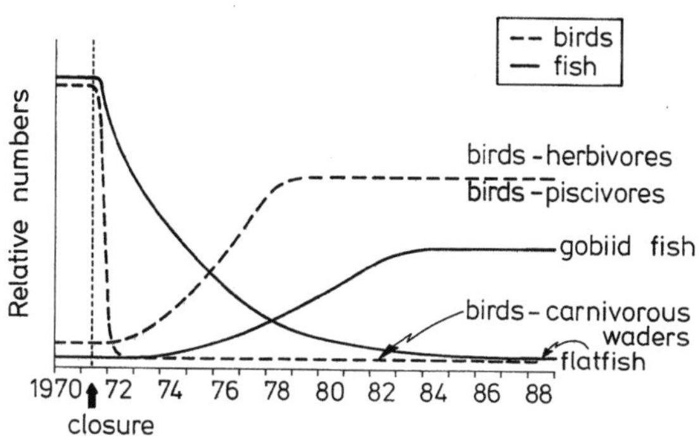

Fig. 20. Long-term changes in relative numbers of dominant groups of primary and secondary consumers in Grevelingen lagoon, following the closure of the estuary in 1971 (derived from Nienhuis, 1985). Only qualitative trends are given; annual fluctuations have been left out.

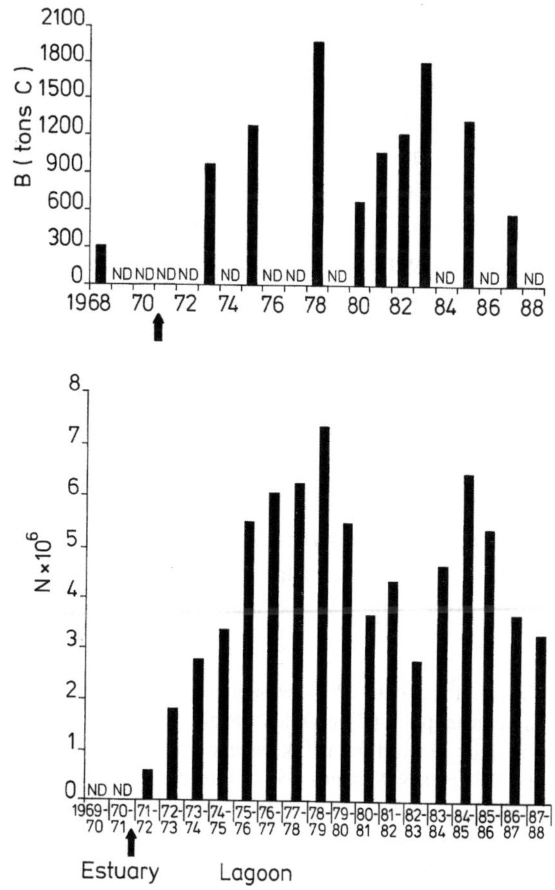

Fig. 21. Upper panel: Long-term trend in integrated biomass (above-ground peak standing stock in July–August) of eelgrass, *Zostera marina*, in entire Grevelingen lagoon, in tons organic carbon, over the period 1968 to 1987 (data from P.H. Nienhuis). ND = no data collected. Arrow indicates closure of Grevelingen estuary in 1971. Lower panel: Numbers of annual (July to July) bird-days (= number of birds times the number of days the birds are present) of (mainly) migratory herbivorous waterfowl in Grevelingen lagoon, over the period 1971 to 1988 (derived from Slob, 1989). ND = no data collected.

change in numbers is related to the much higher transparency of the water (the birds are catching their prey 'on sight'), and the availability of considerable amounts of prey. The prey taken by grebes and mergansers are usually small: 60% of the fresh weight of the stomach contents of these birds consisted of gobiid fish, as was shown by Doornbos (1984) in 1981 and 1982.

Migratory herbivorous birds (geese, ducks, mute swan [*Cygnus olor*], and coot [*Fulica atra*]) took also great advantage of the new situation in Grevelingen lagoon after May 1971 (Fig. 20), which brought them food and shelter (Slob, 1989).

Eelgrass, *Zostera marina*, a rooting submerged phanerogam found an open niche – the sheltered subtidal sand flats – after May 1971. The species colonized the lagoon from the original pre-1971 spots, reached a maximum distribution in 1978 and showed a fluctuating pattern thereafter (Fig. 21), owing to complex causes beyond the scope of this review (cf. Nienhuis, 1983). *Zostera marina* is by far the dominant macrophyte food source in Grevelingen lagoon. The maximum above-ground biomass of *Zostera marina* in July/August, over the period 1973 to 1988, showed a significant correlation with the number of bird-days of herbivorous birds (*Fulica atra, Anas penelope, Anas plathyrhynchos, Anas crecca, Branta bernicla, Cygnus olor*) in the subsequent autumn and winter (Fig. 22). Though the birds only consume 4% of the annual above-ground primary production of eelgrass (Nienhuis & Groenendijk, 1987), that is 11% of the peak standing crop in August, the correlation between the numbers of birds present and their obligate food source is significant. Obviously a large part of the peak standing crop is not available to the birds because it is beyond the reach of their bills, or because the timing of the birds does not match with maximum standing stock of the macrophytes: when most

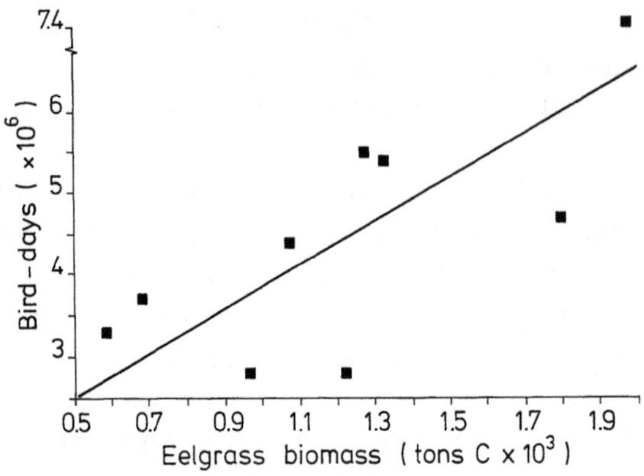

Fig. 22. Relation between maximum above-ground biomass of *Zostera marina*, Grevelingen lagoon in July/August and the number of bird days of herbivorous waterfowl in Grevelingen lagoon in subsequent autumn and winter over the period 1973 to 1987 (data derived from Fig. 21); $n = 9$; $r = 0.74$; $p < 0.025$.

Table 3. Production (P = net or gross primary production or value in between, depending on methods used; Pn = net production) and consumption (C) in Dutch estuaries and former estuaries (in g C m^{-2} y^{-1}); primary production data, see Fig. 5; consumption data: Dollard from Rijkswaterstaat (1985) and Baretta & Ruardy (1988); Wadden Sea from de Wilde & Beukema (1984) and Beukema & Nienhuis (1985); Grevelingen from de Vries *et al.* (1988a) and Doornbos (1987); Veerse Meer from Coosen *et al.* (1990).

		Ems-Dollard	Wadden Sea	Grevelingen	Veerse Meer
Primary	– phytoplankton (P)	100	200	190	240
Producers	– microphytobenthos (P)	50	100	75	80
	– macrophytobenthos (P)	<5	<5	50	120
Primary	– macrozoobenthos (C)	30	200	118	120
Consumers	– id. (Pn)	3	20	14	12
	– birds (herbivores) (C)	–	–	2	5
Secondary	– fish + invertebrates (C)	4	4	3	3
Consumers	– birds (carnivores) (C)	1.7	2	±1	2
	– man (C)	≪1	1	≪1	≪1

birds arrive later in the year, the larger part of the seagrass biomass has already decomposed.

The estuarine foodweb in perspective

Finally two questions about the functioning of the estuarine foodweb will be answered here: (1) What is the carrying capacity of the Dutch estuaries? (Carrying capacity is the potential ability of the ecosystem to offer food and living space [habitat] to the aggregated populations.) (2) What are the bottle necks in the dissipation and transformation processes of primary organic material through the estuary?

Four Dutch estuaries have been studied in enough detail to speculate on the above questions (Table 3). In the Ems-Dollard net production of macrozoobenthos is estimated at 3 g C m^{-2} y^{-1} (Baretta & Ruardy, 1988). This implies a consumption of 30 g C m^{-2} y^{-1}, to sustain this production, assuming an ecological efficiency (P/C × 100) of 10%. Valiela (1984) suggests that the ecological efficiency is often lower than earlier assumed (cf. De Wilde & Beukema, 1984). Table 3 shows that the supply of food amply covers the demand. The extremely low macrozoobenthos production might be explained by the low salinity and high turbidity of Ems-Dollard; species such as *Mytilus edulis* and *Cerastoderma edule*, which make a large contribution to marine estuaries like Wadden Sea and Oosterschelde, are far less common in Ems-Dollard.

A bottle neck is emergent in the transfer from the primary to the secondary consumers. The integrated consumption by carnivorous invertebrates (fish and birds) is 5.7 g C m^{-2} y^{-1} and that is more than the annual resource of macrozoobenthos offered (3 g C m^{-2} y^{-1}). The carnivores may be food limited in Ems-Dollard and competition between fish, larger invertebrates and birds is likely to occur.

In the Wadden Sea estuary net production of macrozoobenthos increased from 11 g C m^{-2} y^{-1} in 1970 to 18 g C m^{-2} y^{-1} in the early eighties, as calculated from maximum late summer biomass data (De Wilde & Beukema, 1984). Assuming an ecological efficiency of 10%, this means a consumption of 180 gc m^{-2} y^{-1}. When the additional detritus source, a low quality food, is not taken into account, it appears that phytoplankton and microphytobenthos together are hardly filling the demands of the primary consumers, taking into account that besides macrozoobenthos also zooplankton and benthic meiofauna are demanding their share, roughly another 100 g C m^{-2} y^{-1} (De Wilde & Beukema, 1984).

A bottle neck is revealed in the transfer from primary consumers to secondary consumers. The 20 g C m^{-2} y^{-1} macrozoobenthos production potentially covers the demands of the carnivores (7 g C m^{-2} y^{-1}; Beukema & Nienhuis, 1985), but a large part of this food source is not available for predators, because the density is too low to be detected as a proper food source, or because it is distributed in the sublittoral zone, being not accessible for waders. Besides that, the production estimate for macrozoobenthos is based on maximum late summer data. In late winter when biomass is at its lowest, food may be a serious limiting resource for migratory waders, especially during severe winters.

Regarding mussel and cockle fisheries, man acts as a competitor for waders during winter. In this context Lambeck (1989) reported serious constraints on food resources of wintering waders during the severe winters of 1985–1986 and 1986–1987 in Oosterschelde estuary. Manipulations by the managers of the storm surge barrier, leading to inaccessibility of the feeding grounds, together with an overexploitation by man of the natural cockle banks resulted in mass killings among wintering oystercatchers.

In the Grevelingen lagoon macrozoobenthos production may be limited by a restricted supply of microalgae (Table 3), which in turn are nitrogen-limited (De Vries & Hopstaken, 1984). The supply of phytoplankton in eutrophicated Veerse Meer seems more luxurious (Table 3) and is supposed to be not fully exploited by the benthic filter feeders, considering the large production of phytoplankton and the smaller net production of macrozoobenthos in this system.

An important difference between the two tidal

estuaries Dollard and Wadden Sea on the one hand and non-tidal Grevelingen and Veerse Meer on the other, is the exploitation of large quantities of macrophytes by herbivorous birds in the latter systems. In both lagoons, however, birds only consume roughly 5% of the annual macrophyte production (cf. 14% consumption calculated by Coosen *et al.* (1990) but presumably based on too low macrophyte production estimates).

It is striking that among the secondary consumers the levels of organic carbon consumption –man excluded– are almost the same, when tidal estuaries are compared with non-tidal lagoons, notwithstanding the fact that the consumer populations show large qualitative differences (Table 3). Carnivorous birds in Grevelingen consume mainly small fish (approximately 60% of the available biomass in 1981–1982; Doornbos, 1984). Almost 20% of the macrozoobenthos production may potentially be consumed by fish and larger invertebrates (Doornbos, 1987). In Veerse Meer the consumption by fish and invertebrates is unknown. The much higher consumption by carnivorous birds in Veerse Meer (Coosen *et al.*, 1990), as compared to Grevelingen (Doornbos, 1987), is explained by the fact that Veerse Meer harbours several times higher densities of zoobenthos-eating diving ducks (Meire *et al.*, 1989). The question whether the carrying capacity of macrozoobenthos in both lagoons fills the needs of the secondary consumers, seems to be answered positively, although carnivorous birds may be food-limited owing to the unpredictable year to year variations in the production of small fish.

In the Delta area of Rhine and Meuse any possible potential effect of eutrophication has completely been masked by the effects of the enclosure of several estuaries in the period 1961 to 1987. All examples from Grevelingen lagoon about long-term changes are clearcut, following an absolute and sudden change in environmental circumstances. Things are different when eutrophication is occurring, a slow and gradual process, loading the estuary with nutrients, presumably giving rise to higher primary production and – far less convincing – to higher secondary production and biomass (cf. examples from the Wadden Sea). Does a higher secondary production, i.e. more food for waders, also lead to a larger population of migratory birds? Or stated otherwise: Dutch coastal eutrophication – is it enough or already too much? (Beukema & Cadée, 1987).

The counteracting negative aspects of chronic eutrophication are all too obvious in many estuarine areas in the world (*e.g.* Baltic, Rosenberg, 1985; northern Adriatic, Sfriso *et al.*, 1987; Black Sea, Bologa, 1989). These include anaerobiosis, toxic algal blooms, mass kills of benthic and epibenthic animals and changes in species patterns. Non-tidal areas with a long residence time of the water mass and prevailing sedimentation and stratification appear to be most vulnerable to eutrophication. Once the unwanted negative effects of eutrophication emerge, the process is very difficult to stop. Subsequent delivery of nutrients from underwater sediments may extend the process for many years. As long as we do not know in detail the long-term effects of eutrophication on estuarine ecosystem functioning, the precautionary principle (Portman, 1989) should be supported, avoiding excessive inputs and reducing future loadings.

Acknowledgements

The author thanks Dr P.G.Harrison (Univ. British Columbia, Vancouver, Canada), Dr J.de Leeuw (DIHO, Yerseke), Dr R.H.D. Lambeck (DIHO, Yerseke), Dr J.J. Beukema (NIOZ, Texel), Dr V.N. de Jonge (DGW, Groningen), Dr K. Essink (DGW, Groningen) for critically reading and improving the manuscript. Communication no. 519 of NIOO, Yerseke, The Netherlands.

References

Bakker, J.P., J. de Leeuw, K.S. Dijkema, P.C. Leendertse, H.H.T. Prins & J. Rozema, 1993. Salt marshes along the coast of The Netherlands. Hydrobiologia 265: 73–95.

Baretta, J. & P. Ruardy (eds), 1988a. Tidal flat estuaries. Springer-Verlag, Berlin: 1–353.

Baretta, J.W. & P. Ruardy, 1988b. Enhanced benthic-pelagic

coupling in the western Wadden Sea model. In EON-projectgroep: The ecosystem of the western Wadden Sea: field research and mathematical modelling. EMOWAD II – NIOZ-report 1988–11: 105–111.

Batje, M. & H. Michaelis, 1986. *Phaeocystis pouchetii* blooms in the Eastfrisian coastal waters, German Bight, North Sea. Mar. Biol. 93: 21–27.

Beukema, J.J., 1974. Seasonal changes in the biomass of the macrozoobenthos of a tidal flat area in the Dutch Wadden Sea. Neth. J. Sea Res. 8: 94–107.

Beukema, J.J., 1976. Biomass and species richness of the macrobenthic animals living on the tidal flats of the Dutch Wadden Sea. Neth. J. Sea Res. 10: 236–261.

Beukema, J.J., 1982. Biologische produktie in de ondiepe, rijke randen van de zee. In W.J. Wolff (ed.), Wadden, duinen, delta. Pudoc, Wageningen: 32–53.

Beukema, J.J., 1989. Long-term changes in macrozoobenthic abundance on the tidal flats of the western part of the Dutch Wadden Sea. Helgoländ. Meeresunters. 43: 405–415.

Beukema, J.J. & G.C. Cadée, 1986. Zoobenthos responses to eutrophication of the Dutch Wadden Sea. Ophelia 26: 55–64.

Beukema, J.J. & G.C. Cadée, 1987. De eutrofiëring van ons kustwater: genoeg of al te veel? Vakbl. Biol. 67: 153–157.

Beukema, J.J. & P.H. Nienhuis, 1985. Processen in oecosystemen. In K.Bakker *et al.* (eds), Inleiding tot de oecologie. Bohn, Scheltema & Holkema, Utrecht: 323–348.

Billen, G. & C. Lancelot, 1987. Modelling benthic microbial processes and their role in nitrogen cycling of temperate coastal ecosystems. In H.T. Blackburn & J. Sorensen (eds), Nitrogen in coastal marine environments. Scope Wiley, New York: 341–378.

Blackburn, T.H. & K. Henriksen, 1983. Nitrogen cycling in different types of sediments from Danish waters. Limnol. Oceanogr. 28: 477–493.

Bokhorst, M., 1988. Inventarisatie van een aantal abiotische factoren in de Westerschelde. DIHO Studentenversl. D6-1988: 1–22.

Bologa, A.S., 1989. Present status of seaweed production along the Romanian Black Sea shore. Vie Milieu 39: 105–109.

Brockmann, U., G. Billen & W.W.C. Gieskes, 1988. North Sea nutrients and eutrophication. In W.Salomons, B.L.Bayne, E.K.Duursma & U.Förstner (eds), Pollution of the North Sea. An assessment. Springer-Verlag, Berlin/ Heidelberg: 348–389.

Broekhuysen, G.J., 1935. The extremes in percentages of dissolved oxygen to which the fauna of a Zostera field in the tide zone at Nieuwediep can be exposed. Arch. Neerl. Zool. 1: 339–346.

Cadée, G.C., 1980. Reappraisal of the production and import of organic carbon in the western Wadden Sea. Neth. J. Sea Res. 14: 305–322.

Cadée, G.C., 1984. Has input of organic matter into the western part of the Dutch Wadden Sea increased during the last decades? Neth. Inst. Sea Res. Publ. Ser. 10: 71–82.

Cadée, G.C., 1986. Increased phytoplankton primary production in the Marsdiep area (western Dutch Wadden Sea). Neth. J. Sea Res. 20: 285–290.

Cadée, G.C.& J. Hegeman, 1974a. Primary production of phytoplankton in the Dutch Wadden Sea. Neth. J. Sea Res. 8: 240–259.

Cadée, G.C. & J. Hegeman, 1974b. Primary production of benthic microflora living on tidal flats in the Dutch Wadden Sea. Neth. J. Sea Res. 8: 260–291.

Cadée, G.C. & J. Hegeman, 1977. Distribution of primary production of the benthic microflora and accumulation of organic matter on a tidal area, Balgzand, Dutch Wadden Sea. Neth. J. Sea Res. 11: 24–41.

Cadée, G.C. & J. Hegeman, 1986. Seasonal and annual variation in *Phaeocystis pouchetii* (Haptophyceae) in the westermost inlet of the Wadden Sea during the 1973 to 1985 period. Neth. J. Sea Res. 20: 29–36.

Cadeé, G.C. & J. Hegeman, 1991. Historical phytoplankton data of the Marsdiep. Hydrobiol. Bull. 24: 111–118.

Cameron, W.M. & D.W. Pritchard, 1963. Estuaries. In M.N. Hill (ed.), The Sea II. Wiley, New York: 306 –324.

Colijn, F., 1983. Primary production in the Ems-Dollard estuary. Ph.D. Thesis, University Groningen. Van Denderen, Groningen: 1–123.

Coosen, J., P. Meire, J.J. Stuart & J.Seys, 1990. Trophic relationships in brackish Lake Veere: the role of macrophytes. In M. Barnes & R.N. Gibson (eds), Trophic relations in the marine environment; Proceed. 24th Europ. Mar. Biol. Symp., Aberdeen Univ. Press: 404–423.

Coosen, J. & A.C. Smaal, 1985. Jaargemiddelde biomassa en activiteit van de dominante bodemdieren in de Oosterschelde. BALANS, Deltadienst, Middelburg, DIHO, Yerseke; Interimreport 1985–12: 1–82.

Daemen, E.A.M.J., 1985. Literatuuronderzoek met betrekking tot de oecologie van het Veerse Meer. Delta Consult, Kapelle: 1–116.

Dankers, N., M. Binsbergen, K. Zegers, R. Laane & M. Rutgers van der Loeff, 1984. Transportation of water, particulate and dissolved organic and inorganic matter between a salt marsh and the Ems-Dollard estuary, the Netherlands. Estuar. Coast. Shelf Sci. 19: 143–165.

Day, J.W. Jr., C.A.S. Hall, W.M. Kemp & A. Yanez-Arancibia (eds), 1989. Estuarine Ecology. Wiley, New York: 1–558.

De Jong, S.A., P.A.G. Hofman, A.J.J. Sandee & E.J. Wagenvoort, 1990. Primary production of benthic microalgae in the Oosterschelde estuary (SW Netherlands). DGW, Middelburg; DIHO Yerseke, BALANS report 43: 1–262.

De Jonge, V.N., 1990. Response of the Dutch Wadden Sea ecosystem to phosphorus discharges from the river Rhine. Hydrobiologia 195: 49–62.

De Jonge, V.N. & K. Essink, 1991. Long-term changes in nutrient loads and primary and secondary producers in the Dutch Wadden Sea. In M. Elliott & J.-P. Ducrotoy (eds), Estuaries and coasts: spatial and temporal intercomparisons. Olsen & Olsen, Fredensborg: 307–316.

De Jonge, V.N., K. Essink & R. Boddeke, 1993. The Dutch Wadden Sea: a changed ecosystem. Hydrobiologia 265: 45–71.

De Jonge, V.N., & L.A. Villerius, 1989. Possible role of carbonate dissolution in estuarine phosphate dynamics. Limnol. Oceanogr. 34: 332–340.

De Ruyter, W.P.M., L. Postma & J.M. de Kok, 1987. Transport Atlas of the Southern North Sea. Rijkswaterstaat Tidal Waters Division, Delft Hydraulics, The Hague: 1–34.

De Vlas, J., 1979. Annual food intake by plaice and flounder in a tidal flat area in the Dutch Wadden Sea, with special reference to consumption of regenerating parts of macrobenthic prey. Neth. J. Sea Res. 13: 362–393.

De Vries, I. & C.F. Hopstaken, 1984. Nutrient cycling and ecosystem behaviour in a salt-water lake. Neth. J. Sea Res. 18: 221–245.

De Vries, I., F. Hopstaken, H. Goossens, M. de Vries, H.de Vries & J. Heringa, 1988a. GREWAQ: an ecological model for Lake Grevelingen. Report T-0215-03, Rijkswaterstaat DGW, The Hague, Delft Hydraulics, Delft, 159 pp. + 83 pp.

De Vries, I., W. van Raaphorst & N. Dankers, 1988b. Extra voedingsstoffen in zee: gevolgen, voordelen, nadelen. Landschap 5: 270–285.

De Wilde, P.A.W.J. & J.J. Beukema, 1984. The role of zoobenthos in the consumption of organic matter in the Dutch Wadden Sea. Neth. Inst. Sea Res. Publ. Ser. 10: 145–158.

Dethlefsen, V., 1989. Fish in the polluted North Sea. Dana 8: 109–129.

Dethlefsen, V. & H. von Westernhagen, 1983. Oxygen deficiency and effects on bottom fauna in the eastern German Bight 1982. Meeresforsch. 30: 42–53.

Dijkema, K.S., 1987. Changes in salt-marsh area in the Netherlands Wadden Sea after 1600. In A.H.L. Huiskes, C.W.P.M. Blom & J. Rozema (eds), Vegetation between land and sea. Dr W. Junk Publishers, Dordrecht: 42–49.

Dijkzeul, A., 1982. De waterkwaliteit van de Rijn in de periode 1970–1981. RIZA Report 82-061: 1–112.

Doornbos, G., 1984. Piscivorous birds on the saline lake Grevelingen, The Netherlands: abundance, prey selection and annual food consumption. Neth. J. Sea Res. 18: 457–479.

Doornbos, G., 1987. The fish fauna of Lake Grevelingen (SW Netherlands). Ph.D. Thesis, University Amsterdam, DIHO, Yerseke: 1–169.

EON-projectgroep, 1988. The ecosystem of the western Wadden Sea: field research and mathematical modelling. NIOZ Report 1988–11: 1–139.

Essink, K., 1985. Monitoring of mercury pollution in Dutch coastal waters by means of the teleostean fish Zoarcus viviparus. Neth. J. sea Res. 19: 177–182.

Fairbridge, R.W., 1980. The estuary: its definition and geodynamic cycle. In E. Olauson & I. Cato (eds), Chemistry and biochemistry of estuaries. Wiley, New York: 1–35.

Fenchel, T.M. & B.B.-Jorgensen, 1977. Detritus food chains of aquatic ecosystems: the role of bacteria. In M. Alexander (ed.), Advances in microbial ecology, Vol I. Plenum, New York: 1–57.

Fenchel, T.M. & R.J. Riedl, 1970. The sulfide system: a new biotic community underneath the oxidized layer of marine sand bottoms. Mar. Biol. 7: 255–268.

Green, J., 1968. The biology of estuarine animals. Sidgwick & Jackson, London, 401 pp.

Groenendijk, A.M., 1984. Primary production of four dominant salt marsh angiosperms in the South-West Netherlands. Vegetatio 57: 143–152.

Hannewijk, A., 1988. De verspreiding en biomassa van macrofyten in het Veerse Meer, 1987. DIHO Rapp. Verslag. 1988–2: 1–25.

Herman, P.M.J. & H. Hummel, 1989. Project group Westerschelde. In C.H.R. Heip & E.S. Nieuwenhuize (eds), Progress Report Delta Institute for Hydrobiological Research, Dutch Royal Academy of Sciences, Amsterdam: 51–52.

Herman, P.M.J. & H. Scholten, 1990. Can suspension-feeders stabilise estuarine ecosystems? In M. Barnes & R.N. Gibson (eds), Trophic relations in the marine environment; Proceed. 24th Europ. Mar. Biol. Symp., Aberdeen Univ. Press: 104–116.

Hickel, W., E. Bauerfeind, U. Niermann & H. von Westernhagen, 1989. Oxygen deficiency in the southeastern North Sea: sources and biological effects. Ber. Biol. Anst. Helgoland 4: 1–148.

Huiskes, A.H.L., 1988. The salt marshes of the Westerschelde and their role in the estuarine ecosystem. Hydrobiol. Bull. 22: 57–63.

Hummel, H., A. Meyboom & L. de Wolf, 1986. The effects of extended periods of drainage and submersion on condition and mortality of benthic animals. J. Exp. Mar. Biol. Ecol 103: 251–266.

Hummel, H., G. Moerland & C. Bakker, 1988. The concomitant existence of a typical coastal and a detritus food chain in the Westerschelde estuary. Hydrobiol. Bull. 22: 35–41.

Kaplan, W., I. Valiela & J.M. Teal, 1979. Denitrification in a salt marsh ecosystem. Limnol. Oceanogr. 24: 726–734.

Kelderman, P., 1980. Phosphate budget and sediment-water exchange in Lake Grevelingen (SW Netherlands). Neth. J. Sea Res. 14: 229–236.

Kelderman, P., 1985. Nutrient dynamics in the sediment of Lake Grevelingen (SW Netherlands). Ph.D. Thesis, University of Groningen. Delft University Press: 1–84.

Kuipers, B.R., P.A.W.J. de Wilde & F. Creutzberg, 1981. Energy flow in a tidal flat ecosystem. Mar. Ecol Prog. Ser. 5: 215–221.

Laane, R.W.P.M., H. Etcheber & J.C. Relexans, 1987. Particulate organic matter in estuaries and its ecological implication for macrobenthos. Mitt. Geol.-Palaont. Inst. Univ. Hamburg, SCOPE/UNEP Sonderband Heft 64: 71–91.

Lambeck, R.H.D., 1989. Beroven de Deltawerken steltlopers van hun voedsel? Vogels 10: 8–12.

Lambeck, R.H.D. & E.B.M. Brummelhuis, 1985. Een bestandsopname in voorjaar 1984 van het macrozoobenthos in het Grevelingenmeer. DIHO Rapp. Versl. 1985–4: 1–28 (unpublished report).

Lambeck, R.H.D. & R. Pouwer, 1986. Een bestandsopname in voorjaar 1985 van het macrozoobenthops in het Grevelingenmeer, en enige notities over lange-termijn ontwikkelingen. DIHO Rapp. Versl. 1986–5: 1–40 (unpublished report).

Lambeck, R.H.D. & G. de Smet, 1987. Een bestandsopname in voorjaar 1986 van het macrozoobenthos in het Grevelingenmeer.DIHO Rapp. Versl. 1987–4: 1–38 (unpublished report).

Lambeck, R.H.D., E.G.J. Wessel & A. Hannewijk, 1989. Macrozoobenthos in het Grevelingenmeer: een bestandsopname in voorjaar 1988. DIHO Rapp. Versl. 1989–5: 1–46 (unpublished report).

Law, C.S. & N.J.P. Owens, 1990. Denitrification and nitrous oxide in the North Sea. Neth. J. Sea Res. 29: 65–74.

Meire, P.M. & J. Coosen, 1985. Benthos and bird communities in a changing estuary. Proceed. Progress in Belgian Oceanographic Research, Brussels 1985: 292–303.

Meire, P.M., J. Seys, T. Ysebaert, P.L. Meininger & H.J.M. Baptist, 1989. A changing Delta: effects of large coastal engineering works on feeding ecological relationships as illustrated by waterbirds. In J.C. Hooghart & C.W.S. Posthumus (eds), Hydro-ecological relations in the Delta waters of the south-West Netherlands. TNO Com. Hydrol. Res. Proceed. Inform. 41: 109–146.

Nelissen, P.H.M. & J. Stefels, 1988. Eutrophication in the North Sea, a literature survey. NIOZ-report 1988–4: 1–100.

Newell, R.C. & J.G. Field, 1983. The contribution of bacteria and detritus to carbon and nitrogen flow in a benthic community. Mar. Biol Lett.4: 23–36.

Nienhuis, P.H. (ed.), 1985. Het Grevelingenmeer. Van estuarium tot zoutwatermeer. Natuur en Techniek, Maastricht, Brussel, 177 pp.

Nienhuis, P.H., 1983. Temporal and spatial patterns of eelgrass (*Zostera marina* L.) in a former estuary in the Netherlands, dominated by human activities. Mar. Techn. Soc. J. 17: 69–77.

Nienhuis, P.H., 1989a. Eutrophication of estuaries and brackish lagoons in the South-West Netherlands. In J.C. Hooghart & C.W.S. Posthumus (eds), Hydro-ecological relations in the Delta waters of the South-West Netherlands. TNO Com. Hydrol. Res. Proceed. Inform.41: 49–70.

Nienhuis, P.H., 1989b. Introduction – Research group Estuarine Foodchains. In C.H.R. Heip & E.S. Nieuwenhuize (eds), Progress Report 1988, Delta Institute for Hydrobiological Research, Dutch Royal Academy Sciences, Amsterdam: 8–9.

Nienhuis, P.H. & E.A.M.J. Daemen, 1985. De biomassa en produktie van het fytobenthos. In P.B.M. Stortelder (ed.), De koolstofbalans in de Oosterschelde. RWS-Deltadienst, Middelburg, The Hague, DIHO, Yerseke. Interimreport BALANS 1985–11: 67–79.

Nienhuis, P.H. & A.M. Groenendijk, 1986. Consumption of eelgrass (*Zostera marina* L.) by birds and invertebrates: an annual budget. Mar. Ecol. Progr. Ser. 29: 29–35

Officer, C.B., T.J. Smayda & R. Mann, 1982. Benthic filter-feeding: a natural eutrophication control. Mar. Ecol. Progr. Ser. 9: 203–210.

Portman, J.E., 1989. The chemical pollution status of the North sea. Dana 8: 95–108.

Postgate, J.R., 1984. The sulphate-reducing bacteria. Cambridge Univ. Press, UK, 208 pp.

Projectgroep Balans, 1988. Voedsel in de Oosterschelde. Rijkswaterstaat – DGW, Middelburg and DIHO, Yerseke, 53 pp.

Rachor, E. & H. Albrecht, 1983. Sauerstoff-Mangel im Bodenwasser der deutschen Bucht. Veröff. Inst. Meeresforsch. Bremerh. 19: 209–227.

Radach, G. & J. Berg, 1986. Trends in den Konzentrationen der Nahrstoffe und des Phytoplanktons in der Helgolander Bucht (Helgoland Reede Daten). Ber. Biol. Anstalt Helgoland 2: 1–63.

Reise, K., 1985. Tidal flat ecology. An experimental approach to species interactions. Ecological Studies 54: 1–191, Springer-Verlag, Berlin.

Reise, K., 1984. Indirect effects of sewage on a sandy tidal flat in the Wadden Sea. Neth. Inst. Sea Res. Publ. Ser. 10: 159–164.

Richardson, K., 1989. Algal blooms in the North Sea: the good, the bad and the ugly. Dana 8: 83–93.

Ridderinkhof, H., 1990. Residual currents and mixing in the Wadden Sea. Ph.D. Thesis, State University Utrecht. Publ. 12 BEWON: 1–91.

Riegman, R., F. Colijn, J.F.P. Malschaert, H.T. Kloosterhuis & G.C. Cadeée, 1990. Assessment of growth rate limiting nutrients in the North Sea by the use of nutrient-uptake kinetics. Neth J. Sea Res. 26: 53–60.

Rijkswaterstaat, 1985. Biological research Ems-Dollard estuary. RWS Commun. The Hague 40: 1–182.

Rijkswaterstaat, 1989. Water voor nu en later; Derde nota waterhuishouding, samenvatting. Rijkswaterstaat, The Hague, 66 pp.

Rijkswaterstaat, 1990a. Beleidsanalyse waterbeheer Veerse meer. Nota AX-89 + append. RWS-DZ, Middelburg.

Rijkswaterstaat, 1990b. 1990 Interim report on the quality status of the North Sea; Third Intern. Conference on the protection of the North Sea, March 7–8, The Hague. Rijkswaterstaat, The Hague, 48 pp.

Rosenberg, R., 1985. Eutrophication – the future marine coastal nuisance? Mar. Poll. Bull. 16: 227–231.

Scholten, H., O. Klepper, P.H. Nienhuis & M. Knoester, 1990. Oosterschelde estuary (SW Netherlands): a self-sustaining estuary? Hydrobiologia 195: 201–215.

Sfriso, A., A. Marcomini & B. Pavoni, 1987. Relationships between macroalgal biomass and nutrient concentrations in a hypertrophic area of the Venice Lagoon. Mar. Environm. Res. 22: 297–312.

Slob, G.J., 1989. 15 jaar vogelontwikkelingen in het afgesloten

Grevelingenbekken. Staatsbosbeheer, Goes: 1–50, 50 appendices.

Smaal, A.C., M. Knoester, P.H. Nienhuis & P.M. Meire, 1990. Changes in the Eastern Scheldt ecosystem induced by the Delta works. In M. Elliott & J.-P. Ducrotoy (eds), Estuaries and coasts: spatial and temporal intercomparisons. Olsen & Olsen, Fredensborg: 379–384.

Sörensen, J., 1984. Seasonal variation and control of oxygen, nitrate, and sulfate respiration in coastal marine sediments. In M.J.Klug & C.A. Reddy (eds), Current perspectives in microbial ecology. Amer. Soc. Microbiol., Washington, DC: 447–453.

Stronkhorst, J., R.N.M. Duin & H.A. Haas, 1985. Primaire produktie-onderzoek in het Veerse Meer (1982–1983). Nota DDMI 85.01, Middelburg: 1–52.

Tijssen, S.B. & A.J. van Bennekom, 1976. Lage zuurstofgehaltes in het water op het Balgzand. H2O 9(2): 28–31.

Valiela, I., 1984. Marine ecological processes. In D.E. Reichle (ed.), Springer advanced texts in life sciences. Springer, New York, 546 pp.

Van Buuren, J.T., 1988. De Noordzee en eutrofiëring. H_2O 21 (20): 591–595.

Van der Veer, H.W., W. van Raaphorst & M.J.N. Bergman, 1988. Eutrophication of the Dutch Wadden Sea – external nutrients loadings of the Marsdiep and Vliestroom basins. In EON-projectgroup: The ecosystem of the western Wadden Sea: field research and mathematical modelling. EMOWAD II – NIOZ report 1988: 113–122.

Vegter, F. & P.R.M. de Visscher, 1987. Nutrients and phytoplankton primary production in the marine tidal Oosterschelde estuary (The Netherlands). Hydrobiol. Bull. 21: 149–158.

Veldhuis, M.J.W., 1987. The ecophysiology of the colonial alga *Phaeocystis pouchetii.* Ph.D. Thesis, Univ. Groningen, Van Denderen Groningen, 127 pp.

Vosjan, J.H., 1987. A sketchy outline of the fate of organic matter in the Dutch Wadden Sea. Hydrobiol. Bull. 21: 127–132.

Weichart, G. 1986. Nutrients in the German Bight, a trend analysis. Deutsch. Hydrogr. Zeit. 39: 197–206.

Wellershaus, S., 1984. Seston particles in the Weser estuary. Neth. Inst. Sea Res. Publ.Ser. 10: 107–111.

Wetsteyn, L.P.M.J., J.C.H. Peeters, R.N.M. Duin, F. Vegter & R.P.M. de Visscher, 1990. Phytoplankton primary production and nutrients in the Oosterschelde (The Netherlands) during the pre-barrier period 1980–1984. Hydrobiologia 195: 163–177.

Wetsteyn, L.P.M.J. & L. Peperzak, 1988. Abiotische variabelen, primaire produktie en kwalitatieve fytoplanktonsamenstelling en -successie in de Oosterschelde in 1987. Nota GWAO-88.1009. Rijkswaterstaat DGW, Middelburg: 1–59.

Witte, J.IJ. & J.J. Zijlstra, 1984. The meiofauna of a tidal flat in the western part of the Wadden Sea and its role in the benthic ecosystem. Mar. Ecol. Prog. Ser. 14: 129–138.

Wollast, R., 1988. The Scheldt estuary. In W. Salomons, B.L. Bayne, E.K. Duursma & U. Förstner (eds), Pollution of the North Sea – an assessment. Springer-Verlag, Berlin: 183–193.

Hydrobiologia **265**: 45–71, 1993.
E. P. H. Best & J. P. Bakker (eds), Netherlands-Wetlands.

The Dutch Wadden Sea: a changed ecosystem

V.N. de Jonge[1], K. Essink[1] & R. Boddeke[2]
[1]*Rijkswaterstaat, Tidal Waters Division (DGW), P.O. Box 207, 9750 AE Haren (GN), The Netherlands*
[2]*The Netherlands Institute for Fishery Investigations (RIVO), P.O. Box 68, 1970 AB IJmuiden, The Netherlands*

Key words: Dutch Wadden Sea, ecosystem, coastal engineering, dredging, fisheries, management

Abstract

Since 1600 the surface area of the Dutch Wadden Sea decreased considerably by successive reclamations of saltmarshes. In 1932 the Zuiderzee (3200 km^2) was closed off from the Wadden Sea causing in the remaining part an increase in tidal range and current velocities. In 1969 the Lauwerszee (91 km^2) was closed off and turned into a freshwater lake as well. Man's use of the Wadden Sea changed simultaneously. Dredging in harbours and shipping routes, and extraction of sand and shells became common practice. Extraction of sand increased manifold between 1960 and 1985. These activities did contribute to the turbidity of the Wadden Sea water. Discharge of nitrogen and phosphorus compounds into the western Wadden Sea increased also manifold since 1950, causing an increase in phytoplankton production, duration of phytoplankton blooms, and intertidal macrozoobenthic biomass. Loads of metals and organochlorine contaminants entering the Wadden Sea were hard to estimate.

Fisheries changed drastically since the 1930's. Fishing for 'Zuiderzee' herring came to an end shortly after closing off the Zuiderzee. The anchovy fishery ceased in the 1960', that for flounder in 1983. Undersized brown shrimps were fished until 1971. Selective shrimptrawls and sorting devices with flushing seawater were introduced to reduce mortality among young flatfish and shrimps. Oysters became extinct in the 1960's due to over-exploitation of the natural beds. Production of mussels increased more than ten times between 1950 and 1961 due to 'culturing', catches of cockles increased slowly between 1955 and 1984. Whelks were fished until 1970.

The most important changes in the biotic system of the Wadden Sea, increased production of microalgae and intertidal macrozoobenthos, can be attributed to increased nutrient loads. Eutrophication provided ample food supply for mussels which are harvested mainly by man and eider duck, and may have caused also increased growth in juvenile plaice. Increased turbidity may have impaired life conditions for adult dab, and have prevented also recovery of sublittoral eelgrass beds after their disappearance in the 1930's due to the 'wasting disease'. Increased turbidity in the Wadden Sea is probably caused by closing off the Zuiderzee (1932), a significant increase of dredge spoil disposal near Hoek van Holland between 1970 and 1983, and a more than 10-fold increase of mussel culturing in the Wadden Sea since 1950. Stocks of several bird species breeding in the Wadden Sea area suffered great losses in the early 1960's due to a pesticide accident. Most of the breeding populations have recovered. PCB's caused a reproduction failure among harbour seals in the 1970's.

Since 1980 official Dutch policy aims at multiple use of the Wadden Sea, with emphasis on protection and restoration of the natural environment. The 3rd Water Management Plan (1989) aims at a development of the Wadden Sea ecosystem towards the situation of ca. 1930, i.e. without undoing present sea dikes and reclaimed polders. Management of the Dutch Wadden Sea will therefore have to focus mainly on reduction of eutrophication, pollution and turbidity. Some management options are discussed.

1. Introduction

The Wadden Sea is a shallow inshore area with a surface of roughly 6,000 km^2 extending from The Netherlands to Denmark. The Dutch part of the Wadden Sea with an area of approximately 3,000 km^2 is situated between the North Sea and the mainland (Fig. 1). It is separated from the North Sea by a series of barrier islands and from the mainland by dikes.

The Wadden Sea is a highly dynamic area. The dynamics are caused by many factors. The most striking factors are tide and climate (wind and temperature). Freshwater discharges and the outflow of the River Rhine cause changes in salinity. Together with the freshwater, particulates and dissolved chemical compounds are carried to the area. Part of these substances comes from weathering processes and part from human activity. These compounds are nutrients, like phosphorus and nitrogen, but also contaminants, like heavy metals, PCB's, TBT, HCH and many others. Some dynamics in the Wadden Sea are directly caused by human activities. Examples are the mussel, cockle and shrimp fisheries. Extraction of bottom deposits and dumping of harbour spoil also contribute to this. Finally, engineering works can serve as an example of human inter-

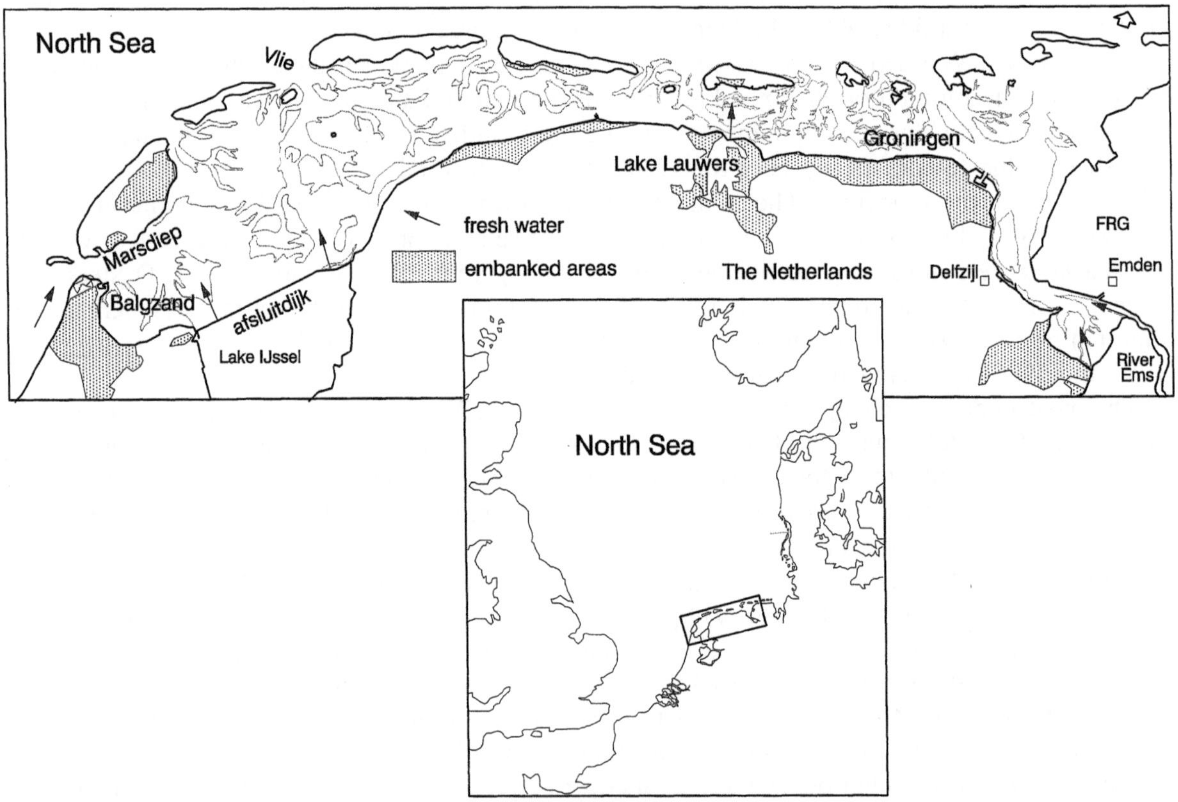

Fig. 1. Map of the Dutch Wadden Sea and Ems Estuary with the main tidal channels (dotted lines), areas embanked since 1600 (after: Dijkema, 1987), and the main routes of fresh water to the Wadden Sea (arrows).

ference which may result in dramatic changes in the dynamics of a shallow coastal area like the Wadden Sea.

In the last 50 years important changes occurred in the abiotic environment of the Wadden Sea, ranging from increased input of nutrients and contaminants to loss of extensive areas due to embankments and related changes in hydraulic processes.

In this paper some of the major changes in the abiotic environment and in the biocoenosis (species composition as well as production) of the Dutch Wadden Sea will be described. The interrelationships between the environmental changes, the role of human activities and the biota will be discussed, and some alternatives for management of the Wadden Sea environment will be presented.

2. Characteristics and functioning of the ecosystem

2.1. Geomorphology

The structure of the Wadden Sea area is the result of different geological processes, such as tectonic subsidence, compaction of the substratum and eustatic rise of the sea level (Veenstra, 1980). Recently, local subsidence due to gas exploitation occurs.

In the past, a girdle of barrier sand banks was formed in front of the coastline. The banks were formed by sand transport due to wind induced waves, and to a lesser extent to water currents (Postma, 1980). When these banks reached the high water level the wind became an important factor in the displacement of the sand in eastern direction, resulting in the formation of beaches and dunes. These morphological entities grew into islands. Fine materials (mud) accreted on sheltered localities behind the dunes. Finally, embankments gave the islands their present shape.

Approximately 50% of the present area of the Wadden Sea consists of intertidal flats. In this paper two main areas of the Dutch Wadden Sea are distinguished, viz. the western Wadden Sea (mainly the Marsdiep tidal basin) and the eastern Wadden Sea (the area between Lake Lauwers and Ems Estuary).

2.2. Physical processes

The tide in the Wadden Sea is driven by the tide in the North Sea. The tide in the Southern Bight of the North Sea is dominated by a semi-diurnal lunar tide, and generates the vertical water movement in the tidal inlets. This results in strong tidal currents with complicated residual patterns in the different tidal basins in the Wadden Sea (Zimmerman, 1976; Eisma, 1980). These currents are responsible for the transport of dissolved and particulate substances.

Due to several inputs of fresh water (cf. Fig. 1), the Wadden Sea is an estuarine environment (cf. Nienhuis, 1992). An important indirect source of fresh water is the northward residual current along the Dutch coast (Van Veen, 1938) transporting a mixture of Atlantic water entering through the Straits of Dover and fresh water from the Rivers Scheldt, Meuse and Rhine. Fresh water enters the Wadden Sea also from the River Ems and through a number of small canals (approx. 20 $m^3 s^{-1}$ annually), from Lake Lauwers (20–50 $m^3 s^{-1}$ annually) and from Lake IJssel (250–670 $m^3 s^{-1}$ annually). The fresh water in the Dutch Wadden Sea mainly originates from the River Rhine. The discharge varies, due to variations in precipitation and in the melting of snow in the Swiss Alps.

2.3. Sedimentological processes

The Wadden Sea imports particles from the North Sea through the tidal inlets. Estimates of sediment import vary considerably. McCave (1973) and Eisma (1981) gave figures of 0.4 and 1.6×10^6 tonnes respectively of mud (silt) per annum for the entire Dutch Wadden Sea. For the Marsdiep tidal inlet alone, an annual import of approximately 7×10^6 tonnes sediment (silt and sand) was calculated by us from the data of Glim *et al.* (1987). Based on data of Van Heuvel (1991) and Reenders & Van der Meulen (1972) a total annual sediment input in the Ems estuary (ca. 460 km^2) of ca. 5.6×10^6 tonnes could be calculated of which ca. 1.5×10^6 tonnes consists of mud (fraction <55 μm). Recently, De Jonge (1992) gave approximately the same figure (1.4×10^6 tonnes of mud). This value is considerably higher than the values given by the other authors.

Together with inorganic sediment also organic particulate material is imported into the Wadden Sea (Postma, 1954). The input of organic matter and particulate phosphorus into the Marsdiep tidal basin was estimated at approx. 250 gC m^{-2} and 6 gP m^{-2} in 1970 (De Jonge & Postma, 1974). There are also indications that at present the western Wadden Sea is not an organic carbon importing system but an exporting system (Baretta & Ruardij, 1989). Very fine particulates, colloids and dissolved substances are thought to be transported in seaward direction (Postma, 1967).

2.4. Supply and cycling of nutrients and contaminants

Nutrients reach the Wadden Sea from different sources of fresh water (see above) and in different forms. For the present study we distinguish total-P, total-N, phosphate and total dissolved inorganic nitrogen. The changes in the annual loads, together with the fresh water discharge, are given in Fig. 2. All curves clearly demonstrate the manifold increase in nutrient loads from the 1950's (or early 1970's) onwards.

It is conspicuous that the phosphate loads from the River Rhine increased steadily and drop again after 1981 (Fig. 2a). For total phosphorus the load levels off since ca. 1965.

The phosphorus loads from Lake IJssel increased significantly since the early seventies. The data indicate a possible decrease since 1984, three years later than in the River Rhine itself (Fig. 2b). The nutrient loads from Lake Lauwers show an increase till 1979, and a stabilization thereafter (Fig. 2c).

In all fresh water bodies (Lake IJssel, Lake Lauwers and River Ems) discharging on the Dutch Wadden Sea phosphorus is the potential limiting factor for the development of phytoplankton (De Jonge & Essink, 1991).

Various polluting substances reach the Wadden Sea directly (*e.g.* by pipelines) or indirectly (with the fresh water from rivers and canals). Effects of organic wastewater discharge are known from the Ems estuary (BOEDE, 1985) and from the eastern Wadden Sea (Essink, 1978). In the Wadden Sea these effects are mainly of local importance (Essink, 1984).

The rivers Rhine and Ems are important sources of a variety of contaminants that eventually can reach the Wadden Sea. Reliable data on the amounts of contaminants entering the Wadden Sea from these sources are not available, mainly because the transport from Hoek van Holland northward along the Dutch coast can hardly be quantified.

The contaminant load of the river Rhine is rather well documented. The annual loads of mercury and cadmium showed a significant decrease in 1970–1975 and 1978–1983 respectively (Van Meerendonk *et al.*, 1988). In the western Dutch Wadden Sea also a decrease in mercury contents in eelpout (*Zoarces viviparus*) could be demonstrated (Essink, 1988). A decrease in pollution of the Dutch Wadden Sea by cadmium can be inferred from data on chemical monitoring of invertebrate species such as shrimp (*Crangon crangon*) and lugworm (*Arenicola marina*) (Essink, 1989).

Little information is available on changes in the input of organochlorine pesticides, polychlorinated biphenyls (PCB's) and polycyclic aromatic hydrocarbons (PAHs). In the western Dutch Wadden Sea freshwater discharges from Lake IJssel are a significant source of PAHs (Van Meerendonk *et al.*, 1988).

In the 1960's an accidental industrial discharge of telodrin and other pesticides caused an extensive mortality among gulls, terns, eider ducks and spoonbills breeding in the Wadden Sea (Koeman, 1971; Rooth & Jonkers, 1972; Swennen, 1972).

A general picture of nutrient cycling in Dutch

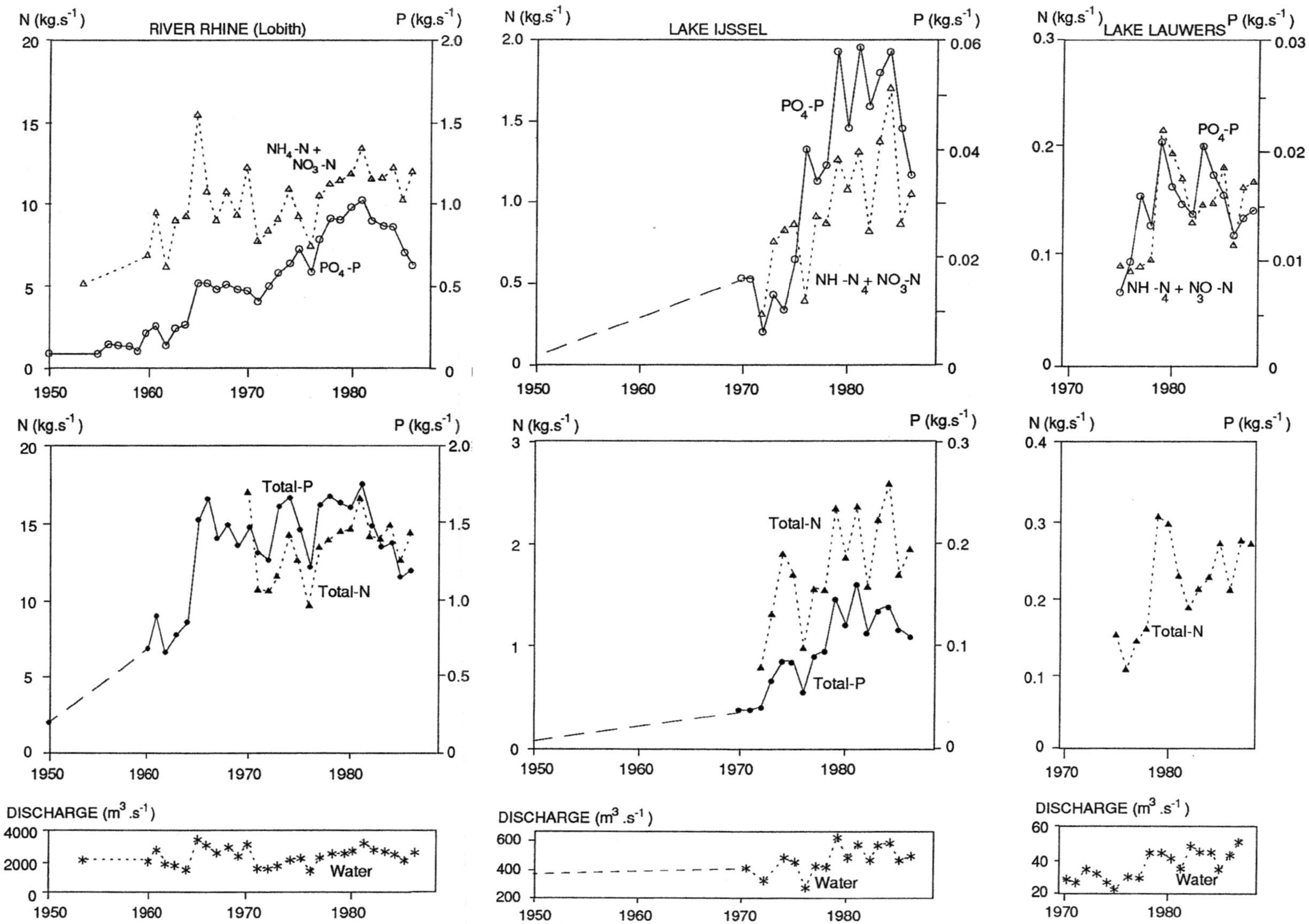

Fig. 2. Time series of annual nutrient loads (kg s^{-1}) and the mean annual water discharge (m^3 s^{-1}) of the River Rhine, Lake IJssel and Lake Lauwers. Data from Rijkswaterstaat.

estuaries is drawn by Nienhuis (1993). The general concept on the accumulation of sediment (including particulate organic matter) in the Wadden Sea is given above. Part of the imported organic carbon is mineralized in the Wadden Sea resulting in dissolved nutrients. These may be taken up by local phytoplankton. Recent studies have shown the direct importance for the western Wadden Sea of the phosphorus nutrient input from Lake IJssel (Van Raaphorst & Van der Veer, 1990; De Jonge, 1990) (Fig. 3). The supply of phosphorus from the different sources, the transport routes of this nutrient and the transformations of the different phosphorus compounds are shown in the diagram of Fig. 3.

It is assumed that the nutrient supply of the eastern Wadden Sea depends largely on the mineralization of organic matter from the North Sea and/or the adjacent Wadden Sea basins. For the Ems estuary nutrients are derived from the rivers Westerwoldsche Aa and Ems and the coastal zone of the North Sea (De Jonge & Essink, 1991).

The cycling of contaminants is schematized in the diagram of Fig. 4. The figure clearly illustrates that the levels of contaminants depend on different equilibria between water, sediment and organisms. The equilibrium constants are mainly dependent on the properties of the different substances or carriers (Boon & Duinker, 1986).

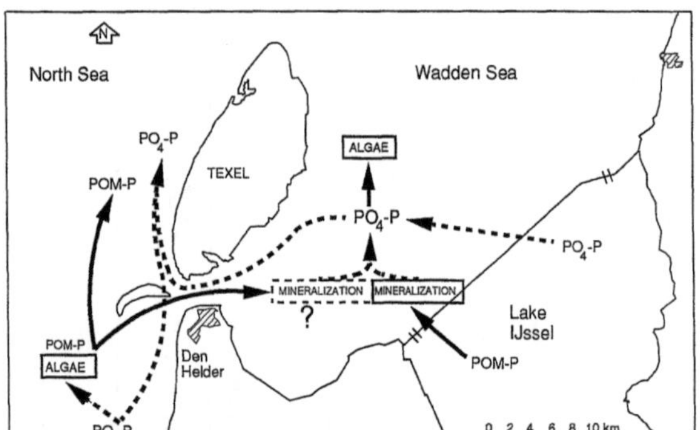

Fig. 3. Diagram representing the main route of phosphorus and the main conversion of the different phosphorus components in the western Dutch Wadden Sea. From: de Jonge (1990).

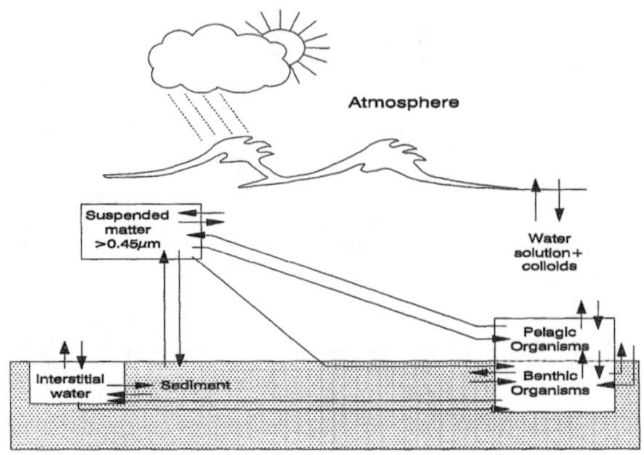

Fig. 4. Interactions between different marine environmental compartments with regard to contaminants. After: Boon & Duinker (1986).

2.5. Food web

The food web of the Wadden Sea does not differ from that of other brackish ecosystems. Primary production occurs by phytoplankton and phytobenthos in which benthic diatoms play a dominant role. The annual primary production varies dramatically possibly mainly as a result of changes in nutrient supply (De Jonge, 1990). The lowest estimate of phytoplankton primary production (ca. 40 gC m^{-2} a^{-1}) is from the late 1940's and early 1950's (Postma, 1954). The largest value (520 gC m^{-2} a^{-1}) comes from Cadée (1986, Fig. 12). A considerable part of the primary production in the Wadden Sea is used by macrozoobenthos. The standing stock of the macrozoobenthos also changes significantly from year to year. However, on the long run an increasing trend in macrozoobenthos biomass is observed which remarkably agrees with the trend in primary production (De Jonge & Essink, 1991). Lowest macrozoobenthos production was ca. 20 g m^{-2} a^{-1} AFDW while the highest was 85 g m^{-2} a^{-1} AFDW (Beukema and Cadée, 1986).

Other groups of organisms have not been studied as thoroughly as the algae and macrozoobenthos. A conservative carbon budget for the western Wadden Sea was made by Baretta & Ruardij (1989) (Table 1). For a more detailed picture of the functioning of the food web the reader is referred to Nienhuis (1992).

Table 1. Annual carbon budget of the western Wadden Sea for 1986. From: Baretta & Ruardij (1989).

Import/production	(10^3 ton C a^{-1})		Export/consumption
Input Lake IJssel	133	146	Output North Sea
Pelagic primary production	610	342	Pelagic consumption
Benthic primary production	53	109	Intertidal consumption
From carbon pool	25	192	Subtidal consumption
		32	Buried
Total	821	821	

3. Human activities

3.1. Coastal engineering

Since 1600 many areas in the Wadden Sea were embanked or otherwise separated from the sea (Fig. 1). During the 20th century small parts of the Wadden Sea were embanked in the province of Groningen, south of the island of Ameland, and in the Balgzand area. Although these activities temporarily reduced the total surface of the salt-marshes, accretion caused the formation of new ones (Dijkema, 1987). The closure of the Lauwerszee in 1969 resulted in a loss of another 91 km^2.

The construction of the 'Afsluitdijk', closing off the Zuiderzee in 1932, decreased the total area of the Wadden Sea with 3,200 km^2, a greater tidal range (Fig. 5), current velocities increasing 10–25% (Thijsse, 1972) and a much steeper salinity gradient (Fig. 17). Between 1932 and ca. 1950 a very rapid accretion in part of the western Wadden Sea took place (Glim *et al.*, 1987).

Before 1953 natural materials (*e.g.* Vilvordian limestone and basalt) were used as a cover of sea-dikes providing a suitable substrate and hiding places for plants and animals. Since the start in 1953 of a large-scale reconstruction of sea-dikes concrete and bitumen were used to cover the sea-dikes removing all cavities, cracks and ridges, releasing a.o. toxic polyaromatic hydrocarbons (PAHs). A general impoverishment or even disappearance of the characteristic dike communities of flora and fauna has been the result (Den Hartog, 1959, 1961).

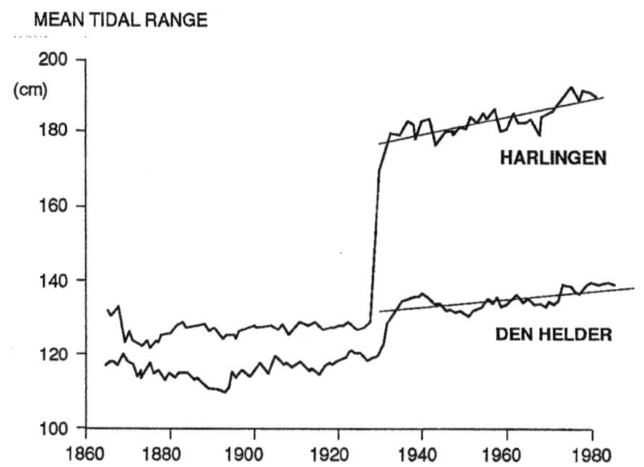

Fig. 5. Changes in the mean tidal range (cm) near Harlingen and Den Helder (western Dutch Wadden Sea). Data from: Rijkswaterstaat.

3.2. Dredging

Dredging of shipping channels and dumping of harbour dredge spoil does increase turbidity in an estuary (De Jonge, 1983; De Jonge & Essink, 1991). In the Wadden Sea 2–3 × 10^6 m^3 of sediment are dredged annually. The majority of this (ca. 2 × 10^6 m^3) is dredged in harbours of the Wadden Sea. The dredged sediment is dumped in the Wadden Sea. As the sediment in Wadden Sea harbours is contaminated with metals (*e.g.* zinc) and organic contaminants (*e.g.* tributyltin) possibilities to store contaminated dredge spoil are presently investigated.

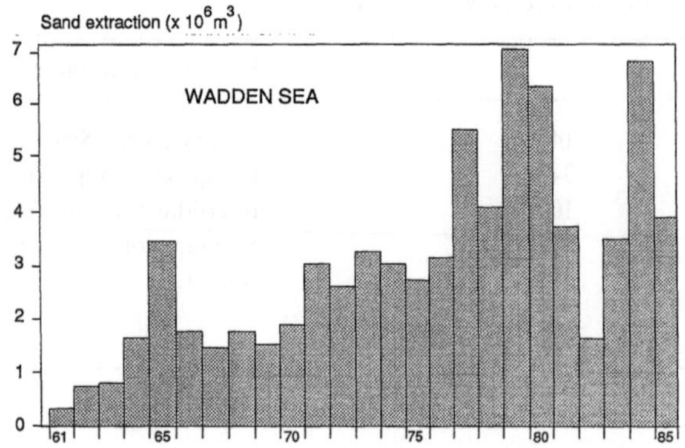

Fig. 6. Quantities of sand extracted anually from the Wadden Sea. Data from: Rakhorst (1982) and Rijkswaterstaat.

3.3. *Extraction of deposits*

In the Wadden Sea shells, clay, and sand are extracted (Rakhorst, 1982). The quantities of sand extracted are high (Fig. 6), and environmental effects of this deposit extraction can be serious. Most of the extraction activities are carried out in the main channels close to or in the tidal inlet, leading to hydraulic changes that extend over nearly the entire tidal basin. The extraction of sand also causes morphological anomalies in the system resulting in a changed erosion-sedimentation cycle and increased suspended matter concentrations (De Jonge, 1983). Because these activities are carried out on many localities, they are considered as one of the causes for the increased suspended matter concentrations in parts of the Wadden Sea (Fig. 7; see also De Jonge, 1986).

3.4. *Fisheries*

The fisheries on fish and shrimps in the Dutch Wadden Sea have changed drastically since the 1930's when the fisheries in this area had a true artisanal character. A wide variety of species was fished with various gear.

Before the Zuiderzee was closed off in 1932, an extensive fishery on 'Zuiderzee herring' (*Clupea harengus*) was carried out in the western Wadden Sea, the Zuiderzee herring being a race spawning

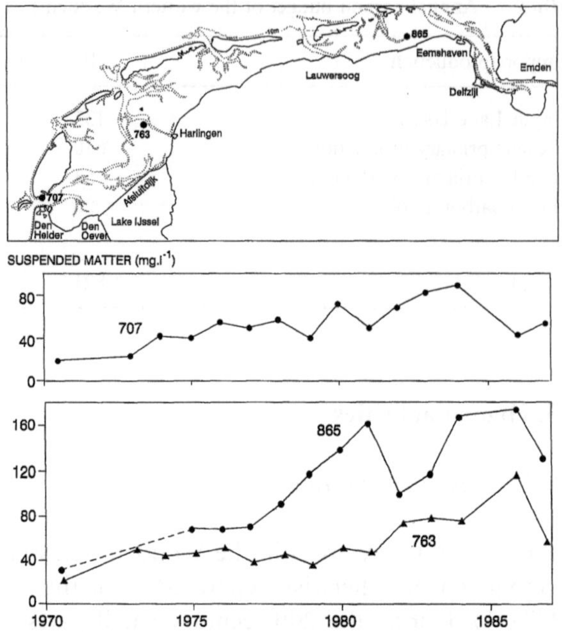

Fig. 7. Time series of annual mean suspended matter concentrations at three locations in the Wadden Sea. Data from: Rijkswaterstaat.

in March–April in the eastern part of the Zuiderzee. Before 1932, catches of herring were approx. 10,000 tonnes/year in the Zuiderzee, and 2500 tonnes in the western Wadden Sea (Fig. 8; Boddeke, 1991). After a temporary increase to an average catch of 14,000 tonnes/year in 1934–1936, specific herring fishery disappeared from the western Wadden Sea.

Nowadays, the Dutch Wadden Sea is still an important nursery area for herring from the central North Sea and the English Channel. Larvae arrive in the Wadden Sea in February–March and April–May, respectively. At the end of the summer the juveniles return to the North Sea (Van Beek, 1992).

The most valuable pelagic species, the anchovy (*Engraulis encrasicolus*), was caught till 1959 in quantities comparable to those obtained from the Dutch Wadden Sea before the closing off of the Zuiderzee (Fig. 9). Specific fishery for this species ceased in the 1960's. Anchovies, are still present in the Wadden Sea in summer, but recent recordings of spawning do not exist (Boddeke, 1992).

Recently, fishery for smelt (*Osmerus eperlanus*) carried out with pelagic trawls and trammle nets

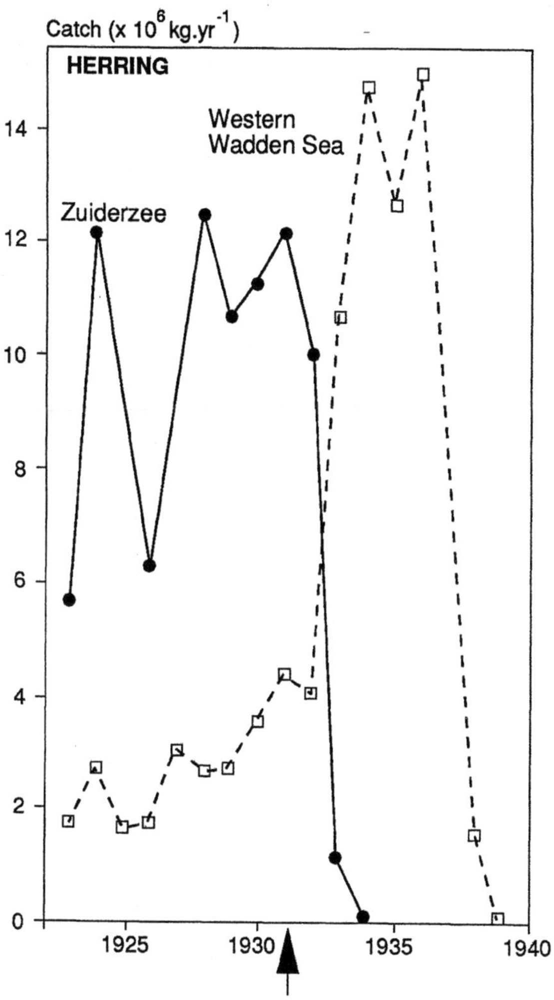

Fig. 8. Annual catch of herring in the Zuiderzee and in the western Dutch Wadden Sea. Closure of the Zuiderzee is indicated (↑). After: Boddeke (1991).

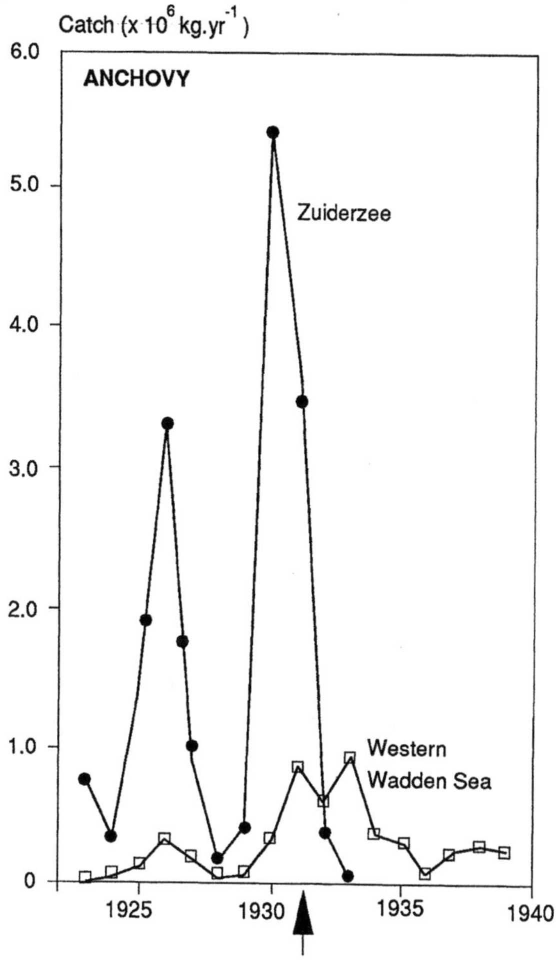

Fig. 9. Annual catch of anchovy in the Zuiderzee and in the western Dutch Wadden Sea. Closure of the Zuiderzee is indicated (↑). After: Boddeke (1991).

near the Afsluitdijk sluices during winter became of some importance, with landings of 73 tonnes in 1987. Before 1940, no fishing for smelt occurred in the Dutch Wadden Sea.

In the 1930's, a rather large fishery for flounder (*Platichthys flesus*) existed in the Dutch Wadden Sea, carried out with fixed nets and hooklines. Flounder, landed and sold alive, was a popular fish for human consumption for which relatively high prices were paid. Landings from the western Dutch Wadden Sea were 269 tonnes/year on average in 1930–1938. At present, landings of flounder are negligible. In 1983, the latest year of which data are available, landings were less than 1 tonne. This decline of the landings seems to be caused mainly by the decreased demand for flounder (Boddeke, 1992).

During the 1960's the total annual catch of rays (mainly *Raja clavata*) in the Wadden Sea amounted to 13–121 kg. After 1966 no rays were brought to the fish auction any more.

Fishing for eel (*Anguilla anguilla*), using mainly various types of fyke nets, did not change substantially with time. Concomitantly to the development of the mussel culture in the western Wadden Sea in the 1950's, landings of eel increased (see 4.5) (Boddeke, 1967b).

In the Dutch Wadden Sea the fishery for shrimps (*Crangon crangon*) has been subject to a remarkable development during the last 50 years. Before 1940, fishing for shrimps for human consumption (> 52 mm total length) was hampered by slack demand, forcing fishermen to sell their catch (undersized *and* consumption size) to the fishmeal industry. In the eastern Wadden Sea fishing with otter trawls and fixed nets was even entirely for industrial purposes. In the 1950's this situation started to change. Otter trawls were replaced by beamtrawls of which the groundrope was provided with wooden or rubber rollers to avoid the by-catch of bottom material. Fixed nets disappeared. Landing of undersized shrimps got prohibited in 1971.

In 1962, Dutch fishery research started an extensive programme to save the by-catch of undersized shrimps and flatfish. At first, this led to the development of a selective trawl from which flatfish escape. Later on, this development was followed by the introduction of fish-friendly sorting devices on board replacing the old shaking sieves. Later, these new sieves were provided with basins with running seawater to receive the catch, and with conveyor belts to transport the catch onto the sieves. From these sieves undersized shrimps and fish are returned to sea automatically by flushing seawater (Boddeke, 1989a). By these improvements, shrimp fishing in Dutch coastal waters has become very selective. The strong decrease in the losses of undersized shrimps has considerably improved the position of the fishery for shrimps in the Dutch Wadden Sea, which in the 1970's seemed to be doomed to disappear.

The Wadden Sea share of the shrimps landed from the Dutch coastal area was only 20% in 1976–1977, but increased till 50% in 1986–1987. At present, approx. 90 vessels are licensed to fish for shrimps in the Dutch Wadden Sea. This number is not expected to increase in future.

Over-exploitation of natural oyster (*Ostrea edulis*) beds has probably led to the extinction of this species in the Dutch Wadden Sea in the 1960's (Postuma & Rauck, 1978).

The annual production of mussels (*Mytilus edulis*) in the Wadden Sea increased from 6,000 tons in 1949 to 67,000 tons in 1962, after which the landings fluctuated at a somewhat lower level (Fig. 10). In 1985–87, the average annual landing was ca. 56,000 tons.

In the early 1950's fisheries on the edible cockle (*Cerastoderma edule*) started. Between 1955 and 1973, annual catches slowly increased to approx. 2×10^6 kg fresh meat. Between 1975 and 1984, catches reached an average level of 9×10^6 kg yr^{-1} (De Vlas, 1987). The cockle makes up ca. 16% of the total macrozoobenthos biomass in the western Wadden Sea tidal flats (Beukema, 1976). Large fluctuations occur in the cockle stocks due to their strongly varying success of spatfall, and mortality during severe winters (Beukema, 1979, 1984; De Vlas, 1987). In years with a dense cockle population the cockle fisheries removes only a few percents of the stock. However, in years with a low population density up to 40% is removed (De Vlas, 1987).

Since historic times a specific fisheries for whelks (*Buccinum undatum*) existed in the western Wadden Sea, carried out mainly by fishermen from Texel and Wieringen. In 1949–1952 average landings amounted to 65 tonnes yr^{-1}. A gradual decline occurred since then. In 1965–1969 only 26 tons were landed annually, after which this species disappeared from fisheries statistics. Although the whelk fishery regularly suffered from slack demand and low prices, the decrease in

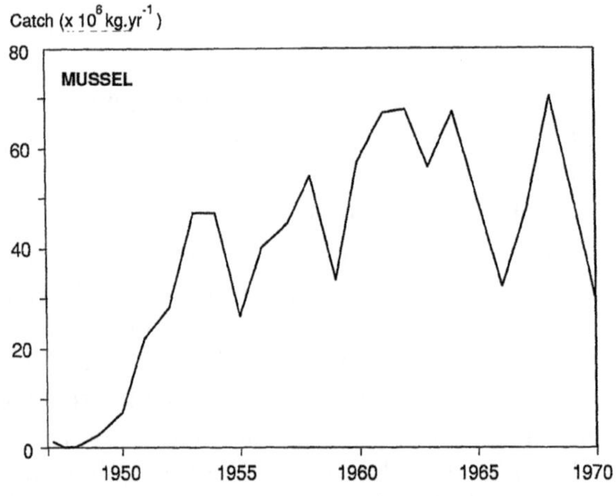

Fig. 10. Annual catch of mussels in the Wadden Sea. After: Boddeke (1991).

landings seems to reflect a decline of the stock in the western Wadden Sea. Nowadays, commercial fishing for whelks would be impossible due to the virtual absence of the species in this area. In the Eastern Scheldt whelk fishing disappeared as well, together with the whelks (R. Dijkema, pers. comm.).

At the tidal flats lugworms (*Arenicola marina*) are collected manually and mechanically as bait for sports fishery. The effects of this activity on bottom fauna and bottom structure are local, and temporary only 1% of the population of lugworms is harvested (Van den Heiligenberg, 1987)

4. Changes in the biotic system

4.1. Flora

Previously, it was generally assumed that the nutrient supply for algal growth in the Wadden Sea was obtained from mineralization of North Sea suspended matter accumulated in the Wadden Sea (cf. De Jonge & Postma, 1974). Recently however, indications were found that an important part of the nutrient supply is obtained from Lake IJssel (Cadée, 1984; Van Raaphorst & Van der Veer, 1990; De Jonge, 1990).

Phytoplankton

Primary production values for the western Wadden Sea vary considerably (Fig. 11). Unfortunately, such data are not available for the eastern Wadden Sea. The first estimate of phytoplankton primary production in the western Wadden Sea (Postma, 1954: 20–40 $gCm^{-2} yr^{-1}$) indicates the production to be significantly lower than at present (Cadée, 1986; Veldhuis *et al.*, 1988: 150–520 $gC\ m^{-2}a\ yr^{-1}$). Microphytobenthos primary production also shows considerable variation over time. Phytoplankton primary production increased significantly as did the production of microphytobenthos.

Little is known about changes in the seasonal occurrence of phytoplankton species. A statistically significant increase in the blooming period of

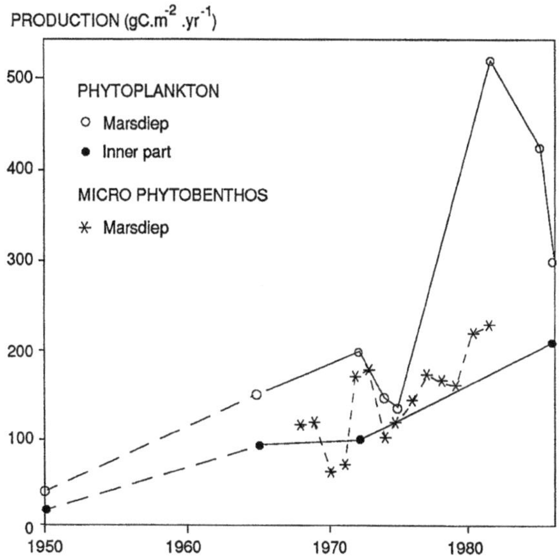

Fig. 11. Annual primary production of the phytoplankton and the microphytobenthos ($gC\ m^{-2}\ yr^{-1}$) in the Marsdiep tidal basin. Data from: Postma (1954), Postma & Rommets (1970), Cadée & Hegeman (1974, 1979), Cadée (1986), Veldhuis *et al.* (1988), and Beukema & Cadée (1987).

Phaeocystis pouchetii (*Haptophyceae*) from 20 days in the early 1970's to more than 60 days in the mid 1980's was reported by Cadée & Hegeman (1986), although recently Cadée and Hegeman (1991) referred to records of a relatively long blooming period of Phaeocystis in 1897 and 1899. Several algae, viz. *Dinophysis acuminata*, are known to produce toxic excretion products. This species is responsible for diarrethic poisoning among consumers of bivalves (Kat, 1987). The blooming often occurs at the end of the summer. Although indications exist for an increased frequency of blooms of *Dinophysis acuminata*, start and distribution of these blooms are still unpredictable.

Green macroalgae

Very conspicuous in the Wadden Sea are *Enteromorpha* and *Ulva* species. In the Wadden Sea four *Ulva* species can be found: *Ulva lactuca*, *U. curvata*, *U. rigida* and *U. scandinavica* (Koeman & Van den Hoek, 1981). Before the closure of the Zuiderzee *Ulva* was a common genus in eelgrass

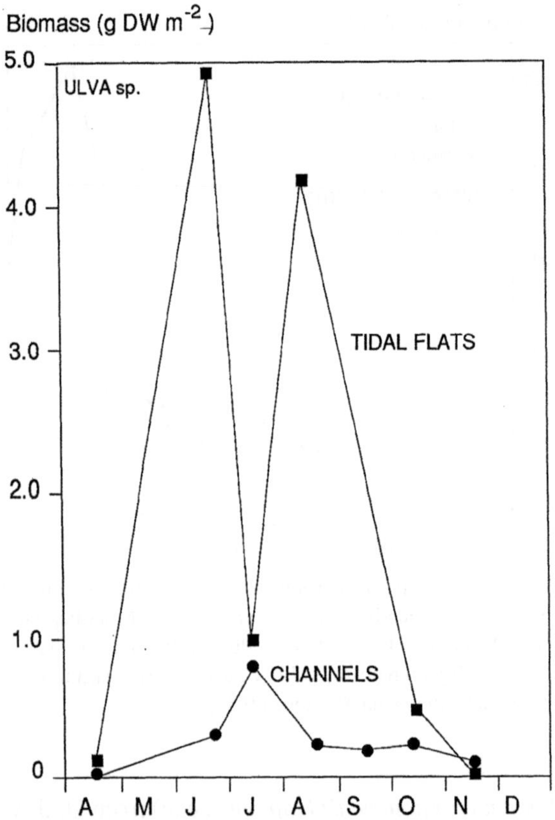

Fig. 12. Seasonal fluctuation in the biomass (g dry weight) of *Ulva* species on a tidal flat and in the adjacent tidal channel in the western Dutch Wadden Sea in 1988. From: De Jonge & Peletier, unpubl.

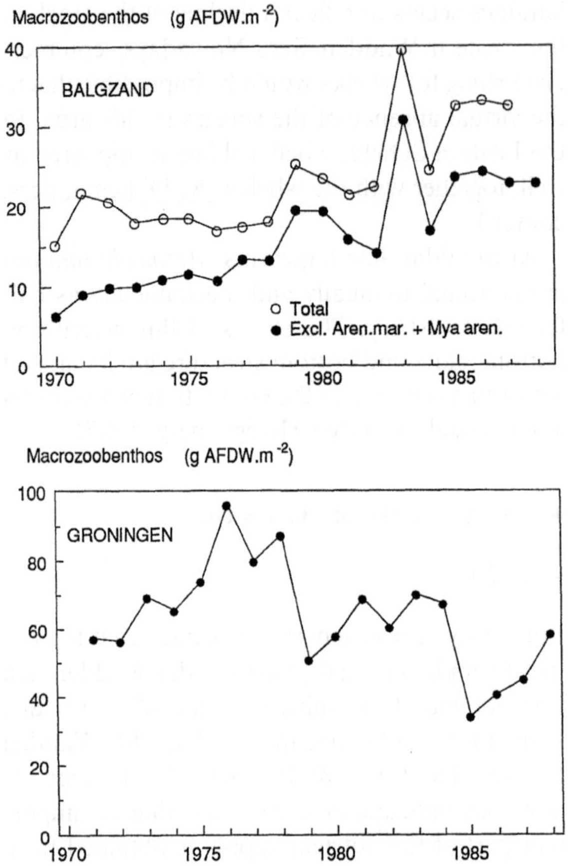

Fig. 13. Time series of the biomass of the macrozoobenthos in the western (Balgzand; data from Beukema) and in the eastern (Groningen) Dutch Wadden Sea. At Groningen no data on *Arenicola marina* and large *Mya arenaria* were available.

beds (Van Goor, 1919). *Ulva* starts to develop in April, dependent on the water temperature. It reaches maximal biomasses in July–August and dies at the end of the growing season when temperatures decrease. Recent data show that the biomass of this species may vary with a factor 15 within the course of a year (Peletier, 1992). Development of *Ulva* in the Wadden Sea seems to depend largely on the course of the summer. Bright and calm summers are favourable for its development. In 1988 and 1989, both years having a bright summer, *Ulva* was notably present. Mussel beds are an important substrate for *Ulva* species. In June many of the plants become afloat, mainly due to the activities of the mussel farmers starting to harvest mussels with dredge nets. Currents determine whether *Ulva* can be found on the flats or in the tidal channels (Fig. 12). In the Dutch Wadden Sea there are indications for an increase in the mean annual biomass of *Ulva* over the last 30 years.

Enteromorpha species can occur on dikes and sandy flats. On the sandy flats in spring *Enteromorpha radiata* can develop extensive vegetations with a coverage up to 100%. However, presence of only one plant per m^2 is a common feature (Van den Hoek *et al.*, 1979).

Phanerogams

Repeated endikements over consecutive centuries have reduced the total surface of saltmarshes in the Dutch Wadden Sea. Especially the mainland saltmarshes diminished, whereas the island saltmarshes were not greatly reduced (Fig. 1; Bakker *et al.*, 1993). Presently, saltmarshes make up only ca. 3% of the surface of the Wadden Sea, whereas originally about 15% was covered with saltmarshes (Dijkema, 1987).

The deliberate introduction by man of cord-grass (*Spartina anglica*) in the 19th century has led to reduced variety in the vegetation of the lower saltmarsh, it has partially replaced *Zostera* and algal communities of high mud flats, and caused the disappearance in the Netherlands of the native *Spartina maritima* (Westhoff, 1985).

Phanerogams living in the Wadden Sea are the eelgrasses *Zostera marina* and *Zostera noltii*. Before the construction of the Afsluitdijk in 1932 beds of *Zostera marina* covered up to 150 km^2 of the sublittoral in the transition zone between the Zuiderzee and the western Wadden Sea. In the littoral zone both species occurred. *Zostera marina* in the littoral zone differed from the subtidal one by the presence of broader leaves and rhizomes. The subtidal eelgrass disappeared in the early thirties due to an infection caused by *Labyrinthula macrocystis*. This infection occurred in the entire Dutch-German-Danish Wadden Sea. The species recovered, but never came back in the Dutch part of the Wadden Sea, most likely as a result of the changed hydrological conditions due to the construction of the Afsluitdijk. Together with the sublittoral eelgrass beds several associated species disappeared, *e.g.* the fishes *Syngnathus typhle* and *Spinachia spinachia*, the gastropod *Rissoa membranacea*, the cephalopod *Sepia officinalis*, and the epiphytic algae *Fosliella lejolisii* and *Myriotrichia clavaeformis* (Van Goor, 1919; Den Hartog, 1983). The littoral eelgrass beds also decreased. Since the early seventies the total surface as well as the vegetation density of these beds declined. In Table 2 the decrease in eelgrass vegetation is illustrated for the period 1920–32 to 1988. At present, *Zostera marina* as well as *Zostera noltii* have disappeared from the western Dutch Wadden Sea almost completely (De Jong & De Jonge, 1989; Giesen, 1989; De Jonge & De Jong, 1992). In the eastern Dutch Wadden Sea some populations of *Z. noltii* are still present.

4.2. Fauna

Macrozoobenthos

Some changes occurred in the species composition of the macrozoobenthos over time. Many alien species reached the Dutch marine and brackish environment (Den Hartog & Van der Velde, 1987). Among those immigrants are the clams *Mya arenaria* and *Petricola pholadiformis*, and, recently, the American jack-knife clam *Ensis directus* (Essink, 1985, 1986).

Mya arenaria was introduced into European waters already in the 16–17th century and has become an integrated part in the Wadden Sea benthic community. Small specimens serve as food for shore crabs (*Carcinus maenas*), while siphons of larger animals are being cropped by flatfish (De Vlas, 1979) and wading birds (Zwarts & Wanink, 1984). *Ensis directus* has developed intertidal populations of low densities (a few per m^2), and is consumed by oystercatcher (*Haematopus ostralegus*) (Swennen *et al.*, 1985).

In the Dutch Wadden Sea with its large proportion of intertidal flats macrozoobenthos is a quantitative important element in the ecosystem.

Table 2. Surface covered and biomass of eelgrass in the western Dutch Wadden Sea for different periods. After: Van den Hoek *et al.* (1979) and De Jong & De Jonge (1989).

Period		1920–1932	1950–1960	1972–1973	1988
Surface	(km^2)	150	4–4.5	1.6	2
Biomass	(g C m^{-2})	85–350	ca. 10	4.5	?
Total biomass	(tonnes C)	12700–52500	40–45	7	?

An annual secondary production of ca. 30 g ash-free dry weight m^{-2} (De Wilde & Beukema, 1984) supports large numbers of demersal fish as well as birds (Wolff & Smit, 1984). During the last two decades in the western Wadden Sea intertidal macrozoobenthic biomass and production have increased significantly (Fig. 13) (Beukema & Cadée, 1986; Beukema, 1989). This increase is well correlated with the increased phytoplankton primary production due to increased phosphate loads from Lake IJssel (De Jonge, 1990). In the eastern Wadden Sea intertidal macrozoobenthic biomass did not show an increase in the same period (Fig. 13). This is in agreement with the virtual absence of increased eutrophication in that area (De Jonge & Essink, 1991).

An important change in the benthic system is induced by the change from free fishing to 'culturing' of mussels in the western Dutch Wadden Sea since 1950. The annual production of mussels strongly increased from 6,000 tonnes in 1949 to about 67,000 tonnes in 1962, and fluctuates around a somewhat lower level since then (Dankers, 1987). Most of the culture lots are situated in the subtidal area. They are seeded with young mussels that are fished from wild intertidal and subtidal mussel beds. Dekker (1989) estimates that in the subtidal of the western Wadden Sea (tidal channels excluded) mussels make up about 65% of the macrozoobenthic biomass. In the intertidal *Mytilus edulis* contributes about 23% to the biomass (Beukema, 1976).

Shrimps

Since time immemorial, the stock of brown shrimps in the Wadden Sea has been subject to a wide variety of changes (Boddeke, 1978). Changes occurring after 1960 have been mainly advantageous for the brown shrimp stock (see: 5.2 and Boddeke, 1989b). However, the increase in terms of catches of consumption shrimps has been modest and restricted to August–November, due to the very high natural mortality of brown shrimps, in particular in October–March (Boddeke, 1967a, 1989a).

Fishes

Before 1962, dab (*Limanda limanda*) including large numbers of adults, was the dominating flatfish species in the western Wadden Sea all year round. Numbers present in this area during summer decreased rapidly between 1962 and 1965. In winter adult dab still enter through the tidal inlets. Since 1965, juvenile plaice (*Pleuronectes platessa*) dominates the flatfish stock in this area (Boddeke, 1967b).

The thornback ray *Raja clavata* used to be a rather common species in the Wadden Sea (Witte & Zijlstra, 1978). Although no good quantitative data are available the abundance of *R. clavata* seems to have decreased significantly in the 1950's. The decrease of *R. clavata* in the Wadden Sea followed the decrease of rays in the south-eastern North Sea, which is most likely caused by increased and more efficient demersal fishing.

High abundances of eel (*Anguilla anguilla*), eelpout (*Zoarces viviparus*), butterfish (*Pholis gunellus*), bull rout (*Myoxocephalus scorpius*), and sole (*Solea solea*) were observed to be related to mussel beds and culture lots, the area of which increased considerably after 1950. Mussel beds provide food and shelter for these species.

The sea stickleback (*Spinachia spinachia*) has become rare in the Dutch Wadden Sea since the disappearance of sublittoral eelgrass vegetations. No deep-snouted pipefish (*Syngnathus typhle*) have been recorded since 1935, probable also due to the disappearance of eelgrass (Witte & Zijlstra, 1978).

Birds

During the 20th century the population size of gulls (*Larus ridibundus*, *L. canus*, *L. argentatus*) in western Europe has increased considerably. In the Dutch Wadden Sea an increase of the breeding population at the Wadden islands has been observed (Goethe, 1980a, 1980b, 1980c), notwithstanding a temporary decrease during the 1960's due to pesticide mortality (Swennen, 1982). See for further information Wolff (1993).

The eider (*Somateria mollissima*) breeds on the Dutch Wadden Sea islands since 1906. Since 1925, the breeding population increased to approx. 6,000 pairs in 1960 (Teixeira, 1979). During the 1960's the population decreased till 1,000 pairs in 1965 due to a pesticide caused mortality (Swennen, 1972). Since then the breeding population increased again till 5,000–7,000 pairs in 1979–83 (SOVON, 1987) (Fig. 14). Increased protection of breeding areas and reduction of shooting has contributed to the increase of the breeding population in western Europe. The Dutch Wadden Sea, and the western part in particular, has a high significance for non-breeding and migratory eiders. In April–June 10,000–15,000 non-breeding birds are present, the numbers wintering in this area are estimated at 100,000–200,000 (SOVON, 1987).

In the Dutch Wadden Sea the food of eiders consists for 40% of mussels, for 40% of cockles, and for 20% of other prey (Swennen, 1976). In January 1987 almost 150,000 eiders were counted in the Dutch Wadden Sea. In that same month 37% of all eiders in the entire Wadden Sea were present near mussel culture plots (Swennen, Nehls & Laursen, 1989). Eiders normally have a diving depth of 3–6 meters (Voous, 1960). Therefore, sublittoral mussel culture lots with abundant loose mussels are very suitable feeding grounds. In the western Wadden Sea eiders harvest an

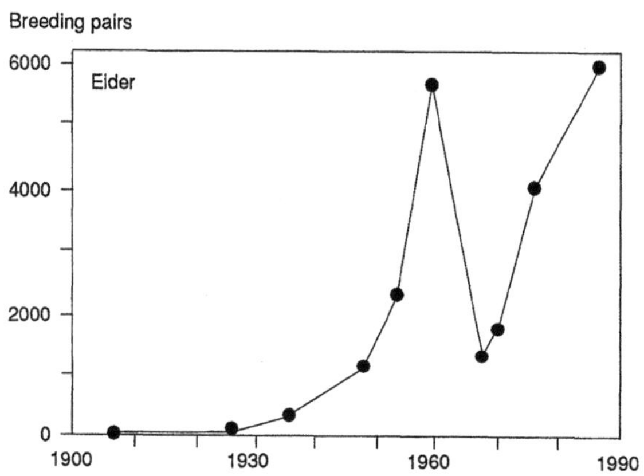

Fig. 14. Number of breeding pairs of the eider (*Somateria mollissima*) in The Netherlands. After: Vertegaal & Van der Salm (1988).

estimated 30,000 tons of mussels from culture lots annually (Swennen, Nehls & Laursen, 1989). This is approx. 50% of the annual yield (tons) of the mussel farmers, but the mussels consumed by eiders are on average smaller than those of normal human consumption size.

The mass wintering of eiders in the Dutch Wadden Sea is known since systematic bird counting started in the 1960's. Before that time the numbers wintering were estimated at several thousands (SOVON, 1987). The large difference between this estimate and present numbers suggests a rapid increase of wintering birds already before 1960, i.e. at the time of the extension of the mussel culture in the Wadden Sea.

The oystercatcher (*Haematopus ostralegus*) is a common breeding bird around the Dutch Wadden Sea. At present, about half of the Dutch population breeds along the coast, whereas the other half breeds inland in the provinces Friesland and Noord-Holland (SOVON, 1987). In earlier times breeding areas were concentrated along the coasts only. The total breeding population in the Netherlands has increased from 2,000–6,000 pairs in 1930 to 80,000–100,000 pairs in 1987 (Vertegaal & Van der Salm, 1988).

The increase in the breeding population in the Dutch Wadden Sea area runs parallel with the increase in intertidal macrozoobenthic production in the western half of the Dutch Wadden Sea (Beukema & Cadée, 1986; Beukema, 1989). The increased breeding in inland grassland seems to be caused by an increase of prey abundance due to increased use of manure (Vertegaal & Van der Salm, 1988).

For dark-bellied brent geese (*Branta bernicla bernicla*) the Dutch Wadden Sea is an important wintering area. The international population seems to have decreased severely in the 1930's. However, accurate countings (also to confirm the assumption that this was caused by the almost complete disappearance of eelgrass meadows) are lacking (Voous, 1960). Dark-bellied brent geese feed mainly on salt marsh plants (Bakker *et al.*, 1991), *Zostera marina*, *Z. noltii*, and on *Ulva* and *Enteromorpha* species, and, in recent times, on grass.

In the 1920's, brent geese concentrating on eelgrass beds in the western Wadden Sea were subject to professional shooting with special large guns. In The Netherlands regular shooting ceased in February 1950, in Denmark not earlier than 1972 (Madsen, 1986).

In 1955–1957 the population of dark-bellied brent geese was estimated at 15,500 birds. From then on numbers increased rapidly, especially after the prohibition of brent geese hunting in Denmark in 1972 (Fig. 15). The population size stabilized at a level of 120,000–200,000 birds during the 1980's. The increased supply of *Ulva* and *Enteromorpha* (see above) is favourable for brent geese, in particular in autumn. In March–May when numbers of brent geese in the Dutch Wadden Sea reach their annual peak, supply of green algae, however, is low. Brent geese then feed largely on grass.

Mammals

The Dutch Wadden Sea has to date been an important biotope for the common or harbour seal (*Phoca vitulina*). Numbers amounted to 2,700–ca. 3,000 animals in 1940–1950 (Fig. 16). Numbers decreased strongly to ca. 1,200 in 1960, followed by a temporary recovery of numbers between 1960 and 1964 (to ca. 1500 animals) presumably caused by the stopping of seal hunting in The Netherlands (Reijnders, 1981). Numbers decreased again to less than 500 in the early 1970's. Decreased reproduction success due to pollutants

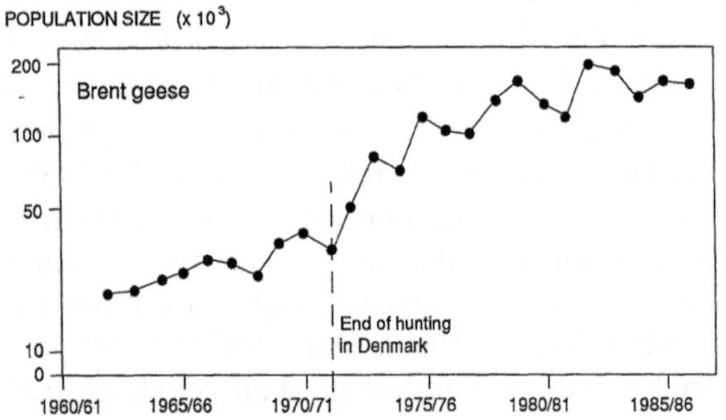

Fig. 15. Development of the population size in winter of the dark-bellied brent geese (*Branta bernicla bernicla*). After: Vertegaal & Van der Salm (1988).

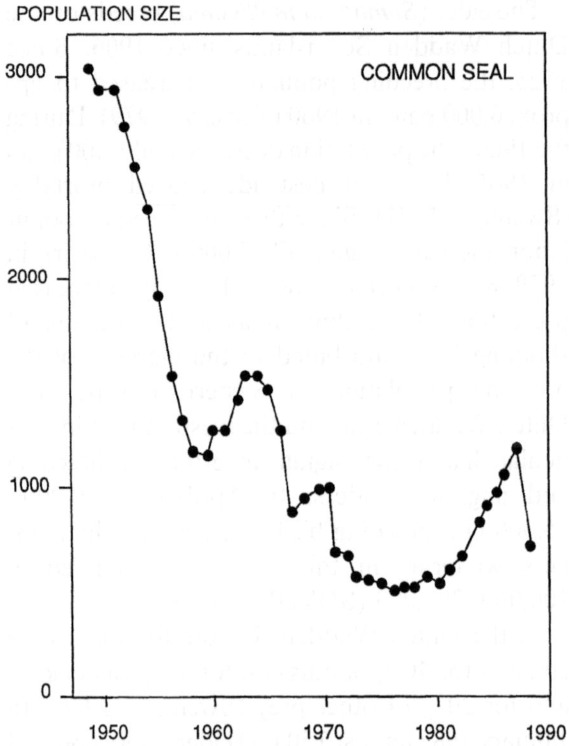

Fig. 16. Population size of the common seal (*Phoca vitulina*) in the Dutch Wadden Sea. After data from Reijnders.

(*e.g.* PCB's) is considered the major cause of the gross decrease of the population from 1965 (1500 animals) towards the 1970's (ca. 500 animals) (Reijnders, 1980, 1986). From 1980 onwards the stock recovered, reaching a level of ca. 1100 specimens in 1987. This was caused by (a) increased reproduction success, (b) increased numbers of rehabilitated young seals, and (c) increased immigration of seals from the German Wadden Sea (Reijnders, 1985).

In 1988 the NW European seal population suffered serious losses due to a fast spreading infection by *Canine distemper* virus (Osterhaus & Vedder, 1988). This caused the Dutch seal population to decrease to ca. 450 animals in 1989.

5. Causes of changes in the ecosystem

5.1. *Physical parameters*

Due to the construction of the Afsluitdijk in 1932, the salinity of the western Dutch Wadden Sea decreased on average. Before 1932 a gradual salinity gradient existed from the mouth of the River IJssel to the tidal inlets in the Wadden Sea (Fig. 17). After the closure of the Zuiderzee this gradient was built up over a much shorter distance between the sluices in the Afsluitdijk and the tidal inlets (Rakhorst, 1983; Van der Hoeven, 1982).

At least four possible causes of increased turbidity in the western Dutch Wadden Sea (cf. Fig. 7) can be mentioned. Firstly, the construction of the Afsluitdijk causing an increase in tidal current velocities resulting in higher suspended matter concentrations. Secondly, dredging for sand and shells in and near tidal inlets also may have caused higher suspended matter concentrations (cf. De Jonge, 1983). Thirdly, dredging of the Rotterdam shipping route (Eurogeul and Nieuwe Waterweg) increased considerably from ca. 2×10^6 m^3 before 1960 to ca. 17×10^6 m^3 yr^{-1} in the 1980's (cf. De Jonge & Essink, 1992). A considerable amount of this dredged and dumped

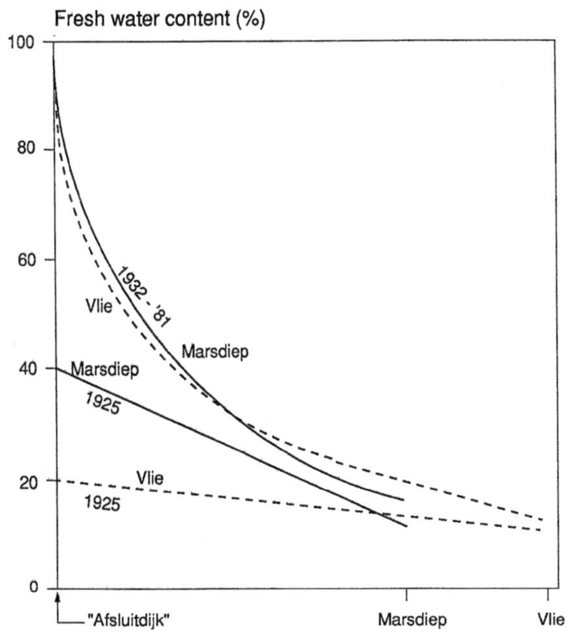

Fig. 17. The mean salinity gradient between the (position of) the Afsluitdijk and the tidal inlets Marsdiep (———) and Vlie (-------) in the western Dutch Wadden Sea before (1925) and after (average over 1932–1981) closing off the Zuiderzee. After: Van der Hoeven (1982).

(near Hoek van Holland) material is subject to further transport (Louisse, 1987), and may end up in the western Dutch Wadden Sea (Scholten *et al.*, 1988). This may explain the good correlation between the suspended matter concentrations in the Marsdiep tidal inlet and the amounts of dredge spoil dumped at the Loswal Noord dumping site near Hoek van Holland (Fig. 18). Fourthly, the strong increase in mussel culturing between 1953 and 1962 may have led to increased turbidity due to increased sublitoral deposits of silt, feces and pseudofeces, which subsequently resuspended as the mussels are being relaid and fished from the culture lots.

5.2. *Nutrients and productivity*

Increased nutrient loads due to discharge of fresh water into the Wadden Sea (Fig. 2b) can lead to increased productivity of the Wadden Sea ecosystem. In the Marsdiep tidal basin the annual primary production, and the mean annual chlorophyll-*a* concentration seems to be mainly fueled by the nutrient loads from Lake IJssel (Fig. 19). On an annual basis phosphate appears to be a critical nutrient. Both annual phytoplankton production and microphytobenthos develop proportional to the discharge of phosphate (cf. De Jonge, 1990). This correlation clearly holds for the Mars-

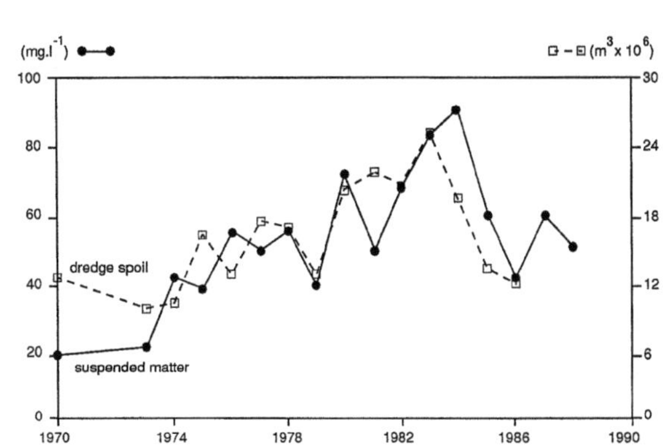

Fig. 18. Time series of mean annual amount dredge spoil dumped at Loswal Noord (near Hoek van Holland) dumping site (m^3) and mean annual suspended matter concentration in Marsdiep tidal inlet (mg l^{-1}). After: De Jong & De Jonge (1989).

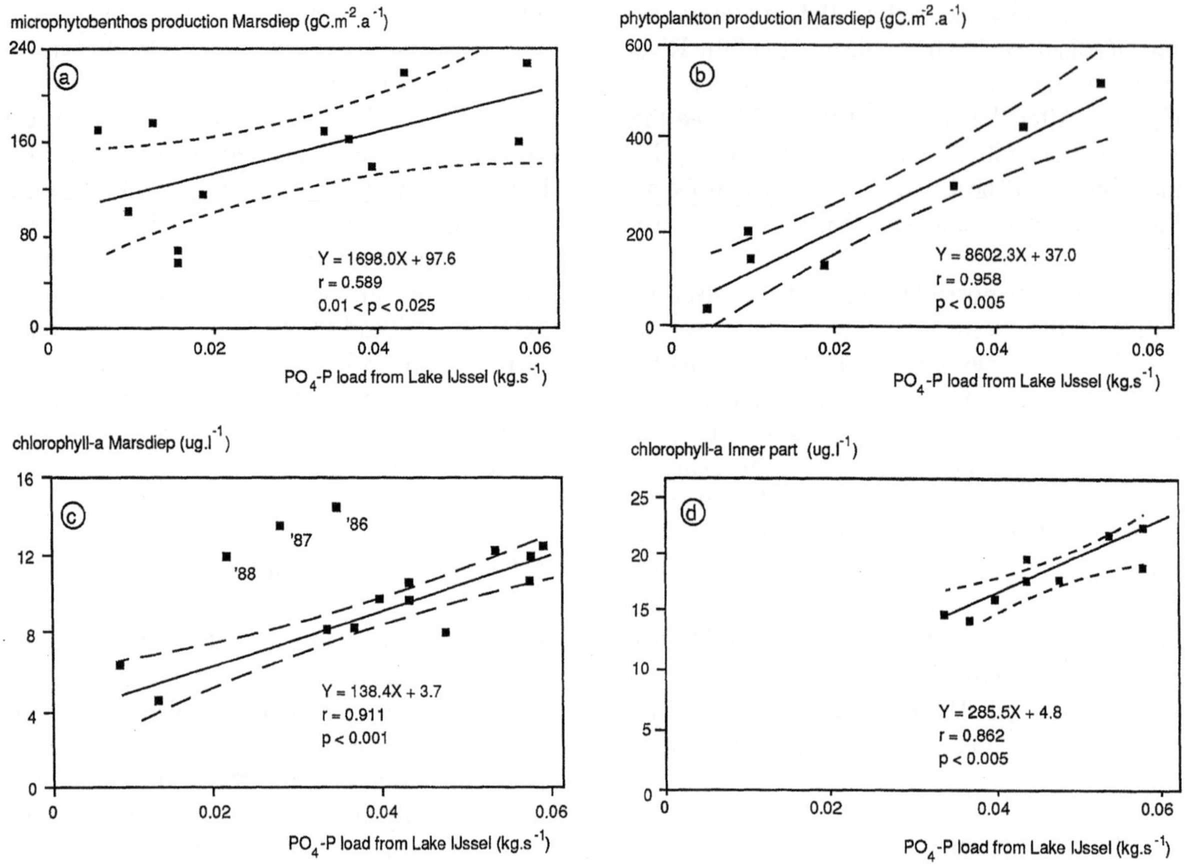

Fig. 19. Relation between phosphate loads and the annual primary production of microphytobenthos (a) and phytoplankton (b) in the Marsdiep tidal inlet. Relation between phosphate loads and the mean chlorophyll-*a* concentration in the Marsdiep tidal inlet (c) and the inner part of the Marsdiep tidal basin (d).

diep tidal inlet. In the inner part of the Marsdiep tidal basin the increase in the primary production was less significant, probably due to higher turbidity (Fig. 11).

In the eastern Wadden Sea available chlorophyll-*a* data suggest that primary production did not increase. This is in agreement with the much smaller discharge of nutrients in that area (see Fig. 2c).

Often a positive correlation is assumed between eutrophication and the production of *Ulva* sp. and of phytoplankton, toxin producing species included (Piriou, 1985; Marchetti *et al.*, 1989). In the Ythan Estuary (Scotland) a correlation was found between nutrient loads (especially nitrogen) and biomass of *Enteromorpha* sp. (Raffaelli *et al.*, 1989). In the Wadden Sea, evidence for such a correlation is lacking for *Ulva*. For phytoplankton in the western Wadden Sea the relationship is shown in Fig. 19b. A causal relationship between presence of *Dinophysis acuminata* and eutrophication is still unclear (Kat, 1987).

In the western Wadden Sea, an increase in abundance and biomass of intertidal macrozoobenthos has been related to eutrophication (Beukema & Cadée, 1986; Beukema, 1989; De Jonge, 1990). In the eastern Wadden Sea, the macrozoobenthos biomass did not change significantly (Fig. 13). This is in accordance with the virtual absence of eutrophication in that area (see above), and confirms the conclusion by Beukema & Cadée that eutrophication is the main factor responsible for increased macrozoobenthos production in the western Wadden Sea. This increase includes species, such as blue mussel (*Mytilus edulis*) and cockle (*Cerastoderma edule*), which are harvested both by man and birds.

In the 1980's the Dutch Wadden Sea produced about 20% of the recruitment to the North Sea plaice stock, and ca. 8% to the North Sea sole stock. Rauck & Zijlstra (1978) showed that from 1955 to 1973 the average length of juvenile plaice in their first winter increased with approx. 1 cm. This may indicate better feeding conditions due to eutrophication. Since then, growth has not changed (Van Beek *et al.*, 1989). Also for sole (*Solea solea*) an increase in growth rate in the sixties and early seventies has been reported (De Veen, 1976; Van Beek, 1988).

Between 1962 and 1965, the Dutch Wadden Sea, especially the western part, changed from an area dominated by adult dab into an environment in which juvenile plaice is the most common flatfish. Dab is a visual predator preying upon brown shrimps and other invertebrates in the area. Increased turbidity may have impaired its life conditions in the western Wadden Sea.

5.3. *Stocks and fisheries*

The increase in commercial cockle landings in the Wadden Sea until 1984 (see above) was caused by both an increased fishing effort and an increased cockle stock (Beukema & Cadée, 1986). During 1985–87 the landings decreased sharply, due to high natural mortality during three successive severe winters. Fluctuations in cockle production are not uncommon (Dijkema, 1988). The recent decrease in cockle landings may therefore be temporary. Depletion of cockle stocks by continued fishing in lean years (cf. De Vlas, 1987) will possibly affect the natural consumers of cockles such as the oystercatcher (*Haematopus ostralegus*) and eider (*Somateria mollissima*). Since 1987, cockle fishing in The Netherlands has been strictly regulated. The total number of vessels has been fixed at 37; the annual number of fishing days per vessel was 33 in 1989 (between mid-August and December). In 1990 court judged the fishermen to be free to fish every part of the Wadden Sea with the result that in 1991 there were no cockles available neither for man nor for the birds. In spring 1991 the breeding success of the eiders was ca. 10% the value of the previous years.

In the 1950's organized mussel farming created many mussel culture lots in subtidal areas of the western Dutch Wadden Sea. Mussel farmers collect small mussels mainly in the sublittoral, seed them on these lots, and regularly remove starfishes (*Asterias rubens*), being a significant predator of mussels (Dijkema & Van Stralen, 1989), and *Ulva*. This practice has increased the mussel production at the culture lots. Parallel to this development mussels in wild banks showed an increase since the 1960's. Apparently, increased eutrophication provided ample supply of food for these molluscs. Since the 1980's, however, a decline occurred. In 1990 no wild banks were available anymore due to the seed fisheries of the mussel farmers. Consequently, oystercatchers and eiders suffered from food shortage.

Musselbeds also provide food and shelter for various fish species such as sole (*Solea solea*) eelpout (*Zoarces viviparus*) and bull rout (*Myoxocephalus scorpius*) in (Table 3). The attractiveness of musselbeds for these fish species consists of the availability of food (sole: *Nereis succinea*; viv. blenny: amphipods and tiny mussel seed; bull rout: amphipods) and shelter (viv. blenny, bull rout).

In the period 1949–52 the average annual landing of shrimps (*Crangon crangon*) from the Wadden Sea amounted to 1,268 tons of shrimps for

Table 3. Numbers of fish caught in 30-minute hauls with a commercial shrimper along and on wild musselbanks in the western Dutch Wadden Sea in 1967.

Species	Along mussel bank		On mussel bank	
	May	July	May	July
Dab	48	180	103	158
Plaice	2554	466	500	1040
Flounder	157	20	121	160
Sole	44	58	238	265
Eelpout	62	30	1187	1252
Bull rout	13	8	201	148

consumption, and 4,140 tons of undersized shrimps. In the period 1985–88 only consumption shrimps were landed at an annual average of 3,206 tons. The increased yield of consumption shrimps is larger than expected on basis of increased fishing efficiency and protection of undersized shrimps only (Boddeke, 1967, 1989a). This implies that since the fifties the Wadden Sea environment has changed in favour of the brown shrimps.

5.4. Turbidity and eelgrass

De Jonge (1986) and Giesen (1990) showed that in the western Wadden Sea Secchi disc readings as indication for water clearness have decreased from approx. 2.5 m in 1930 to 1.2 m in the late 1970's) (Fig. 20). For the tidal flats in the western Wadden Sea the Secchi disc values were ca. 1.8 m in the 1930's and ca. 0.6 m in the 1970's. When the present light conditions in the western

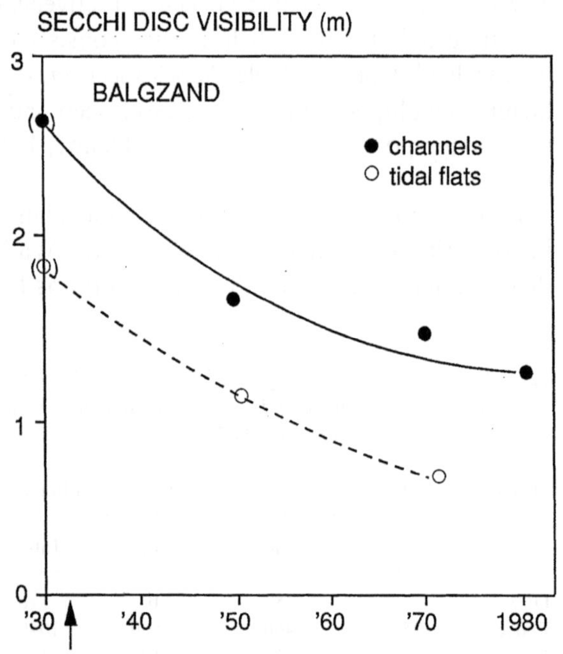

Fig. 20. Average Secchi disc visibilities in tidal channels and above tidal flats in the Balgzand area between 1930 and 1980. Closure of the Zuiderzee is indicated (↑). After: De Jonge (1986) and Giesen (1990).

Wadden Sea are used in the simulation model for growth of *Zostera marina* in the saline Lake Grevelingen (southwest of The Netherlands) (Verhagen & Nienhuis, 1983), the tentative conclusion may be drawn that eelgrass is unable to live neither in the subtidal areas of the western Dutch Wadden Sea nor in most intertidal areas due to increased turbidity (Giesen, 1990). Epiphytes may cover the leaves of eelgrass reducing growth by prevention of light penetration (Den Hartog, 1983). Increased eutrophication leads to enhanced epiphytal growth (Den Hartog, 1987; Kemp *et al.*, 1983), which may even result in eelgrass mortality (Neverauskaus, 1987).

5.5. Contaminants

Several pollutants such as mineral oil, metals and organochlorine pesticides end up in the Wadden Sea. These contaminants may have direct as well as indirect effects on organisms. Indirect effects are mostly related to accumulation in the food web and conversion (*e.g.* methylation of mercury). Bird mortality due to oil pollution is an example of a direct effect (cf. Swennen & Spaans, 1970).

Data on contaminant levels in benthic invertebrates and fish in the Dutch Wadden Sea show that the western Wadden Sea and the Ems Estuary are more polluted than the intermediate part (cf. Essink, 1989). Although the levels of contaminants found in benthic evertebrates differ greatly, no indications are available of a possible detrimental effect on their populations. Effects at a low trophic level however, have been demonstrated in experiments with enclosed plankton communities (Kuiper, 1982). Everaarts *et al.* (1979) demonstrated sublethal effects of copper and zinc at concentrations locally present in sea water: impairment of the oxygen binding capacity of haemoglobin in the lugworm (*Arenicola marina*).

In 1978, after a period of mercury pollution being about twice as high in the Ems Estuary as compared with the western Wadden Sea survival of fry of eelpout (*Zoarces viviparus*) born of females from the Ems Estuary was little, but significantly less than survival of fry produced by females from the western Wadden Sea (Essink, 1988, 1989).

Accumulated pesticides (telodrin and dieldrin) caused a severe decline in numbers of breeding pairs of several bird species in the early 1960's (Koeman, 1971; Rooth & Jonkers, 1972; Swennen, 1972). Since then, numbers have largely recovered (SOVON, 1987).

Polychlorinated biphenyls (PCB's) taken up with the food (fish from the Wadden Sea) were proven to be responsable for a reproduction failure among harbour seals (Reijnders, 1986).

6. Management

Management of man's activities in the Wadden Sea has been a matter of concern for many years (*e.g.* Wolff, 1976; Wolff & Zijlstra, 1980; Zijlstra, 1972). In 1980, the Dutch government and parliament agreed upon a policy with respect to the protection and management of the Wadden Sea (Anonymous, 1980). In subsequent years an overall management strategy was accepted, and several management plans dealing with nature management, shipping, coastal protection, water quality management, prevention of oil pollution, etc. were produced (Anonymous, 1987, 1988; Vreugdenhil, 1984). In 1981, large parts of the Dutch Wadden Sea were protected by the Dutch Nature Conservation Act, and in 1984 the entire Dutch Wadden Sea was officially added to the list of internationally important wetlands under the Ramsar Convention. Recently, three new government policy plans were accepted by the Dutch parliament. The first two (Anonymous, 1989a, 1989c) focus on management of the (natural) environment; the third one (Anonymous, 1989b) deals with integrated management of watersystems. Among the policy objectives indicated above statements are found, such as "The government policy with respect to the Wadden Sea is protection, conservation, and, where necessary, restoration of the Wadden Sea as a natural environment" (Anonymous, 1980), "Protection against pollution", "Restoration of [.....] eelgrass beds in the Wadden Sea", "Future ecological functioning of water systems, permitting lasting multiple use by man" (Anonymous, 1989b). At the same time the Third Water Management Plan (Anonymous, 1989b) aims at the Wadden Sea being primarily a natural environment, in which there is limited room for fisheries, extraction of sand and shells, and recreational activities.

Official Dutch policy clearly points at multiple use of the Wadden Sea, with emphasis on protection, and restoration of the natural environment. What has to be pursued in management of the Wadden Sea? A pristine Wadden Sea, such as existed in the Middle Ages? Unfortunately, we do not know very well what the Wadden Sea looked like at that time. Moreover, do we want to undo the reclaimed polders, which now are valuable agricultural areas? Do we want to undo the construction of the Afsluitdijk? More realistic, but still extreme in some respects, the Third Water Management Plan aims at a development of the Wadden Sea ecosystem towards the situation of ca. 1930, i.e. without undoing of the Afsluitdijk, other sea dikes and reclaimed polders. This means that, without treating the subject exhaustively, management of the Wadden Sea has to focus on reduction of eutrophication, pollution, and turbidity.

Eutrophication

Reduction of eutrophication and chemical pollution will lead to quite different and partly even unexpected changes in the Wadden Sea ecosystem. Reduction of eutrophication will lead to a decrease in the productivity, which of course has its pros and cons. Among the cons are a reduced macrozoobenthic production, which serves as food for birds, demersal fishes and brown shrimps, and a reduced carrying capacity for bivalve stocks (mussels, cockles) that can be exploited by man. Among the pros are a reduction in algal blooms, such as of *Phaeocystis pouchetii*, and possibly a change in the phytoplankton species composition in favour of diatoms.

Pollution

Reduction in chemical pollution in the Wadden Sea will, by definition, improve the quality of the ecosystem. A more effective system of collection of bilge-water etc. in harbours around the Wadden Sea will certainly contribute to a reduction in chronic oil pollution. However, for most contaminants reduction in pollution of the Wadden Sea can only be achieved by a reduction in their emissions at the sources. As many pollutants reaching the Wadden Sea originate from sources located in the catchment area of the River Rhine, such a reduction can only be achieved by coordinated international measures. Following previous agreements by the International Commission for Protection of the River Rhine against Pollution (1987), and the Second North Sea Ministers Conference (1987), recently, at the Third North Sea Ministers Conference (1990) it was agreed to reduce further emissions of phosphorus, nitrogen, cadmium, lead, mercury, dioxins, pesticides and other noxious compounds to 50 or 70% by 1995. PCB's will be banned and destroyed by 1995. Effects of these measures, however, will take some time. As shown in Fig. 19c, a reduction of phosphate load to the western Wadden Sea in the years 1986–88 did not lead to reduced phytoplankton biomass. It is assumed that this is caused by subsequent delivery of phosphorus accumulated in the sediments.

Hydro-morphology

Reduction of turbidity as far as determined by increased current velocities due to the construction of the Afsluitdijk is not a realistic option. Some reduction of turbidity may be achieved by a proper regulation of dredging and extraction of sand and shells. To be able to do so a thorough study of hydromorphological processes is necessary, aimed at indicating locations where extraction of sediments or dredging will not, or only to a small extent, lead to an increased erosion-sedimentation cycle (cf. De Jonge, 1983). Mussels cultured on subtidal culture lots will, by their nature, accumulate silt, pseudofeces and feces. These deposits will be resuspended as soon as the mussels are harvested, or by temporarily increased water turbulence due to wind stress. The more frequent these events of resuspension occur, the more sediment will be resuspended, leading to increased turbidity. From this it follows that, theoretically, a reduction of turbidity may be obtained by reducing the resuspension frequency of deposits in mussel beds. Whether this will be a real option or not has to be shown by a proper study. However, even if it would be possible to manage the resuspension frequency of deposits in subtidal mussel culture beds, that would not include the subtidal wild mussel beds, and, therefore, be not very effective.

Decreased turbidity would increase the chances for restoration of eelgrass (*Zostera marina*) beds in the western Dutch Wadden Sea. Once re-established eelgrass beds will act as sediment traps, reducing water turbidity.

Finally, if it can be demonstrated that dumping of dredged materials at the Loswal Noord dumping site affects the suspended matter concentrations in the western Dutch Wadden Sea reduction of turbidity in the Wadden Sea may require alternative strategies of dredging and dredge spoil disposal along the Dutch coast at considerable distances from the Wadden Sea.

Above, only some general lines for the management of the Dutch Wadden Sea are given. Elaborating these would require a thorough knowledge of the often complex multiple causal relationships, as well as a good insight into the various, often technical, possibilities for management. Many of the causes of the changes in the Wadden Sea ecosystem described in this paper are not known well enough, making further research necessary. Moreover, management of the Wadden Sea requires a good fit to the delicately balanced policy with respect to the multiple and lasting use of this internationally valued wetland. Recently, as a first step towards a new management of the Dutch Wadden Sea a Wadden Action Plan has been formulated by the Ministry of Public Works and Water Management (Anonymous, 1990).

7. Conclusions

At present, the Wadden Sea ecosystem is more eutrophic and more turbid than in the past. It is also more productive than before, which has resulted in increased biomass of algae and macrozoobenthos, well growing plaice in relative high densities, a high mussel and cockle production. Part of this high production is consumed by birds, supporting increased populations (eider, oystercatcher). In the western Dutch Wadden Sea eiders have become important share-holders in the Dutch mussel industry.

The increased productivity in this area caused some species to become more abundant (*e.g.* juvenile plaice and eel), while at the same time other species decreased significantly (*e.g.* dab and anchovy). Some species even disappeared (almost) completely due to the disappearance of the special habitat formed by the sublittoral eelgrass vegetation. Among these species are fishes, a gastropod, a caphalopod and epiphytic algae.

In the late 1980's the loads of nutrients decreased dramatically causing a decrease of the phytoplankton production. Macrozoobenthos changed dramatically due to man's influence. In 1990 cockles and wild mussel banks were completely removed by fishermen.

It is very difficult to assess the contribution of different causal factors, such as eutrophication, changed hydrography (due to coastal engineering) and pollution, in the observed ecological changes. Qualitative ecosystem models can be helpful in solving this problem.

References

Anonymous, 1980. Nota over de hoofdlijnen van de ontwikkeling van de Waddenzee. Ministerie van Volkshuisvesting, Ruimtelijke Ordening en Milieubeheer. Tweede Kamer, 1980–1981, 13 933 nr. 53(in Dutch).

Anonymous, 1987. Waddenzee: Beheersplan Ontgrondingen, Beheersplan Scheepvaart, Beheersplan Kabels en Buisleidingen, Beheersplan Buitendijkse Gronden, Beheersplan Kustbescherming, Beheersplan Bestrijding Verontreiniging door Olie. Waddencomité Rijkswaterstaat, Leeuwarden (in Dutch).

Anonymous, 1988. Beheersplan Natuur. Ministerie van Landbouw en Visserij, The Hague, 60 pp. (in Dutch.)

Anonymous, 1989a. Nationaal Milieubeleidsplan. Ministerie van Volkshuisvesting, Ruimtelijke Ordening en Milieubeheer, Ministerie van Economische Zaken, Ministerie van Landbouw en Visserij, ministerie van Verkeer en Waterstaat. Tweede Kamer, 1988–1989, 21 137, nrs. 1–2 (in Dutch).

Anonymous, 1989b. Derde Nota Waterhuishouding. Ministerie van Verkeer en Waterstaat, Ministerie van Volkshuisvesting, Ruimtelijke Ordening en Milieubeheer, Ministerie van Landbouw en Visserij. Tweede Kamer, 1988–1989, 21 250, nrs. 1–2 (in Dutch).

Anonymous, 1989c. Natuurbeleidsplan. Ministerie van Landbouw en Visserij, The Hague, 179 pp. (in Dutch)

Anonymous, 1990. Wadden Actie Plan; Discussienota. Ministerie van Verkeer en Waterstaat, Rijkswaterstaat/DGSM, 46 pp. (in Dutch).

Bakker, J.P., J. de Leeuw, K.S. Dijkema, P.C. Leendertse, H.H.T. Prins & J. Rozema, 1993. Salt marshes along the coast of The Netherlands. Hydrobiologia 265: 73–95.

Baretta, J.W. & P. Ruardij, 1989. Carbon flows in the western Wadden Sea: model calculations. Scient. Mar. 53: 523–529

Beek, F.A. van, 1988. On the growth of sole in the North Sea. I.C.E.S., C.M./G: 24 (mimeo).

Beek, F.A. van, 1992. De vissen in de Waddenzee en estuaria. In W.J. Wolff (ed.), Handboek voor Natuurbeheer. Waddenzee en estuaria. PUDOC, Wageningen (in press) (in Dutch).

Beek, F.A. van, A.D. Rijnsdorp & R. de Clerck, 1989. Monitoring juvenile stocks of flatfish in the Wadden Sea and the coastal areas of the southern North Sea. Helgol. Meeresunters. 43: 461–477.

Beukema, J.J., 1976. Biomass and species richness of the macrobenthic animals living on the tidal flats of the Dutch Wadden Sea. Neth. J. Sea Res. 10: 236–261.

Beukema, J.J., 1979. Biomass and species richness of the macrobenthic animals living on a tidal flat area in the Dutch Wadden Sea: effects of a severe winter. Neth. J. Sea Res. 13: 203–223.

Beukema, J.J., 1984. Zoobenthos survival during severe winters at high and low tidal flats in the Dutch Wadden Sea. In J.S. Gray & M.E. Christiansen (eds), Marine biology of polar regions and effects of stress on marine organisms. John Wiley, Chichester: 351–361.

Beukema, J.J., 1989. Long-term changes in macrozoobenthos abundance on the tidal flats of the western Wadden Sea. Helgol. Meeresunters. 43: 405–415.

Beukema, J.J. & G.C. Cadée, 1986. Zoobenthos responses to eutrophication of the Dutch Wadden Sea. Ophelia 26: 55–64.

Beukema, J.J. & G.C. Cadée, 1987. De eutrofiëring van ons kustwater: genoeg of al te veel? Vakblad v. Biologen 67: 153–157 (in Dutch).

Boddeke, R., 1967a. Den Oever–Harlingen. Een vergelijking van twee garnalenhavens. Visserij 20: 159–165 (in Dutch).

Boddeke, R. 1967b. Visserij-biologische veranderingen in

de westelijke Waddenzee. Visserij 20: 213–222 (in Dutch).

Boddeke, R., 1978. Changes in the stock of Brown Shrimp (*Crangon crangon* L.) in the coastal area of the Netherlands. Rapp. P.-v. Réun. Cons. int. Explor. Mer 172: 239–249.

Boddeke, R., 1989a. Management of the brown shrimp (*Crangon crangon*) stock in Dutch coastal waters. In J.F. Caddy (ed.), Marine invertebrate fisheries: their assessment and management. J. Wiley & Sons, New York: 35–62.

Boddeke, R., 1989b. Beschrijving van de populaties van de garnaal (*Crangon crangon* (L.)) langs de Nederlandse kust en in de estuaria. In Ecologisch profiel vissen, Rijkswaterstaat, Dienst Getijdewateren.

Boddeke, R., 1992. Visserij in de estuaria. In W.J. Wolff (ed.), Handboek voor Natuurbeheer. Waddenzee en Estuaria. PUDOC, Wageningen (in press) (in Dutch).

BOEDE, 1985. Biological Research Ems-Dollard Estuary. Rijkswaterstaat Communications no. 40, The Hague, 182 pp.

Boon, J.P. & J.C. Duinker, 1986. Monitoring of cyclic organochlorines in the marine environment. Environ. Monit. Assess. 7: 189–208.

Cadée, G.C., 1984. Has input of organic matter into the western part of the Dutch Wadden Sea increased during the last decades? In R.W.P.M. Laane & W.J. Wolff(eds), The role of organic matter in the Wadden Sea. Neth. Inst. Sea Res. Publ. Ser. 10: 71–82.

Cadée, G.C., 1986. Increased phytoplankton primary production in the Marsdiep area (western Dutch Wadden Sea). Neth. J. Sea Res. 20: 285–290.

Cadée, G.C. & J. Hegeman, 1974. Primary production of phytoplankton in the Dutch Wadden Sea. Neth. J. Sea Res. 8: 240–259.

Cadée, G.C. & J. Hegeman, 1979. Phytoplankton primary production, chlorophyll and composition in an inlet of the western Dutch Wadden Sea (Marsdiep). Neth. J. Sea Res. 13: 224–241.

Cadée, G.C. & J. Hegeman, 1986. Seasonal and annual variation in *Phaeocystis pouchetii* (Haptophyceae) in the westernmost inlet of the Dutch Wadden Sea during the 1973 to 1985 period. Neth. J. Sea Res. 20: 29–36.

Cadée, G.C. & J. Hegeman, 1991. Historical phytoplankton data of the Marsdiep. Hydrobiol. Bull. 24: 111–118.

Dankers, N., 1987. Some ecological effects of the mussel culture in the Dutch Wadden Sea. In S. Tougaard & S. Asbirk (eds), Proceedings of the 5th International Wadden Sea Symposium, Esbjerg, Denmark, Sept. 29–Oct. 3, 1986. The National Forest and Nature Agency and The Museum of Fisheries and Shipping, Esbjerg: 201–213.

Dekker, R., 1989. The macrozoobenthos of the subtidal western Dutch Wadden Sea. I. Biomass and species richness. Neth. J. Sea Res. 23: 57–68.

Dijkema, K.S., 1987. Changes in salt-marsh area in the Netherlands Wadden Sea after 1600. In A.H.L. Huiskes, C.W.P.M. Blom & J. Rozema (eds), Vegetation between land and sea. Dr W. Junk Publishers, Dordrecht: 42–49.

Dijkema, R., 1988. De Nederlandse kokkelvisserij in 1987. Achtergronden en suggesties voor toekomstig beheer. Rijksinstituut voor Visserijonderzoek, IJmuiden. RIVO rapport AQ 88–01 (in Dutch).

Dijkema, R. & M. van Stralen, 1989. Mussel cultivation in the Netherlands. World Aquaculture 20(4): 56–68.

Eisma, D., 1980. Natural forces. In K.S. Dijkema, H.-E. Reineck & W.J. Wolff (eds), Geomorphology of the Wadden Sea area. Balkema, Rotterdam: 20–31.

Eisma, D., 1981. Supply and deposition of suspended matter in the North Sea. Spec. Publs. Int. Ass. Sediment 5: 415–428.

Essink, K.,1978. The effects of pollution by organic waste on macrofauna in the eastern Wadden Sea. Neth. Inst. Sea Res. Publ. Ser. 1: 1–135.

Essink, K., 1984. The discharge of organic waste into the Wadden Sea – local effects. Neth. Inst. Sea Res. Publ. Ser. 10: 165–177.

Essink, K., 1985. On the occurrence of the American jackknife clam *Ensis directus* (Conrad, 1843) (Bivalvia, Cultellidae) in the Dutch Wadden Sea. Basteria 49: 73–80.

Essink, K., 1986. Note on the distribution of the American jack-knife clam *Ensis directus* (Conrad, 1843) in N.W. Europe (Bivalvia, Cultellidae). Basteria 50: 33–34.

Essink, K., 1988. Decrease of mercury pollution in the Dutch Wadden Sea and Ems Estuary. Mar. Pollut. Bull. 19: 317–319.

Essink, K., 1989. Chemical monitoring in the Dutch Wadden Sea by means of benthic invertebrates and fish. Helgol. Meeresunters. 43: 435–446.

Everaarts, J.M.N., W.K. Osborne & E. Verhaaf, 1979. Effect of copper and zinc on some oxygen binding properties of the haemoglobin of the polychaete *Arenicola marina*. Neth. J. Sea Res. 13: 571–580.

Giesen, W., 1990. Wasting disease and present eelgrass condition. Report of the Laboratory of Aquatic Ecology, Catholic University Nijmegen, The Netherlands, 138 pp.

Glim, G.W., G. Kool, M.F. Lieshout & M. de Boer, 1987. Erosie en sedimentatie in de binnendelta van het zeegat van Texel 1932–1982. Rijkswaterstaat, Directie Noord-Holland, Report ANWX 87.H201, 33 pp. and 22 figures (in Dutch).

Goethe, F., 1980a. Black-headed gull (*Larus ridibundus* L.). In C.J. Smit & W.J. Wolff (eds), Birds of the Wadden Sea. Balkema, Rotterdam: 219–229.

Goethe, F., 1980b. Common gull (*Larus canus* L.). In C.J. Smit & W.J. Wolff (eds), Birds of the Wadden Sea. Balkema, Rotterdam: 229–238.

Goethe, F., 1980c. Herring gull (*Larus argentatus* Pontoppidan). In C.J. Smit & W.J. Wolff (eds), Birds of the Wadden Sea. Balkema, Rotterdam: 238–250.

Goor, A.C.J. van, 1919. Het zeegras (*Zostera marina* L.) en zijn beteekenis voor het leven der visschen. Rapp. Verh. Rijksinst. Viss. onderz. I: 415–498 (in Dutch).

Hartog, C. den, 1959. The epilithic algal community occurring along the coast of the Netherlands. Wentia 1: 1–241.

Hartog, C. den, 1961. Verarming en verrijking van de algen-

flora der Nederlandse kust. Het Zeepaard 21: 108–114 (in Dutch).
Hartog, C. den, 1983. Structural uniformity and diversity in *Zostera*-dominated communities in western Europe. Mar. Techn. Soc. J. 17: 6–14.
Hartog, C. den, 1987. 'Wasting disease' and other dynamic phenomena in *Zostera* beds. Aquat. Bot. 27: 3–14.
Hartog, C. den & G. Van der Velde, 1987. Invasions by plants and animals into coastal, brackish and fresh water in the Netherlands. Ecology Proceedings C 90: 31–37.
Heiligenberg, T. van den, 1987. Effects of mechanical and manual harvesting of lugworms, *Arenicola marina* L., on the benthic fauna of tidal flats in the Dutch Wadden Sea. Biol. Conserv. 39: 165–177.
Heuvel, Tj. van, 1991. Sediment transport in het Eems-Dollard estuarium, volgens de methode McLaren. Rijkswaterstaat, Dienst Getijdewateren, Nota GWWS-91.002, 25 pp. (in Dutch).
Hoek, C. van den, W. Admiraal, F. Colijn & V.N. de Jonge, 1979. The role of algae and seagrasses in the ecosystem of the Wadden Sea: a review. In W.J. Wolff (ed.), Flora and vegetation of the Wadden Sea. Balkema, Rotterdam: 9–118.
Hoeven, P.C.T. van der, 1982. Observations of surface watertemperature and salinity, State Office of Fishery Research (RIVO): 1860–1981, Scientific report W.R. 82–8, De Bilt, 118 pp.
Jong, D.J. de & V.N. de Jonge, 1989. Zeegras. Een ecologisch profiel en het voorkomen in Nederland. Rijkswaterstaat, Tidal Waters Division, The Hague. Report GWAO-89.1003, 34 pp.
Jonge, V.N. de, 1983. Relations between annual dredging activities, suspended matter concentrations, and the development of the tidal regime in the Ems estuary. Can. J. Fish. Aquat. Sci. 40 (Suppl. 1): 289–300.
Jonge, V.N. de, 1986. Extraction of deposits in the Dutch Wadden Sea: a possible cause of increased turbidity. Rijkswaterstaat, Tidal Waters Division, Groningen. Report No. GWAO-86.453.
Jonge, V.N. de, 1990. Response of the Dutch Wadden Sea ecosystem to phosphorus discharges from the River Rhine. Hydrobiologia 195: 49–62.
Jonge, V.N. de, 1992. Physical processes and dynamics of microphytobenthos in the Ems estuary. Ph.D. Thesis, Rijksuniversiteit Groningen, 176 pp.
Jonge, V.N. de & K. Essink, 1991. Long-term changes in nutrient loads and primary and secundary producers in the Dutch Wadden Sea. p. 307–316. In M. Elliott & J.-P. Ducrotoy (eds): Estuaries and coasts: Spatial and temporal intercomparisons, Olsen & Olsen, Fredensborg, Denmark.
Jonge, V.N. de & K. Essink, 1992. Scheepvaart. De baggerproblematiek en andere effekten van scheepvaart op het estuariene ecosysteem. In W.J. Wolff (ed.), Handboek voor Natuurbeheer, Waddenzee en Estuaria, PUDOC, Wageningen, (in press) (in Dutch).
Jonge, V.N. de & D.J. de Jong, 1992. Role of tide, light and fisheries in the decline of *Zostera marina* L. in the Dutch Wadden Sea. Neth. Inst. Sea Res. Publ. Ser., No. 20-1992: 161–176.
Jonge, V.N. de & H. Postma, 1974. Phosphorus compounds in the Dutch Wadden Sea. Neth. J. Sea Res. 8: 139–153.
Kat, M., 1987. Diarrhetic mussel poisoning. Measures and consequences in the Netherlands. Rapp. P-v. Réun. Cons. int. Explor. Mer 187: 83 – 88.
Kemp, W.M., W.R. Boynton, R.R. Twilley, J.C. Stevenson & J.C. Means, 1983. The decline of submerged vascular plants in Upper Chesapeake Bay: summary of results concerning possible causes. Mar. Tech. Soc. J. 17: 78–89.
Koeman, J.H., 1971. Het voorkomen en de toxicologische betekenis van enkele chloorkoolwaterstoffen aan de Nederlandse kust in de periode van 1965 tot 1970. Ph.D. Thesis, Rijksuniversiteit Utrecht (in Dutch).
Koeman, R.P.T. & C. van den Hoek, 1981. The taxonomy of *Ulva* (Chlorophyceae) in The Netherlands. Br. Phycol. J. 19: 9–53.
Kuiper, J., 1982. The use of enclosed plankton communities in aquatic ecotoxicology. Fate and effects of mercury, cadmium and selected aromatic organics in a marine model ecosystem. Ph.D. Thesis, Landbouwhogeschool Wageningen.
Louisse, C.J., 1987. Aard en omvang van het slibtransport gebeuren op Loswal Noord. Rijkswaterstaat, Tidal Waters Division, The Hague, report GWAO-87. 002. (in Dutch).
Madsen, J., 1986. Danske rastepladser for gæse. Gåsetaellinger 1980–1983. Miljøministeriet, Fredningsstyrelsen, Kopenhagen, 114 pp. (in Danish).
Marchetti, R., A. Provini & G. Crosa, 1989. Nutrient load carried by the River Po into the Adriatic Sea, 1968–87. Mar. Poll. Bull. 20: 168–172.
McCave, N., 1973. Mud in the North Sea. In D. Goldberg (ed.), North Sea Science, NATO North Sea Science Conference, Cambridge, London: 75–100.
Meerendonk, J.H. van, G.M. Janssen & B. Frederiks, 1988. De aanvoer van voedingstoffen en microverontreinigingen naar de Waddenzee en Eems-Dollard. Rijkswaterstaat, Tidal Waters Division, Haren, Nota nr. GWWS-88.002, 84 pp. (in Dutch).
Neverauskaus, V.P., 1987. Monitoring seagrass beds around a sewage sludge out fall in South Australia. Mar. Pollut. Bull. 18: 158–164.
Nienhuis, P.H., 1993. Nutrient cycling and foodwebs in Dutch estuaries. Hydrobiologia 265: 15–44.
Osterhaus, A.D.M. & E.J. Vedder, 1988. Identification of a virus causing recent seal death. Nature 335: 20.
Peletier, H., 1992. Measurements of biomass, pigments and photosynthesis of *Ulva* sp. in the western Wadden Sea in 1988–91. Rijkswaterstaat, Tidal Waters Division. Report DGW-92.005 (in Dutch).
Piriou, J-Y, 1985. Les marées vertes sur le littoral Breton, bilan 1985. Rapport Direction de l'environnement et de recherches oceanografiques, IFREMER – Centre de Brest SDP, 79 pp. (in French).
Postma, H., 1954. Hydrography of the Dutch Wadden Sea. Archs. néerl. Zool. 10: 405–511.

Postma, H., 1967. Sediment transport and sedimentation in the estuarine environment. In G.H. Lauff (ed.), Estuaries. Am. Assoc. Adv. Sci.: 158–179.

Postma, H., 1980. Sediment transport and sedimentation. In E. Olausson & I. Cato (eds), Chemistry and biochemistry of estuaries. John Wiley, New York: 153–186.

Postma, H. & J.W. Rommets, 1970. Primary production in the Wadden Sea. Neth. J. Sea Res. 4: 470–493.

Postuma, K.H. & G. Rauck, 1978. The fishery in the Wadden Sea. In N. Dankers, W.J. Wolff & J.J. Zijlstra (eds), Fishes and fisheries of the Wadden Sea. Balkema, Rotterdam: 139–157.

Raaphorst, W. van & H. van der Veer, 1990. The phosphorus budget of the Marsdiep tidal basin (Dutch Wadden Sea) in the period 1950–1985: importance of the exchange with the North Sea. Hydrobiologia 195: 21–38.

Raffaelli, D., S. Hull & H. Milne, 1989. Long-term changes in nutrients, weed mats and shorebirds in an Estuarine System. Cah. Biol. Mar. 30: 259–270.

Rakhorst, H.D., 1982. Inventarisatie onderhoudsbaggerwerk en slibstortingen. Rijkswaterstaat, Directie Waterhuishouding en Waterbeweging, Adviesdienst Hoorn, Nota WWKZ-82.H001. (in Dutch).

Rakhorst, H.D., 1983. Zoutgehaltemetingen Waddenzee. Enige bewerkingen. Rijkswaterstaat, Adviesdienst Hoorn, Nota WWKZ-83.H003, 19 pp. (in Dutch).

Reenders, R. and D.H. van der Meulen, 1972. De ontwikkeling van de Dollard over de periode 1952–1969/'70. Rijkswaterstaat, Direktie Groningen, afd. Studiedienst, nota 72.1, 23 pp and 14 figures (in Dutch).

Reijnders, P.J.H., 1980. Organochlorine and heavy metal residues in harbour seals from the Wadden Sea and their possible effects on reproduction. Neth. J. Sea Res. 14: 30–65.

Reijnders, P.J.H., 1981. Management and conservation of the harbour seal, *Phoca vitulina*, population in the international Wadden Sea area. Biol. Conserv. 19: 213–221.

Reijnders, P.J.H., 1985. Waarom onze zeehondenpopulatie 1984 toch heeft gehaald. Waddenbulletin 20: 79–81 (in Dutch).

Reijnders, P.J.H., 1986. Reproductive failure in common seals feeding on fish from polluted coastal waters. Nature 324: 456–457.

Reise, K., 1990. Historische Veränderungen in der Ökologie des Wattenmeeres. Rheinisch-Westfälische Akademie der Wissenschaften, Vorträge Nr. 382. Westdeutscher Verlag: 35–55.

Rooth, J. & D.A. Jonkers, 1972. The status of some piscivorous birds in the Netherlands. TNO-Nieuws 27: 551–555.

Scholten, M.C.T., C.J.M. Kramer, H. van het Groenewoud, W. C. de Kock, C.T. Bowmer & M. van Veen, 1988. De ecologische effecten van havenspecieberging op Loswal Noord voor het waddenecosysteem. Hoofdgroep Maatschappelijke Technologie TNO, Delft, Report nr.: R 88/279. (in Dutch).

SOVON, 1987. Atlas van de Nederlandse vogels. SOVON, Arnhem. (in Dutch)

Swennen, C., 1972. Chlorinated hydrocarbons attacked the Eider population in the Netherlands. TNO-Nieuws 27: 556–560.

Swennen, C., 1976. Populatie-structuur en voedsel van de Eidereend *Somateria m. mollissima* in de Nederlandse Waddenzee. Ardea 64: 311–371 (in Dutch).

Swennen, C., 1982. De vogels langs onze kust. In W.J. Wolff *et al.* (eds), Wadden, duinen, delta. PUDOC, Wageningen: 78–100.

Swennen, C. & A.L. Spaans, 1970. De sterfte van zeevogels door olie in februari 1969 in het Waddengebied. Het Vogeljaar 18: 233–245 (in Dutch).

Swennen, C., M.F. Leopold & M. Stock, 1985. Notes on growth and behaviour of the American razor clam *Ensis directus* in the Wadden Sea and the predation on it by birds. Helgol. Meeresunters. 39: 255–261.

Swennen, C., G. Nehls & K. Laursen, 1989. Numbers and distribution of eiders *Somateria mollissima* in the Wadden Sea. Neth. J. Sea Res. 24: 83–92.

Teixeira, R.M. (ed.), 1979. Atlas van de Nederlandse broedvogels. Vereniging tot Behoud van Natuurmonumenten, 's-Graveland. (in Dutch).

Thijsse, J. Th., 1972. Een halve eeuw Zuiderzeewerken 1920–1970. Tjeenk Willink, Groningen, 469 pp. (in Dutch).

Veen, J.F. de, 1976. On changes in some biological parameters in the North Sea sole (*Solea solea* L.). J. Cons. int. Explor. Mer 37: 60–90.

Veen, Joh. van, 1938. Water movements in the Straits of Dover. J. Cons. perm. int. Explor. Mer 13: 7–36.

Veenstra, H.J., 1980. Introduction to the geomorphology of the Wadden Sea area. In K.S. Dijkema, H.-E. Reineck and W.J. Wolff (eds), Geomorphology of the Wadden Sea area. Balkema, Rotterdam: 8–19.

Veldhuis, M.J.W., F. Colijn, L.A.H. Venekamp & L.A. Villerius, 1988. Phytoplankton primary production and biomass in the western Dutch Wadden Sea (The Netherlands); a comparison with an ecosystem model. Neth. J. Sea Res. 22: 37–49.

Verhagen, J.H.G. & P.H. Nienhuis, 1983. A simulation model of production, seasonal changes in biomass and distribution of eelgrass (*Zostera marina*) in Lake Grevelingen. Mar. Ecol. Progr. Ser. 10: 187–195.

Vertegaal, C.T.M. & J.N.C. van der Salm, 1988. Ecologische profielen: kusten zeevogels. 6. Scholekster *Haematopus ostralegus*. Advies- en Onderzoeksbureau voor Duinbeheer Duin + Kust. Leiden: 39–48 (in Dutch).

Vlas, J. de, 1987. Annual food intake by plaice and flounder in a tidal flat area in the Dutch Wadden Sea, with special reference to consumption of regenerating parts of macrobenthic prey. Neth. J. Sea Res. 13: 117–153.

Vlas, J. de, 1987. Effects of the cockle fisheries on the macrobenthos in the Wadden Sea. In S. Tougaard & S. Asbirk (eds), Proceedings of the 5th International Wadden Sea Symposium, Esbjerg, Denmark, Sept. 29–Oct. 3, 1986. The

National Forest and Nature Agency and The Museum of Fisheries and Shipping, Esbjerg: 215–227.

Voous, K.H., 1960. Atlas van de Europese vogels. Elsevier, Amsterdam/Brussel, 284 pp. (in Dutch).

Vreugdenhil, D., 1984. Management of the Dutch Wadden Sea. Wat. Sci. Tech. 16: 485–495.

Westhoff, V., 1985. Nature management in coastal areas of Western Europe. Vegetatio 62: 520–532.

Wilde, P.A.W.J. de & J.J. Beukema, 1984. The role of the zoobenthos in the consumption of organic matter in the Dutch Wadden Sea. Neth. Inst. Sea Res. – Publ. Ser. 10: 145–148.

Witte, J.Y. & J.J. Zijlstra, 1978. The species of fish occurring in the Wadden Sea. In N. Dankers, W.J. Wolff & J.J. Zijlstra (eds), Fishes and fisheries of the Wadden Sea. Balkema, Rotterdam: 10–19.

Wolff, W.J. (ed.), 1976. Proceedings of the Conference of Wadden Sea Experts held at the Island of Schiermonnikoog, The Netherlands, 26–28 November 1975. Government Publishing Office, The Hague, 96 pp.

Wolff, W.J., 1993. Netherlands-Wetlands. Hydrobiologia 265: 1–13.

Wolff, W.J. & J.J. Zijlstra, 1980. Management of the Wadden Sea. Helgol. Meeresunters. 33: 596–613.

Wolff, W.J. & C.J. Smit, 1984. The Dutch Wadden Sea. In P.R. Evans, J.D. Goss-Custard & W.G. Hale (eds), Coastal waders and winterfowl in winter. Cambridge Univ. Press: 238–252.

Zimmerman, J.T.F., 1976. Mixing and flushing of tidal embayments in the western Dutch Wadden Sea. Ph.D. Thesis, Free University Amsterdam, 439 pp.

Zwarts, L. & J. Wanink, 1984. How oystercatchers and curlews successively deplete clams. In P.R. Evans, J.D. Goss-Custard & W.G. Hale (eds), Coastal waders and winterfowl in winter. Cambridge Univ. Press: 69–83.

Zijlstra, J.J., 1972. On the importance of the Wadden Sea as a nursery area in relation to the conservation of the southern North Sea fishery resources. Symp. zool. Soc. Lond. 29: 233–258.

Hydrobiologia **265**: 73–95, 1993.
E. P. H. Best & J. P. Bakker (eds), Netherlands-Wetlands.

Salt marshes along the coast of The Netherlands

J. P. Bakker[1], J. de Leeuw[2], K. S. Dijkema[3], P. C. Leendertse[4], H. H. T. Prins[5] & J. Rozema[4]
[1]*Laboratory of Plant Ecology, University of Groningen, P.O. Box 14, 9750 AA Haren, The Netherlands*
[2]*Delta Institute for Hydrobiological Research, Vierstraat 28, 4401 EA Yerseke, The Netherlands (present address: Division of Vegetation and Agricultural Sciences ITC, P.O. Box 6, 7500 AA Enschede, The Netherlands)*
[3]*Research Institute for Nature Management, P.O. Box 59, 1790 AB Den Burg (Texel), The Netherlands*
[4]*Department of Ecology and Ecotoxicology, Free University, De Boelelaan 1089, 1081 HV Amsterdam, The Netherlands*
[5]*Zoological Laboratory, University of Groningen, P.O. Box 14, 9750 AA Haren, The Netherlands (present address: Department of Nature Conservancy, Agricultural University of Wageningen, Ritzema Bosweg 32A, 6703 AZ Wageningen, The Netherlands)*

Key words: exploitation, fauna, management, nitrogen cycle, vegetation, salt marsh, sea level rise

Abstract

The area of salt marshes does no longer increase. The recent erosion coincides with a rise in MHT-level in the last 25 years. Despite the decrease in area, sedimentation continues, especially in the lower salt marsh, which acts as a sink of nitrogen. Assimilation and mineralization of nitrogen are in balance in most plant communities along the gradient from lower to higher salt marshes. Mineralization of nitrogen increases towards the higher salt marsh, whereas the above-ground production and the mean nitrogen content of plants decrease. There is a positive correlation between quality of food plants in salt marshes and breeding success of Brent geese in the arctic tundra.Sedimentation on mainland salt marshes can compensate for the expected sea level rise. This is not the case for island salt marshes, if the relative sea level rise is more than 0.5–1.0 cm yr^{-1}. The natural succession on salt marshes results in an accumulation of organic material, which is related to the dominance of single plant species. It is not clear to which extent this process is enhanced by eutrophication from acid deposition and seawater.Human exploitation of unprotected salt marshes is old and heavy in the system of mound settlements. Reclamation rates by dikes in the last centuries were higher than the rate of area increase.Grazing by cattle as a management practice results in both a higher plant species-richness and community diversity than abandoning; hay-making is intermediate, but shows less structural diversity than grazing with low stocking density. The invertebrate fauna is favoured by a short period of abandoning, but eventually characteristic salt marsh invertebrates are replaced by inland species. Many bird species prefer grazed salt marshes.The final section gives some perspectives. Provided that no further embankments take place the optimal nature management option for plants and animals is a vegetation pattern, which includes areas with a low canopy (grazed) and areas with a tall canopy.

System characteristics

Area increase and sedimentation in salt marshes

Conditions for the formation of salt marshes are best on a gently sloping shoreline with little wave energy and sufficient sediment supply (Dijkema, 1987). Surface elevation, tidal amplitude, and drainage must be sufficient to allow periods of soil aeration necessary for plant growth (Armstrong *et al.*, 1985). Pioneer plants (*Salicornia* *dolichostachya** and *Spartina anglica*) and further sedimentation create an environment which promotes a closed coverage of perennial halophytic plants. Around Mean High Tide (MHT)-level *Puccinellia maritima* reaches sufficient abundance to enhance maximal sedimentation.

The total area of salt marshes in the Dutch Wadden Sea area is presently 7,300 ha, including 1,900 ha pioneer zone (Dijkema *et al.*, 1990). The average seaward expansion along the mainland coast was 8.2 m yr^{-1} for Friesland and 4.7 m yr^{-1} for Groningen for the period 1960–1985. These rates are similar to those observed in German mainland salt marshes, namely, 7.1 m yr^{-1} for the period 1930–1978 and 4.7 m yr^{-1} for the period 1940–1979 (Dieckmann, 1988). The seaward expansion of the western section of the Friesland mainland coast only took place in recent years (Fig. 1), whereas it was constant for the entire period (> 1 m yr^{-1}) in the eastern section. Almost all seaward expansion of the central section of Friesland and the western section of the Groningen coast occurred in the 1960's. The marsh area stabilized later on and erosion occurred during the past 10 years. The eastern section of the Groningen coast shows a net reduction in salt marsh area after 1960. The total area of the mainland saltmarshes increased by 1,000 ha in

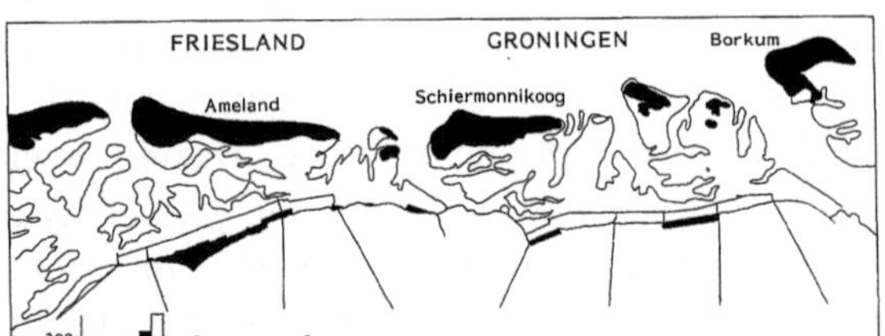

Fig. 1. Seaward expansion of the contourline of NAP + 1 m and of the salt marsh edge in the Netherlands Wadden Sea (after Bouwsema *et al.*, 1986).

* Nomenclature of taxa follows Van der Meijden *et al.* (1983), that of syntaxa Westhoff & Den Held (1969).

the period 1960–1978, whereas a reduction of 380 ha took place in the period 1978–1988 (P.Bouwsema, pers.comm.).

Sedimentation was higher from 1960 to 1968, shortly after the construction of the sedimentation fields, than in later years (1968–1987) in the Wadden Sea mainland coastal area. The sedimentation balances differ in the various sections along the coast (Fig. 2). In the sections of west and east Friesland and in west Groningen the balance is positive relative to MHT (0.9–1.7 cm yr^{-1} on average), whereas in the sections of central Friesland and east Groningen a negative balance is found in the pioneer zone (−0.8 cm yr^{-1}) for the period 1978–1987. The sedimentation is highest between MHT-level (MHT = NAP + 1.25 m) and 20 cm above MHT-level and decreases again at higher elevation due to less frequent tidal flooding, namely, 0.7–2.0 cm yr^{-1} (at NAP + 1.30 m) to 0.6 cm yr^{-1} (at NAP + 2.00 m) in mainland salt marshes and 0.4 cm yr^{-1} (at NAP + 1.30 m) to 0 cm yr^{-1} (at NAP + 2.00 m) in barrier island marshes (Dijkema *et al.*, 1990) (NAP = Nieuw Amsterdams Peil; similar to Ordnance Date).

The sedimentation rate is affected by grazing as demonstrated in the mainland salt marsh of the Leybucht in Germany. In the *Puccinellia maritima*-zone (NAP + 1.40 m) a sedimentation of 2.3 cm yr^{-1} was found in the ungrazed area and 1.7 cm yr^{-1} in the grazed site. No differences were found between grazed and ungrazed sites in the *Festuca rubra*-zone at NAP + 1.8 m (1.6 cm yr^{-1}) and in the *Salicornia dolichostachya*-zone at NAP + 1.1 m (1.8 cm yr^{-1}) (Andresen *et al.*, 1990).

The recent erosion of salt marshes in the Wadden Sea area might be caused by the increase in MHT-level, namely, 0.44 cm yr^{-1} in the Dutch Wadden Sea from 1960 to 1986, (Dijkema *et al.*, 1990) and 0.46 cm yr^{-1} in the adjacent German Wadden Sea from 1954 to 1986 (Jensen *et al.*, 1988). The higher flooding frequency might eventually result in cliff formation along the salt marshes (Dijkema *et al.*, 1990).

Most estuarine salt marshes in the south-west of the Netherlands are eroding. The high rates of

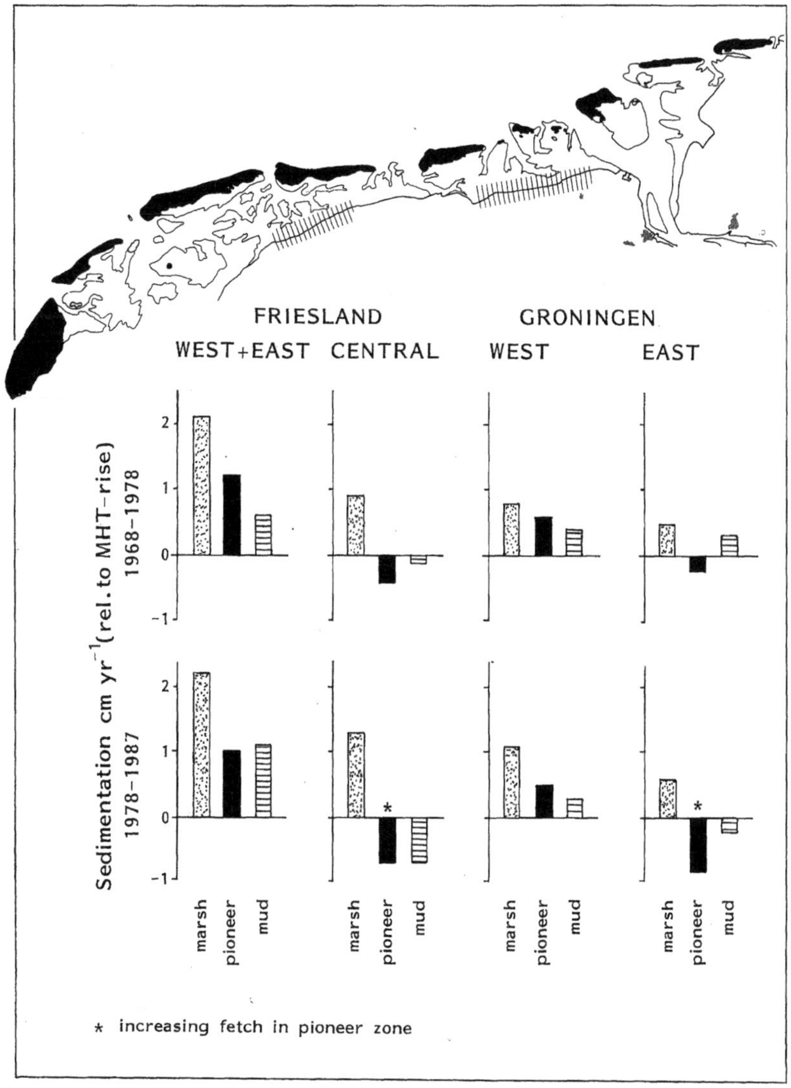

Fig. 2. Sedimentation relative to MHT-rise in mainland salt marshes and mud flats in the Netherlands Wadden Sea for the period after the construction of sedimentation fields (after Dijkema *et al.*, 1989).

cliff erosion (up to 4 m yr^{-1}) in the eastern part of the Westerschelde have been related to increased tidal stream velocities caused by intensive dredging. The area increase slowed down in the 1930' in the Oosterschelde, and erosion took place over the last decades. The cliff length of the marsh-mud transition increased from 60% in the period 1950 to 1970 to 90% in the early 1980's. This has been attributed to increased tidal amplitude and stream velocities caused by various human activities. The construction of the Oosterschelde storm surge barrier caused a reduction in the tidal amplitude and it was therefore expected that the rate of cliff erosion would decrease. To date, the rate has, however, remained high (De Jong, 1991).

Geochemical cycles

System behaviour

Import–export

Geochemical cycles in salt marshes are open systems due to import and export of large amounts of organic material and nutrients. Transport of particulate and dissolved matter between salt marshes and the sea takes place by tidal water. Salt marshes are often regarded as sinks for organic material, nutrients and heavy metals since sedimentation of particulate matter takes place. The net direction of transport of materials between salt marshes and the sea is, however, complicated. In several studies different methodological approaches have been applied to estimate the net exchange and the variation in results can partly be ascribed to the approach used, especially with respect to the hydrographical balance (Wolaver *et al.*, 1983; Dankers *et al.*, 1984). Much of the variability among various marshes may be due to factors related to geographic position, marsh morphology and hydrology (Whiting *et al.*, 1987).

Phosphate is generally exported from the salt marsh on an annual basis, whereas organic phosphorus is imported (Jordan *et al.*, 1983). In the Dutch Ems-Dollard estuary also a small export of phosphate of 2.8 kg P ha^{-1} yr^{-1} was found (Dankers *et al.*, 1984). A net overall export of nitrogen, ranging from 7 to 111 kg N ha^{-1} yr^{-1} is often found (Rozema & Leendertse, 1991). The role of salt marshes in the nitrogen cycle is mainly that of nitrogen processor since oxidized forms, like nitrate, are generally exported, and reduced forms, like ammonia and dissolved organic nitrogen, are imported (Abd. Aziz & Nedwell, 1986). In a mesohaline marsh, in contrast, a large annual import of nitrogen was found (Wolaver *et al.*, 1983). The annual import of nitrate in and export of ammonia out of the Ems-Dollard salt marsh as measured by Dankers *et al.* (1984) is in agreement with the conclusions of Abd. Aziz & Nedwell (1986). The large input of particulate matter in the Ems-Dollard salt marsh could, however, mean an input of particulate nitrogen into the marsh and hence a net input of nitrogen.

Many authors suggest that the quantification of nitrogen fluxes during storm tides may be important for the annual balance, but such measurements are rarely done because of practical problems. Dankers *et al.* (1984) estimated the export of coarse debris during storm tides to be low as compared to the primary production of the salt marsh.

Effects of age of salt marshes

Valiela & Teal (1979) expected that relatively old salt marshes, like at their study site, eventually reach a steady state in which the import and export of nitrogen are nearly balanced. Relatively young marshes like those in the Wadden Sea may, therefore, act differently i.e. as a sink for nitrogen as a whole (Leendertse, 1989).

Effects of zonation of plant communities

The objective of transport studies such as the afore-mentioned is to quantify the import and export for a whole salt marsh. They consist, however, of different plant communities according to the abiotic zonation. To understand the mechanisms of the production and decomposition phenomena of salt marshes, it is necessary to have knowledge of the plant community level. Jordan *et al.* (1983), for example, found differences in the annual nitrogen budget of higher and lower parts of a brackish salt marsh, namely, a net import of 9 kg N ha^{-1} yr^{-1} in the lower marsh and a net export of 28 kg N ha^{-1} yr^{-1} in the higher marsh.

Nitrogen cycle

Net exchange between salt marsh and sea, assimilation, decomposition and mineralization

Nitrogen plays an important part in the functioning of salt marshes. It can control processes like the productivity of primary and secondary producers, and of decomposition rates of organic material (Teal, 1986). This became clear from a study of the N cycle of the Great Sippewisset Marsh (USA). N was imported by sea water, dry and wet deposition, groundwater, sedimentation and fixation. It was exported by sea water and denitrification. Considerable assimilation and mineralization occurred in the marsh itself. A British study on the annual cycling of nitrogen in a *Puccinellia maritima* community indicated, for instance, that assimilation (225 kg N ha^{-1} yr^{-1}), and decomposition and mineralization (195 kg N ha^{-1} yr^{-1}) were high compared to the net import and export (7 kg N ha^{-1} yr^{-1}), but about half as high as the total import and export (500 kg N ha^{-1} yr^{-1}), while aerial deposition was considered negligible (Fig. 3; Abd. Aziz & Nedwell, 1986).

Mineralization usually outweighs humification in salt marshes (Beeftink, 1966). This is illustrated by the following calculations. The soil nitrogen pool of sandy and clay-rich salt marshes in northern Germany ranged from 5,000 to 12,000 kg ha^{-1}; humus accumulation from 1,700 to 12,400 kg ha^{-1} (sedimentation rate of 1 cm yr^{-1} assumed; Scherfose, 1987). The N mobility, i.e. the ratio mineral N: total N, in salt marshes is high (2.6 to 8.7%; Scherfose, 1987) compared to that in inland ecosystems (1 to 4%; Alexander, 1977).

Aerial deposition

Although aerial deposition was considered negligible (Abd. Aziz & Nedwell, 1986), it can be considerable, at least 43 kg N ha^{-1} yr^{-1} (22 kg NO_x and 21 kg NH_y ha^{-1} yr^{-1}) in the Netherlands in 1987 (Additioneel Programma Verzuringsonderzoek, 1989). If this deposition had been the same

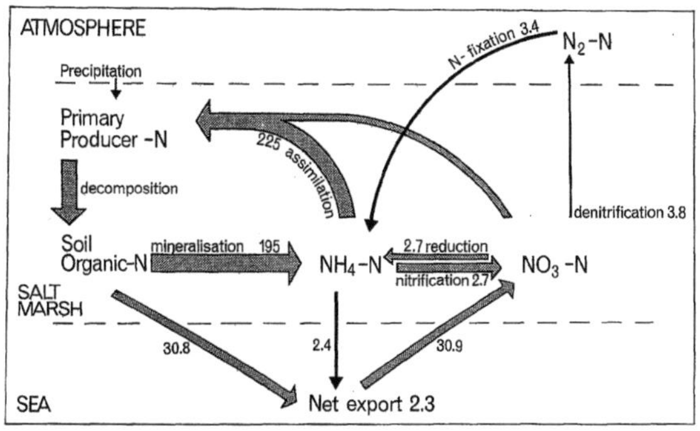

Fig. 3. The annual nitrogen budget for a *Puccinellia marritima* community in eastern England. Fluxes of nitrogen are shown as kg N ha^{-1} yr^{-1} (after Abd. Aziz & Nedwell, 1986).

in the British study, it would have accounted for 22% of the mineralized nitrogen in the *Puccinellia maritima* community.

Hay-making and grazing

The removal of nitrogen by hay-making and grazing is often not taken into account in many budget studies. The amount of nitrogen removed by haymaking on the island of Schiermonnikoog can range from 16 to 131 kg N ha^{-1} yr^{-1} in various plant communities (Table 1). The export of nitrogen by grazing, however, cannot merely be derived from the utilized biomass, since utilization includes both foraging and trampling. Foraging can amount to 30% of the herbage utilized by cattle and 50% for that by geese (Looijen & Bakker, 1987). This implies that the amount of nitrogen removed by grazing ranges from 10–70 kg N ha^{-1} yr^{-1} for the combined effects of cattle and geese-grazing (Table 1). Thus, going back to the British study (Abd. Aziz & Nedwell, 1986) cattle can remove about 40 kg N ha^{-1} yr^{-1} and geese 30 kg N ha^{-1} yr^{-1} from a *Puccinellia maritima* community.

Nitrogen is not removed by herbivores, but in fact reallocated. It reaches the soil again as dung and contributes as such to the nitrogen cycle. The mineral cycling of nitrogen, can even up to tenfold be stimulated (Floate, 1970). The above-ground plant biomass was demonstrated to increase by geese-grazing, possibly due to the rapid recycling of nitrogen from faeces (Fig. 4; Hik & Jefferies, 1990).

Balancing items of nitrogen budgets

Mineralization of nitrogen strongly varies with zonation and different plant communities (Table 2; Scherfose, 1987), and generalization of data is therefore risky. The nitrogen assimilation and mineralization in the British *Puccinellia maritima* community (Abd. Aziz & Nedwell, 1986), a Danish *Halimione portulacoides* community (Henriksen & Jensen, 1979) and a German *Festuca*

Table 1. The removal of plant nitrogen by hay-making and grazing, calculated from the mean biomass removed by anual cutting, the biomass utilized by cattle and geese, and the mean nitrogen contents of the plant communities. Biomass utilization by grazing was measured as the difference between herbage under exclusion cages and in open plots according to Looijen & Bakker (1987). Trampling effects were estimated as 70% of utilization by cattle and 50% of utilization by geese (Looijen & Bakker, 1987). –: not measured.

Plant community Schiermonnikoog The Netherlands	Hay-making		Cattle grazing		Geese grazing		Mean nitrogen content (%)	Total grazing corrected for trampling. Nitrogen removed kg ha^{-1} yr^{-1}
	Mean biomass removed ($g\ m^{-2}$)	Nitrogen removed (kg ha^{-1} yr^{-1})	Biomass utilized ($g\ m^{-2}$)	Nitrogen removed (kg ha^{-1} yr^{-1})	Biomass utilized ($g\ m^{-2}$)	Nitrogen removed (kg ha^{-1} yr^{-1})		
Puccinellia maritima	–	–	415	145	181	63	3.5	70
Juncus gerardii	311	84	174	47	38	10	2.7	18
Festuca rubra	273	55	144	29	75	15	2.0	15
Armeria maritima	113	16	229	32	–	–	1.4	10
Juncus maritimus	355	60	1054	180	–	–	1.7	60
Elymus pycnanthus	266	38	–	–	–	–	1.4	–
Artemisia maritima	413	131	–	–	–	–	3.2	–

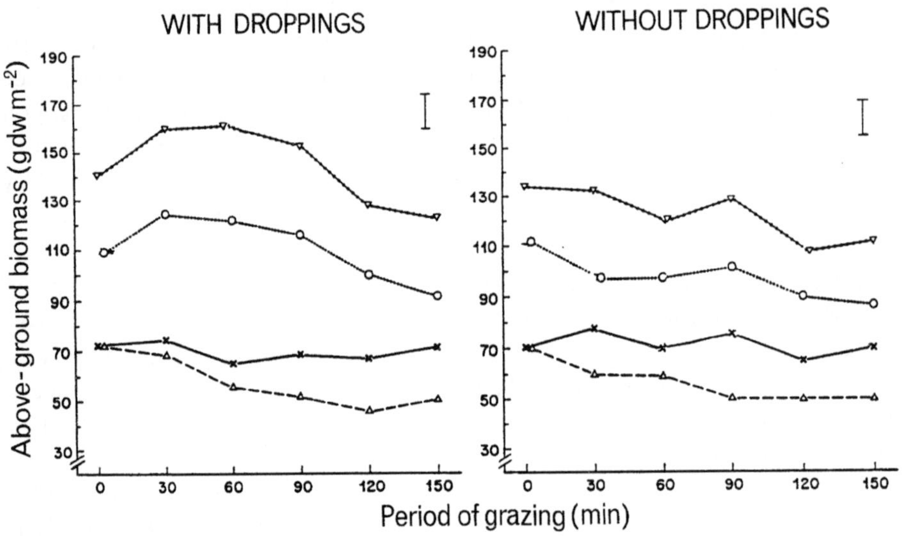

Fig. 4. The above-ground plant biomass after different periods of grazing with droppings left and with the droppings removed (after Hik & Jefferies, 1990).

rubra community (Zimmek, 1975) balance more or less. A shortage of mineralized nitrogen was, however, estimated in a German *Puccinellia maritima* community which was possibly replenishned by bird droppings and groundwater from the higher parts of the marsh (Zimmek, 1975). The nitrogen assimilation by roots was quantified in the British *Puccinellia maritima* community (Abd. Aziz & Nedwell, 1986). This might be important for the balance between mineralization and assimilation, but it is not clear whether it is taken into account in the other studies.

Heavy metals

The exchange of heavy metals between sea and salt marsh has not been measured as intensively as the exchange of nutrients. The retention of heavy metals is mainly due to the anoxic soil con-

Table 2. Standing crop, including accumulated litter, of salt marsh plant communities and the mineralization and decomposition rates in different salt marsh zones and a beach plain. (1) Groenendijk, 1987. (2) Henriksen & Jensen, 1979; (3) Abd.Aziz & Nedwell, 1979; (4) Zimmek, 1975; (5) Scherfose, 1987; (6) Rozema, 1978; (7) Bakker, 1989; (8) Looijen & Bakker, 1987. –: not measured.

Salt marsh zone	Plant community	Inundation frequency ($N\ yr^{-1}$)	Standing crop + litter ($gwd\ m^{-2}$)	Net mineralization rate ($kg\ N\ ha^{-1}$)		Decomposition rate (% cellulose disappeared)
				Per year	Per growing season	Per growing season
Pioneerzone	Salicornia europaea	350–600	–	–	50 (5)	–
	Spartina anglica	350–600	970 (1)		35 (5)	–
Lower salt marsh	Puccinellia maritima	130–350	360 (1)	195 (3); 50 (4)	60 (5)	96 (8)
	Halimione portulacoides	130	1300 (1)	110 (2)	90 (5)	–
Mid salt marsh	Juncus gerardii	40	400–800 (7)	–	70 (5)	76 (8)
	Artemisia maritima	30–160	1100–1300 (7)	–	110 (5)	–
Upper salt marsh	Juncus maritimus	12–60	800–1500 (7)	–	75 (5)	93 (8)
	Elymus pycnanthus	20–40	1300–1500 (7)	–	–	–
	Festuca rubra	12–30	400–1000 (7)	150 (4)	100 (5)	58 (8)
	Armeria maritima	30–80	200–400 (7)	–	55 (5)	15 (8)
Beach plain		2–5	–	15 (6)	–	–

ditions in salt marshes. The sediments contain high levels of sulfide under reduced conditions and lead-, zinc-, copper- and cadmium-ions form insoluble and relatively stable sulfides under these conditions. Budget studies for lead and cadmium were carried out by applying contaminated fertilizers to the salt marsh and estimating the retention (Banus *et al.*, 1975). Lead and cadmium were largely retained in the salt marsh soil (96% and 35–80%, respectively). Other data included retention of 20–35% for cadmium, 20–50% for chrome and 20–45% for zinc (Giblin *et al.*, 1983). The oxidation of the soil by plant roots may reduce its retention for heavy metals. The exchange of heavy metals between sea and salt marsh will also depend on the amount of metals taken up by plants and animals. Transport of metal-containing plant material to the sea means export from salt marsh sediments.

Original and unexploited plant communities

The tidal marsh landscapes which formerly extended all along the whole Dutch coast have changed dramatically in (pre)historic times. The investigation of plant remains such as pollen and macro-fossils from prehistoric settlements yielded evidence concerning these habitats.

Habitats far from the sea were supplied with fresh water from the hinterland. A continuous sea level rise took place from the last glacial period onwards. Natural drainage became blocked and an extensive accumulation of organic material took place as far as the fringes of the Pleistocene cover sands. Both fen-peat (*Magnocaricion*, *Phragmition*, *Alnion glutinosae*) and bog-peat (*Sphagnion*, *Ericion tetralicis*) developed, depending on the regional and local hydrology (Fig. 5). Most peat was covered by clay during transgression periods, vast areas from the middle ages onwards (Behre, 1979). One example of peat which has not been covered, is located outside the dike and is therefore floating, is in the Jadebusen in Germany (Schwimmendes Moor) (Wiermann, 1965; Luck, 1976). River bank forests (*Fraxino-Ulmetum*) have been recorded from the early subboreal period (ca. 4500 B.C.) until the Iron Age (ca. 500 B.C.) on the elevated levees deposited by water along the lower parts of rivers, for instance

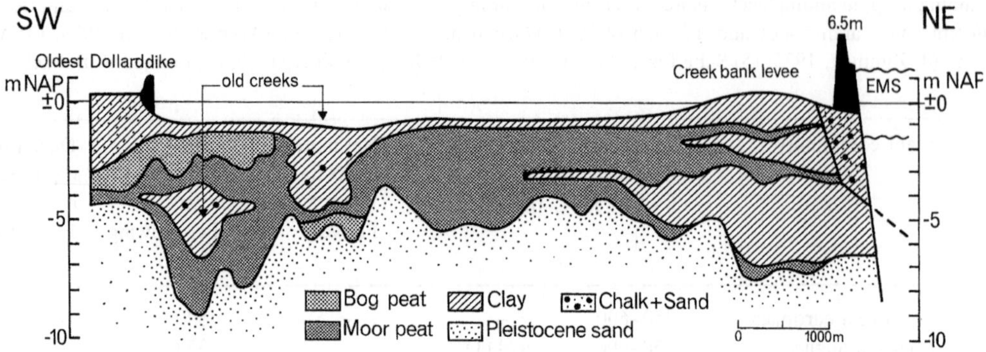

Fig. 5. Transect through the marshes between the river Ems and the Dollard in north-west Germany (after Behre, 1979).

the Ems in Germany. Mainly *Phragmition* and *Bidention* communities and some *Magnocaricion* occurred in the neighbourhood of the levees in the period from 700 B.C. until 300 A.D. as examples of natural treeless communities (Behre, 1979).

The transition zone between fresh and salt tidal water landscapes must have been represented by plant communities indicating brackish conditions. Indeed brackish elements of fresh water *Phragmition* communities (especially *Scirpus maritimus*) and of salt water *Armerion maritimae* communities (especially *Eleocharis palustris* ssp. *uniglumis*) have been found (Behre, 1979).

All the aforementioned plant communities were not protected by dikes. Salt marsh communities are the only unprotected ones at present. In order to understand the synecology of the former salt marsh plant communities, they are based on the present knowledge of plant communities. Starting from the tidal flats, a characteristic zonation in the natural halosere of salt marsh communities is found on the sandy barrier islands of the Wadden Sea (Westhoff & Den Held, 1969; Beeftink, 1977A; Dijkema, 1983). The *Salicornietum strictae* and *Spartinetum townsendii* occur on tidal flats with daily submergence in the pioneer zone. The lower salt marsh, between mean high water (MHW) and spring tide level is characterized by the *Puccinellion maritimae* with the *Puccinellietum maritimae* in depressions, the *Halimionetum portulacoidis* on creek bank levees and the *Plantagini-Limonietum* in the intermediate positions. The *Armerion maritimae* occurs on the mid and upper salt marsh from just below the spring tide level up to the level reached by winter storm floods. The *Juncetum gerardii* is found on the clay soils, whereas the *Artemisietum maritimae* occurs on sand ridges and creek bank levees. The *Puccinellio-Spergularion salinae* community can develop in places with irregularly changing conditions such as occasionally stagnant water. The *Atriplici-Agropyretum pungentis* can develop higher on the salt marsh after the deposition of drift material. *Lolio-Potentillion* communities can be found on the upper slightly desalinated salt marsh (communities of *Agrostis stolonifera* and *Trifolium fragiferum*) and on adjacent low dunes (community of *Ononis spinosa* and *Carex distans*). A more ephemerous community can also be found at the transition of wet/salt and dry/fresh, viz. the *Saginion maritimae*. The relationship of the halosere on mainland salt marshes to the inundation frequency is shown in Fig. 6.

Plant biomass characteristics

The above-ground standing crop shows a large variation in the different plant communities over the salt marsh zones and from year-to-year (Table 2). These differences can partly be attributed to various amounts of accumulated litter in the biomass samples. Since the standing crop and the amount of litter depend on the relative rates of production and decomposition, it seems necessary to focus on these processes.

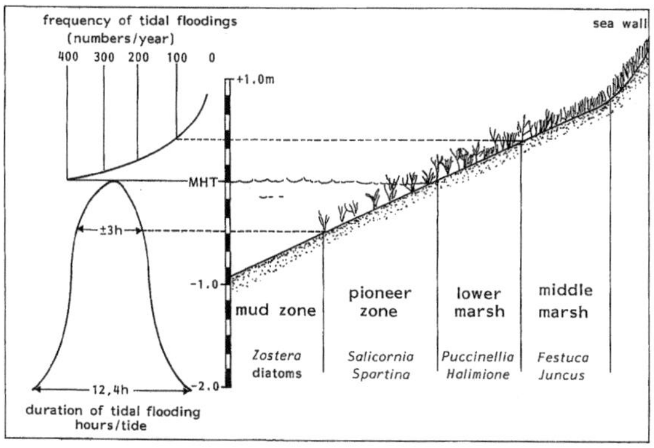

Fig. 6. Zonation of mainland salt marshes in relation to duration and frequency of tidal floodings (after Erchinger, 1985).

The primary production in salt marshes is extremely high. Above-ground production in Dutch salt marshes varies widely: from 300–400 g m^{-2} in *Juncus gerardii* and *Festuca rubra* communities on higher salt marshes in the Wadden Sea area to over 1,500 g m^{-2} in *Spartina anglica* stands on low marshes in the south-west of the Netherlands and in *Juncus maritimus* communities on Schiermonnikoog (Table 3). This suggests a generally negative relationship between the above-ground production of salt marsh plant communities and their frequency of inundation (Table 3).

A high spatial and temporal variation within plant communities is, however, found. A large range (100–1,000 g m^{-2}) in production estimates was found in *Spartina anglica* communities in the south-west of the Netherlands (De Leeuw & Buth, 1991). Large year-to-year fluctuations in standing crop occurred on Schiermonnikoog (in the northern Netherlands) in annually cut areas in various plant communities on the mid- and high-salt marsh with no litter present (De Leeuw *et al.*, 1990). These communities showed, however, a synchronous pattern in peak above-ground biomass over a 13-year period. These fluctuations could be explained by the rainfall deficit in the growing season, but not by the inundation frequency (Fig. 7). The soil salinity was positively and the soil moisture content was negatively cor-

Table 3. Annual above-ground, below-ground and the ratio below-gr.: above-gr. production for Dutch salt marsh plant communities. Above-ground production estimated according to Smalley, 1959; below-ground production according to the change in biomass technique (Dahlman & Kucera, 1965). (1) Groenendijk, 1984; (2) Groenendijk & Vink-Lievaart, 1987; (3) Ketner, 1972; (4) Looijen & Bakker, 1987; Wolff *et al.*, 1980; (6) A. Fuchs, unpubl.

Salt marsh zone	Plant community	Locality	Above-ground production (gwd m^{-2} yr^{-1})	Below-ground production (gwd m^{-2} yr^{-1})	Ratio Below-gr. above-gr.
Pioneerzone	*Spartina anglica*	Oosterschelde	1650 (1)	6044 (2)	3.7
	Spartina anglica	Oosterschelde	1057 (6)		
	Spartina anglica	Oosterschelde	1030 (5)		
Lower salt marsh	*Puccinellia maritima*	Schiermonnikoog	767 (4)		
	Puccinellia maritima	Terschelling	497 (3)		
	Puccinellia maritima	Oosterschelde	992 (6)		
	Puccinellia maritima	Oosterschelde	837 (5)		
	Halimione portulacoides	Oosterschelde	1400 (1)	7977 (2)	5.7
	Halimione portulacoides	Oosterschelde	575 (6)		
Mid salt marsh	*Triglochin maritima*	Oosterschelde	70 (1)	3475 (2)	5.0
	Plantago maritima	Terschelling	514 (3)	1138 (3)	2.2
	Juncus gerardii	Terschelling	344 (3)	601 (3)	1.7
	Juncus gerardii	Schiermonnikoog	429 (4)		
Upper salt marsh	*Elymus pycnanthus*	Oosterschelde	900 (1)	4421 (2)	4.9
	Elymus pycnanthus	Oosterschelde	1008 (5)		
	Festuca rubra	Schiermonnikoog	388 (4)		
	Juncus maritimus	Schiermonnikoog	1707 (4)		

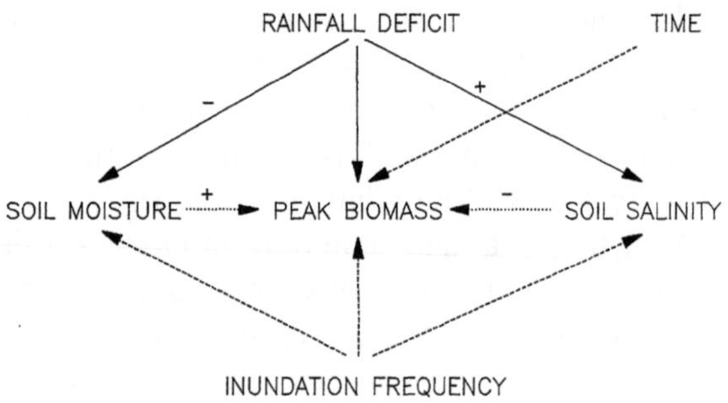

Fig. 7. Factors affecting peak biomass accumulation; ——— = significant correlation, ------ = no significant correlation, ········ = suggested relation (after De Leeuw *et al.*, 1990).

related with the rainfall deficit in a two-year period in these mid- and high-salt marsh communities. The production of lower salt marsh ecosystems, in contrast, is probably not affected by the rainfall deficit during the growing season (De Leeuw *et al.*, 1991).

Below-ground production generally outweighs above-ground production in salt marshes. It also varies widely: from 300–1,000 g m^{-2} (Ketner, 1972) to 3,500–8,000 g m^{-2} (Groenendijk, 1987). The high below-ground production estimates in the Oosterschelde estuary (Table 3) are similar to those reported from north American marshes (Groenendijk, 1987).

The question arises why lower salt marsh halophytes, in particular, invest so much in below-ground production. It was demonstrated that the above-ground production of several *Spartina alterniflora* marshes varied widely, whereas the below-ground production remained nearly constant (Teal, 1986). A larger part of the total production was invested below-ground in low productive marshes than in high productive ones. The grass might first allocate sufficient assimilates to below-ground structures to gain the nutrients required and subsequently invests the excess into photosynthetic parts (Teal, 1986). The data in Table 3 show, in contrast, the opposite. The plant communities with the higher above-ground production studied by Groenendijk (1987) invested a larger proportion of the total production into below-ground structures than the lower productive communities studied by Ketner (1972). Reason for this may be that the roots of the salt marsh species have other functions apart from the acquisition of nutrients. They anchor the plants in the sometimes unstable sediments and protect the substrate from erosion (Van Eerdt, 1985). It may be assumed that the latter function of roots is more important in the lower than in the higher salt marsh.

High decomposition rates in salt marsh ecosystems have been reported from both the Wadden Sea area (Table 2) and the south-west of the Netherlands (Hemminga & Buth, 1991). Decomposition rates (% weight loss day^{-1}) estimated with litter bag techniques for different plant parts of a number of salt marsh species in the south-west of the Netherlands varied between 0 and 0.55%. These data were used to obtain a ranking of factors according to the extent in which they induce variations in the rates of decomposition. Tissue composition was the most important factor in the explanation of this variation, in particular the content of lignin in different plant parts and species. Environmental heterogeneity came next, while the decomposition stimulating effect of the larger macrofauna was the least important factor. The effects of a number of environmental variables was studied using the cotton strip method. The water, silt and salt content of the soil and the coverage by the vegetation were most important. Fifty-one and 56% of the variance in loss of tensile strength of the strips in respectively spring and summer was explained by these factors (Hemminga & Buth, 1991).

Geese grazing: from salt marshes to the arctic tundra

The relationship between Brent geese and food plants in the spring staging area, and their breeding success in the arctic tundra, is shown in Fig. 8.

The preference of geese for grazed salt marshes can be attributed to the quality of forage i.e. the high ratio of green to dead biomass in the *Puccinellia maritima*, the *Juncus gerardii* and the *Festuca rubra* communities (Bakker, 1989). De-

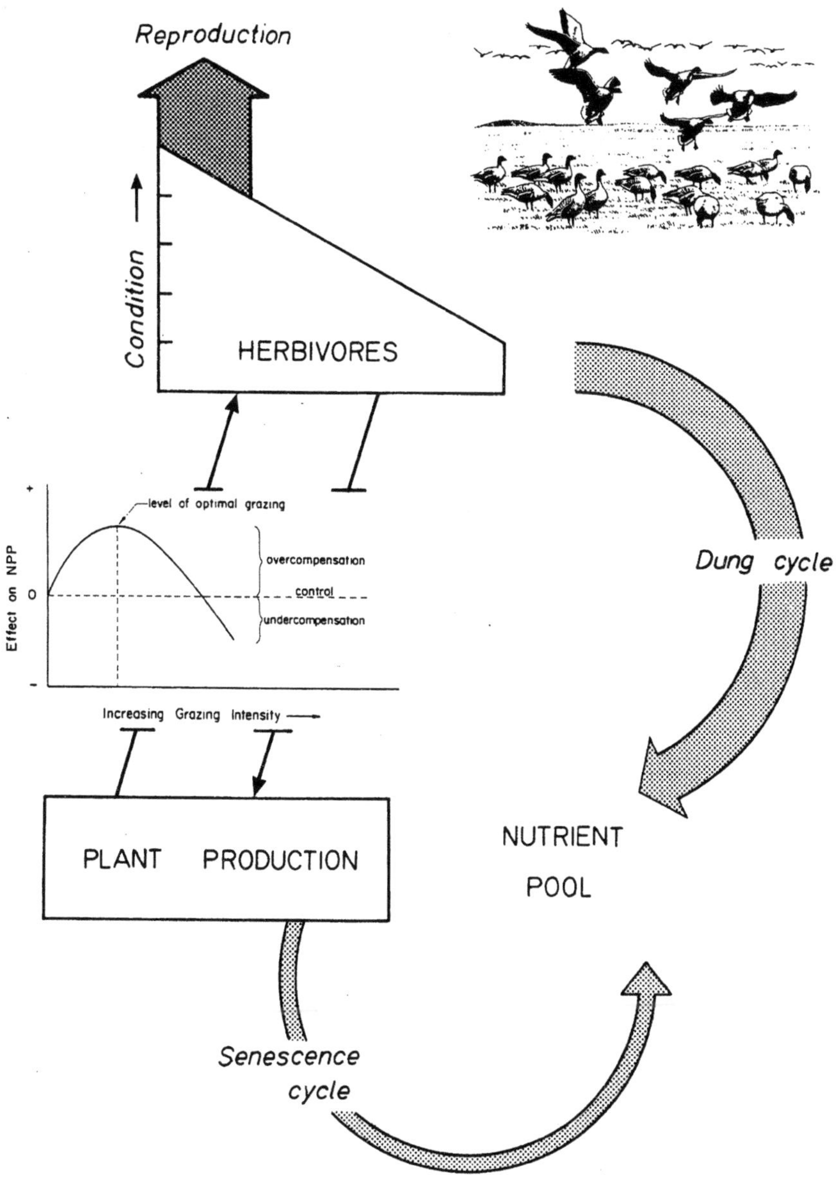

Fig. 8. Flow diagram of herbivore/plant interactions according to the grazing optimization hypothesis (NNP = net above-ground primary production). The relative importance of the dung cycle as contrasted to the return of nutrients via senescence, is symbolized by the thickness of the arrows.

tailed measurements revealed a nitrogen content of 4.2% in *Festuca rubra* in February, of 4.2% in Barnacle geese-grazed and of 3.2% in ungrazed plants until the end of April, suggesting that grazing kept the nitrogen content higher in grazed than in ungrazed plants and that geese so to speak manipulate the quality of their food supply by repeated grazing (Drent & Prins, 1987).

The food quality in the spring staging areas in the salt marshes is important since they serve as a source to gain weight before migration and to hatch their young. Successful parents of Barnacle geese, i.e. females accompanied by young in the autumn had a higher body mass before departure in the spring than unsuccessful parents (Ebbinge *et al.*, 1982). The share of juveniles in the total population in a year-to-year comparison increase with increasing food quality (Drent & Prins,

1987). Food supply, which depends on the position of individuals in the flock, also influenced the yearly breeding success of individual Brent geese. Goose visitations of the salt marsh of the island of Schiermonnikoog followed a pulsating pattern with massive visits every 3–5 days (Prins *et al.*, 1980). Each flock visit the entire harvestable fraction was removed and it was demonstrated that the birds at the rear of the flock already faced a heavily depleted food supply, especially of the preferred food plant in this zone, *Plantago maritima*. The reason for the strong preference for this species over the alternative food plant (*Puccinellia maritima*) is the far higher harvesting potential for geese. It was demonstrated that half of all *Plantago* feeding time was spent by only 12% of the birds, whereas 27% of the birds obtained no *Plantago* at all and had to feed on *Puccinellia* (Drent & Prins, 1987).

Threats

Effects of sea level rise and climatic change

Sea level rise

The relative sea level rise in the Wadden Sea area consists of several components. The eustatic sea level rise is 0.10–0.12 cm yr^{-1} over the period 1880–1980 (Gornitz & Lebedeff, 1987), the local tectonic subsidence is estimated at 0–0.04 cm yr^{-1} over the period 1926–1987 (P. Noomen, in prep.). Subsidence due to gas extraction along the Groningen coast was 0.1–0.2 cm yr^{-1} over the period 1964–1987. The mean relative sea level rise, calculated from the eustatic sea level rise and the tectonic subsidence can be estimated at 0.10–0.16 cm yr^{-1}. This value is somewhat smaller than that registered by tide gauges for the Dutch Wadden Sea over the period 1933–1980, viz. 0.15–0.17 cm yr^{-1} (Van Malde, 1984). In an other section on sedimentation in salt marshes, the rise of MHT level of 0.44 cm yr^{-1} was mentioned, but that rise is independent of the relative sea level rise.

The problem in predicting the survival potential of salt marshes regarding the future sea level rise lies in the uncertainty of the expected rise value. The globally expected eustatic mean sea level rise is 0.5 cm yr^{-1} until the year 2025, but the standard deviation equals the expected rise (J.Oerlemans, in Dijkema *et al.*, 1990). Moreover, there are no predictions of future MHT levels. One has to reckon with an extra subsidence of 0.1–0.6 cm yr^{-1} until 2025 for the Groningen coast and of 0.8–1.5 cm yr^{-1} for the island of Ameland as a result of gas extraction. The MHT rise has to be considered in relation to both the salt marsh zone itself and to the tidal flat transition in front of it. It seems that the potential loss of salt marsh area through erosion from the seaward edge does not depend on the sedimentation processes on the salt marsh itself, but on that in the pioneer zones in front of the marsh (Dijkema *et al.*, 1988; Boorman *et al.*, 1989).

Erosion

Sea edge erosion may be expected for more than half of the Dutch man-made salt marshes. It will result in large scale cliff formation, provided that the present sedimentation deficit in the pioneer zone continues. Cliffs show a horizontal retreat of 0.02–0.4 m yr^{-1} along the Groningen foreland, 1.9 m yr^{-1} along the Ems estuary (De Glopper, 1981), and 6 m yr^{-1} on mainland salt marshes in Germany (Erchinger, 1987). All barrier islands in the Wadden Sea have a mean cliff erosion of 0.5 m yr^{-1} (Ehlers, 1988).

Sedimentation

The present sedimentation rate on mainland salt marshes is high enough to compensate for a future relative sea level rise of 1–2 cm yr^{-1}, which is more than the expected global sea level rise. This is not true for the higher marsh zones which have a lower sedimentation rate. The sedimentation rate will, however, increase as a result of the increase in the number of tidal floodings due to a sedimentation deficit to that typical of the lower marsh zone (De Glopper, 1981; Bouwsema *et al.*, 1986). An increased sea level rise will, therefore, probably not affect the mainland salt marshes by regressive succession, i.e., that the plant communities of the higher salt marsh will be replaced by communities of the lower marsh zone.

The barrier island type of salt marshes in the Wadden Sea area will eventually disappear if the relative sea level rise is more than 0.5–1.0 cm yr^{-1}. However, there is some evidence for a balance between the present MHT rise of 0.44 cm yr^{-1} and sedimentation within these marshes (Dijkema *et al.*, 1990).

At a long-term sedimentation deficit in the marsh zone as a result of the sea level rise, the survival of the salt marsh depends on the response of the vegetation to the increase in frequency and duration of floodings. For instance regressive succession to plant communities of lower salt marsh zones or to bare tidal flats would locally occur on the barrier island of Ameland, even if the increased sedimentation is taken into account (Dankers *et al.*, 1987). Sedimentation may, however, increase during years with raised MHT levels due to a delay in the regressive succession of the vegetation and thus may compensate for (part of) the sea level rise (Dijkema *et al.*, 1990). More insight is needed on the sedimentation balance and vegetation dynamics in the pioneer zone as a function of year-to-year fluctuations in MHT, wave energy levels, vegetation structure, drainage, and soil fertility. The process of sedimentation itself as well as its effects upon the succession on sites with low and high sedimentation rates should be studied.

Climatic change

Anthropogenic impact on the global atmosphere results in an increase of CO_2, and possibly temperature and ultra-violet radiation. Specific effects on salt marsh plants can be the following. The increase in atmospheric CO_2 will generally favour carbon assimilation of C3-species (Brouns, 1988; Rozema *et al.*, 1991). Interaction with salinity is to be expected in such a way that salt marsh plants will improve their water use efficiency and will tolerate higher levels of salinity.

The expected rise in temperature will, however, be primarily more advantageous to C4-species, *e.g. Spartina anglica* in Dutch coastal salt marshes and *Spartina maritima* in southern Europe.

Increased UV-B radiation will adversely affect many plant species in salt marshes, dicotyledons perhaps more severely than monocotyledons. The combined effects of global change factors have to be studied intensively in the near future.

Eutrophication and toxification

Natural and cultural eutrophication

The succession towards dominance of one or few plant species on various sites over the salt marsh zones can be a result of natural eutrophication i.e. the accumulation of litter and the subsequent decomposition and mineralization of the organic material. Eutrophication of the salt marsh by atmospheric deposition and flooding water may accelerate this process.

Severe nutrient loading has occurred in salt marshes during recent decades originating from anthropogenic activities. It is estimated that 208,000 tons of nitrogen and 18,500 tons of phosphorus enter the Dutch Waddens Sea every year (Van Meerendonk *et al.*, 1988). The Scheldt estuary has an annual load of 9,500 tons of nitrogen and 790 tons of phosphorus (Nienhuis, 1989; Herman *et al.*, 1991). The water of the western Wadden Sea contained three times more phosphorus in 1970 than in 1950 (De Jonge & Postma, 1974). It can, therefore, be expected that salt marshes in both the Wadden Sea and Westerschelde are confronted with increased amounts of nutrients brought in by tidal waters (Leendertse, 1989).

Atmospheric deposition of nitrogen has also increased greatly in the last decades: it presently amounts to 43 kg N ha^{-1} yr^{-1} (Additioneel Programma Verzuringsonderzoek, 1989).

The application of nitrogen (the limiting factor) fertilizers usually increases the production of the vegetation and decreases the number of plant species (Tyler, 1976; Beeftink, 1977B; Jefferies & Perkins, 1977; Valiela, 1984; Asjes, 1988). The application of fertilizers and grazing on a higher salt marsh will eventually cause the replacement of halophytic plant communities by common inland communities (Bakker, 1989).

The increased nitrogen input might accelerate the spread of fast growing plant species, and eventually facilitate their dominance in unexploited sites: *Elymus pycnanthus* on salt marshes and *Calamagrostis canescens* in beach plains. The mechanisms behind this acceleration are yet unknown. To predict the effects on salt marshes of nutrient input by sediments as a result of sea level rise, of an envisaged reduction in the nutrient input from atmospheric deposition and of nutrients dissolved in sea water, and of the impact of management regimes like grazing and cutting, scenarios have to be developed. These scenarios should include the existing knowledge on the balance of organic matter and nutrients and the subdivision over different compartments in various plant communities over the zonation, and can be used to estimate the effects of changes in various factors incorporated in these scenarios.

Toxification

Salt marsh plants take up heavy metals from the soil (Beeftink *et al.*, 1982). The uptake differs between species and is also strongly influenced by soil conditions like pH, redox potential and sulfide content. So far no obvious effects of heavy metals on production and germination in salt marsh plants have been found in Dutch field studies (Huiskes & Rozema, 1988).

Management

Human exploitation of the unprotected coastal area

People living at the fringes of the Pleistocene Plateau in the northeastern part of the Netherlands, in territories adjacent to navigable parts of little rivers which flowed into the coastal areas, are assumed to have been the first to exploit the salt marshes. Competition between the arable and pastoral component of the agricultural system might have arisen due to sand-drifts and peat-formation on the Plateau, which forced the herdsmen to move farther away from the villages (Van Gijn & Waterbolk, 1984). The suitability of the salt marshes for grazing might have resulted in 'transhumance', i.e. the movement of herds of domestic livestock to complementary seasonal pastures in the period 700–500 B.C. (Waterbolk, 1988). This system must have changed gradually into year-round occupation of the salt marshes.

The forests on elevated levees along the lower part of the rivers in the fresh water landscape were cleared very early and had completely vanished in the Roman Period in northwestern Germany. The treeless anthropogenic replacement communities *Molinio-Arrhenatheretea* grassland communities, *Chenopodietea* arable field communities and *Plantaginetea* treading communities were found from 700 B.C. onwards (Behre, 1979).

Farm-steads were built initially (ca. 600 B.C.) on the salt marsh, but later also on mounds of salt marsh sods because of transgression periods from ca. 500 B.C. onwards. These mounds gradually grew higher through the accumulation of waste and dung and, therefore, particularly through macro-fossils, became excellent records of former settlement and land use.

The dwelling places were situated at the highest points of the salt marsh viz. salt marsh ridges and creek bank levees. These were also the only places to grow arable crops (toponyme: 'valge') and were preferred as cattle-grazing areas. It has been demonstrated that crops grown from species found in the mounds can be cultivated on the upper salt marsh in the Netherlands (Van Zeist *et al.*, 1976) and in Germany (Körner-Grohne, 1967). Moreover, ploughing marks have been discovered on former salt marshes (Haarnagel, 1979). The salt marsh depressions between the creek bank levees were mainly used for haymaking (toponyme: 'meede') and sheep-grazing (Miedema, 1983).

In a salt marsh area of 150 km^2 northwest of the city of Groningen, the Netherlands, most mounds were only inhabited for a few centuries but the area was, as a whole, inhabited continuously from the early Iron Age (ca. 600 B.C.) onwards (Waterbolk, 1976; Miedema, 1983). Each farm-stead needed food for seven to eight people to be realised on 5 ha of arable field including the dwelling place, and 30 ha of grassland (20 cattle, 5 horses, 20 sheep/goats) (Haarnagel, 1979; Miedema, 1983). Not the whole area which was suitable for agricultural practices was indeed needed during the early Iron Age (Table 4). Only a minor part (ca. 600 ha) of the inferior lower salt marsh had to be exploited. During the late Iron Age half of the area suitable as arable land was needed and ca. 7,000 ha of the lower salt marsh was exploited as grassland. During the Roman Period all the suitable grassland was exploited and there was probably even a shortage of 700 ha which was possibly compensated for by stubble grazing of the arable fields (Miedema, 1983).

The salt marsh plant communities in the northern part of the Netherlands and adjacent Germany from the time of the mound settlements were derived from palaeo-ecological samples (see the section on original and unexploited plant communities) (Körner-Grohne, 1967; Van Zeist, 1974; Behre, 1979; 1985). The lower salt marsh contained the *Puccinellietum maritimae* and the mid salt marsh the *Juncetum gerardii*. It was assumed that only the *Juncetum gerardii leontodontetosum autumnalis* with *Lolio-Potentillion* and *Saginion maritimae* elements occurred on the upper salt marsh. It must, therefore, be assumed that the salt marshes were heavily exploited (Waterbolk, 1976), as Miedema (1983) had already suggested.

The exploitation of salt marshes continued. Hay-making has gradually decreased and nowadays the most widespread use is open-range grazing for domestic animals. Seventy% of the northwestern European salt marshes are still exploited, but the area of abandoned salt marshes increases rapidly (Bakker, 1989).

Protection of coastal areas by dikes

The only defence against inundations were centuries long the mounds. Different settlements did not cooperate in the building of dikes as protection against the sea. A start was, most likely, made ca. 1000 A.D. in the northern Netherlands (Waterbolk, 1976). It started in adjacent Germany in the 11th century (Behre, 1985). The first low dikes enclosed cultivated salt marsh areas around one or two mounds. A closed sea-defence probably existed in Groningen and Friesland around 1100 A.D., featuring dikes with a maximum height of 2.50 m. Vast salt marsh areas, however, remained unprotected by dikes and had, therefore, no owners.

During the 12th century Cistercian monks founded the Klaarkamp Abbey near Rinsumageest in the neighbourhood of Dokkum in Friesland. They built dikes and reclaimed the unprotected salt marsh (Reitsma & Bakker, 1986).

Later on the sea reclaimed large parts of the embanked areas during flood disasters. In these flooded areas as well as in other areas outside the existing reclamations, sedimentation allowed the

Table 4. Area suited and needed for arable field and grassland on the salt marsh northwest of the city of Groningen during three settlement periods (after Miedema, 1983).

	Area suited		Area needed	
	Arable field	Grassland	Arable field	Grassland
Early iron age	730	5400	160	1200
Late iron age	1940	10950	1000	6630
Roman period	1940	10950	1980	11660

growth of new foreland (for instance Middelzee, Lauwerszee, Fivel, Dollard in the Netherlands and Leybucht, Harlebucht, Jadebusen in adjacent Germany). Over the last three or four centuries, the Wadden Sea has gradually diminished in size through successive reclamation of the newly accreted salt marshes for agricultural purposes. The continuing sedimentation, however, always produced new salt marshes.

Nowadays, the growth of salt marshes is much slower and occurs in the Wadden Sea along the mainland coast under controlled conditions. The reclamation rates in the last centuries were higher than the rate of area increase of salt marsh. To compare salt marsh areas in different years and regions, the acreage of salt marsh areas are related to the extent of their tidal basins (catchment area of a tidal inlet), resulting in a proportional share. Dijkema (1987) proposed to take the proportional share of salt marshes between 1600 and 1800 as a standard for 'naturalness' of the Dutch Wadden Sea, since a rough balance between the rate of area increase and the rate of embankment was found in that period. The mainland salt marshes of the eastern Dutch Wadden Sea area cover nowadays 50 km^2 less than expected from this standard (Fig. 9). This is only partly compensated for by the island salt marshes which cover 18 km^2 more than calculated based on the standard (the latter is due to the positive effect of sand dikes on the salt marsh growth). With unaltered area increase and without taking sea level rise into account (and no embankment) the 'naturalness' standard for the mainland salt marshes will be reached again in 70 to 135 years.

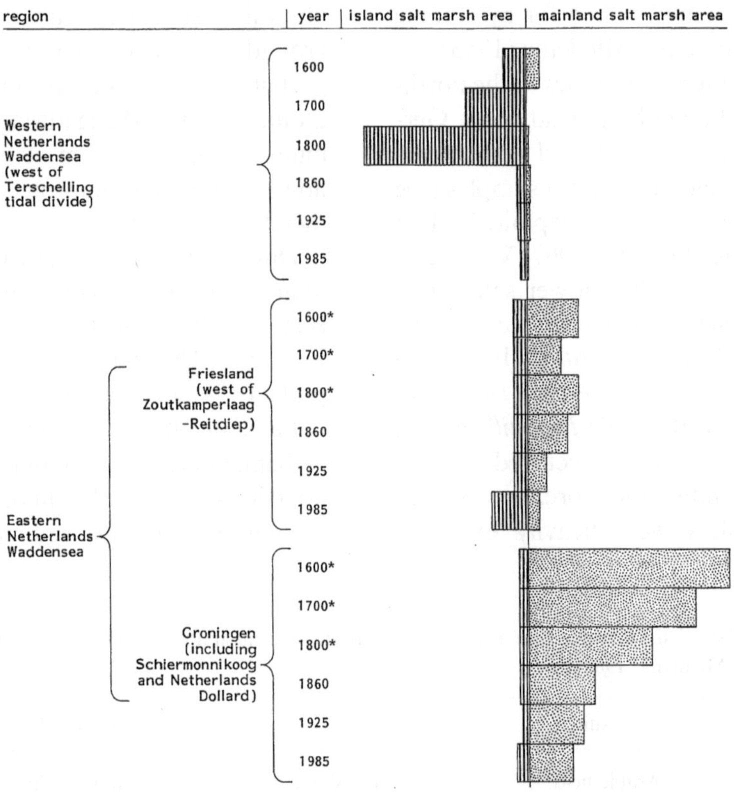

Fig. 9. Salt marsh area in the Netherlands Waddens Sea after 1600, in percentages of the total tidal area. The former island of Huisduinen has been included. Data for 1985 including the Slufter on Texel, but excluding summer polders and mainland pioneer vegetation. ------- = mean salt marsh percentage for the years 1600, 1700 and 1800 (from Dijkema, 1987, courtesy of Kluwer Academic Publishers).

Effects of management on salt marsh plant communities

Soil aeration is a key-factor in the initial stages of salt marsh formation, namely, in the pioneer zone and low salt marsh. Soil salinity becomes more important with increasing elevation. The zonation of plant species along the gradient from tidal flats towards upper salt marsh is, however, also determined by other than abiotic factors. Competitive interactions between plant species are important in the mid- and high salt marsh in determining the zonation pattern (Snow & Vince, 1984; Bertness & Ellison, 1987). It is strongly influenced by management practices, for instance grazing and hay-making.

Cessation of grazing often caused dominance of one or a few plant species. Which species reaches dominance depends on the tidal level and salinity of the sea water, *e.g. Puccinellia martitima* in England (Ranwell in Beeftink, 1977b), *Festuca rubra* or *Elymus repens* in the Baltic (Schmeisky, 1977), *Phragmites australis* in the Baltic (Siira, 1970), *Elymus pycnanthus* on the barrier islands in the Netherlands (Bakker, 1989) and Germany (Andresen *et al.*, 1990), and *Atriplex prostrata* and *Elymus repens* in mainland salt marshes of the Netherlands (Dijkema, 1983).

A successional scheme was derived from long-term studies of vegetational change after abandoning a previously grazed salt marsh, and after resumption of grazing of an abandoned salt marsh on the island of Schiermonnikoog in the Netherlands (Fig. 10). After the resumption of grazing, the return to the initial successional stage, generally took 5–10 years. After the cessation of the grazing regime it often takes 10–20 years to reach a new equilibrium (Bakker, 1989). *Elymus pycnanthus* outcompeted most other species. The number of plant communities as well as the number of species in each community decreased after cessation of grazing. A number of rare species, viz., *Bupleurum tenuissimum*, *Cicendia filiformis*, *Cochlearia anglica*, *Parapholis strigosa* (Wolff, 1988), are threatened in this way.

Grazing of a previously abandoned salt marsh effectuated a larger species-richness than a hay-

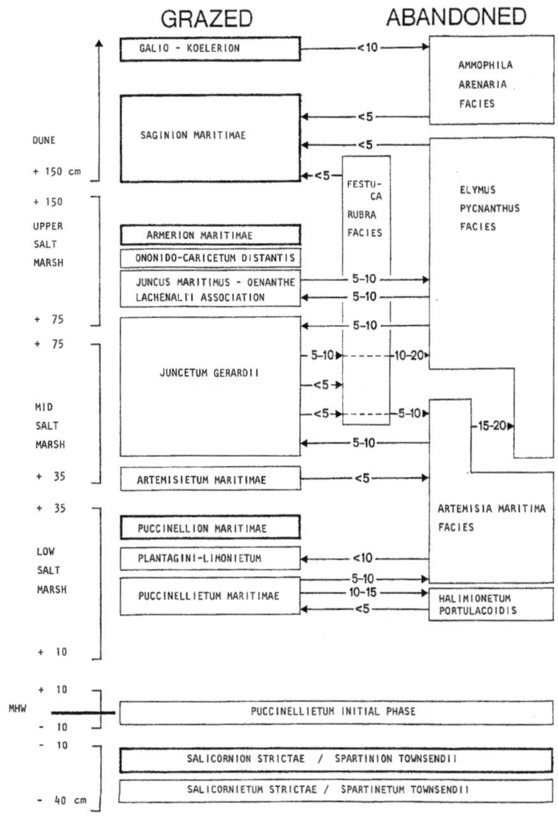

Fig. 10. The position of syntaxa with respect to MHW and the number of years it requires to change after the cessation or the resumption of heavy grazing by 1.6 heifers ha^{-1} (from Bakker, 1989, courtesy of Kluwer Academic Publishers).

making regime (Bakker, 1989). This may be because grazing facilitates the establishment of more species by the creation of open spaces, while hay-making causes the formation of dense turf dominated by *Festuca rubra*. The species-richness on a grazed lower salt marsh is, however very low, because of the complete destruction of the topsoil by heavy grazing (Bakker, 1989).

Grazing with moderate stocking density creates a pattern of closely grazed areas and lightly grazed patches. The comparison of above-ground standing crop in the grazed and abandoned plant communities on the island salt marsh of Schiermonnikoog (Fig. 11) shows different rates of herbage utilization of the various plant communities. These differences emerged soon after the resumption of grazing, and maintained themselves. The lightly grazed patches range from one

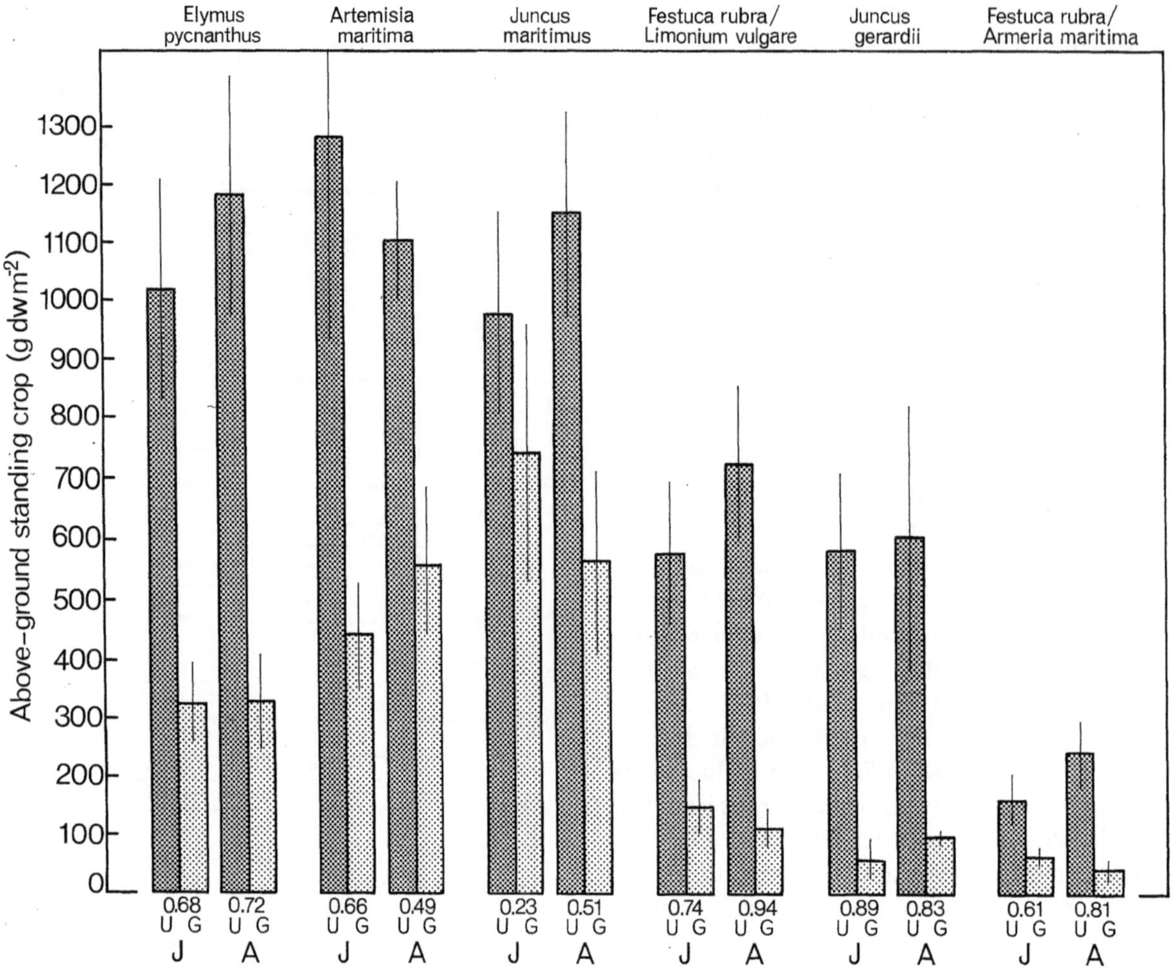

Fig. 11. Average above-ground standing crop including litter (± SE) in ungrazed (first bar) and grazed (second bar) salt marsh plant communities in June and in August. The fraction utilized, i.e. difference between grazed and ungrazed,is indicated below the bars (from Bakker, 1989, courtesy of Kluwer Academic Publishers).

ha to small isolated clumps of *Juncus maritimus* (Bakker, 1989). It was found in a mainland salt marsh that the height of the canopy increased from the dike towards the tidal flats in a grazed salt marsh. No such a trend was found in the abandoned site, which indicates a decrease of grazing impact towards the tidal flats (Andresen *et al.*, 1990).

Effects of management on fauna

Invertebrates

The invertebrate species diversity was highest in the *Puccinellia maritima* and the *Armerion martimae* communities of the salt marshes in northern Germany (Rahmann *et al.*, 1987). The number of individuals decreased in the following sequence: abandoned for 10 years > abandoned for 1 year > cut in July > nearly 2 cattle ha^{-1} > 5 sheep ha^{-1}. The numbers of species and faunal groups were highest in the abandoned area, with a dominance of cicades (*Cicadina*), bugs (*Heteroptera*) and snails (*Gastropoda*). The grazed areas were very poor with only beetles (*Coleoptera*) and bugs.

The effects on the macrofaunal invertebrates of cessation of cattle grazing were studied three years after the withdrawal of the grazers (Irmler & Heydemann, 1985, 1986). The number of species did not change, but that of individuals belonging to trampling-sensitive species increased in the abandoned area, namely, springtails (*Collembola*), spiders (*Aranea*) and species feeding on tall plants or parts of them, namely, butterflies (*Lepidoptera*), aphids (*Aphinidae*), bees (*Apidae*) and hoverflies (*Syrphidae*), whereas the coprophagous species and those feeding on algae in the bare soil decreased.

The cessation of grazing caused litter accumulation and hence, the food web shifted from predominance by herbivorous animals to a food web dominated by detrivorous animals (Andresen *et al.*, 1990). On the most heavily grazed site the macrofaunal community of the lower salt marsh spread to the higher salt marsh. On the not grazed area the *Elymus pycnanthus* community had spread after eight years. Characteristic salt marsh invertebrates had disappeared within 10 years and were replaced by non-characteristic inland species (Andresen *et al.*, 1990).

Birds

The effect of grazing on the function of salt marshes as a foraging or resting area for birds during high tides and for breeding was studied in the German Wadden Sea area (Schultz, 1987). The increase in the number of breeding Oystercatchers (*Haemotopus ostralegus*) was related to the increased grazing of salt marshes, whereas the number of Redshanks (*Tringa totanus*) decreased. The avifauna was in this case negatively influenced by the frequency with which the birds were disturbed by flooding, the trampling of nests and pulli, and the increased stocking density. The vegetation composition and structure was considered less important. Foraging Barnacle (*Branta leucopsis*) and Brent geese (*B. bernicla*) preferred, however, salt marshes grazed by live-stock.

Perspectives for maintenance, management and restoration

Best strategies for maintenance of the salt marsh ecosystem in the Wadden Sea area are being discussed for a long time (Ovesen, 1990). The management practices in Germany have focused on coastal protection and agricultural practices with hardly any attention for natural processes. Large areas of salt marshes are designed nature reserve recently, and the major part is abandoned now. This is in contrast to the Dutch and Danish management, where extensive grazing is the predominant management regime.

Coastal protection in Germany included until recently facilitation of a maximal sedimentation in man-made salt marshes. The marsh was considered to be heavily drained and grazed since ungrazed salt marshes were perceived to erode more easily. An opposite management was chosen by nature conservation authorities, namely abandonment, which resulted in a decrease in plant and animal species-richness.

The Duch salt marshes are nowadays marginal agricultural areas and it is, therefore, difficult to have them grazed by domestic live-stock. One of the private organizations for nature conservation in the Netherlands brings Limousin cattle to the salt marsh for the summer period, whereas the cattle are brought to the Drenthe Plateau for the winter period and then forage on hay cut from inland nature reserves. It seems to be the beginning of a modern 'transhumance'.

The second trilateral working conference on salt marsh management in the international Wadden Sea area held in October 1989 at Rømø, Denmark, agreed on a number of recommendations (Ovesen, 1990), of which some important ones are listed below:

1. Land reclamation activities should not be carried out in areas where natural sedimentation is taking place considering the fact that erosion and sedimentation are natural processes. The potential expansion of salt marsh areas, by the local removal of summer dikes, should be considered.
2. Protection measures for existing salt marshes should be applied in a differentiated and specific manner. Attention should be paid, in particular to the pioneer zone of the mainland salt marshes to prevent cliff formation becoming a

widespread process. The grazing of salt marshes for coastal protection is unnecessary.

3. Drainage should be based on natural processes and not on man-made technical means.
4. Vegetation management by grazing is to be preferred and should be performed on large areas of 100 ha or more to produce a mosaic of different vegetation types. An advocated maximum stocking density is one head of cattle ha^{-1} or two sheep with their lambs, grazing by horses should be avoided due to turf destruction and associated erosion risk.
5. The vegetation pattern should include areas with short vegetation cover (grazed) and areas with long vegetation cover (ungrazed or very lightly grazed). The determination of the ratio of grazed and ungrazed areas should be based on the results of grazing experiments.

References

Abd. Aziz, S.A. & D.B. Nedwell, 1979. Microbial nitrogen transformations in the salt marsh environment. In R.L. Jefferies & A.J. Davy (eds), Ecological processes in coastal environments. Blackwell Scientific Publ., Oxford: 385–398.

Abd. Aziz, S.A. & D.B. Nedwell, 1986. The nitrogen cycle of an east coast U.K. salt marsh. II. Nitrogen fixation, Nitrification, denitrification, tidal exchange. Estuar. coast. shelf Sci. 22: 689–704.

Additioneel Programma Verzuringsonderzoek, 1989. Evaluatierapport verzuring nr. 00–06. Rijksinst. Volksgezondheid Milieuhygiëne, Bilthoven.

Alexander, M., 1977. Introduction to soil microbiology. Wiley, New York.

Andresen, H., J.P. Bakker, M. Brongers, B. Heydemann & U. Irmler, 1990. Long-term changes of salt marsh communities by cattle grazing. Vegetatio 89: 137–148.

Armstrong, W., E.J. Wright, S. Lythe, T.J. & Gaynard, 1985. Plant zonation and the effects of the spring-neap tidal cycle on soil aeration in a Humber salt marsh. J. Ecol. 73: 323–339.

Asjes, J., 1988. The effect of fertilizer-application on interacting Puccinellia maritima and Spartina anglica. In M.C.T. Scholten & H. Schat (eds), The competitive ability of Puccinellia maritima. TVC-TNO, Delft: 19–21.

Bakker, J.P., 1989. Nature management by grazing and cutting. Kluwer Academic Publishers, Dordrecht.

Banus, M.D., I. Valiela & J.M. Teal, 1975. Lead, zinc, and calcium budgets in experimentally enriched salt marsh ecosystems. Estuar. coast. mar. Sci. 3: 421–430.

Beeftink, W.G., 1966. Vegetation and habitat of salt marshes and beach plains in S.W. Netherlands. Wentia 15: 83–108.

Beeftink, W.G., 1977a. The coastal salt marshes of Western and Northern Europe: an ecological and phytosociological approach. In V.J. Chapman (ed.), Wet coastal ecosystems. Elsevier, Amsterdam: 109–155.

Beeftink, W.G., 1977b. Salt marshes. In R.S.K. Barnes (ed.), The coastline. Wiley, New York: 93–121.

Beeftink, W.G., J. Nieuwenhuize, M. Stoeppler & C. Mohl, 1982. Heavy-metal accumulation in salt marshes from the Western and Eastern Scheldt. Sci. Total Environm. 25: 199–223.

Behre, K.E., 1979. Zur Rekonstruktion ehemaliger Pflanzengesellschaften an der Deutschen Nordseeküste. In O. Wilmanns & R. Tüxen (eds), Werden und Vergehen von Pflanzengesellschaften. Cramer, Vaduz: 181–214.

Behre, K.E., 1985. Die ursprungliche Vegetation in den deutschen Marschgebieten und deren Veränderung durch prähistorische Besiedlung und Meeresspiegelbewegungen. Verh. Gesellsch. für Oekologie 13: 85–96.

Bertness, M.D. & A.M. Ellison, 1987. Determinants of pattern in a New England salt marsh plant community. Ecol. Monogr. 57: 129–147.

Boorman, L.A., J.D. Goss-Custard & S. Mc Grorty, 1989. Climatic change, rising sea level and the British coast. HMSO, London. Natural Environm. Res. Council. ITE research publication 1: 1–24.

Bouwsema, P.J.H., Bossinade, K.S. Dijkema, J.W.Th.M. Van Meegen, R. Reenders & W. Vrieling, 1986. De ontwikkeling van de hoogte en van de omvang van de kwelders in de landaanwinningswerken in Friesland en Groningen. Rijkswaterstaat Dir. Groningen, Nota ANA-86.05, RIN-report 86/3: 1–58.

Brouns, J.J.W.M., 1988. The impact of elevated carbon dioxide levels on marine and coastal ecosystems. NIOZ-report 1988–7, RIN-report 88–58, Texel: 1–101.

Dahlman, R.C. & C.L. Kucera, 1965. Root productivity and turnover in native prairie. Ecology 46: 84–89.

Dankers, N., M. Binsbergen, K. Zegers, R. Laane & M. Rutgers Van Der Loeff, 1984. Transportation of water, particulate and dissolved organic and inorganic matter between a salt marsh and the Ems-Dollard estuary, the Netherlands. Estuar. coast. shelf Sci. 19: 143–165.

Dankers, N., K.S. Dijkema, G. Londo & P.A. Slim, 1987. De ecologische effecten van bodemdaling op Ameland. RIN-report 87/14: 1–90.

De Glopper, R.J., 1981. De snelheid van de opslibbing en van de terugschrijdende erosie op de kwelders langs de noordkust van Friesland en Groningen. In 50 jaar onderzoek. Flevobericht 163. Rijksdienst voor de IJsselmeerpolders, Lelystad: 43–51.

De Jong, Z., 1991. Mudflat salt marsh boundary changes after construction of the Eastern Scheldt storm surge barrier. Progress report DIHO 1990, Kon. Ned. Acad. v. Wetensch., in press.

De Jonge, V.N. & H. Postma, 1974. Phosphorus compounds in the Dutch Wadden Sea. Neth. J. Sea Res. 8: 139–153.

De Leeuw, J. & G.J. Buth, 1991. Spatial and temporal variation in peak standing crop of European tidal marshes. In M. Elliott & J.P. Ducrotoy (eds), Estuaries and Coasts: Spatial and temporal intercomparisons. Olsen & Olsen, Fredensborg, in press.

De Leeuw, J., H. Olff & J.P. Bakker, 1990. Year-to-year variation in peak above-ground biomass of six salt marsh angiosperm communities as related to rainfall deficit and inundation frequency. Aquatic Botany 36: 139–151.

De Leeuw, J., A. Van Den Dool, W. De Munck, J. Nieuwenhuize & W.G. Beeftink, 1991. Factors influencing the soil salinity regime along an intertidal gradient. Estuar. coast. shelf Sci. 32: 87–97.

Dieckmann, R., 1988. Entwicklung der Vorländer an der nordfriesischen Festlandküste. Wasser und Boden 3: 146–150.

Drent, R. H. & H.H.T. Prins, 1987. The herbivore as a prisoner of its food supply. In J. Van Andel, J.P. Bakker & R.W. Snaydon (eds), Disturbance in Grasslands. Dr W. Junk Publishers, Dordrecht: 131–147.

Dijkema, K.S., 1983. The salt-marsh vegetation of the mainland coast, estuaries and Halligen. In K.S. Dijkema & W.J. Wolff (eds), Flora and Vegetation of the Wadden Sea islands and coastal areas. Balkema, Rotterdam: 302–312.

Dijkema, K.S., 1987. Changes in salt-marsh area in the Netherlands Wadden Sea after 1600. In A.H.L. Huiskes, C.W.P.M. Blom & J. Rozema (eds), Vegetation between land and sea. Dr W. Junk Publishers, Dordrecht: 42–49.

Dijkema, K.S., J. Van Den Bergs, J.H. Bossinade, P. Bouwsema, R.J. De Glopper & J.W.Th.M. van Meegen, 1988. Effecten van rijzendammen op de opslibbing en de omvang van de vegetatiezones in de Friese en Groninger landaanwinningswerken. Rijkswaterstaat Dir. Groningen, Nota GRAN 1988-2010. RIN-report 88/66, RIJP-report 1988-33 Cbw: 1–119.

Dijkema, K.S., J.H. Bossinade, P. Bouwsema & R.J. De Glopper, 1990. Salt marshes in the Netherlands Wadden Sea: rising high tide levels and accretion enhancement. In J.J. Beukema, W.J. Wolff & J.J.W.H. Brouns (eds), Expected effects of climatic change on marine coastal ecosystems. Kluwer Academic Publishers, Dordrecht: 173–188.

Ebbinge, B., A. St. Joseph, P. Prokosh & B. Spaans, 1982. The importance of spring staging areas for arctic-breeding geese wintering in Western Europe. Aquila 89: 249–258.

Ehlers, J., 1988. Morphologische Veränderungen auf der Wattseite der Barriere-Inseln des Wattenmeeres. Die Küste 47: 3–30.

Erchinger, H.F., 1985. Dünen, Watt und Saltwiesen. Der Niedersachsiche Minister für Ernährung, Landwirtschaft und Forsten, Hannover: 1–59.

Erchinger, H.F., 1987. Salzwiesenbildung und -erhaltung – Lahnungsbau und Begrupping für den Küstenschutz. In N. Kempf, J. Lamp & P. Prokosch (eds), Salzwiesen: Geformt von Küstenschutz, Landwirtschaft oder Natur? WWF-Deutschland, Tagungsbericht 1, Husum: 279–296.

Floate, M.J.S., 1970. Mineralization of nitrogen and phosphorus from organic materials and animal origin and its significance in the nutrient cycle in grazed uplands and hill soils. J. Br. Grassl. Soc. 25: 295-302.

Giblin, A.E., I. Valiela & J.M. Teal, 1983. The fate of metals introduced into a New England salt marsh. Water Air Soil Pollut. 20: 81–98.

Gornitz, V. & S. Lebedeff, 1987. Global sea-level changes during the past century. In D. Nummedal, O.H. Pilkey & J.D. Howard (eds), Sea-level fluctuations and coastal evolution. Soc. Econ. Paleontol. and Mineralog. Special Publ. 41, Tulsa, Oklahoma: 3–16.

Groenendijk, A.M., 1984. Primary production of four dominant salt-marsh angiosperms in the SW Netherlands. Vegetatio 57: 143–152.

Groenendijk, A.M., 1987. Ecological consequences of a storm surge barrier in the Oosterschelde salt marshes. Ph.D. Thesis, Univ. Utrecht.

Groenendijk, A.M. & M.A. Vink-Lievaart, 1987. Biomass and production on a Dutch salt marsh: emphasis on the below-ground component. Vegetatio 70: 21–27.

Haarnagel, W., 1979. Die Grabung der Feddersen Wierde. Feddersen Wierde 2: 1–364.

Hemminga, M.A. & G.J.C. Buth, 1991. Decomposition in salt marsh ecosystems of the SW Netherlands: the effects of biotic and abiotic factors. Vegetatio 92: 73–83

Henriksen, K. & A. Jensen, 1979. Nitrogen mineralization in a salt marsh ecosystem dominated by Halimione portulacoides. In R.L. Jefferies & A.J. Davy (eds), Ecological processes in coastal environments. Blackwell Scientific Publ., Oxford: 373–384.

Herman, P.M.J., H. Hummel, M. Bokhorst & A.G.A. Merks, 1991. The Westerschelde: Interaction between eutrophication and chemical pollution? In M. Elliott & J.P. Ducrotoy (eds), Estuaries and Coasts: Spatial and temperoral intercomparisons. Olsen & Olsen, Fredensborg, in press.

Hik, D.S. & R.L. Jefferies, 1990. Increases in the net aboveground primary production of a salt marsh forage grass: a test of the predictions of the herbivore-optimization model. J. Ecol. 78: 180–195.

Huiskes, A.H.L. & J. Rozema, 1988. The impact of anthropogenic activities on the coastal wetlands of the North Sea. In W. Salomons, B.L. Bayne, E.K. Duursma & U. Förstner (eds), Pollution of the North Sea: an assessment. Springer-Verlag, Berlin: 455–473.

Irmler, U. & B. Heydemann, 1985. Der Einfluss der Rinder Beweidung auf die Struktur der Salzwiesen-Biozönose. Verh. Gesellsch. Oekologie 13: 71–76.

Irmler, U. & B. Heydemann, 1986. Die ökologische Problematik der Beweidung von Salzwiesen an der Niedersäsichen Küste – am Beispiel der Leybucht. Naturschutz Landschaftspflege Niedersachsen Beiheft 15.

Jefferies, R.L. & N. Perkins, 1977. The effects on the vegetation of the additions of inorganic nutrients to salt marsh soils at Stiffkey, Norfolk. J. Ecol. 65: 867–882.

Jensen, A., K. Henriksen & M.B. Rasmussen, 1985. The distribution and interconversions of ammonium and nitrate in the Skallingen salt marsh (Denmark) and their exchange with the adjacent coastal water. Vegetatio 62: 357–366.

Jensen, J., H.E. Mügge & G. Visscher, 1988. Untersuchungen zur Wasserstandentwicklung in der Deutschen Bucht. Die Küste 47: 135–161.

Jordan, T. E., D.L. Correll & D.F. Whigham, 1983. Nutrient flux in the Rhode River: tidal exchange of nutrients by brackish marshes. Estuar. coast. shelf Sci. 17: 651–667.

Ketner, P., 1972. Primary production of salt-marsh communities on the island of Terschelling in the Netherlands. Ph.D. Thesis, Univ. Utrecht.

Körner-Grohne, U. 1967. Geobotanische Untersuchungen auf der Feddersen Wierde. Feddersen Wierde 1: 1–358.

Leendertse, P.C., 1989. De effekten van verontreiniging (eutrofiëring, microverontreinigingen en olie) van het Waddenzeewater op kweldervegetaties. Report Rijkswaterstaat Dir. Getijdewateren and Dept. Ecology and Ecotoxicology, Free Univ. Amsterdam.

Looijen, R.C. & J.P. Bakker, 1987. Utilization of different salt marsh plant communities by cattle and geese. In A.H.L. Huiskes, C.W.P.M. Blom & J. Rozema (eds), Vegetation between land and sea. Dr W. Junk Publishers, Dordrecht: 52–64.

Luck, G., 1976. Waddengebied van Nedersachsen. In H.J. Abrahamse, W. Joenje & N. Van Leeuwen-Seelt (eds), Waddenzee. Landelijke Ver. tot Behoud van de Waddenzee, Harlingen/Natuurmonumenten, 's-Graveland: 63–75.

Miedema, M., 1983. Vijfentwintig eeuwen bewoning in het terpen gebied van NW van Groningen. Ph.D. Thesis, Univ. Amsterdam.

Nienhuis, P.H., 1989. Eutrophication of estuaries and lagoons in the south-west Netherlands. In J.C. Hooghart & C.W.S. Posthumus (eds), Hydro-ecological relations in the Delta-waters of the south-west Netherlands. TNO committee on Hydrological Research, The Hague: 49–70.

Ovesen, C.H., 1990. Salt marsh management in the Wadden Sea Region. Proceedings of the Second Trilateral Working Conference on salt marsh management in the Wadden Sea Region. The Ministry of the Environment/The National Forest and Nature Agency, Hørsholm.

Prins, H.H.T., R.C. Ydenberg & R.H. Drent, 1980. The interacting of Brent Geese (*Branta bernicla*) and sea plantain (*Plantago maritima*) during spring staging: field observations and experiments. Acta Bot. Neerl. 29: 585–596.

Rahmann, M., H. Rahmann, N. Kempf, B. Hoffmann & H. Gloger, 1987. Auswirkungen unterschiedlicher landwirtschaftlicher Nutzung auf die Flora und Fauna der Salzwiesen an der ostfriesischen Wattenmeerküste. Senckenb. Marit. 19: 163–193.

Reitsma, D.T. & J.P. Bakker, 1986. Landgebruik op Schiermonnikoog, vroeger en nu. Waddenbulletin 21: 94–100.

Rozema, J., 1978. On the ecology of some halophytes from a beach plain in the Netherlands. Ph.D. Thesis, Univ. Amsterdam.

Rozema, J., F. Dorel, R. Janissen, G.H. Lenssen, R.A. Briekman, W. Arp & B.G. Drake, 1990. Effect of elevated atmospheric CO_2 on growth, photosynthesis and water relations of salt marsh grass species. Aquat. Bot. 39: 45–55.

Rozema, J. & P.C. Leendertse, 1991. Natural and man-made stresses in coastal wetlands. In J. Rozema & J.A.C. Verkleij (eds), Ecological responses to environmental stresses. Dr W. Junk Publishers, Dordrecht: 220–234.

Scherfose, V., 1987. Pflanzensoziologische und ökologische Untersuchungen in Salzrasen der Nordseeinsel Spiekeroog. II. Bodenchemische Untersuchungen, Stickstoff-Netto-Mineralization und Salzbelastung. Tuexenia 7: 173–198.

Schmeisky, H., 1977. Der Einfluss von Weidetieren auf Salzpflanzen-gesellschaften an der Ostsee. In R. Tüxen (ed.), Vegetation und Fauna. Cramer, Vaduz: 481–498.

Schultz, W., 1987. Einfluss der Beweidung von Salzwiesen auf die Vogelfauna. In N. Kempf, J. Lamp & P. Prokosch (eds), Salzwiesen: Geformt von Küstenschutz, Landwirtschaft oder Natur? Tagungsbericht I der Umweltstiftung WWF-Deutschland, Husum: 255–270.

Siira, J., 1970. Study on the ecology of the seashore meadows of the Bothnian Bay with special reference to the Liminka area. Aquilo Ser. Bot. 9: 1–109.

Smalley, A.E., 1959. The growth cycle of Spartina and its relation to the insect populations in the marsh. Proc. Salt Marsh Conf. March 1958, Sapelo Island Georgia.

Snow, A.A. & S.W. Vince, 1984. Plant zonation in an Alaskan salt marsh II. An experimental study of the role of edaphic conditions. J. Ecol. 72: 669–684.

Teal, J.M., 1986. The ecology of a regularly flooded salt marsh of New England: a community profile. U.S. Fish & Wildl. Service, Biol. Rep. 85 (7.4). 61 pp.

Tyler, G., 1976. On the effects of phosphorus and nitrogen, supplied to Baltic shore-meadow vegetation. Bot. Notiser 120: 433–447.

Valiela, I., 1984. Marine ecological processes. Springer Verlag, New York, 546 pp.

Valiela, I. & J.M. Teal, 1979. The nitrogen budget of a salt marsh ecosystem. Nature 280: 652–656.

Van der Meijden, R., E.J. Weeda, F.A.C.B. Adema & G.J. De Jonckheere, 1983. Flora van Nederland. Wolters-Noordhoff, Groningen.

Van Eerdt, M.M., 1985. The influence of vegetation on erosion and accretion in salt marshes of the Oosterschelde, the Netherlands.Vegetatio 62: 367–375.

Van Gijn, A.L. & H.T. Waterbolk, 1984. The colonization of the salt marshes of Friesland and Groningen: the possibility of a transhumant prelude. Palaeohistoria 26: 101–122.

Van Malde, J., 1984. Voorlopige uitkomsten van voortgezet onderzoek naar de gemiddelde zeeniveaus in de Nederlandse kustwateren. Rijkswaterstaat, Dir. Waterhuishouding en Waterbeweging. Nota WW-WH 84.08: 1–20.

Van Meerendonk, J.H. , G.M. Janssen & B. Frederiks, 1988. De aanvoer van voedingsstoffen en microverontreinigingen naar de Waddenzee en Eems-Dollard. Min. v. Verkeer en Waterstaat, Rijkswaterstaat, GWWS 88–002.

Van Zeist, W., 1974. Palaeobotanical studies of settlement sites in the coastal area of the Netherlands. Palaeohistoria 16: 223–371.

Van Zeist, W., T.C. Van Hoorn, S. Bottema & H. Woldring, 1976. An agricultural experiment on the unprotected salt marsh. Palaeohistoria 18: 11–153.

Waterbolk, H.T., 1976. Oude bewoning in het Waddengebied. In J. Abrahamse, W. Joenje & N. Van Leeuwen-Seelt (eds), Waddenzee. Landelijke Ver. tot Behoud van de Waddenzee, Harlingen: 210–22.

Waterbolk, H.T., 1988. Zomerbewoning in het terpengebied? In M. Bierma, A.T. Clason, E. Kramer & G.J. De Langen (eds), Terpen en wierden in het Fries-Groningse Kustgebied, Wolters-Noordhoff/Forsten, Groningen: 1–19.

Westhoff, V., 1987. Salt marsh communities of three West Frisian islands, with some notes on their long-term succession during half a century. In A.H.L. Huiskes, C.W.P.M. Blom & J. Rozema (eds), Vegetation between land and sea. Dr W. Junk Publishers, Dordrecht: 16–40.

Westhoff, V. & A.J. Den Held, 1969. Plantengemeenschappen in Nederland. Thieme, Zutphen.

Whiting, G.J., H.N. McKellar Jr., B. Kjerfve & J.D. Spurrier, 1987. Nitrogen exchange between a southeastern USA salt marsh ecosystem and the coastal ocean. Mar. Biol. 95: 173–182.

Wolaver, T.G., J.C. Zieman, R. Wetzel & K.L. Webb, 1983. Tidal exchange of nitrogen and phosphorus between a mesohaline vegetated marsh and the surrounding estuary in the lower Chesapeake Bay. Estuarine coastal shelf Sci. 16: 321–332.

Wiermann, R., 1965. Moorkundliche und vegetationsgeschichtliche Betrachtungen zum Aussendeichsmoor bei Schestedt (Jadebusen). Ber. Deutsche Bot. Ges. 78: 269–278.

Wolff, W.J., 1988. De internationale betekenis van de Nederlandse natuur. RIN-report 88/32.

Wolff, W.J., M.J. van Eeden & E. Lammens, 1980. Primary production and import of particulate matter on a salt marsh in the Netherlands. Neth. J. Sea Res. 13: 242–255.

Zimmek, G.E., 1975. Die Mineralstickstoff-Versorgung einiger Salzrasen-Gesellschaften des Graswarders vor Heiligenhafen. Ph.D. Thesis, Univ. Göttingen.

Hydrobiologia **265**: 97–128, 1993.
E. P. H. Best & J. P. Bakker (eds), Netherlands-Wetlands.

The rivers Rhine and Meuse in The Netherlands: present state and signs of ecological recovery

W. Admiraal[1], G. van der Velde[2], H. Smit[3] & W.G. Cazemier[4]
[1]*National Institute of Public Health and Environmental Protection, P.O. Box 1, 3720 BA Bilthoven, The Netherlands;* [2]*Laboratory of Aquatic Ecology, Catholic University of Nijmegen, Toernooiveld, 6525 ED Nijmegen, The Netherlands;* [3]*Institute for Inland Water Management and Waste Water Treatment, P.O. Box 17, 8200 AA Lelystad, The Netherlands;* [4]*Netherlands Institute for Fisheries Research, P.O. Box 68, 1970 AB IJmuiden, The Netherlands*

Key words: River Rhine, River Meuse, river ecosystems, ecological rehabilitation, pollution abatement

Abstract

The ecosystems of the rivers Rhine and Meuse have suffered drastic environmental changes, for example because of the regulation of the stream bed and the construction of weirs and dams. Furthermore, discharges of industrial and municipal waste water have caused the water quality of these rivers to deteriorate; this problem became acute in the sixties and seventies. Recently some chemical parameters of water quality have improved in the Rhine, and as a consequence some aquatic communities are showing signs of recovery after decades of severe degradation. This paper describes the present state of the aquatic communities in the Dutch part of the rivers Rhine and Meuse, using published observations on plankton, macrophytes, invertebrates, and fish. The sparse information on the food chain in these rivers is summarized. The main channels of the Rhine and Meuse are characterized by a dense plankton that develops rapidly in the nutrient-rich river waters. The stream beds, now dominated by wave-exposed sand and gravel, have a sparse fauna and flora. The river banks, mostly consolidated by blocks of stone, offer a substratum for numerous benthic organisms, particularly now that the water quality has improved. The floodplain waters and old river channels harbour a flora and fauna rich in species. The degree of water exchange with the river is crucial for the ecological development of the river and its backwaters. Today the freshwater tidal reaches of the rivers occupy a very restricted area, and only remnants of the previously abundant vegetation of rushes are found.

Losses in the numbers of animal and plant species, notably those specific to rivers, are evident, but over the last 15 years several species have returned. Allochthonous species (exotics), including crustaceans and molluscs, have also settled in the Rhine and Meuse. Fish species characteristic of these rivers, such as river lamprey, sea trout, barbel, and flounder, have recently been observed in appreciable numbers.

The Rhine Action Programme provides a framework for the large-scale rehabilitation of the Rhine. Experiments on re-stocking the Rhine with Atlantic salmon and on the ecological rehabilitation of floodplains are being carried out on the assumption that there will be a further reduction of the pollution load. A similar programme is required for the Meuse.

Introduction

The rivers Rhine and Meuse play a vital role in the hydrology of The Netherlands (TNO, 1986). The ecosystems of these two major rivers with their extensive floodplains and delta regions once encompassed a considerable part of The Netherlands, but since the Middle Ages, when the inhabitants started to build dikes, they have been subjected to human interference and have altered greatly. For centuries the rivers have been intensively exploited for various purposes; inevitably, conflicts of interests arose. By the end of the 19th century the Rhine was being intensively fished, and in 1885 problems with the exploitation of salmon stocks led to the 'salmon treaty' between the Rhine states (Smit & de Jong, 1989). Hydrographic changes, such as the damming of Rhine tributaries in Germany, contributed to the decline of migratory fish species. The construction of sewer systems in this century and the subsequent massive discharge of organically polluted water into the Rhine and Meuse caused the oxygen conditions of the rivers to deteriorate. Post-war industrial development added a massive load of toxic materials, culminating in an overall degradation of biological conditions in the Rhine over the period 1960–1970 (Wolff, 1978). Since then, the organic load has been effectively reduced in the Rhine, but concentrations of nitrogen and phosphorus in the water of the two rivers are still very high. The input of toxic materials into the Rhine, notably of some of the heavy metals, has also been considerably reduced, under the pressure of international agreements made by the Rhine states. Sanitation measures enabled the biological communities to recover partially in the early eighties (Friedrich & Müller, 1984; Van Urk, 1984b). Today the Rhine and Meuse are highly eutrophic rivers with a strongly modified morphology, carrying a reduced but persisting load of many toxic compounds.

A growing public awareness of the widespread detrimental effects of river water on the North Sea and Wadden Sea and on the quality of drinking water paved the way for the next phase in the rehabilitation of the river. After the Sandoz accident in Basle November 1986 the states bordering the river Rhine agreed on the Rhine Action Programme (IKSR, 1987), which has three main aims: to create conditions which would enable the return of higher organisms (such as the salmon); to safeguard the use of Rhine water for the preparation of drinking water; and to abate the contamination of sediments with toxic compounds. Recently a fourth objective has been added: to fulfil the requirements of the North Sea Action Plan. The Rhine Action Programme was the first pollution abatement programme for the Rhine to include ecological objectives (Smit & van Urk, 1987). So far, a 'Meuse Action Programme' has not been adopted, despite the similarities between the environmental problems in the rivers Rhine and Meuse.

Since ca. 1970 the hydrobiology of several of the Dutch wetlands has been intensively studied and some have been subjected to systems analysis (cf. Nienhuis *et al.*, de Jonge *et al.*, de Haan *et al.*, 1993). The Rhine and Meuse, however, which had deteriorated so much, attracted little interest. However, since the seventies, when these rivers began to show signs of recovery, the research has been intensified, particularly after the project 'Ecological Rehabilitation of the Rhine' started in 1988. This review describes the ecosystems of the rivers Rhine and Meuse in their present state and analyses the trends in their rehabilitation.

Morphology and hydrology

The Rhine and Meuse catchment areas have been described extensively by Friedrich & Müller (1984), Van Urk (1984b), and Descy & Empain (1984). An essential difference between these two rivers is their water source: the Rhine is fed by meltwater from the Alps and by rainwater from its catchment, whereas the Meuse is fed by rainwater only. Hence, the Rhine has a more stable discharge over the seasons, while the Meuse usually shows a distinct summer minimum in its flow.

The water systems of the Rhine and Meuse in The Netherlands can be divided into truly riverine reaches that include the distributaries of the Rhine: the R. Waal, the R. Nederrijn/Lek, and the R. IJssel as well as the slow-flowing delta waters: the former main estuary of the Rhine and Meuse, the Hollands Diep/Haringvliet, and the Rotterdam Waterway (Fig. 1). Each of these waterways is strongly affected by engineering works.

River regulation has strongly modified the river beds in the course of the centuries (Van Urk & Smit, 1989). Figure 2 shows the changes over the last 200 years in a short stretch of the R. Waal resulting from the construction of groynes. The narrowing of the main bed, the 'summer bed', together with the straightening of river bends has increased the scour that deepened the shipping route, resulting in a facilitated water transport and a lowered water level. In the German section of the lower Rhine the water level has fallen by 1 or 2 m in the course of this century (Fig. 3). There is an increased incidence of inundation of the floodplains during discharge peaks in summer

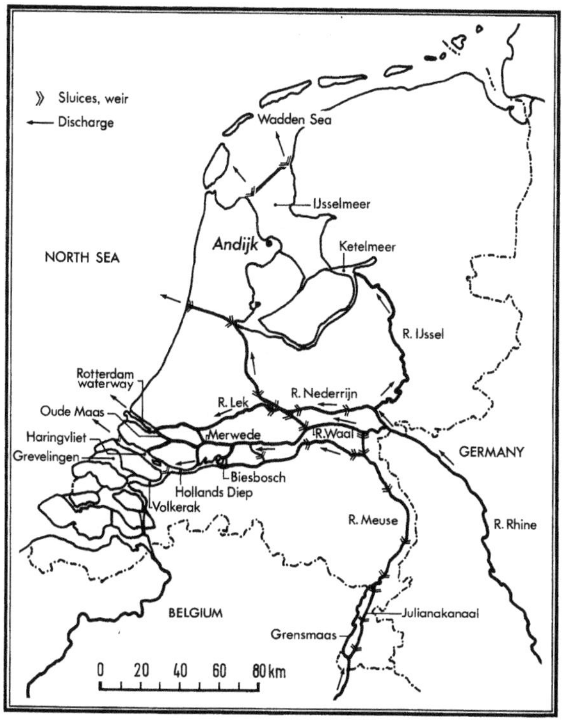

Fig. 1. The lower course of the R. Rhine and Meuse in the Netherlands. The location of the main engineering works is indicated. After TNO (1986).

(Brock *et al.*, 1987), probably because of accelerated water transport from the catchment.

The weirs in the River Nederrijn/Lek (Fig. 1) were primarily constructed to ensure a minimum river flow in the R. IJssel. They are closed at low water discharge. The weirs in the R. Meuse and a lateral canal (the Julianakanaal) were primarily constructed to allow shipping to pass during low summer discharge (Fig. 1).

Until very recently all these works were executed without regard to their ecological consequences. In 1970, for instance, the former main estuary of the Rhine, the Haringvliet, was dammed and a large intertidal marsh, the Biesbosch, lost its unique character (cf. Van Urk, 1984). The tidal amplitude in another freshwater tidal area, the Oude Maas, was reduced from ca. 1.9 m to ca. 1.1 m after the Haringvliet was closed. As a result the only freshwater tidal marsh is now restricted to the banks of the Oude Maas (south of Rotterdam) and the western stretch of the R. Lek (Fig. 4).

The freshwater basin of the Hollands Diep/Haringvliet receives about 35% of the Rhine water; the discharge of the Haringvliet sluices depends on the discharge of the Rhine and on the water flow in the Rotterdam Waterway needed to limit the intrusion of salt water. Hence the residence time of Rhine water in the Hollands Diep/Haringvliet is very variable. It averages between 1 and 2 weeks.

The R. IJssel discharges about 15% of the Rhine water into the IJsselmeer via the Ketelmeer. These basins were created by the closure of the Zuiderzee in 1932 and the construction of polders. The Rhine water in the lake has a residence time of 0.5 year and affects the quality of the water in the western Wadden Sea (cf. de Jonge *et al.*, 1993), and in shallow lakes in the northern part of the Netherlands (cf. de Haan *et al.*, 1993), as well as the drinking water inlet at the village of Andijk (Province of North Holland).The rivers have numerous backwaters in the 'winter bed'. The backwaters situated in the river forelands are normally flooded by river water when water levels are high in winter or spring, and occasionally also in summer. Smits (1989) made an inventory

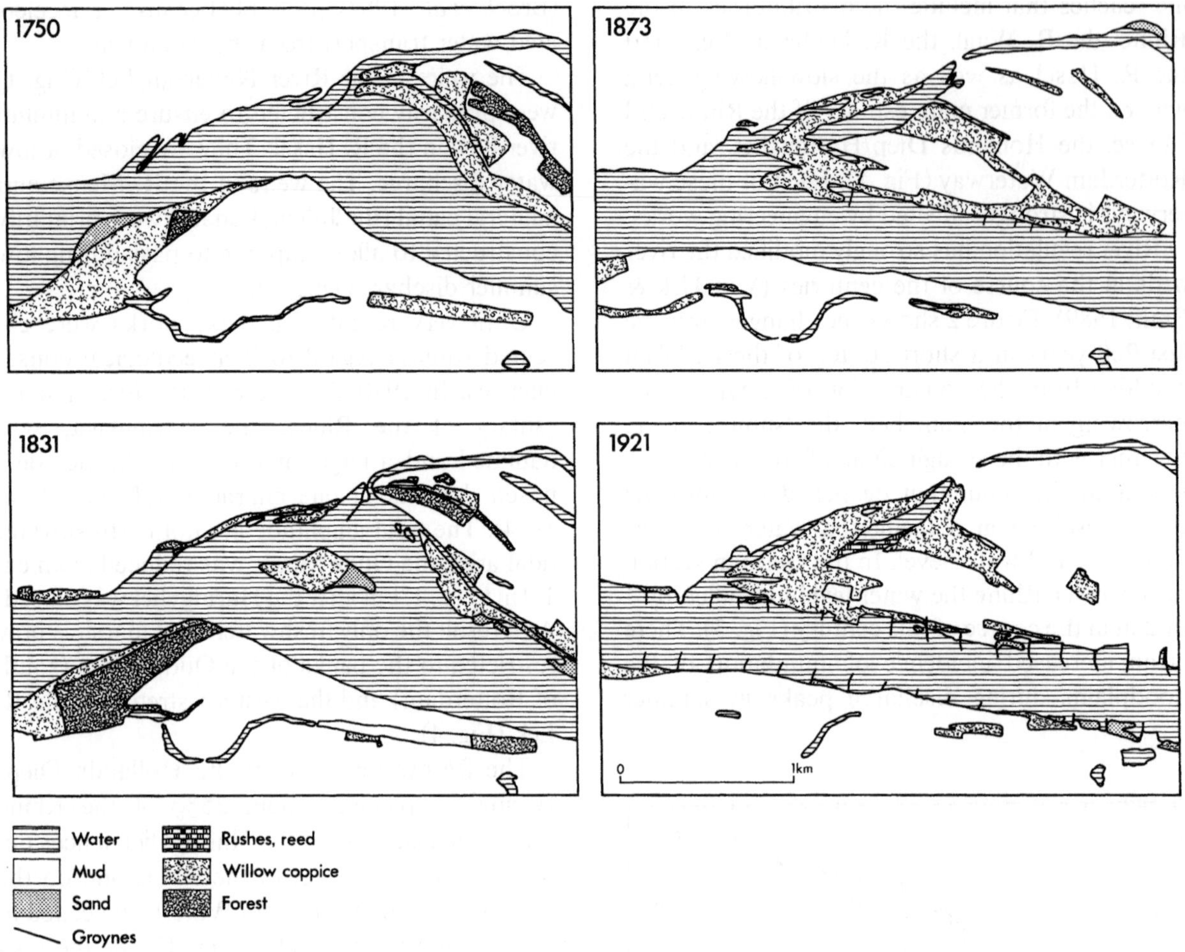

Fig. 2. Development of the morphology of a short stretch of the R. Waal (river km 899–901) since 1750. After Van Urk & Smit (1989).

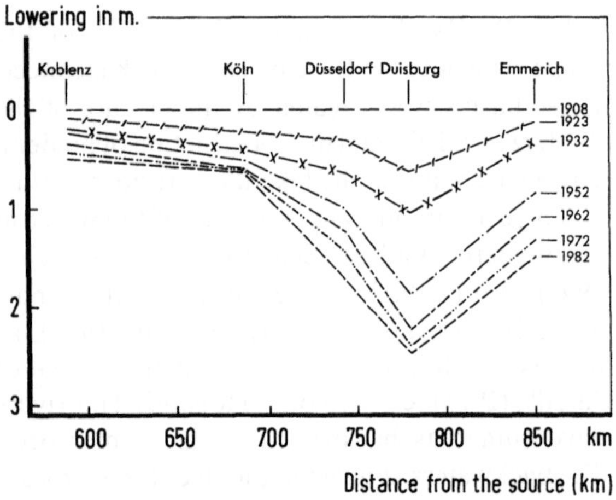

Fig. 3. Lowering of the mean water level in the Middle and Lower Rhine since 1908. The large decrease near Duisburg is partly due to mining. After Smit & de Jong (1989).

of the backwaters in the river forelands of the Rhine and Meuse. He found 170 abandoned meanders, 76 of which are connected with the river. He also described 47 breakthrough ponds (15 still connected with the river), 50 small rivers and brooks, 84 clay pits (12 still connected with the river), 163 sand and gravel pits (81 connected with the river), 101 harbours and 41 ditches and canals. There are also thousands of small areas of water (marshes, shallow ponds, etc.), most of which are temporary.

Shipping on the rivers is intensive (266×10^9 tons in 1986) and is a physical threat to biological communities on the banks. The wave action churns up the river sediments so that coarse-grained 'beaches' prevail between the groynes, as

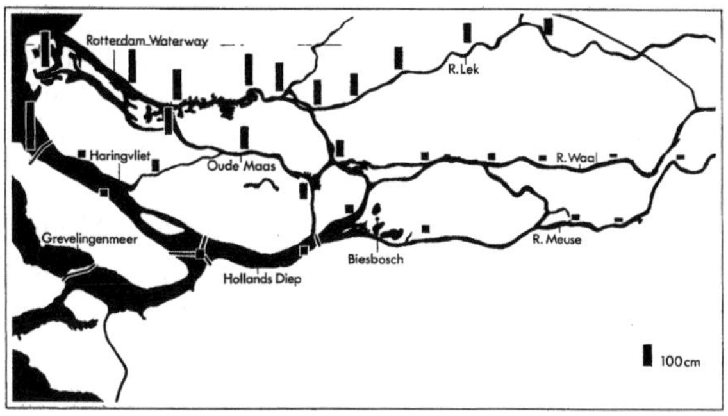

Fig. 4. Tidal amplitude (in m) in the Rhine-Meuse delta after closure of the Haringvliet. After H. Coops, unpublished.

on the R. Waal; in other places rubble is dumped to protect the banks from erosion.

The temporary occurrence of an ice cover on rivers is probably a powerful factor enforcing morphological changes in natural rivers. These effects are probably limited in the Rhine and Meuse because of the thermal pollution. The temperature of the Rhine water is raised by about 2 °C above its natural value. The Meuse waters are also warmed up by several degrees (Descy & Empain, 1984). Klink (1989) has argued that the numbers of days with an ice cover on the Rhine have declined significantly during this century.

Transport and fate of materials

Rivers are typically 'open' systems; the flow of water and the dissolved or particulate substances therein is essentially unidirectional. This is also true for the relatively short pre-estuarine reach of the lower Rhine and Meuse in The Netherlands. It is only in the estuarine reaches of the rivers that materials may accumulate or be transported upstream.

The Rhine carries an average of 3.4 10^9 kg yr^{-1} of suspended matter (Van Urk & Smit, 1989). The concentrations in the Rhine are generally somewhat higher than in the Meuse, but there is a typical difference in the behaviour of silt in the two rivers. The bed load transport in the Rhine may be limited (cf. Van Urk & Smit, 1989), but the compartments of the Meuse between the weirs tend to accumulate silt by sedimentation in the relatively slow-flowing waters. This material may be resuspended during peak discharge and transported downstream. It has been suggested that the transport of suspended matter in the Rhine has decreased, due to the changed morphology, from 4.2 to 3.4 10^9 kg yr^{-1} over the period of 1880–1980 (Van Urk & Smit, 1989).

In the past the suspended matter of the Rhine and Meuse supplemented the estuarine cycle of this material (de Jonge *et al.*, 1993; Nienhuis *et al.*, 1993), but now most of the riverine particulate matter is deposited in the Hollands Diep, the Rotterdam Harbours, the Ketelmeer, and in the IJsselmeer. The disproportional dimensions of water volume and water discharge in the Hollands Diep/Haringvliet has led to a typical accretion pattern, with materials from different periods tending to accumulate in different parts of the basin (Fig. 5).

The load of various substances in the Rhine and Meuse (*e.g.* cadmium or phosphate) dominates the budgets of these substances for The Netherlands; a significant fraction of the material is retained by the settling river sediment (De Kruijf *et al.*, 1988). Heavy metals and refractive organic pollutants have been accumulating in the sediment for decades, and large deposits are now present (Van Broekhoven, 1987). It is not known whether these toxic materials have been effectively immobilized or whether they are available

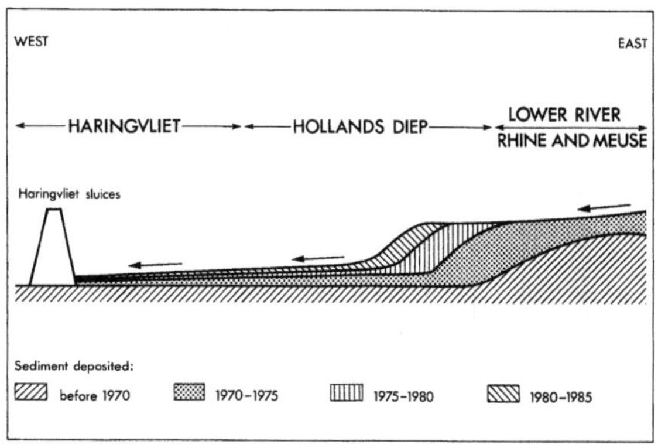

Fig. 5. Schematic representation of the sedimentation of riverine silt in the Hollands Diep and Haringvliet, recent sediments tending to be deposited further downstream. After Rijkswaterstaat (1987).

to organisms when they are re-exposed by scouring.

The concentrations of most heavy metals in the Rhine water fell appreciably over the period 1975–1985 (Van Broekhoven, 1987; Fig. 6), but the values are still much higher than the background levels (Salomons & Förstner, 1984). The concentrations of metals in the Meuse water do not show the consistent decrease found in the Rhine water. Concentrations of copper (Fig. 6) have fallen to levels similar to those in the Rhine, but concentrations of chromium and cadmium (especially the latter) have recently begun to rise. The annual load of cadmium carried by the Meuse now exceeds that of the much larger Rhine (Van Vuuren, 1989).

Oxygen concentrations in the Rhine and Meuse are shown in Fig. 7. In recent years the average oxygen concentrations in the Rhine and Meuse have been equivalent to a saturation level of slightly over 80%. The Rhine has recovered from periods with low oxygen levels in the sixties and

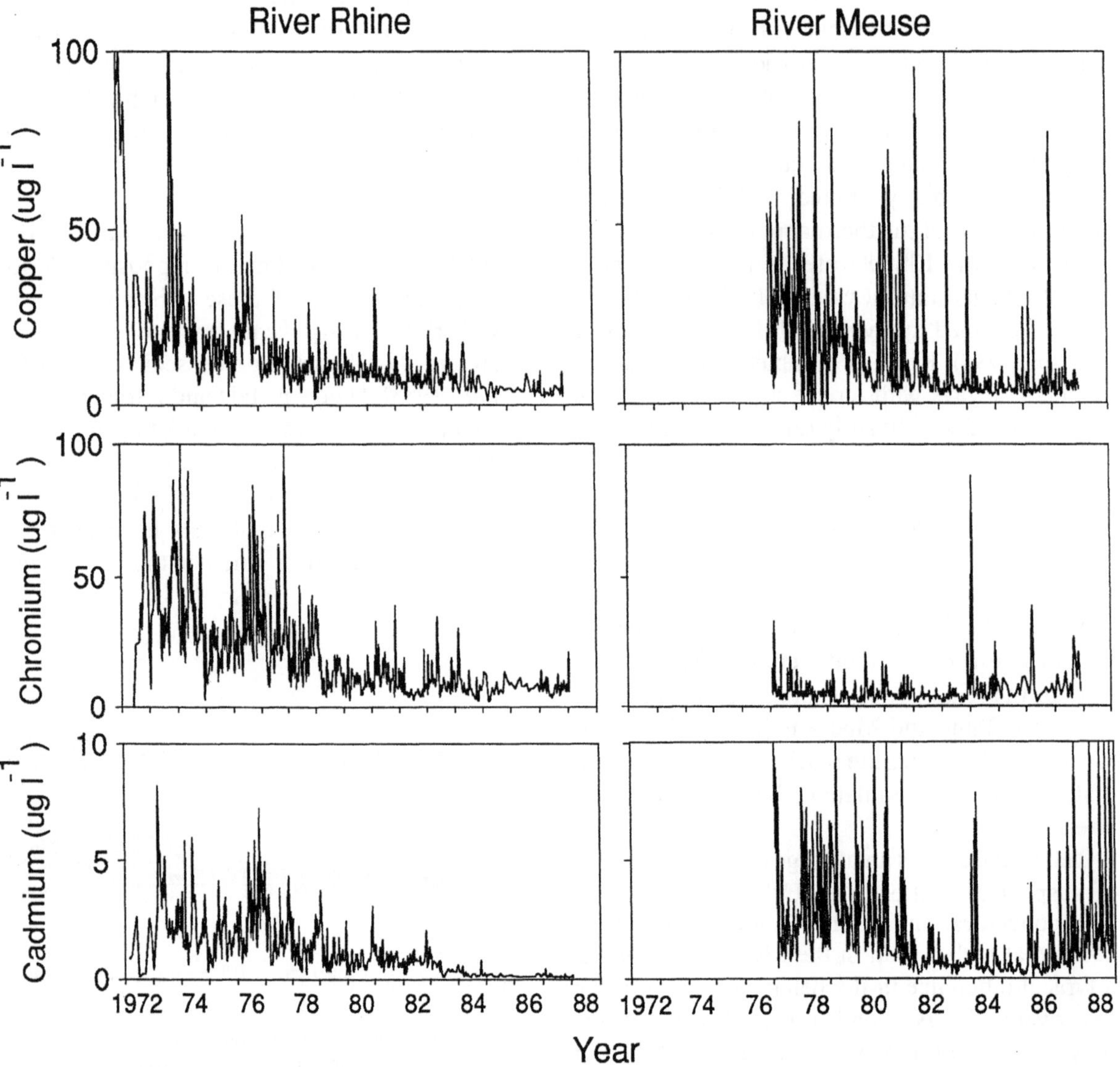

Fig. 6. Concentrations of total copper, total chromium, and total cadmium in the water of the Rhine and Meuse in 1972–1988. Data from Cooperating Drinking Water Companies (RIWA).

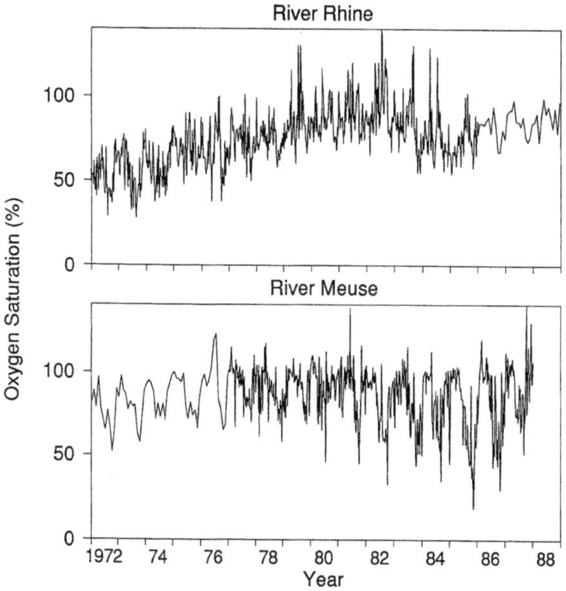

Fig. 7. As Fig. 6 for oxygen saturation.

seventies. The Meuse shows considerable summer depressions in oxygen concentration.

At present the concentrations of nitrogen compounds in the Rhine are higher than in the Meuse, but the opposite is true for the phosphate concentrations (Fig. 8). In the Rhine, phosphate concentrations have fallen consistently over the last decade, probably as a result of measures reducing the production and emission of phosphates. No such consistent trend is seen in the Meuse.

Chloride concentrations in the two rivers are different (Fig. 9): the continuously high levels in the Rhine are due to the ongoing discharges from the French potassium mines, together with the increasing discharge of salt drainage water from coal mining in Germany. In autumn, chloride concentrations in both rivers are relatively high because of the low water discharge.

Concentrations of organic micropollutants in the Rhine and Meuse differ as well. In recent years, cholinesterase-inhibiting compounds have amounted to 0.5–3 $\mu g\,l^{-1}$ in the Meuse and to 0.2–2 $\mu g\,l^{-1}$ in the Rhine, whereas the levels of absorbable organic halogen compounds in the Rhine are twice those in the Meuse (40–50 $\mu g\,l^{-1}$ versus 10–20 $\mu g\,l - 1$; data from the Cooperating Drinking Water Companies, RIWA). There is a similar difference between the rivers in the extractable and volatile fractions of halogen compounds.

The concentrations shown in Figs 6–8 were measured at the Dutch/German and Dutch/Belgian borders. Some of the values may decrease in the Dutch reaches of the rivers, due to adsorption, evaporation, or degradation. In the Dutch reach of the Meuse the long residence time of the water may allow water quality to improve despite the local discharges. De Zwart & Folkerts (1990) indicated that the toxicity of the concentrated organic fraction in Rhine water decreased during transport in the Netherlands. However, the concentrations of a few substances such as cadmium rise in the Rotterdam Waterway, as a consequence of discharges in the Rotterdam harbour.

The river forelands of the Rhine and Meuse are polluted by heavy metals and organic micropollutants such as PCB's. The deposition of contaminated silt after flooding adversely affects soil quality: the most frequently flooded soils have the highest concentrations of zinc, cadmium, copper, lead, cobalt, nickel and chromium in their top layers (Demon & Van Broekhoven, 1989), whereas erosive sediments show lower values. The Rhine Action Programme aims at halving the concentrations of pollutants in river water, including suspended materials, by 1995, the reference year being 1985. In most cases the contaminated soil in the river forelands is covered by or mixed with less contaminated silt, but this process of 'dilution' takes many decades, especially if the summer dikes remain intact. In forelands with summer dikes the sedimentation rate is estimated at 3 mm per year, based on measurements after floods in 1987 and 1988. After removal of the summer dikes this rate is estimated to increase to at least 10 mm per year.

Plankton

Peelen (1975) and Friedrich & Müller (1984) analysed trends in plankton development in the Rhine and Meuse on the basis of older, mostly qualitative observations. Peelen (1975) concluded

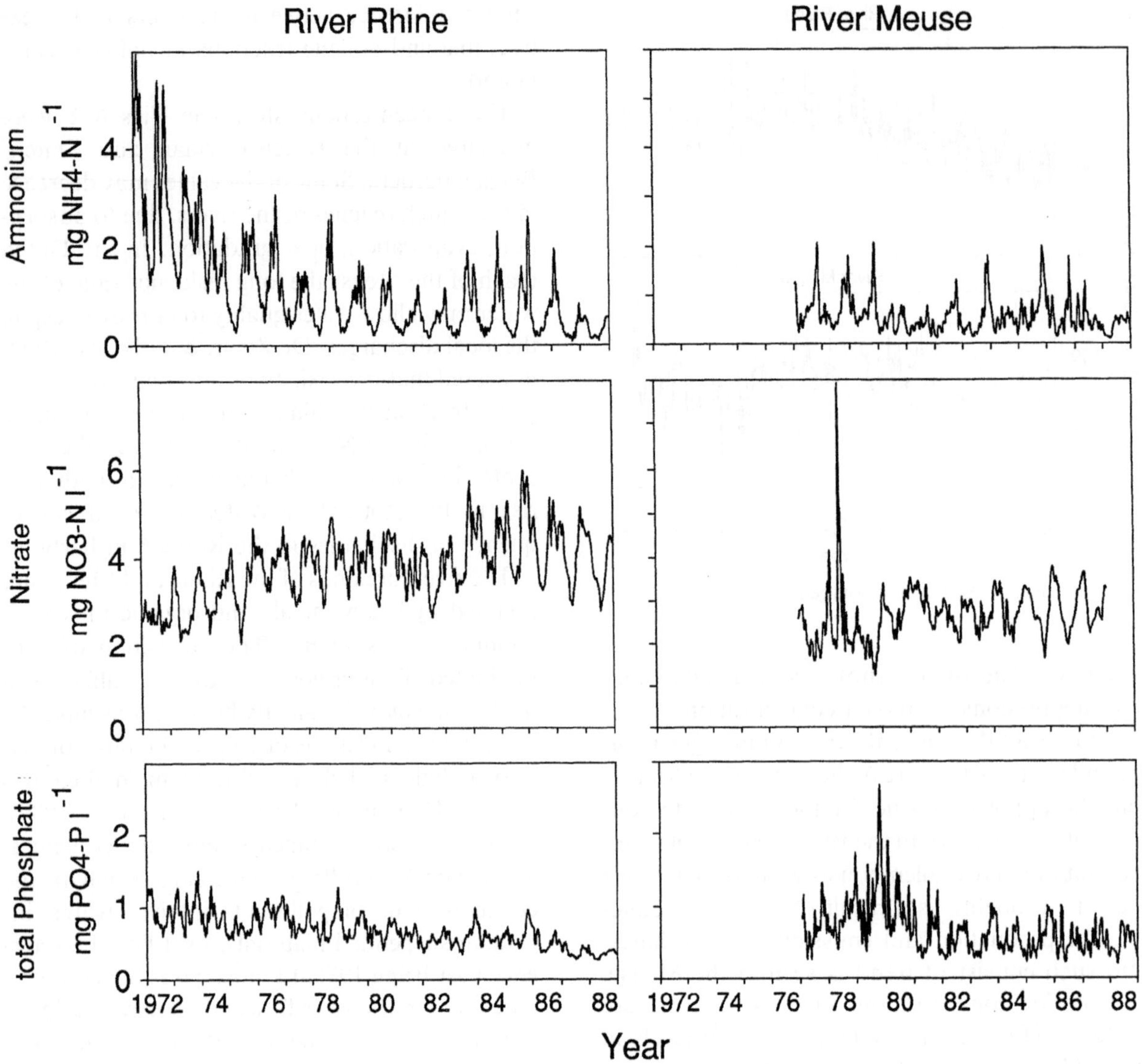

Fig. 8. As Fig. 6 for nitrogen compounds and total phosphate.

that between 1916 and 1972 the saprobic index of the phyto- and zooplankton did not shift significantly. However, it was clear that the plankton of both rivers was more sparse when it was investigated by De Lint (1916, in the Rhine, quoted in Peelen, 1975) and Romijn (1918, in the Meuse, quoted in Peelen, 1975) than in the period 1966–1972 (Peelen, 1975). According to Lauterborn (1939, in Peelen, 1975) plankton numbers in the Rhine increased rapidly after about 1890 (cf. Friedrich & Müller, 1984; Friedrich, 1990). In 1918 the Meuse was still free flowing. After the weirs were constructed in The Netherlands (from the twenties onwards) the residence time of the water increased; this, in addition to the increasing nutrient load, will have stimulated plankton development. Similar conditions may have stimulated the plankton input to the Rhine from tributaries. Increased development of plankton has also been reported for the regulated Rivers Ruhr (Nusch, 1978) and Neckar (Backhaus & Kemball, 1978).

The massive occurrence of planktonic microalgae, notably diatoms, in recent years has become

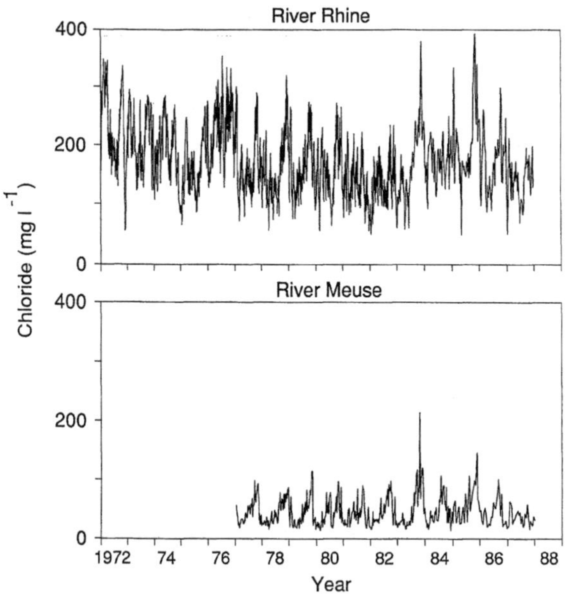

Fig. 9. As Fig. 6 for chloride.

clear in a palaeolimnological study of sediment cores from the Rhine area (A. Klink, unpublished), which showed that sediment-inhabiting pennate diatoms dominate the spectrum of diatom frustules in the sediments deposited in the 19th century, whereas planktonic centric species are numerically dominant in sediment layers deposited in the course of the 20th century.

A few ubiquitous species nowadays dominate the phytoplankton of the Rhine and Meuse. They include *Stephanodiscus hantzschii* Grün., *Melosira granulata* Ehr., and *Cyclotella meneghiniana* Kütz. (Friedrich & Müller, 1984; Descy, 1987). These species are also very common in other eutrophic rivers (*e.g* the R. Thames, Lack, 1971). Friedrich & Müller (1984) suggested that, in addition to eutrophication, the input of salt into the Rhine has contributed to the abundance of indifferent algal species.

In 1977 the *Daphnia* toxicity test showed that Rhine water in The Netherlands was acutely toxic (Slooff, 1983; Slooff *et al.*, 1985). Therefore, it seems likely that the zooplankton development in the river must also have suffered. Phytoplankton development in the Rhine between the Ruhr area and the German/Dutch border was interrupted in 1983 by the local input of toxic materials (Friedrich & Viehweg, 1984). Recently, Tubbing & Admiraal (1991) and Tubbing *et al.* (1992) have indicated that the present pollution of the Rhine is still affecting the growth of bacterioplankton and the photosynthesis of the phytoplankton.

Nowadays, the Rhine and the Meuse have similar densities of phytoplankton and zooplankton (De Ruyter van Steveninck *et al.*, 1990b). However, in the Dutch reaches of the Rhine the zooplankton is in an early stage of development (Admiraal *et al.*, 1990a; De Ruyter van Steveninck *et al.*, 1990a, b), whereas in the Meuse at the Belgian/Dutch border a fully developed zooplankton with numerous rotifers and crustaceans may be present (Fig. 10).

The phytoplankton, dominated by diatoms and green algae (Friedrich & Müller, 1984; Peelen, 1975), develops in the middle reach of the river, probably by selective development of inocula from tributaries and backwaters. Friedrich & Viehweg (1984) have demonstrated the growth of phytoplankton in the Rhine upstream of The Netherlands (cf. Franz, 1990), but the chlorophyll concentrations in the Netherlands tend to be more or less stable in the pre-estuarine river sections (Van

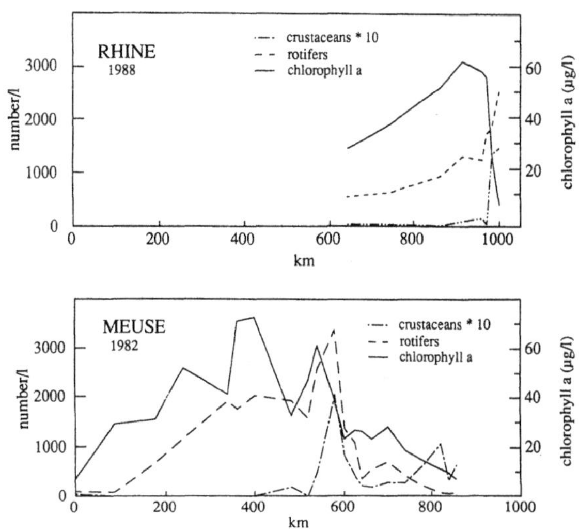

Fig. 10. Development of the plankton during transport of water in the R. Rhine (in May 1988) and Meuse (in Sept 1982). Density of phytoplankton measured as chlorophyll-*a*; density of zooplankton as numbers per l. Horizontal scale: km from the source. From De Ruyter van Steveninck *et al.* (1990b).

Urk, 1984a) or decrease in the sedimentation areas. However, trends in the dissolved silicate concentrations measured at various stations in the Rhine and Meuse indicate rapid growth of phytoplankton (certainly of diatoms) during transport in The Netherlands (Admiraal *et al.*, 1990b; De Ruyter van Steveninck *et al.*, 1990b).

The development of zooplankton (Fig. 11) follows that of the phytoplankton through the seasons. In addition to the true zooplankton, large numbers of the planktonic larvae of the zebra mussel *Dreissena polymorpha* (Pallas) have been found in the Rhine (De Ruyter van Steveninck *et al.*, 1990a; Borcherding & de Ruyter van Steveninck, 1992) and in the Meuse (RIWA, 1988).

Zooplankton develops in higher numbers and densities in standing waters than in the river.

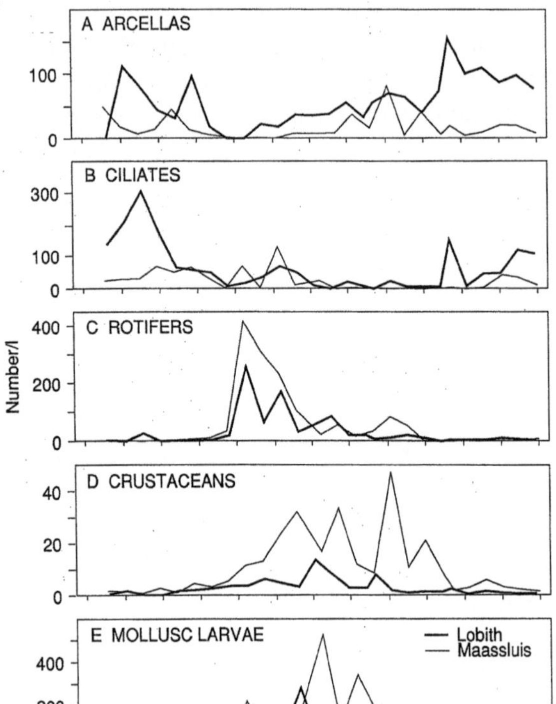

Fig. 11. Seasonal changes in zooplankton densities at two stations on the Dutch part of the R. Rhine. Data 1987. Station Lobith is at the German-Dutch border; Station Maassluis is downstream of Rotterdam (De Ruyter van Steveninck *et al.*, 1990a).

Wibaut-Isebree Moens (1965) found that gravel pits along the Meuse contain higher biomasses of zooplankton, mainly *Cyclops* sp., than the river. There are probably also typical differences between the phytoplankton of the river and the backwaters. Thus Leentvaar (1966) found *Ceratium hirundinella* (O.F. Müller) Schrank in the breakthrough ponds; this dinoflagellate dominates where there is a deep thermocline.

Vegetation

The vegetation along the rivers and in the backwaters has changed since 1950 (Maenen, 1989, Van den Brink *et al.*, 1991a). Most of the man-made banks of the Rhine and Meuse are now totally unsuitable for macrophytes (Leemans & Reiling, 1986; Leemans, 1989). The numbers of species of aquatic macrophytes and helophytes in the floodplain waters have declined dramatically, probably because summer floods are more common and the average water level is lower and hence the floodplain drier (Van Urk & Smit, 1989). Janse (1986) and Van den Brink *et al.* (1991a) classified the backwaters according to inundation frequency. Preliminary results (Fig. 12) indicate that the more frequent the flooding, the fewer the taxa in the vegetation (as well as in the macroinvertebrate fauna). This agrees with earlier findings by Van der Voo & Westhoff (1961), Van Donselaar (1961), Van de Steeg (1984), Brock *et al.* (1987), and Van der Sman *et al.* (1988). Species predominating in situations of infrequent inundation, such as *Stratiotes aloides* L., *Ranunculus lingua* L. and *Sparganium erectum* L. have declined.

The increased salinity of the river water has also contributed to the decline of formerly common species (cf. Table 1). Higher salinities together with the increased fluctuations in water-level and the increased turbulence of the water (shipping), may have contributed to the decline of the pondweed *Potamogeton nodosus* Poiret. This species was known from several localities in the River Waal, but has now disappeared. It is still present at some localities in the River Meuse,

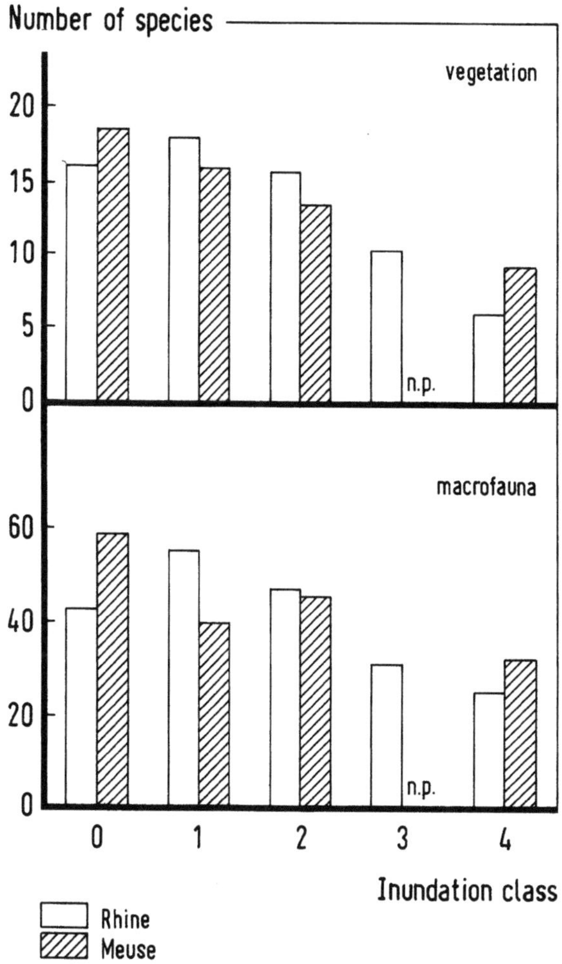

Fig. 12. Mean numbers of aquatic and helophyte plant species and macroinvertebrate species in riverine backwaters subjected to different regimes of inundation. Class 0: not inundated; class 1: <2 inundations per year; class 2: 2–20 inundations per year; class 3: 21–40 inundations per year; class 4: >40 inundations per year. n.p.: not present. Source: Van den Brink & van der Velde (1989).

near Kampen in the mouth of the River IJssel (Maenen, 1989) and in the Biesbosch (Smit & Coops, 1988). In contrast, *Potamogeton pectinatus* L., a species tolerant of high salinities, organic pollution and turbulence, has expanded in a number of areas in the Meuse and lower Rhine.

The distribution of nymphaeid and other macrophytes in abandoned meanders was studied 30 years ago by Van der Voo & Westhoff (1961); they frequently found species such as *Nuphar lutea* (L.) Sm. and *Nymphoides peltata* (Gmelin) O. Kuntze. The increased frequency of summer inundations has negatively affected the abundance of these plant species, and in particular that of *Nymphoides* (Brock *et al.*, 1987).

In the tidal part of the lower River Rhine there are vestiges of marsh vegetation on the banks of the River Oude Maas (Fig. 13; WOM, 1977); most of this vegetation disappeared from the delta after the Haringvliet was dammed in 1970. The richness of this vegetation is well known from the former intertidal marshes of the Biesbosch (Zonneveld, 1960; Van Urk,1984b). The rush species *Scirpus triqueter* L. still occurs in a few small localities along the R. Oude Maas. This species may be considered as an endangered component of the Rhine ecosystem. Bulrush (*Scirpus lacustris* L.) and sea clubrush (*Scirpus maritimus* L.) may grow higher up on the banks; the former in silty to sandy sites and the latter in exposed sandy sites (Coops & Smit, 1988). Two subspecies of the bulrush can be found, the freshwater bulrush (*S. l.* spp. *lacustris* L.) and the brackish water bulrush, *S. l.* spp. *tabernaemontani* (Gmelin) Syme. *S.* × *carinatus* Smith, a hybrid between *S. lacustris* and *S. triqueter*, occurs together with *S. lacustris*.

People have long exploited this vegetation zone, cultivating bulrush (*Scirpus lacustris*). In 1988 0.19 km^2 was still being cultivated by two growers, compared with about 2 km^2 in the sixties. In their stands *S. lacustris* spp. *lacustris, S. lacustris ssp tabernaemontani*, and *S.* × *carinatus* co-exist with *Callitriche stagnalis* Scop. and *Polygonum hydropiper* L. (Anonymous, 1977). The dried stems are used to weave chair seats and as a material to stopper whiskey vats. This extensive 'wetland farming' is a good example of a sustainable exploitation of the natural resources in the Rhine (Smit & Coops, 1991).

The characteristic mixed community of reed (*Phragmites australis* (Cav.)) Trin ex Steudel and *Caltha palustris* L. occurs higher up on the banks. This vegetation is limited to areas were the reed is commercially cut, which is still done in several areas, although the economic prospects are bleak. *Caltha palustris* is well adapted to the intertidal environment. In contrast to the normal form, the variety growing along the R. Oude Maas pro-

Table 1. Classification of macrophyte species according to their occurrence in gradients of salinity and inundation in the R. Rhine and Meuse. After Maenen (1989).

Highest salinity in meq $\cdot l^{-1}$ \ Highest fluctuation in water level in m.	4.00–7.52	7.52–8.75	8.75–9.41
12.2–15.9	*Alisma plantago-aquatica* *Callitriche obtusangula* *Callitriche stagnalis* *Carex disticha* *Glyceria fluitans* *Juncus articulatus* *Potamogeton lucens* *Potamogeton nodosus* *Potamogeton perfoliatus* *Sium latifolia*	*Callitriche platycarpa* *Nymphoides peltata* *Typha angustifolia* *Typha latifolia*	*Eleocharis acicularis* *Scirpus maritimus*
15.9–17.9	*Carex acutiformis* *Nasturtium officinale* *Potamogeton trichoides* *Sagittaria sagittifolia* *Sparganium erectum* *Spirodela polyrhiza*	*Acorus calamus* *Alisma lanceolatum* *Caltha palustris* *Epilobium hirsu-stum* *Glyceria maxima* *Lysimachia vulgaris* *Myosotis laxa* *Nuphar lutea* *Phragmites australis* *Potamogeton crispus* *Potamogeton pusillus* *Rumex conglomeratus* *Rumex hydrolapathum* *Scirpus lacustris* *Scirpus sylvaticus* *Senecio paludosus* *Sparganium emersum* *Veronica anagallis-aquatica* *Veronica beccabunga*	*Butomus umbellatus* *Eleocharis palustris* *Galium palustre* *Iris pseudacorus* *Juncus bufonius* *Lycopus europaeus* *Mentha aquatica* *Thalictrum flavum*
17.9–19.2		*Ceratophyllum demersum* *Elodea nuttallii* *Myriophyllum spicatum* *Oenanthe aquatica* *Scutellaria galericulata* *Senecio congestus* *Zannichellia palustris*	*Carex acuta* *Lemna minor* *Lythrum salicaria* *Myosotis palustris* *Phalaris arundinacea* *Polygonum amphibium* *Potamogeton pectinatus* *Rorippa amphibia* *Veronica catenata*

duces additional roots at the internodes and is called *C. p.* var. *araneosa* van Steenis, because these extra roots resemble a spider. The rooted nodes promote vegetative reproduction. Many marsh plants can be found in places where the reed is not cut: *Angelica archangelica* L., *Leucojum*

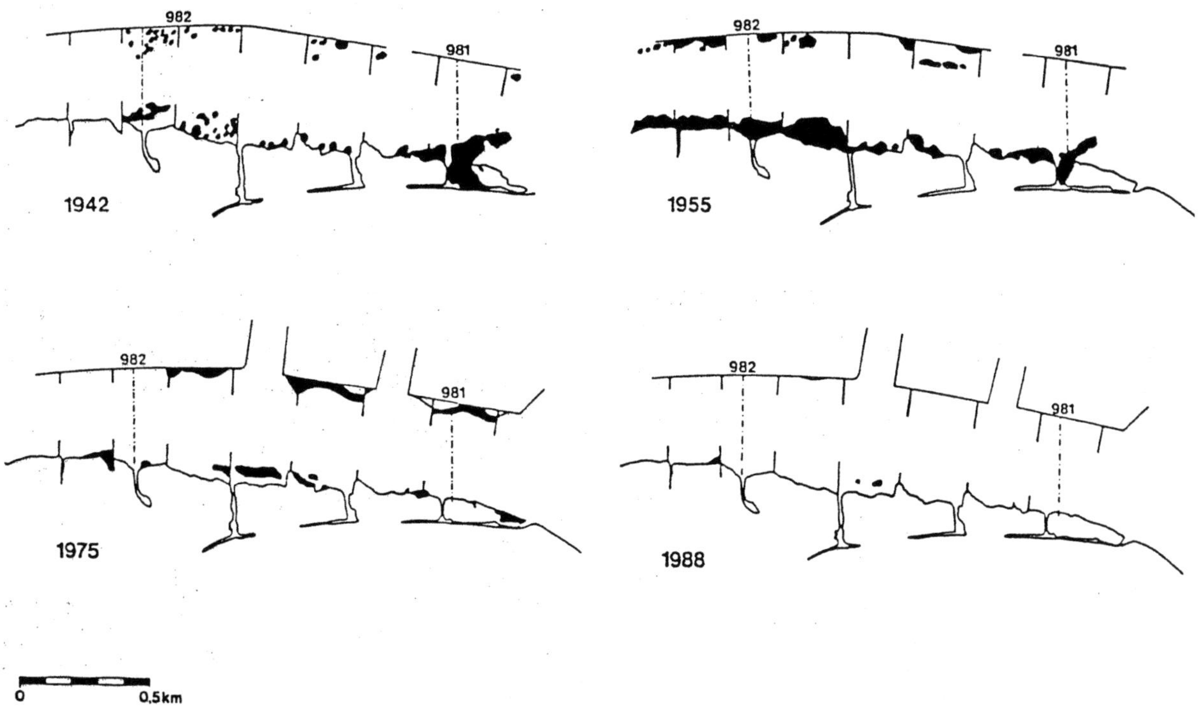

Fig. 13. Development of bulrush vegetation (black) on the banks of the R. Oude Maas (Rhine km 981–982) in the period 1942–1988. From Smit & Coops (1990).

aestivum L. and *Senecio fluviatilis* Wallr. are typical. A well developed willow zone is found higher up on the banks, with *Salix alba* L., *S. viminalis* L., *S. dasyclados* Wimmer and *S. fragilis* L. as the most important species. Willow shoots used to be a useful crop, but its importance has greatly declined because of the hard labour involved and low prices obtained for this material.

Macro-invertebrates

The large Dutch rivers have lost most of their characteristic macroinvertebrate species, although a relatively rich fauna was present until 1940 (Redeke, 1948). Figure 14 shows that even before 1940 the numbers of insect species in the Rhine had decreased strongly; this can be attributed to canalization, regulation, increased shipping and input of organic wastewater. Representatives of the Trichoptera, Ephemeroptera, and Plecoptera, which can be regarded as typical elements of the fauna in large European rivers, have almost completely disappeared from the Rhine and Meuse (Smit, 1985; Van den Brink *et al.*, 1989). Burrowing filter-feeding animals like the mayfly larva *Palingenia longicauda* (Ol.) have vanished (Van Urk & Smit, 1989). Many rheophilous species of insects have disappeared because suitable sites for larval development and oviposition have become scarce. The river pearl mussel *Margaritifera auricularia* (Spengler), a species of deep sandy rivers, used to live in the Rhine and Meuse until about a century ago, but it too has become extinct (Kuijper, 1988).

Discharge of industrial waste and organic pollution have caused low oxygen concentrations and high levels of toxic substances (Figs 6 and 7). Thick layers of contaminated silt have been deposited on the river bed in the lower reaches of the Rhine and Meuse, leading to a further decline of characteristic zoobenthos species. At present most rheophilous species occur on the stones of the groynes. The effects of contaminants on the

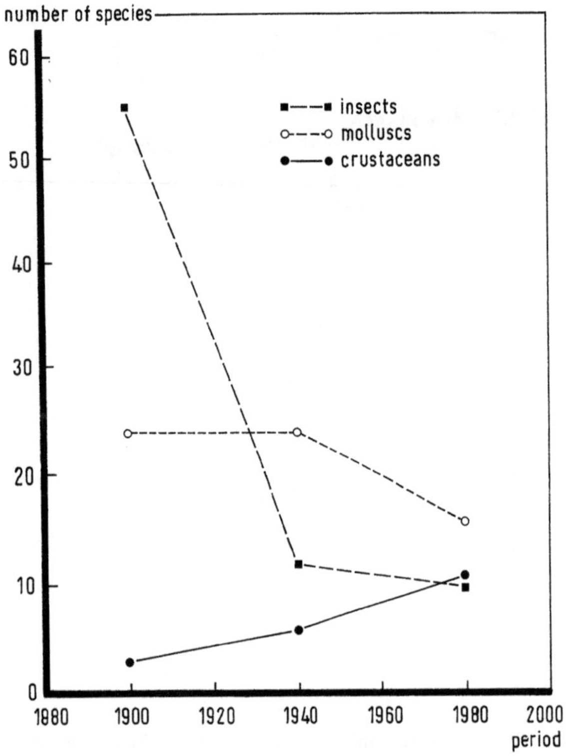

Fig. 14. Changes in the numbers of macroinvertebrate species in the Dutch section of the R. Rhine in the course of this century. Number of insect species excluding chironomids. Based on Van den Brink *et al.* (1990).

invertebrate communities in the sedimentation areas have recently been demonstrated. Van Urk & bij de Vaate (1990) investigated chironomids in a pollution gradient from the polluted Ketelmeer to an adjacent less polluted area. Mean densities and individual weight increased and the incidence of malformations in the larvae decreased along this gradient. A distinct increase in head capsule malformations of chironomid larvae has occurred during this century (Klink, 1986a). Heavy metals and pesticides are considered to be responsible for the severe head capsule malformations, especially in the sedimentation areas (Van Urk & Kerkum, 1986, 1987).

Van Urk & Smit (1989) found very low densities of macrofauna in the sediments of the river Waal and Van Urk *et al.* (1990) found very low densities of chironomids in the R. IJssel. From the presence of exuviae of chironomids, Wilson & Wilson (1985a) concluded that the lower Rhine and the Waal have low proportions of sediment-dwellers, suggesting that the sediments are unable to support a normal chironomid fauna. Accordingly, Klink (1986a) found appreciable numbers of Simuliidae deposited in sediments of the river Rhine dating from the 18th century (numbers of Simuliidae amounting to 0.3 of those of the Chironomidae), whereas at present Simuliidae are virtually absent in the Rhine.

The bed of the River Meuse harbours more macro-invertebrate species than the beds of the Rhine branches. In the Meuse, especially in the section called the Grensmaas, there are more characteristic riverine species than in the Rhine because shipping is less intensive and the morphology is less disturbed (i.e. more of the natural habitats have been preserved: Klink & Moller Pillot, 1982; Smit, 1982; Peeters, 1988). Smit & Gardeniers (1986) found 17 species of macro-invertebrates on the deep bed of the River Meuse, of which 7 were restricted to this habitat.

Although the fauna inhabiting the sandy bottom of the rivers Rhine and Meuse is very poor, new species are occasionally discovered. The chironomid *Lipiniella arenicola* Shilova established itself on sandy flats after the Haringvliet was dammed (Smit *et al.*, 1991), and *Paratendipes intermedius* Mg. was discovered in sandy river sediment of the Rhine in 1989 (H.P.J.J. Cuppen, pers. comm.). Two new fluviatile immigrant mussel species living on sandy and sandy-clayey sediments (*Corbicula* spec.) have recently been discovered in the R. Lek, the Hollands Diep (Bij de Vaate & Greijdanus-Klaas, 1990), and in the R. Waal and Rhine near the German/Dutch border (Bij de Vaate, 1991). A clear indication of the improvement of the circumstances for the sand inhabiting fauna is the recolonization by the mayfly *Ephoron virgo* (Olivier) after almost 50 years of absence. Settlement started in the German part of the R.Rhine and in 1991 it reached the Netherlands section of the river (Van den Brink *et al.*, 1991c).

Although the fauna typical of detrital and clayey deposits has suffered less from pollution than that of sandy sediments, it too has declined. Large mussels (*Anodonta* and *Unio*) were important in this community. In the seventies no large mussels were observed in the river Rhine (Peeters & Wolff, 1973), but since then oxygen conditions have improved and they have been found again. At the very low water levels in 1989 six-year-old specimens of *Anodonta anatina* (L.) and *Unio pictorum* (L.) were found to be abundant in the R. Waal near Nijmegen, and there were smaller numbers of *U. tumidus* (Philipsson). In 1990 the fluviatile species *Pseudanodonta complanata* (Rosm.) has been found in the R. IJssel and the R. Merwede (J.N. de Vries, F.W.B. van den Brink, pers. comm.).

Numerous chironomid species nowadays inhabit the stony substrates in the Rhine (Moller-Pillot & Krebs, 1981; Klink 1982, 1983; Wilson & Wilson, 1985a, b). The caddis flies *Hydropsyche contubernalis* McLachlan and *Ecnomus tenellus* (Ramb.) returned in 1978 and 1980 respectively, after an absence of many years (Fig. 15, Van Urk *et al.*, 1990). The stony embankments of the rivers are also very suitable for some immigrant species such as the zebra mussel *Dreissena polymorpha* from the Ponto-Caspian area, the snail *Potamopyrgus antipodarum* (Gray) from New Zealand, *Physella acuta* (Drap.) from Southern Europe and North Africa, and the North American planarian *Dugesia tigrina* (Girard). These species coexist

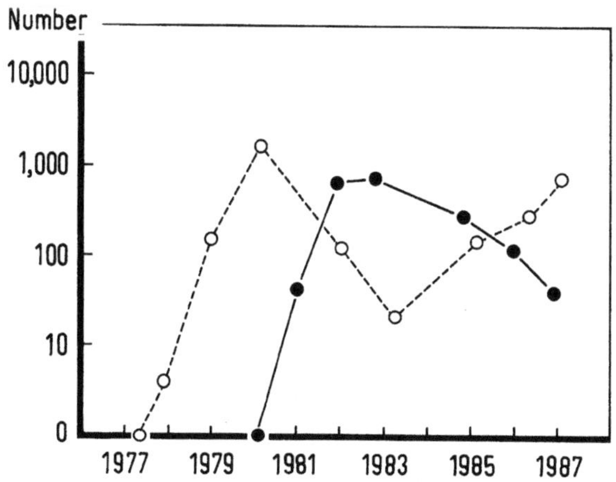

Fig. 15. Recolonization of the R. IJssel by caddis flies. Numbers of larvae of *Hydropsyche contubernalis* (open symbols) and *Ecnomus tenellus* (closed symbols) on 40 stones collected from groynes in the period 1977–1987. Source: Van Urk *et al.* (1990) (quoted in De Wit *et al.*, 1989).

with ubiquitous species like the snails *Radix peregra f. ovata* (O.F. Müller) and *Bithynia tentaculata* (L.) and the river limpet *Ancylus fluviatilis* (O.F. Müller). In 1987 Van den Brink *et al.* (1989) found an immigrant amphipod species, *Corophium curvispinum* Sars, from the Ponto-Caspian area. This species appeared to be very successful and nowadays very high densities of more than 100,000 specimens per m^2 of stone surface occur. Its success is due to a combination of factors, viz. the increased salinity, higher temperatures and the increased food supply. *C. curvispinum* is a filter feeder stimulated by the dense phytoplankton development, caused by eutrophication. The muddy tubes build by this animal species covers the entire surface of solid substrates, thereby diminishing the populations of other animal species occupying this habitat (Van den Brink *et al.*, 1991b, c). *Corophium curvispinum* is still expanding its area in the Netherlands and in 1991 it has already been found in the IJsselmeer, Markermeer and a number of canals (S. Pinkster, pers. comm.).

Klink (1989) indicated that the rise in water temperature due to discharge of cooling water benefits some chironomid species, *e.g. Rheocricotopus chalybeatus* (Edw.), *Nanocladius rectinervis* (K.) and the *Rheotanytarsus* species, all of which are now very abundant. These higher temperatures are also favourable for other invertebrate species, such as the shrimps *Atyaephyra desmaresti* (Millet) and *Palaemon longirostris* H. Milne Edwards (Van den Brink & Van der Velde, 1986a, b), the planarians *Dugesia polychroa* (Schmidt) and *D. tigrina*, the snails *Physella acuta* and *Ferrissia wautieri* (Mirolli), etc. Further research on the distribution of these thermophilous species is necessary in order to predict future changes that might result from a climate-induced temperature rise.

Salt concentrations in Rhine water often exceed 200 mg l^{-1} (Fig. 9) and this explains why mobile macro-invertebrate species formerly restricted to the brackish or tidal parts of the river have been observed as far upstream as the German-Dutch border. When the filtering screens of the cooling-water intakes of some power plants along the rivers were sampled recently it was found that the mudcrab *Rhithropanopeus harrisi* (Gould), an early immigrant from North-America, as well as the white prawn (*Palaemon longirostris*) and the brackish water amphipod *Gammarus zaddachi* Sexton had all extended their range eastwards (Den Hartog *et al.*, 1989). However, another brackish water amphipod *Gammarus tigrinus* Sexton, originating from North America, is able to breed in the river and has caused the numbers of the freshwater amphipod *Gammarus pulex* (L.) to decline. Van den Brink & van der Velde (1986a) and Van der Velde *et al.* (1990) proved that the typical migrating brackish water crustacean species showed a pattern of occurrence similar to that of *G. zaddachi*, with low numbers in summer and high numbers in spring and autumn, whereas river-dwelling species like the North-American crayfish *Orconectes limosus* (Rafinesque), the amphipod *Gammarus pulex*, the isopod *Asellus aquaticus* (L.) and the freshwater shrimp *Atyaephyra desmaresti* showed peaks in numbers in summer as a result of reproduction (see also Van den Brink *et al.*, 1988). Van den Brink & van der Velde and their colleagues did not find the estuarine species as small juveniles or as adults carrying eggs. Furthermore, these species showed very skewed sex ratios in the freshwater section of the river, in contrast to the river-dwelling species.

Both *Palaemon longirostris* and *Atyaephyra desmaresti* have an Atlantic-Mediterranean distribution and are at the northern limit of their range in the Netherlands. The annual records on both species show a negative correlation with the number of frosty days; mild winters favour these species (Van den Brink & van der Velde, 1986a, b).

Fish

Fish populations in the rivers Rhine and Meuse have changed dramatically during the last century. Table 2 shows that about ten indigenous species of fish have become extinct or very rare. Characteristic migrating species such as sturgeon (*Acipenser sturio* L.), houting (*Coregonus oxyrinchus* (L.)), and salmon (*Salmo salar* L.) have died out, and populations of allis shad (*Alosa alosa* (L.)), twaite shad (*Alosa fallax* (Lac.)), sea lamprey (*Petromyzon marinus* L.), and river lamprey (*Lampetra fluviatilis* (L.)) have declined drastically. Other characteristic species, like barbel (*Barbus barbus* (L.)), dace (*Leuciscus leuciscus* (L.)), minnow (*Phoxinus phoxinus* (L.)), and burbot (*Lota lota* (L.)) have also suffered seriously.

For centuries the anadromous species that were caught during their spawning run were fished in the Rhine and Meuse. The highly flourishing fisheries for sturgeon, salmon, sea trout, allis shad, twaite shad and houting largely disappeared in the first half of this century, as has been described by numerous authors (Boddeke, 1974; Bürger, 1926; Cazemier, 1988, Deelder & van Drimmelen, 1959; Denzer, 1966; Van Drimmelen, 1981 and 1987; De Groot 1988, 1989a, b, c, d, 1990a, b; Hoek, 1916; De Jong *et al.*, 1988; Kuhn, 1976; Leentvaar, 1963; Lobregt & van Os, 1977; Lelek & Köhler, 1989; Van Lonkhuyzen & Vonk, 1920; Philippart, 1985, 1988a, b; Redeke, 1941; Van Ruremonde, 1988; Verhey, 1961; Wolff, 1978). In Fig. 16 catches of sturgeon, allis shad, salmon and twaite shad are plotted to demonstrate the decline of the stocks. Despite rescue operations

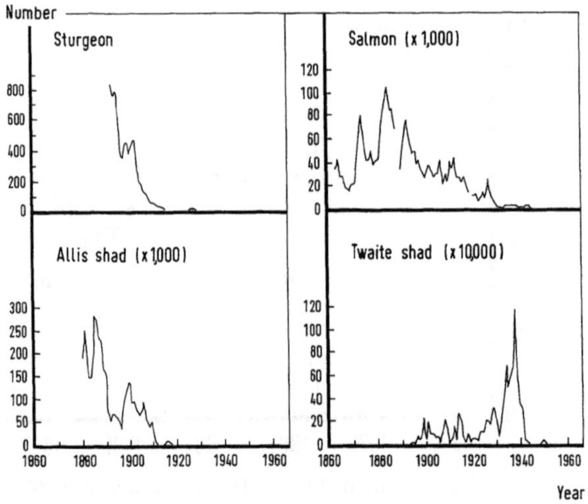

Fig. 16. Landings of four anadromous fish species in the Netherlands in the period 1880–1950. For scientific names see Table 2. Source: fisheries statistics summarized by W.G. Cazemier.

Table 2. Occurrence of fish species in the R. Rhine and R. Meuse in the period 1987–1988. Data for The Netherlands from Cazemier & Heermans (1988); data for Germany from Berg *et al.* (1989), Borchard *et al.* (1986), Hessisches Ministerium (1987), and Lelek (1989); data on the Belgian part of the Meuse according to Philippart *et al.* (1988 a, b)

		A	B	C	D	E	F	G	H	I	J
Petromyzon marinus L.	Sea lamprey	+	+	–	3	3	+	2	–	+	
Lampetra fluviatilis (L.)	River lamprey	+	+	–	4	4	+	2	–	+	
Lampetra planeri (Bloch)	Brook lamprey	+	–	–	1	1	+	1	–	–	
Acipenser sturio L.	Sturgeon	+	+	–	0	0	–	0	–	–	
Anguilla anguilla (L.)	Eel	+	–	+	5	5	+	5	+	+	
Alosa alosa (L.)	Allis shad	+	+	–	2	1	+	1	–	–	
Alosa fallax (Lac.)	Twaite shad	+	+	–	3	2	+	1	–	+	
Salmo salar L.	Atlantic salmon	+	+	–	2	2	?	0	–	–	
Oncorhynchus mykiss (Walbaum)	Rainbouw trout	–	–	–	2	2	+	1	+	–	
Salmo trutta trutta L.	Sea trout	+	+	–	4	4	+	4	+	+	
Salmo trutta trutta morpha fario L.	Brown trout	+	–	–	1	1	+	3	+	–	
Oncorhynchus kisutch (Walbaum)	Coho salmon	–	+	–	2	1	+	1	–	–	a
Salvelinus fontinalis (Mitchill)	American brook trout	–	–	–	1	1	+	2	+	–	
Coregonus albula (L.)	Vendace	+	–	–	0	0	–	0	–	–	
Coregonus lavaretus (L.)	Powan	+	–	–	3	3	+	3	–	–	
Coregonus oxyrinchus (L.)	Houting	+	+	–	0	0	–	–	–	–	
Thymallus thymallus (L.)	Grayling	+	–	–	1	1	+	2	+	–	
Osmerus eperlanus (L.)	Smelt	+	+	–	3	5	+	4	–	+	b
Esox lucius L.	Pike	+	–	–	4	4	+	4	+	–	
Umbra pygmaea (De Kay)	Mud minnow	–	–	–	1	1	–	3	+	–	
Cyprinus carpio (L.)	Carp	–	–	–	4	4	+	4	+	–	
Abramis brama (L.)	Bream	+	–	–	5	5	+	5	+	–	
Abramis bjoerkna (L.)	White Bream	+	–	–	5	4	+	5	+	–	
Alburnoides bipunctatus (Bl.)	Stream bleak (Spirlin)	+	–	–	0	0	+	0	+	–	
Alburnus alburnus (L.)	Bleak	+	–	–	5	5	+	5	+	–	
Aspius aspius (L.)	Asp	–	–	–	1	1	+	2	–	–	
Barbus barbus (L.)	Barbel	+	–	–	3	1	+	3	+	–	
Carassius carassius (L.)	Crucian carp	+	–	–	3	2	+	3	+	–	
Carassius auratus (L.)	Goldfish	–	–	–	1	1	+	1	+	–	
Carassius auratus gibelio (Bloch)	Gibel carp	–	–	–	3	2	+	4	+	–	
Chondrostoma nasus (L.)	Hotu or nase	+	–	–	2	1	+	1	+	–	
Ctenopharyngodon idella (Val.)	Grass carp	–	–	–	2	2	+	2	–	–	
Gobio gobio (L.)	Gudgeon	+	–	–	4	3	+	4	+	–	
Hypophtalmichthys molitrix (Val.)	Silver carp	–	–	–	1	1	+	2	–	–	
Hypophtalmichthys nobilis (Richardson)	Big head	–	–	–	2	2	–	2	–	–	
Leucaspius delineatus (Heckel)	Rain bleak	+	–	–	1	1	+	1	+	–	
Leuciscus leuciscus (L.)	Dace	+	–	–	3	2	+	4	+	–	
Leuciscus idus (L.)	Ide	+	–	–	4	4	+	4	+	–	
Leuciscus cephalus (L.)	Chub	+	–	–	2	1	+	4	+	–	
Phoxinus phoxinus (L.)	Minnow	+	–	–	1	1	+	1	+	–	
Rhodeus sericeus amarus (Bloch)	Bitterling	+	–	–	1	2	+	2	+	–	
Rutilus rutilus (L.)	Roach	+	–	–	5	5	+	5	+	–	
Rutilus erythrophthalmus (L.)	Rudd	+	–	–	4	4	+	4	+	–	

Table 2. (Continued)

		A	B	C	D	E	F	G	H	I	J
Tinca tinca (L.)	Tench	+	–	–	4	3	+	4	+	–	
Vimba vimba (L.)	Vimba	–	–	–	2	1	+	1	–	–	c
Cobitis taenia L.	Spiny loach	+	–	–	3	2	+	3	–	–	
Misgurnus fossilis (L.)	Weather loach	+	–	–	3	3	+	1	+	–	
Noemacheilus barbatulus (L.)	Stone loach	+	–	–	3	1	+	3	+	–	
Silurus glanis L.	Sheatfish	+	–	–	2	2	+	2	+	–	
Ictalurus melas (Raf.)	Black bullhead	–	–	–	1	1	+	2	+	–	
Ictalurus nebulosus (Le S.)	Brown bullhead	–	–	–	2	1	+	2	+	–	
Lota lota (L.)	Burbot	+	–	–	3	1	+	2	+	–	
Gasterosteus aculeatus L.	Three-spined stickleback	+	+	–	4	4	+	4	+	+	
Pungitius pungitius (L.)	Ten-spined stickleback	+	–	–	2	1	+	2	–	–	
Lebistes reticulatus (Peters)	Guppy	–	–	–	1	1	–	1	–	–	
Lepomis gibbosus (L.)	Pumpkinseed	–	–	–	1	2	+	1	+	–	
Perca fluviatilis L.	Perch	+	–	–	5	5	+	5	+	–	
Gymnocephalus cernuus (L.)	Ruffe	+	–	–	5	5	+	5	+	–	
Stizostedion lucioperca (L.)	Pikeperch	–	–	–	5	5	+	5	+	–	
Cottus gobio L.	Bullhead	+	–	–	4	4	+	4	+	–	
Platichthys flesus (L.)	Flounder	+	–	+	4	5	+	4	+	+	

A: Indigenous in The Netherlands
B: Anadromous
C: Catadromous
D: Present frequency in the lower Rhine in The Netherlands (0: extinct: 1: no recent records; 2: occasionally; 3: fairly rare; 4: (locally) common; 5: abundant).
E: Present frequency of occurrence in the IJselmeer and Ketelmeer (classes as under D).
F: Presence in the German part of the river Rhine. –: no records; +: presence established; ?: identification dubious.
G: Presence in the Dutch part of the Meuse classes as under D.
H: Presence in the Belgian part of the Meuse.
I: Presence in Dutch coastal waters.
J: a: record from the tidal zone only; b: mainly as land-locked populations; c: first and only specimen, captured in the R. Nederrijn in 1989 (Cazemier & Heesen, 1989).

by means of fishery regulations and mass releases of hatchery-reared salmon, houting and allis shad, the decline of the migratory fish stocks has continued (De Groot, 1985). The fisheries exploiting the anadromous fish species smelt (*Osmerus eperlanus* (L.)) and twaite shad in the Haringvliet were brought to an end by the closure of the Haringvliet in 1970 (Steinmetz, 1975; Wiegerinck & Heesen, 1987).

In the 19th century the fisheries for eel (*Anguilla anguilla* (L.)), pike (*Esox lucius* L.), barbel (*Barbus barbus* (L.)), chub (*Leuciscus cephalus* (L.)), bream (*Abramis brama* (L.)), roach (*Rutilus rutilus* (L.)), and perch (*Perca fluviatilis* (L.)) were of minor importance in the Dutch rivers. But after the anadromous species had disappeared, the fishermen who remained in the trade concentrated on eel and resident fish, of which pikeperch (*Stizostedion lucioperca* (L.)) became more and more important (De Boer & te Brinke, 1984; Lobregt & van Os, 1977). These fisheries however, have also declined: Table 3 shows the decreasing number of eel fisheries in the Dutch part of the Rhine.

The collapse of fish stocks and fisheries was most probably caused by the reconstruction of the river system plus the deterioration of the water quality. There is ample evidence that river engineering works have had deleterious effects on the

Table 3. Numbers of eel fisheries in the Dutch branches of the R. Rhine. After Cazemier (1988).

Year	Number	Reference
1913	193	Lobregt & van Os, 1977
1949	100	Deelder & van Drimmelen, 1959
1952	76	Lobregt & van Os, 1977
1957	35	Leentvaar, 1963
1957	26	Deelder & van Drimmelen, 1959
1963	16	Leentvaar, 1963

species numbers and abundance (Hoek, 1916; Alabaster, 1985; Klein, 1989; Lelek, 1989; Lelek & Köhler, 1989). The construction of fish passes in combination with almost every weir in the main stream seems to have been insufficient to prevent the decline of migrating fish populations. The water works in general led to the disappearance of specific spawning grounds, feeding biotopes and nursery areas and to the obstruction of migration routes.

The low oxygen concentration in the rivers and the massive discharge of toxic materials in the period of 1960 to 1970 were the most likely causes of the nadir in the number of fish species (Van den Brink *et al.*, 1990; Van der Velde *et al.*, 1990). Slooff (1983) and Slooff *et al.* (1985) demonstrated the abundance of deformations in bream in the Rhine.

Since the water quality of the Rhine began to improve in the mid-seventies, the fish community has been recovering (Willemsen, 1983; Cazemier, 1984; Cazemier, 1988; Lelek, 1989). The present oxygen content of the Rhine water even at low river discharge is suitable for salmonids. The severe pollution of the R. Meuse in the sixties is also decreasing (Philippart *et al.*, 1988a). Despite the continuing problems with *e.g.* high levels of polychlorinated biphenyls (PCB) and mercury in fish (Van der Valk *et al.*, 1989), water quality as a whole has substantially improved and as a consequence the concentrations of several pollutants in fish have fallen (Fig. 17). Some of the characteristic fish species are showing signs of recovery (Hadderingh *et al.*, 1983; Cazemier, 1984; Cazemier & Heermans, 1988; Cazemier, 1989; Heesen, 1989; P.J.M. Bergers, pers. comm., 1989).

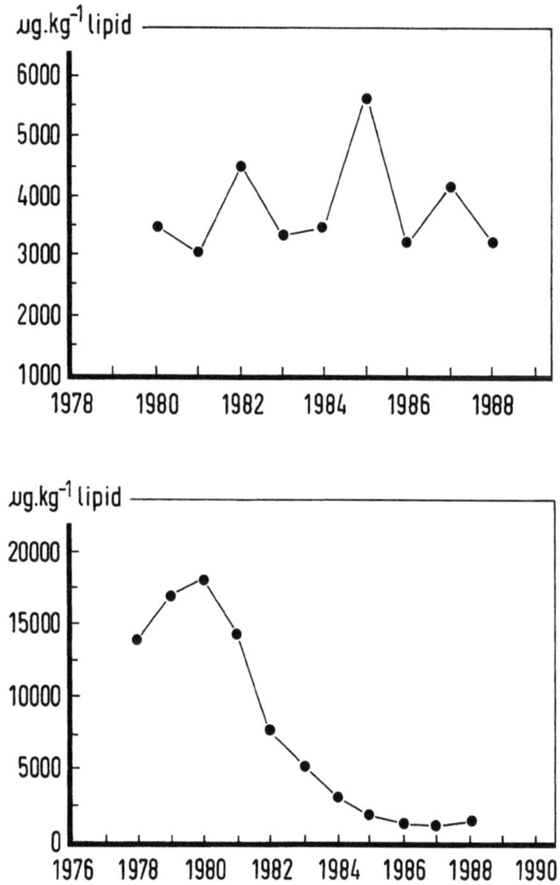

Fig. 17. Development of the concentrations of PCB-153 (upper plot) and HCB (lower plot) in the fat of yellow eel (*Anguilla anguilla*) in the period 1978–1988. Source: Van der Valk *et al.* (1989).

During the sixties and early seventies a number of species, such as river lamprey, sea trout (*Salmo trutta trutta* L.), whitefish (*C. lavaretus* L.), bleak (*Alburnus alburnus* (L.)), barbel, ide (*Leuciscus idus* (L.)), burbot, bullhead (*C. gobio* L.), and flounder (*Platichthys flesus* (L.)) were extremely scarce in the rivers. Now these species are caught regularly at different sites, some in appreciable numbers (Table 4).

In Table 2 a few indigenous species are qualified as 'extinct', 'no recent records', or 'occasionally occurring' in the rivers Rhine and Meuse. The prospects of re-establishing these species in the Dutch parts of the rivers Rhine and Meuse were recently discussed by De Groot (1989a, b, c, d). A natural comeback of sturgeon seems very un-

Table 4. Numbers of some characteristic fish species caught by eight professional fishermen during 1987 and 1988 in different river reaches. Common commercial fish species were not recorded. Numbers have indicative value only, because of different fishing techniques and fishing effort. After Cazemier & Heermans (1988) and Heesen (1989).

Species	River reaches			
	Rijn, IJssel, Nederrijn	Ketelmeer	Maas, Amer	Hollands Diep
Petromyzon marinus	12	5	1	25
Lampetra fluviatilis	15	140	20	50
Salmo trutta trutta	300	650	50	120
Coregonus lavaretus	20	190	50	0
Osmerus eperlanus	0	> 23,000	> 3,900	70
Alburnus alburnus	2,600	> 7,500	> 10,000	> 1,700
Barbus barbus	100	0	15	0
Gobio gobio	4,000	10	500	0
Leuciscus leuciscus	0	1	250	0
Leuciscus idus	500	1,300	2,800	2
Leuciscus cephalus	7	0	700	4
Siluris glanis	15	10	4	0
Lota lota	600	20	1	0
Cottus gobio	6,000	0	500	20
Platichthys flesus	130	5,500	280	17,000

likely even if the water quality improves. There is no evidence that spawning and nursery grounds are available in the lower Rhine or Meuse. Furthermore, the nearest population of sturgeon is a very small one in the R. Garonne (France). A spontaneous revival of a natural population of shad and salmon also seems unlikely, for more or less the same reasons as those mentioned for sturgeon. Even improvement of the water quality and the opening up of the spawning areas for the salmon in the tributaries, the rivers would have to be re-stocked with eggs, parr or smolts. De Groot (1989a, 1990a) has discussed the problems associated with such a reintroduction. The return of the salmon in the Rhine is one of the objectives of the 'Rhine Action Programme'. Therefore, much of the fishery research focuses on this fish. In the Belgian state of Wallonia a programme called Saumon 2000 has also been started for the reintroduction of the salmon in the Meuse drainage basin (Philippart *et al.*, 1988b). The rearing of smolts of salmon and the closely related sea trout started some years ago in Nordrhein-Westfalen (Germany) and in Wallonia, and small numbers have already been introduced into tributaries. Three marked smolts of salmon and eight marked sea trout smolts, stocked in the R. Sieg, have already been captured in Dutch parts of the Rhine. Since the mid-seventies juvenile and adult sea trout have been observed in increasing numbers in the Dutch parts of the Rhine and Meuse, and since 1984 adults of this species have appeared every year in the lower reaches of some Rhine tributaries: the R. Dhünn (Borchard *et al.*, 1986), the R. Sieg (Steinberg, pers. comm.) and the R. Lahn (Brenner, pers. comm.; Klein, 1989). Adult sea trout is also regularly caught in the Belgian part of the Meuse (Philippart, 1985). These phenomena indicate that the water quality of the main rivers may no longer be an obstacle to the migration of juvenile and adult anadromous salmonids. However, Van Brummelen (1989) indicated that the current concentrations of some pollutants in the Rhine may still interfere with the homing behaviour of salmon. Nevertheless in the period 1957–1991 ten adult salmon have been reported from Dutch inland waters and one from the German part of the R. Rhine. Three of these were tagged in Scandinavia; the origin of the other ones is not known. The frequency of the salmon catches tends to increase: seven individuals were caught in 1990 and 1991 together.

Food chains

The trophic relations between organisms are generally decisive factors determining the structure of plant and animal communities. Regrettably, the food chains of the Dutch rivers have only marginally been investigated. The available information on plant/herbivore interaction (Gregory, 1983) and on predator/prey relations in streams (Allan, 1983) mostly pertains to streams smaller than the lower reaches of the Rhine and Meuse. These review papers prompted us to consider the basic structure of the food chains in the large rivers in The Netherlands. Allochthonous material (detritus) is introduced into the lower Rhine and Meuse by the transport of particulates in river water and hardly by direct terrestrial input. Autochthonous production of benthic macrophytes and microphytes, directly exploited by stream herbivores, is restricted to shallow waters near the banks and to floodplain waters with abundant macrophyte growth. As in the smaller streams referred to by Allan (1983), grazing on microphytes, i.e. diatoms or filamentous green algae, is probably essential for the local fauna. A well-developed microphytobenthos is present in the floodplain waters (van der Velde, pers. obs.), in the shallow parts of the Haringvliet (Bijkerk, pers. comm.; De Jong *et al.*, 1989) and on hard substrates on the river banks (Admiraal, pers. obs.). Chironomid communities in shallow parts of the Haringvliet probably feed mainly on benthic microalgae that are stimulated by the light penetrating deeply into the relatively clear water. Well-developed and productive microalgal mats of diatoms, or occasionally desmids, have been found on the shallows of the 'Ventjagersplaat' (Bijkerk, pers. comm.; De Jong *et al.*, 1989). Locally, dense populations of herbivorous nematodes have been found (Bongers & van de Haar, 1990). Despite the local importance of benthic algal production in the Rhine and Meuse, planktonic production at present plays a dominant role. A typical difference between the food chains of small streams (cf. Allan, 1983) and those of the Rhine/Meuse is the occurrence of a dense and productive plankton.

In the early seventies Wolff (1978) observed an incomplete exploitation of the phytoplankton and detritus input by the sparse zooplankton and filter-feeding invertebrates. In that period the deposit-feeding oligochaetes (*e.g. Tubifex tubifex* (Müller)) were the only benthic animals of quantitative importance. This food source was most probably exploited by several fish species (*e.g.* roach, *Rutilus rutilus* (L.)) and diving ducks (Wolff, 1978). Recent research has cast light on other aspects of the food chains. This is discussed below.

Carbon budget

The quantitative basis of the food chains in the main channel has been analysed via a tentative carbon budget of the lower Rhine (Fig. 18; Admiraal & van Zanten, 1988). The *in situ* productivity of phytoplankton and the mineralization rate in water and sediment are high compared to the import of detrital particles (POC). In other words: the turnover of organic material is high (Admiraal & van Zanten, 1988). A significant part of the import and production of organic carbon is deposited in the sedimentation areas (Van Eck, 1982; Admiraal & van Zanten, 1988), where it offers an ample food source to benthic fauna. However, the conditions in the deposits, *e.g.* the superficial oxygen penetration and the liberation of high concentrations of ammonium ions in the sediment (de Jong *et al.*, 1989) may be a selection factor for the nematode fauna (Bongers & van de Haar, 1990). The density of the macro-invertebrates does not reflect the very high input rate of organic material (cf. Fortuin, 1985). Admiraal & Botermans (1989) suggested that the coarse river sediments with their effective water exchange offer favourable conditions for nitrification, which is responsible for the oxidation of the large quantities of ammonium ions discharged by the Rhine.

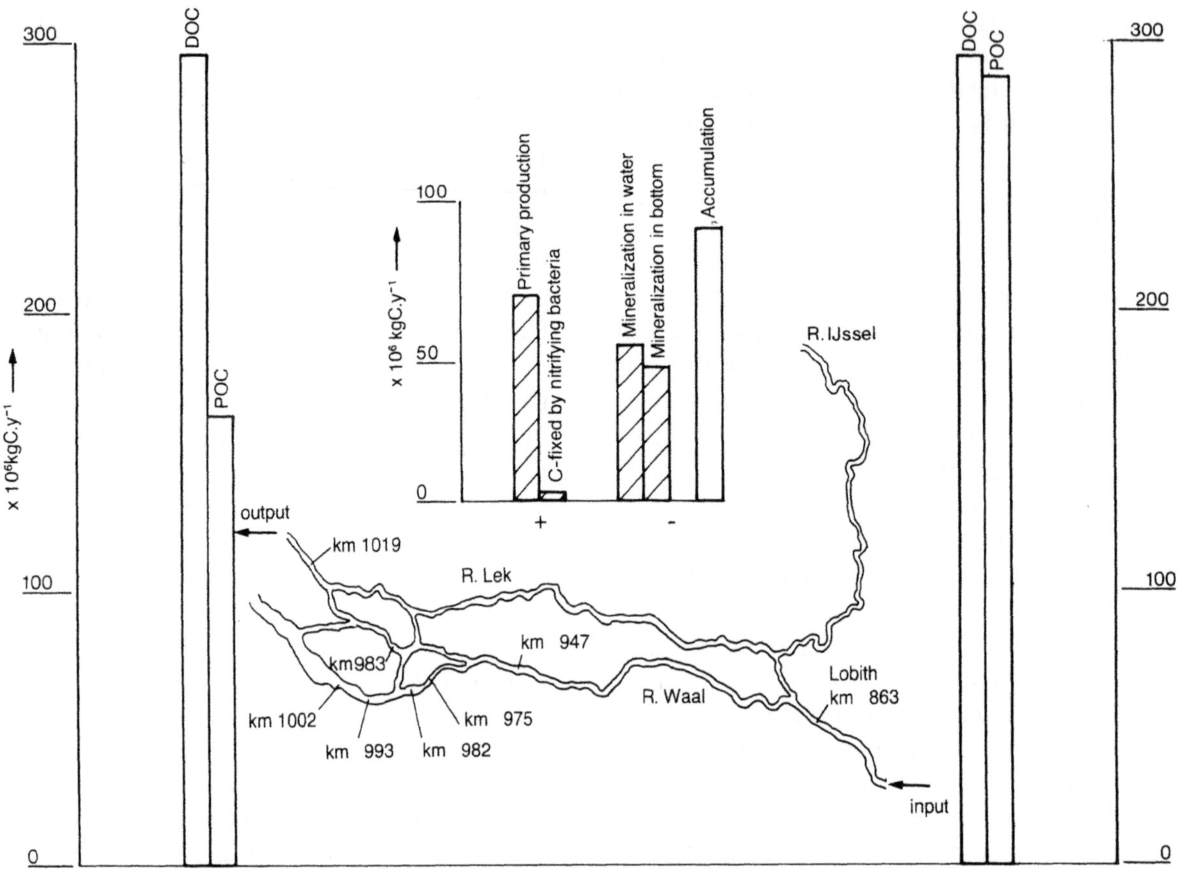

Fig. 18. Summary of input, output, production, mineralization and accumulation of organic carbon in the Dutch section of the lower R. Rhine. POC: particulate organic carbon. DOC: dissolved organic carbon. Source: Admiraal *et al.* (1990).

However, the quantity of organic carbon fixed by nitrifying bacteria is small.

Planktonic food chain

Since the Rhine and Meuse delta forms an important sedimentation basin, we expect that at least part of the phytoplankton losses are caused by settling algae. However, another part of the phytoplankton losses may be attributed to grazing. The depression in numbers of phytoplankton, arcellas, and ciliates in the R. Rhine during the summer is probably caused by the activities of rotifers, crustaceans, and possibly mollusc veliger larvae (De Ruyter van Steveninck *et al.*, 1990a). This is in agreement with the observation that the seasonal increase in zooplankton biomass follows the spring peak in phytoplankton biomass in the German R. Ruhr (a tributary of the Rhine; cf. Nusch, 1978). No quantitative analysis like that established for stagnant waters (cf. de Haan *et al.*, 1993) has been done for the food relations in the river plankton. Yet phytoplankton density is an important variable explaining the nitrogen, carbon, and phosphorus content of suspended matter in the Rhine (Van Eck, 1982; Admiraal *et al.*, 1991). Therefore it seems likely that the prevailing phytoplankton concentrations offer a rich food source for filter-feeding organisms. Smit (unpublished) made preliminary calculations of the production of phytoplankton in the Hollands Diep/Haringvliet and its consumption by the main herbivores, the zooplankton and the abundant zebra mussel (*Dreissena polymorpha* (Pallas)), and indicated that grazing by planktonic and benthic filter feeders matched the strong decreases in phytoplankton density in the river water during its transport in that area. The amphipod *Corophium curvispinum* is nowadays an important filter feeder in the R. Rhine and is likely a significant consumer of phytoplankton (Van den Brink *et al.*, 1991b).

Food of the macrofauna and fish

The food for the macro-invertebrates can be classified into different categories. Cummins (1973) distinguished four functional types among the benthic macrofauna: shredders (chewers and miners), collectors (filter or suspension feeders and sediment or deposit feeders), scrapers (feeding on mineral or organic surfaces), and predators (swallowers and piercers). The diagram in Fig. 19 (after Barnes and Mann, 1980), shows the relationships between these categories, adapted for the Dutch rivers.

Van der Velde *et al.* (1989) analyzed the contents of fish stomachs and found that in the R. Waal perch (*Perca fluviatilis*) had a preference for gammarids, whereas the ruffe (*Gymnocephalus cernuus* (L.)) had a preference for *Hydropsyche* larvae. Pikeperch (*Stizostedion lucioperca*) shifts its diet during its growth in the river (Bergers, pers. comm.). Specimens of perch caught in the backwaters of the Rhine had hardly any gammarids in their stomachs, but specimens caught in the main channel close to the groynes did. The immigrant amphipod species *C. curvispinum* is very numerous and must be considered as a new important food item for fish; it has already been found in the stomachs of ruffe, perch, pikeperch, bullhead and eel.

Although complicated, the food chains involving fish must be worked out, so that the fish populations of the river Rhine can successfully be reestablished within the framework of the Rhine Action Programme.

Birds

The backwaters of the Rhine and Meuse are of great ornithological value. Of the 76 bird species observed there, 36 have been found to breed near these waters. Furthermore, 33 species of waders have been observed (De Soet, 1976).

The rush vegetation and the bare mudflats in the tidal reaches of the Rhine/Meuse are attractive habitats for ducks and geese. During winter the banks remain ice-free longer than other waters, because of the tidal movement. Five to ten per cent of the total numbers of teal (*Anas crecca* L.) occurring in the river delta have been found on the muddy banks of the R. Oude Maas.

The sandy stream bed of the R. Oude Maas contains variable amounts of oligochaetes, chironomids and molluscs. Large numbers of tufted ducks (*Aythya fuligula* L.) have been recorded during the last five winters; many of these diving ducks probably forage on the zebra mussel, *Dreissena polymorpha* (M. van Wouwe, pers. comm.).

Large numbers of various species of ducks and geese overwinter in the Haringvliet (Boudewijn & Mes, 1986). Greyleg goose (*Anser anser* (L.)) and barnacle goose (*Branta leucopsis* (Bechstein)) are mainly found on the former intertidal areas that are now managed as pastures. Wigeon (*Anas penelope* L.) and coot (*Fulica atra* L.) also forage on fields and pastures during winter and use the former estuary to rest. Tufted ducks (*Aythya fuligula*) probably forage on zebra mussels, while pochard (*A. ferina* (L.)), which mainly occur in autumn, forage at night in the nearby saltwater lagoon of Lake Grevelingen (Boudewijn & Kuijpers, 1985). Mute swan (*Cygnus olor* (Gmelin)) forages in summer on filamentous algae and aquatic plants (mainly *Potamogeton pectinatus* and *Zannichellia palustris* L.). Cormorant (*Phalacrocorax carbo* (L.)) and great crested grebe (*Podiceps cristatus* (L.)) are important piscivorous bird species. In the river area cormorants appeared to feed on eleven fish species; important on all localities are eel, bream, roach, ruffe and perch, but cyprinids form the main food (Dirksen *et al.*, 1989a). Cormorants have in-

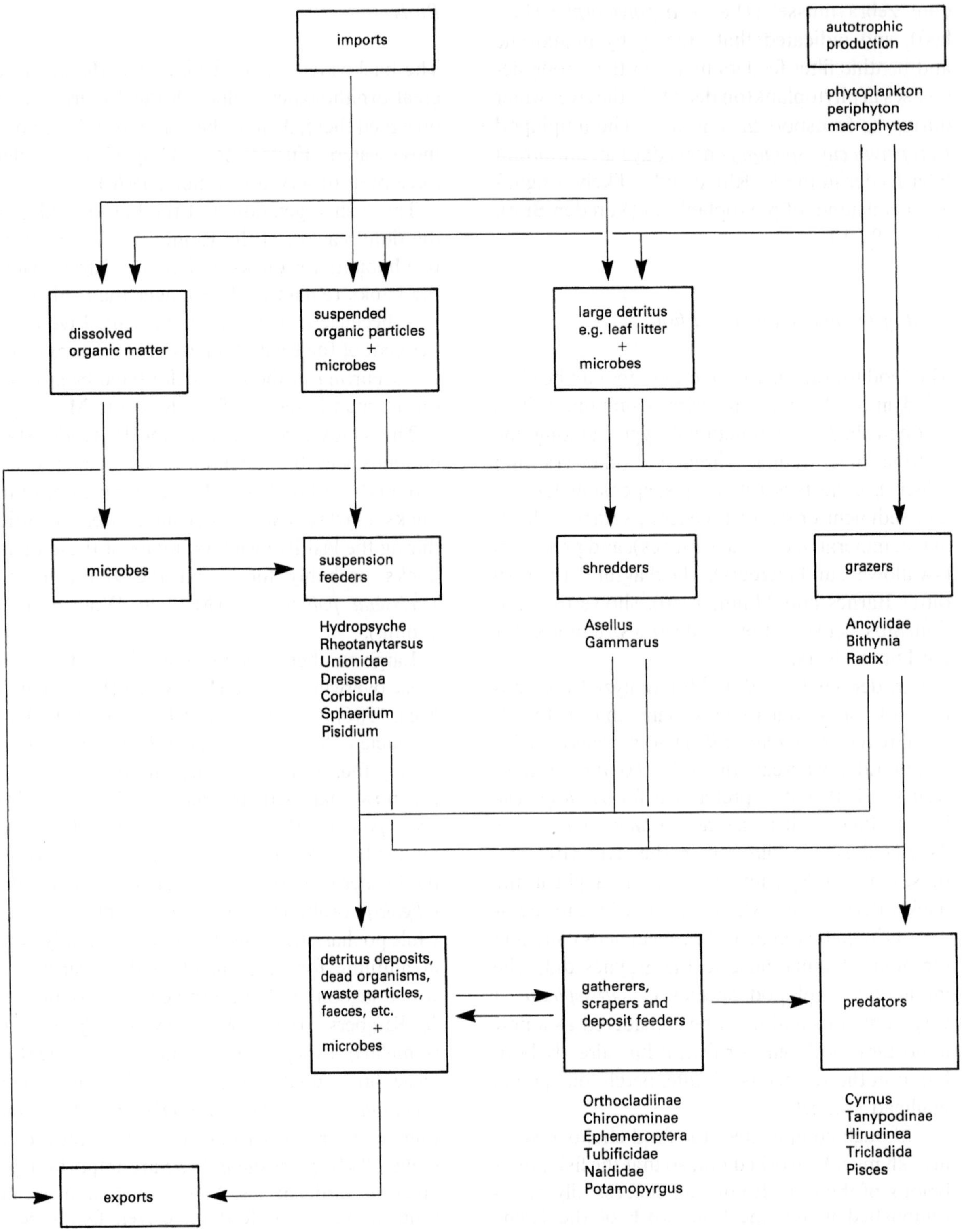

Fig. 19. Food chain of an aquatic ecosystem (after Barnes and Mann, 1980); dominant representatives for the Dutch rivers are indicated (after CUWVO, 1988).

creased in number parallel to their national population and the great crested grebe shows the same trends as in the whole delta (Boudewijn & Mes, 1986).

Recent investigations indicate that the contamination in the sedimentation areas is still interfering with the reproductive success of top carnivores. Tufted ducks (*A. fuligula*) held in captivity for three years were fed zebra mussels from the polluted Haringvliet and from the unpolluted Markermeer (in the IJsselmeer; Marquenie *et al.*, 1986). Compared with the reference group the ducks fed with mussels from the Haringvliet produced fewer and smaller eggs and sat on the nest briefly or not at all. Concentrations of polychlorinated biphenyls (PCB), pesticides and polycyclic aromatic hydrocarbons (PAH) were many times higher in the livers of the contaminated ducks. Dirksen *et al.* (1989b) compared the activities of cormorants (*P. carbo*) near the Rhine and Meuse (in the Biesbosch) and elsewhere. Only 0.55 juveniles per nest flew out in the colony near the Biesbosch, compared with 1.74 and 2.16 juveniles in the two reference colonies. The only large mammal that remained at the top of the food chain, the otter (*Lutra lutra* L.), has now disappeared completely, probably due to pollution (Janse, 1986).

Prospects for ecological recovery

Objectives of the rehabilitation plans for the Rhine and Meuse

The Rhine Action Programme includes ecological objectives (IKSR, 1987); earlier measures for restoring the river were directed mainly towards reducing chemical contamination (Rat von Sachverständigen, 1976; Santema, 1980). The rationale for including biological criteria in the international agreement of the Rhine states is probably the implicit assumption that a 'sensitive' and 'sound' river ecosystem safeguards against uncontrolled contamination of drinking water or an ongoing deposition of polluted sediment. This may be true, but undisturbed river communities no longer exist in the Rhine and Meuse. As we showed above, this is at least partly attributable to environmental conditions other than the emission of toxicants, *e.g.* morphological and hydrographical changes. The absence of pristine river communities has hampered the formulation of suitable biological criteria for rehabilitation. Slooff *et al.* (1985) emphasized the sensitive response of aquatic organisms to existing levels of pollution in the Rhine. Smit (1985) indicated the need to restore part of the morphological diversity in the river bed in addition to reducing toxicant input. In 1987 the salmon was adopted as a symbol for the restoration of the Rhine (IKSR, 1987). Research done by Smit & van Urk (1987) and Vanhemelrijk & van Broekhoven (1990) has indicated ways in which the river communities in the Rhine could develop. Following the approach of Ten Brink *et al.* (1991) they compared recent observations on the abundance of selected plant and animal species (or elements of the flora and fauna) with historical data, and presented a selection of these parameters in the form of an 'amoeba' (Fig. 20). They suggest that the environmental requirements for each of the selected organisms should be defined on the basis of autecological research and an analysis of the population dynamics. This would enable scenarios of ecological rehabilitation to be designed on the basis of a selection of biological parameters. However, there is a large discrepancy between the river communities in the past and at present (Fig. 20) and given the irreversible changes to the river, the rehabilitated communities may differ considerably from the historical ones. This means that the focus on species that characterized the communities in the past (cf. Ten Brink *et al.* (1991) may not be appropriate. Nevertheless, the species symbolic of the Rhine rehabilitation – the salmon – may have a fair chance of being re- established as a self-sustaining population.

Measures and management

It is generally assumed that the Rhine and Meuse will be ecologically rehabilitated if discharges of

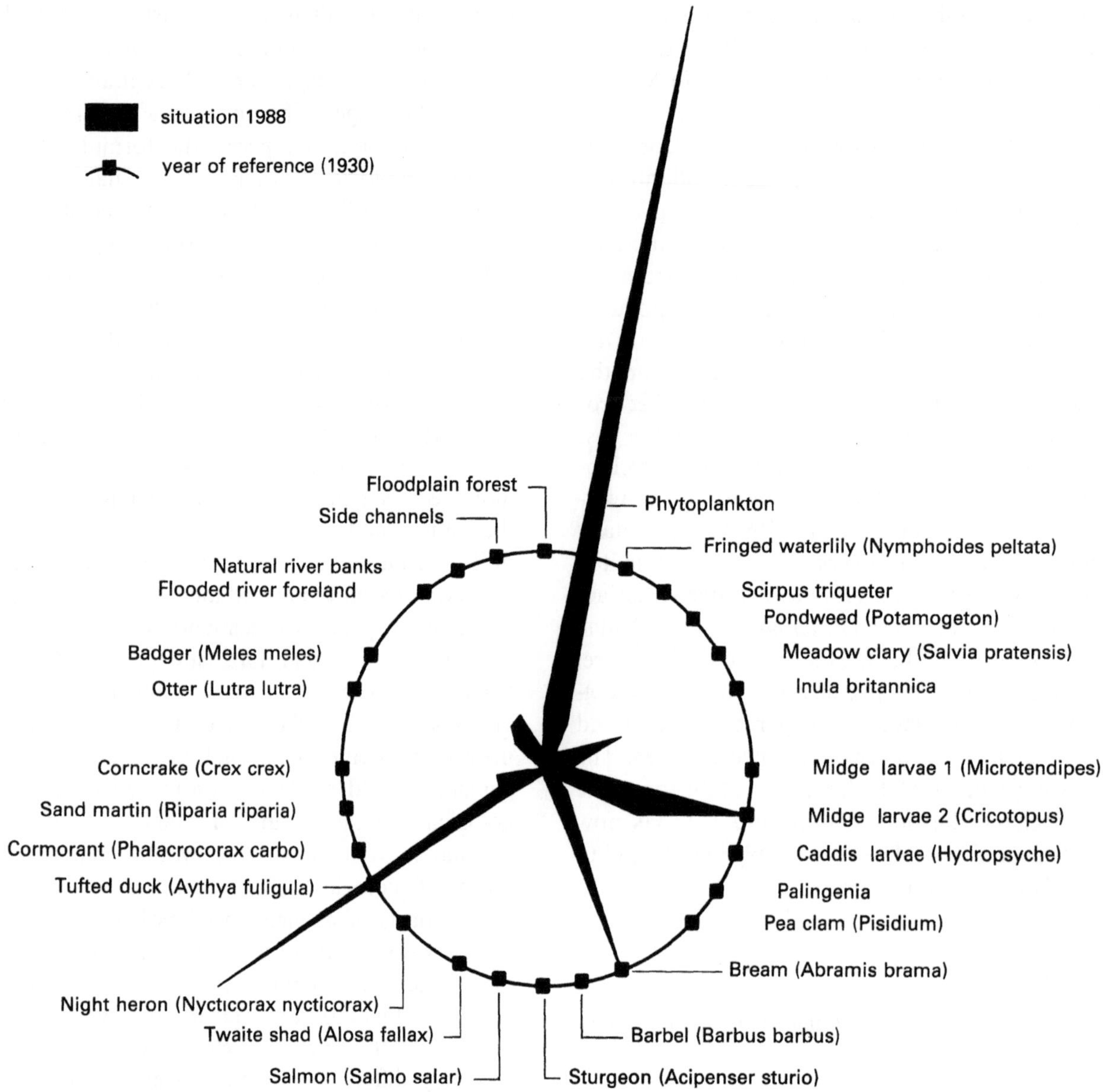

Fig. 20. Radial diagram, comparing the characteristics of the Rhine ecosystem in 1988 with that of ca. 1930. After Vanhemelrijk & Van Broekhoven (1989).

pollutants are reduced and natural habitats are restored (cf. de Wit *et al.*, 1989). So far, the reduction of emissions has been incorporated into the Rhine Action Programme: a 50% reduction in the emissions of about 30 priority substances by 1995 (as compared to 1985) has been arbitrarily chosen. Some of the input of contaminants, however, originates from diffuse sources, *e.g.* from agriculture. This means that effective control must touch on many aspects of environmental policy in the river catchment (cf. Smit & de Jong, 1989). For some of the emissions in the Rhine a 50% reduction can only be an intermediate goal, although any further reduction requires international agreement, or even measures on a river-basin level.

So far no effective remedial action has been undertaken to combat the chemical contamination and the eutrophication of the Meuse. Clearly the international and national coordination of such action is extremely complex: however, lessons can be learned from the organizational structures developed for the R. Rhine and from the international coordination by the International Rhine Commission.

Experiments on improving Rhine habitats locally are currently being discussed or are in progress; examples are presented below. An optimized bank protection policy in the tidal reaches of the river Rhine, now being developed, allocates areas for exploitable marsh vegetation, protects endangered species of flora and fauna and maintains an equilibrium between sedimentation and erosion. Experiments with the construction of breakwaters, placed at some distance from the vegetation, are now in progress; these breakwaters provide an effective shelter against intensive shipping. In the R. Waal the application of willow shoots for bank protection is presently under investigation. It has been suggested that the less artificial flow regime of the water in a morphologically unspoiled part of the Meuse (the Grensmaas) enhances the development of the flora and fauna (Klink, 1986b). The use of the Haringvliet sluices as a storm surge barrier that permits some tidal exchange between the Haringvliet and the North Sea could have far-reaching ecological consequences and is being examined. An optimization of the discharge regime of these and other sluices in favour of migrating fish is currently being tested. Furthermore, effective fish passes will be installed in 1991–1992 at the weirs in the rivers Lek and Meuse and in smaller tributaries.

The flood plains and embankments of the R. Nederrijn (at the 'Blauwe Kamer' and at 'Meinerswijk') have been 'reconstructed' in an attempt to encourage a more natural development in the flood plains. A similar project is being carried out on the R. IJssel ('Duursche Waarden'). As a result of a study on gravel and sand pits connected with the German part of the Rhine just upstream of the Netherlands, Berndt & Neumann (1985) developed recommendations to optimize these backwaters as nature reserves. These initiatives, although aiming at local improvement of the ecological conditions, may form the stepping stones for a more general ecological recovery of the rivers Rhine and Meuse, which will be possible if the policy for reducing discharges succeeds. If we are to follow the changes in water quality and communities, extensive biological monitoring programs are necessary. These are currently in progress.

Acknowledgements

We are indebted to Prof. Dr C. den Hartog, Dr E.D. de Ruyter van Steveninck, and to Dr E.P.H. Best for critical comments on the text, and to the Cooperating Drinking Water companies (RIWA) for the use of their data. J. Burrough-Boenisch corrected the English text, while most of the figures were prepared by J. van Westreenen at the Illustration Department of the Faculty of Natural Sciences of the University of Nijmegen.

References

Admiraal, W. & B. van Zanten, 1988. Impact of biological activity on detritus transported in the lower river Rhine. Freshwat. Biol. 20: 215– 225.

Admiraal, W. & Y.J.H. Botermans, 1989. Comparison of nitrification rates in three branches of the lower river Rhine. Biogeochemistry 8: 135–151.

Admiraal, W., B. van Zanten & E.D. de Ruyter van Steveninck, 1990a. Biological and chemical processes in communities of bacteria, phytoplankton and zooplankton in the lower river Rhine. Limnologie aktuell 1: 151–160. G. Fischer Verlag, Stuttgart.

Admiraal, W., P. Breugem, D.M.L.H.A. Jacobs & E.D. de Ruyter van Steveninck, 1990b. Fixation of dissolved silicate and sedimentation of biogenic silicate in the lower river Rhine during diatom blooms. Biogeochemistry 9: 175–185.

Admiraal, W., D.M.L.H.A. Jacobs, P. Breugem & E.D. de Ruyter van Steveninck, 1991. Effects of phytoplankton development on the elemental composition (C, N, P) of particulate material depositing in the lower River Rhine. Hydrobiologia, in press.

Alabaster, J.S. (ed.), 1985. Habitat modification and freshwater fisheries. Butterworth, London, 278 pp.

Allan, J.D., 1983. Predator-prey relationships in streams. In J.R. Barnes & G.W. Wayne (eds). Stream-ecology. Plenum Press, New York: 191–229.

Anonymous, 1977. De Oude Maas als groene rivier. Rapport Werkgroep Oude Maas. Part I: text (117 pp.). Part II: Appendix.

Backhaus, D. & A. Kemball, 1978. Gewässergüteverhältnisse

und Phytoplanktonentwicklung im Hochrhein, Oberrhein und Neckar. Arch. Hydrobiol. 82: 166–206.

Barnes, R.S.K. & K.W. Mann, 1980. Fundamentals of aquatic ecosystems. Blackwell Scientific Publications, Oxford, 229 pp.

Berg, R., S. Blank & T. Strubelt, 1989. Fische in Baden-Württemberg. Ministerium für Ländlichen Raum, Ernährung, Landwirtschaft und Forsten Baden-Württemberg. Stuttgart, 158 pp.

Berndt, J. & D. Neumann, 1985. Baggerseen mit Verbindung zu Fliessgewässern. Ergebnisse einer ökologischen Studie am Niederrhein. Natur und Landschaft 60: 3–8.

Bij de Vaate, A., 1991. Colonization of the German part of the River Rhine by the Asiatic clam, *Corbicula fluminea* Müller, 1774 (Pelecypoda, Corbiculidae). Bull. zool. Mus. Univ. Amsterdam 13: 13–16.

Bij de Vaate, A. & M. Greijdanus-Klaas, 1990. The asiatic clam, *Corbicula fluminea* (Müller, 1774) (Pelecypoda, Corbiculidae), a new immigrant in the Netherlands. Bull. zool. Mus. Univ. Amsterdam 12: 173–178.

Boddeke, R., 1974. Vissen en vissen. Elsevier, Amsterdam, 235 pp.

Bongers, T. & J. van de Haar, 1990. On the potential of basing an ecological typology of aquatic sediments on the nematode fauna: an example from the river Rhine. Hydrobiol. Bull. 24: 37–45.

Borchard, B., T. Brenner & L. Steinberg, 1986. Fische in Nordrhein-Westfalen. Ministerium für Umwelt, Raumordnung und Landwirtschaft des Landes Nordrhein-Westfalen. Düsseldorf, 127 pp.

Borcherding, J. & E.D. de Ruyter van Steveninck, 1992. Abundance and growth of *Dreissena polymorpha* larvae in the River Rhine during downstream transportation. Limnologie aktuell, G. Fischer Verlag, Stuttgart, in press.

Boudewijn, T. & J.W.M. Kuijpers, 1985. Fourageren de tafeleenden *Aythya ferina* van het Haringvliet in de Grevelingen? Limosa 58: 163–166.

Boudewijn, T. & R. Mes, 1986. De ontwikkeling van de vogelstand in het Hollands-Diep/Haringvlietgebied in de periode 1972–1984 en de invloed van peilbeheer op watervogels. Report Organization 'Ecoland', 152 pp.

Brock, Th.C.M., G. van der Velde & H.M. van de Steeg, 1987. The effects of extreme water level fluctuations on the wetland vegetation of a nymphaeid-dominated oxbow lake in The Netherlands. Arch. Hydrobiol. Beih. 27: 57–73.

Bürger, F., 1926. Die Fischereiverhältnisse im Rhein im Bereich der preussischen Rheinprovinz. Z. Fisch. 24: 217–399.

Cazemier, W.G., 1984. Zeldzame vissen, kreeften en krabben in de binnenwateren. Visserij 37: 117–132.

Cazemier, W.G., 1988. Fish and their environment in large European river ecosystems. The Dutch part of the river Rhine. Science de l'Eau 7: 95–116.

Cazemier, W.G., 1989. Salmoniden in Nederland. Biovisie Magazine 69: 32–35.

Cazemier, W.G. & W. Heermans, 1988. Monitoring van de visstand in het Nederlandse deel van de stroomgebieden van Rijn en Maas in 1987. Report no. BV 88-01 of the Netherlands Institute of Fishery Investigations, IJmuiden, the Netherlands, 31 pp.

Cazemier, W.G. & M.J. Heesen, 1989. First record of *Vimba vimba* (Linnaeus, 1758) (Pisces: Cyprinidae) in the Netherlands. Bull. zool. Mus. Univ. Amsterdam 12: 97–100.

Coops, H. & H. Smit, 1988. Biezen langs de Oude Maas. De Levende Natuur 89: 106–110.

Cummins, K.W., 1973. Trophic relations of aquatic insects. Ann. Rev. Entomol. 18: 183–206.

CUWVO, 1988. Ecologische normdoelstellingen voor Nederlandse oppervlaktewateren. Coordinatiecommissie Uitvoering Wet Verontreiniging Oppervlaktewateren. Report, 154 pp.

De Boer, M. & A.F. te Brinke, 1984. Enkele aspecten van de visstand in de Maas en de Waal. Report no. ZS 84-01 of the Netherlands Institute for Fishery Investigations, IJmuiden, The Netherlands, 21 pp.

Deelder, C.L. & D.E. van Drimmelen, 1959. The decline of the fish stocks in The Netherlands' sections of the rivers Rhine and Meuse. Athens Proceedings of the I.U.C.N. Technical Meeting 4: 185–190.

De Groot, S.J., 1985. Introductions of non-indigenous fish species for release and culture in The Netherlands. Aquaculture 46: 237–257.

De Groot, S.J., 1988. Literatuurstudie naar rekolonisatiemogelijkheden van het stroomgebied van de Rijn door riviertrekvissen en de echte riviervissen. Deel: houtingachtigen. Report no. 88-208 of the Netherlands Institute of Fishery Investigations, IJmuiden, The Netherlands, 23 pp.

De Groot, S.J., 1989a. Idem Deel: zalm. No. 88-205, 86 pp.

De Groot, S.J., 1989b. Idem Deel: forel. No. 88-206, 12 pp.

De Groot, S.J., 1989c. Idem Deel: elft. No. 89-203, 15 pp.

De Groot, S.J., 1989d. Idem Deel: fint. No.89-204, 14 pp.

De Groot, S.J., 1990a. Herstel van riviertrekvissen in de Rijn een realiteit? 1. De Atlantische zalm (*Salmo salar*). De Levende Natuur 91: 82–89.

De Groot, S.J., 1990 b. Herstel van riviertrekvissen in de Rijn een realiteit? 2. De forel (*Salmo trutta trutta*). De Levende Natuur 91: 89–92.

De Haan, H., L. van Liere, Sj. P. Klapwijk & E. van Donk, 1993. The structure and function of fen lakes in relation to water table management in The Netherlands. Hydrobiologia 265: 155–177.

De Jong, S.A., P.A.G. Hofman & A.J.J. Sandee, 1988. Gradiënten van fysisch-chemische en biologische parameters in de Rijn-bodem. Report of the Delta Institute of Hydrobiological Research, Yerseke, the Netherlands, 57 pp.

De Jong, D. (ed.), 1988. Hardinxveld en de riviervisserij. Historische vereniging Hardinxveld-Giessendam, Hardinxveld, 405 pp.

De Jonge, V.N., K. Essink & R. Boddeke, 1993. The Dutch Wadden Sea: a changed ecosystem. Hydrobiologia 265: 45–71.

De Kruijf, H.A.M., E. Jagtman & H. de Haan, 1988. Omgaan

met stofstromen in oppervlaktewateren. In A. Bijlsma (ed.), Integraal waterbeheer, een nieuwe aanpak. NIBI, Utrecht: 57–82.

Demon, J.M.H. & A.L.M. van Broekhoven, 1989. Bodemverontreiniging in de uiterwaarden. Report no. 89.027 of the Inst. Inl. Water Management and Wastew. Treatm., 40 pp.

Den Hartog, C. & G. van der Velde, 1987. Invasions by plant and animals into coastal, brackish and freshwater of the Netherlands. Proceedings Kon. Ned. Ak. Wet. C 90: 31–37.

Den Hartog, C., F.W.B. van den Brink & G. van der Velde, 1989. Brackish-water invaders in the river Rhine. A bioindication for increased salinity level over the years. Naturwissenschaften 76: 80–81.

Denzer, H.W., 1966. Beitrag über die Schädigung der Berufsfischerei am Rhein im Lande Nordrhein-Westfalen (1949–1962) hinsichtlich ihres Umfanges, ihrer Ursachen und ihrer Nachweisbarkeit. Der Fischwirt 16: 253–264.

De Ruyter van Steveninck, E.D., W. Admiraal & B. van Zanten, 1990a. Changes in plankton communities in regulated reaches of the lower river Rhine. Regulated Rivers: Research & Management 5: 67–75.

De Ruyter van Steveninck, E.D., B. van Zanten & W. Admiraal, 1990b. Phases in the development of riverine plankton: examples from the rivers Rhine and Meuse. Hydrobiol. Bull. 24: 47–55.

Descy, J.-P., 1987. Phytoplankton composition and dynamics in the river Meuse (Belgium). Arch. Hydrobiol. Suppl. 78: 225–245.

Descy, J.-P. & A. Empain, 1984. Meuse. In B.A. Whitton (ed.), Ecology of European rivers. Blackwell, Oxford: 1–23.

De Soet, F. (ed.), 1976. De waarden van de uiterwaarden. Report of the Rijksinstituut voor Natuurbeheer. Pudoc, Wageningen, 89 pp.

De Zwart, D. & A.J. Folkerts, 1990. Monitoring the toxicity of organic compounds dissolved in Rhine water. Hydrobiol. Bull. 24: 5–12.

De Wit, J.A.W., W. Admiraal, C. van der Guchte & W.G. Cazemier, 1989. Summary of results and conclusions from the first phase (1988) of the Netherlands research programme 'Ecological Rehabilitation Rhine'. Publication no. 10, 18 pp.

Dirksen, S., T.J. Boudewijn & L.K. Slager, 1989a. Voedselkeus van aalscholvers in zeven Nederlandse broedkolonies in 1987/1988. Bureau Ecoland, Utrecht, the Netherlands. Ecoland-rapport 89–9, 45 pp.

Dirksen, S., T.J. Boudewijn, L.K. Slager & R.G. Mes, 1989b. Broedsucces van aalscholvers in relatie tot vervuiling van het aquatische ecosysteem. Buro Ecoland, Utrecht, the Netherlands. ISBN 90–800259–2–5, 115 pp.

Fortuin, A.W., 1985. Dichtheden en biomassa van de belangrijkste bodemdieren van het Hollands Diep en Haringvliet in 1983. 63 pp. Report 1985–2 of the Delta Institute for Hydrobiological Research, Yerseke, the Netherlands, 63 pp.

Franz, H., 1990. Schwebstoffe im Rhein. Limnologie aktuell 1: 161–180. G. Fischer Verlag, Stuttgart.

Friedrich, G. & D. Müller, 1984. Rhine. In B.A. Whitton (ed.), Ecology of European Rivers. Blackwell, Oxford: 265–315.

Friedrich, G. & M. Viehweg, 1984. Recent developments of the phytoplankton and its activity in the Lower Rhine. Verh. int. Ver. Limnol. 22: 2029–2035.

Friedrich, G., 1990. Das Phytoplankton des Rheins als Indikator. Limnologie aktuell 1: 181–187. G. Fischer Verlag, Stuttgart.

Gregory, S.V., 1983. Plant-herbivore interactions in stream systems. In J.R. Barnes & G.W. Wayne (eds), Streamecology. Plenum Press, New York: 157–189.

Hadderingh, R.H., G.H.F.M. van Aerssen, L. Groeneveld, H.A. Jenner & J.W. van der Stoep, 1983. Fish impingement at power stations situated along the rivers Rhine and Meuse in the Netherlands. Hydrobiol. Bull. 17: 129–141.

Heermans, W., 1988. Overzicht van een aantal barrières en vispassages voor migrerende vis in Nederland, naar de toestand van 1988. Report no. BV 88–03 of the Netherlands Institute for Fishery Investigations, IJmuiden, the Netherlands, 13 pp.

Heesen, M.J., 1989. Monitoring van de visstand in het Nederlandse deel van de stroomgebieden van Rijn en Maas in 1988. Report no. BV 89–01 of the Netherlands Institute for Fishery Investigations, IJmuiden, the Netherlands, 38 pp.

Hessisches Ministerium für Landwirtschaft, Forsten und Naturschutz, 1987. Das Vorkommen der Fische in Fliessgewässern des Landes Hessen. Wiesbaden, 72 pp.

Hoek, P.P.C., 1916. Rapport over den Rijn als zalmrivier, aan de staatscommissie voor het zalmvraagstuk. Report of the Staats commissie voor het zalmvraagstuk. Part 2. Ministerie van Landbouw, Nijverheid en Handel, The Hague: 173–275.

IKSR, 1987. Aktionsprogramm 'Rhein'. Internationale Kommission zum Schütze des Rheins gegen Verunreinigung, Koblenz, F.R.G., 18 pp.

Janse, J.H., 1986. Ecologische waarden van de wateren in het winterbed van de grote rivieren. Report Rijksinstituut voor Natuurbeheer, 114 pp.

Klein, W., 1989. Die Lahn – ein 'reiz'volles Fliessgewässer. Fischökologie Aktuell 1: 17–23.

Klink, A.G., 1982. *Rheopelopia ornata* (Meigen): Description of the metamorphosis and ecology of a river inhabiting Tanypodinae-larva, new to the Dutch fauna (Diptera: Chironomidae). Ent. Ber. (Amst.) 42: 78–80.

Klink, A.G., 1983. *Rheopelopia rhenanus n. sp.*, a common midge of the lithorheophilic fauna in large lowland rivers (Diptera, Chironomidae). Ent. Ber. (Amst.) 43: 136–138.

Klink, A.G., 1986a. Palaeolimnologisch onderzoek naar de geschiedenis van kopafwijkingen bij muggelarven in de grote Nederlandse rivieren. Hydrobiol. Adviesbureau Klink no. 25, 15 pp.

Klink, A.G., 1986b. Literatuuronderzoek naar enige faktoren die invloed hebben op het biologisch herstel van de Grens-

maas. Hydrobiol. Adviesbur. Klink no. 24, Wageningen, 26 pp.
Klink, A.G., 1989. The Lower Rhine: palaeoecological analysis, 183–201. In G.E. Petts (ed.), Historical changes of large alluvial rivers: Western Europe. John Wiley & Sons.
Klink, A.G. & H.K.M. Moller-Pillot, 1982. Onderzoek aan de macroevertebraten in de grote Nederlandse rivieren. Report, 56 pp.
Kuijper, W.J., 1988. Over het vroegere voorkomen van de rivierparelmossel *Margaritifera auricularia* in Nederland. Basteria 52: 133–137.
Kuhn, G., 1976. Die Fischerei am Oberrhein. Heft 83, Hohenheimer Arbeiten, Verlag Eugen Ulmer, Stuttgart, 193 pp.
Lack, T.J.,1971. Quantitative studies on the phytoplankton of the Rivers Thames and Kennet at Reading. Freshwat. Biol. 1: 213–224.
Leemans, J.A.A.M., 1989. Oevertypologie voor de grote rivieren van Nederland. Rijkswaterstaat Rapport no. 4. Projectgroep Milieuvriendelijke Oevers; Rapport 89–3 van de Stichting voor Toegepaste Landschapsecologie, Nijmegen, 86 pp.
Leemans, J.A.A.M. & R. Reiling, 1986. Probleemverkenning naar de landschapsekologische gevolgen van oevererosie langs de Waal. Report, 104 pp.
Leentvaar, P., 1963. Stiefvader Rijn, kunstmoeder Maas. Natura 60: 50–56.
Leentvaar, P., 1966. Presentie en preferentie van plankters in wielen, oude rivierlopen en meren. Water 50: 16–18.
Lelek, A., 1989. The Rhine river and some of its tributaries under human impact in the last two centuries. In D.P. Dodge (ed.), Proc. of the International Large River Symposium. Can. Spec. Publ. Fish. Aquat. Sci. 106: 469–487.
Lelek, A. & C. Köhler, 1989. Zustandsanalyse der Fischartengemeinschaften im Rhein (1987–1988). Fischökologie 1: 47–64.
Lobregt, P. & J. van Os, 1977. De laatste riviervissers. Report Stichting Riviervisserij Nederland, Heerewaarden, 131 pp.
Marquenie, J.M., P. Roele & G. Hoornman, 1986. Onderzoek naar de effecten van contaminanten op duikeenden. Report no. 86/066 of the Dutch organization for applied research TNO, 35 pp.
Maenen, M.M.J., 1989. Water- en oeverplanten in het zomerbed van de grote rivieren in Nederland: voorkomen en relaties met fysisch-chemische parameters in 1988. Report Lab. Aq. Oec. Catholic University of Nijmegen and Publication no. 13. of the project 'Ecological Rehabilitation of the River Rhine', 119 pp.
Micha, J.-C., 1985. Obstacles physiques à la remontée du saumon atlantique dans le bassin mosan en Belgique. In Service de la Pêche de la région Wallone, Compte rendu du colloque 'Réintroduction du saumon atlantique dans la bassin de la Meuse'. Namur, 69–101.
Moller-Pillot, H. & B. Krebs, 1981. Concept van een overzicht van de oekologie van chironomidelarven in Nederland. Report, 41 pp.
Nienhuis, P.N., 1993. Nutrient cycling and foodwebs in Dutch estuaries. Hydrobiologia 265: 15–44.
Nusch, E.A.,1978. Development of planktonic algae in the Ruhr River dependent on nutrient supply, waterflow, irradiance and temperature. Verh. int. Ver. Limnol. 20: 1837–1843.
Peelen, R., 1975. Changes in the composition of the plankton of the rivers Rhine and Meuse in the Netherlands during the last fifty-five years. Verh. int. Ver. Limnol. 19: 1997–2009.
Peeters, E.T.H.M., 1988. Hydrobiologisch onderzoek in de Nederlandse Maas in relatie tot biotopen. Agricultural University of Wageningen, 150 pp.
Peeters, J.C.H. & W. Wolff, 1973. Macrobenthos and fishes of the rivers Meuse and Rhine, The Netherlands. Hydrobiol. Bull. 7: 121–126.
Philippart, J.-C., 1985. Histoire des salmonides migrateurs – saumon de l'Atlantique et truite de mer – dans le bassin du Meuse. In Service de la Pêche de la Région Wallone, Compte rendu du Colloque 'Réintroduction du saumon atlantique dans le bassin de la Meuse'. Namur: 5–47.
Philippart, J.-C., A. Gillet & J.-C. Micha, 1988 a. Fish and their environment in large european river ecosystems. The river Meuse. Science de l'Eau 7: 115–154.
Philippart, J.-C., G. Rimbaud, P. Poncin, E. Baras, C. Blase, J.-C. Micha, A. Gillet & A.P. Jardon, 1988 b. In: Projet AEE Saumon 2000; Rapport final. Convention avec la région Wallone pour l'étude de la faisabilité de la réintroduction du saumon atlantique dans le bassin de la Meuse. Université de Liège, Université de Namur, 114 pp.
Rat von Sachverständigen, 1976. Umweltprobleme des Rheins. Verlag Kohlhammer, Stuttgart, 258 pp.
Redeke, H.C., 1941. Pisces (Cyclostomi-Euichthyes). In H. Boschma (ed.), Fauna van Nederland X. Sijthoff, Leiden, 331 pp.
Redeke, H.C., 1948. Hydrobiologie van Nederland, de zoete wateren. J.C. de Boer, Amsterdam, 580 pp.
Rijkswaterstaat, 1987. De waterbodem van het Noordelijk Deltabekken. Report of the Ministry of Public Works, section Benedenrivieren. Dordrecht, 29 pp.
RIWA, 1988. De samenstelling van het Maaswater in 1985 en 1986. Rapport van de werkgroep waterkwaliteit. Samenwerkende Rijn- en Maaswaterleidingbedrijven, Amsterdam, 212 pp.
Salomons, W. & U. Förstner, 1984. Metals in the hydrocycle. Springer, Berlin, 349 pp.
Santema, P., 1980. The pollution of the river Rhine – the perspective for improvement. In M.J. Stiff (ed.), River pollution control. Horwood, Chichester, UK: 113–144.
Schleuter, A. & T. Tittizer, 1988. Die Makroinvertebratenbesiedlung des Mains in Abhängigheit von der Gewässertiefe und der Korngrösse des Substrates. Arch. Hydrobiol. 113: 133–151.
Slooff, W., 1983. Rijn, Lek, Waal, IJssel en uiterwaarden

onder invloed van ingrepen en verontreinigingen. In G.P. Hekstra & W. Joenje (eds), Rijnwater in Nederland. Oecologische Kring, Arnhem: 13–31.

Slooff, W., J.H. Canton & M.A. van der Gaag, 1985. Betekenis van tien jaar ecotoxicologisch onderzoek aan Rijnwater. H_2O 18: 119–121.

Smit, H., 1982. De Maas. Op weg naar biologische waterbeoordeling van grote rivieren. Report Agricultural University of Wageningen, 100 pp.

Smit, H., 1985. Het ecosysteem van de Nederlandse grote rivieren. De Levende Natuur 86: 162–167.

Smit, H. & H. Coops, 1988. Waterplanten in het Noordelijke Deltabekken. Report no. 88.026x of the Inst. Inl. Water Managem. Wastew. Treatm.

Smit, H. & H. Coops, 1991. Ecological, economic and social aspects of natural and man-made Bulrush (*Scirpus lacustris* L.) wetlands in the Netherlands. Landscape and Urban Planning 20: 33–40.

Smit, H. & J.J.P Gardeniers, 1986. Hydrobiologisch onderzoek in de Maas. Een aanzet tot biologische monitoring van grote rivieren. H_2O 19: 314–317.

Smit, H. & J. de Jong, 1989. Historical and present day management of the river Rhine. In H. Laikari (ed.), River Basin Management 5: 1–10. Advances in Water Pollution Control. Pergamon Press, Oxford.

Smit, H., P. Klaren & M. Snoek, 1991. *Lipiniella arenicola* Shilova (Diptera: Chironomidae) on a sandy flat in the Rhine-Meuse estuary; distribution, population structure, biomass, and production of larvae in relation to periodical drainage. Verh. int. Ver. Limnol. 24: 2918–2923.

Smit, H. & M. Snoek, 1989. Macrofauna in de oeverzone van het noordelijk deltabekken. Biomassa en soortsamenstelling in relatie tot hoogteligging, bodemsamenstelling en verontreiniging. Report no. 89.012 of the Inst. Inl. Water Managem. Wastew. Treatm.

Smit, H. & G. van Urk, 1987. Het herstel van de ecologische waarden van de Rijn: over de zalm en ecologische doelstellingen. H_2O 20: 427–430.

Smits, H., 1989. Locatie en beschrijving van de wateren in het winterbed van de grote rivieren. Report Lab. Aq. Ecol., Catholic University of Nijmegen, The Netherlands, no. 107, 49 pp.

Steinmetz, B., 1975. Resultaten van het visserijkundig onderzoek in het Haringvliet en Hollands Diep in 1974 en 1975. Visserij 28: 474–486.

Ten Brink, B.J.E., S.H. Hosper & F. Colijn, 1991. A quantitative method for description and assessment of ecosystems: the AMOEBE-approach. Mar. Poll. Bull. 23: 265–270.

TNO, 1986. Water in the Netherlands. Netherlands Organisation for Applied Scientific Research TNO, The Hague, 70 pp.

Tubbing, D.M.J. & W. Admiraal, 1991. Sensitivity of bacterioplankton in the river Rhine to various toxicants measured by thymidine incorporation and activity of exoenzymes. Environm. Toxicol. Chem. 10: 1161–1172.

Tubbing, D.M.J., E.D. de Ruyter van Steveninck & W. Admiraal, 1992. Sensitivity of planktonic photosynthesis in the River Rhine to various toxicants. Environm. Toxicol. Water Qual., 7, in press.

Van Broekhoven, A.L.M., 1987. De Rijn in Nederland – Toestand en ontwikkelingen anno 1987. Report no. 87.061 of the Inst. Inl. Water Managem. Wastew. Treatm., 56 pp.

Van Brummelen, T.C., 1989. Chemicals affecting the spawning migration of anadromous fish by causing avoidance response or orientation disability, with special reference to concentrations in the River Rhine. Report of the Univ. of Utrecht, the Netherlands, 77 pp.

Van den Brink, F.W.B. & G. van der Velde, 1986a. Observations on the population dynamics and distribution of the White Prawn *Palaemon longirostris* H. Milne Edwards, 1837 (Crustacea, Decapoda, Natantia) in The Netherlands, with special reference to its occurrence in the major rivers. Arch. Hydrobiol. 107: 465–495.

Van den Brink, F.W.B. & G. van der Velde, 1986b. Observations on the seasonal and yearly occurrence and the distribution of *Atyaephyra desmaresti* (Millet, 1831) (Crustacea, Decapoda, Natantia) in the Netherlands. Hydrobiol. Bull. 19: 193–198.

Van den Brink, F.W.B., G. van der Velde & J.F.M. Geelen, 1988. Life history parameters and temperature-related activity of an American crayfish, *Orconectes limosus* (Rafinesque, 1817) (Crustacea, Decapoda), in the area of the major rivers in The Netherlands. Arch. Hydrobiol. 114: 275–289.

Van den Brink, F.W.B., G. van der Velde & A. bij de Vaate, 1989. A note on the immigration of *Corophium curvispinum* Sars, 1895 (Crustacea: Amphipoda) into the Netherlands via the river Rhine. Bull. zool. Mus. Univ. Amsterdam 11: 211–213.

Van den Brink, F.W.B. & G. van der Velde, 1989. Oecologische waarde en natuurontwikkeling van stagnante wateren in het winterbed van Rijn en Maas. In E.C.L. Marteijn (ed.), Ecologisch Herstel Rijn – Beleid en onderzoek. Symposiumverslag 26 mei 1989. Publications and reports of the project 'Ecological Rehabilitation of the River Rhine' 9: 97–98.

Van den Brink, F.W.B., G. van der Velde & W.G. Cazemier, 1990. The faunistic composition of the freshwater section of the river Rhine in the Netherlands: present state and changes since 1900. Limnologie aktuell 1: 199–216.

Van den Brink, F.W.B., M.M.J. Maenen, G. van der Velde & A. bij de Vaate, 1991a. The (semi-) aquatic vegetation of still waters within the floodplains of the rivers Rhine and Meuse in the Netherlands: historical changes and the role of inundation. Verh. int. Ver. Limnol. 24: 2693–2699.

Van den Brink, F.W.B., G. van der Velde & A. bij de Vaate, 1991b. Amphipod invasion on the Rhine. Nature 352: 576.

Van den Brink, F.W.B., G. van der Velde & A. bij de Vaate, 1991c. Slijkgarnaal bedreigt Rijnfauna. Bionieuws 1, 7.

Van der Sman, A.J.M., O.F.R. van Tongeren & C.P.W.M. Blom, 1988. Growth and reproduction of *Rumex maritimus* and *Chenopodium rubrum* under different waterlogging regimes. Acta Bot. Neerl. 37: 439–450.

Van der Valk, F., H. Pieters & R.C.C. Wegman, 1989. Bioaccumulation in yellow eel (*Anguilla anguilla*) and perch (*Perca fluviatilis*) from the Dutch branches of the Rhine-mercury, organochlorine compounds and polycyclic aromatic hydrocarbons. Publications and reports of the project 'Ecological Rehabilitation of the River Rhine' 7, 53 pp.

Van der Velde, G., F.W.B. van den Brink, M. van der Gaag & P.J.M. Bergers, 1990. Changes in numbers of mobile macro-invertebrates and fish in the River Waal in 1987, studied by sampling the cooling-water intakes of a power plant: first results of a Rhine biomonitoring project. Limnologie aktuell 1: 325–342.

Van der Voo, E.E. & V. Westhoff, 1961. An autecological study of some limnophytes and helophytes in the area of the large rivers. Wentia 5: 163–258.

Van Dessel, B., 1989. Ecologisch herstel van de Rijnmakrofauna. Publications and reports of the project 'Ecological rehabilitation of the River Rhine' 14: 1–109. Rijksinstituut voor Natuurbeheer, 96 pp.

Van de Steeg, H.M., 1984. Effects of summer inundation on flora and vegetation of river foreland in the Rhine area. Acta Bot. Neerl. 33: 365–366.

Van Donselaar J., 1961. On the vegetation of former river beds in the Netherlands. Wentia 5: 1–85.

Van Drimmelen, D.E., 1981. De visserij in het stroomgebied van de Rijn. Report of the Organisation for the Improvement of Inland Fisheries, Nieuwegein, the Netherlands, 23 pp.

Van Drimmelen, D.E., 1987. Schets van de Nederlandse rivier- en binnenvisserij tot het midden van de 20ste eeuw. Report of the Organisation for the Improvement of Inland Fisheries, Nieuwegein, the Netherlands, 128 pp.

Van Eck, G.T.M., 1982. Geochemie van zwevend materiaal en water in het Hollands Diep/Haringvliet. Ph.D. Thesis, Univ. of Utrecht, the Netherlands, 169 pp.

Vanhemelrijk, J.A.M. & A.L.M. van Broekhoven, 1990. Ecologische ontwikkelingsrichting grote rivieren – Aanzet tot quantitatieve uitwerking van ecologische doelstellingen voor de grote rivieren in Nederland. Bijdrage ten behoeve van de Derde Nota Waterhuishouding. Report no. 90.003 of the Inst. Inl. Water Magem. Wastew. Treatm., 98 pp.

Van Lonkhuyzen, J.P. & H. Vonk, 1920. Rapport over de zalmvisscherij op de benedenrivieren en Zuid-Hollandsche stroomen in Nederland. Report Nederlandsche Heidemaatschappij, Arnhem, 15 pp.

Van Ruremonde, R., 1988. Veranderingen van de visfauna in het Nederlandse rivierengebied: een historisch overzicht. Report Lab. Aq. Oec. Catholic University of Nijmegen, 65 pp.

Van Urk, G., 1984a. Algen en de zuurstofhuishouding van de Rijn. H_2O 17: 96–100.

Van Urk, G., 1984b. Lower Rhine-Meuse. In B.A. Whitton (ed.), Ecology of European Rivers. Blackwell, Oxford: 437–468.

Van Urk, G. & F.C.M. Kerkum, 1986. Misvormingen bij muggelarven uit Nederlandse oppervlaktewateren. H_2O 19: 624–627.

Van Urk, G. & F.C.M. Kerkum, 1987. Chironomid mortality after the Sandoz accident and deformities in *Chironomus* larvae due to sediment pollution in the Rhine. Aqua 4: 191–196.

Van Urk, G., F.C.M. Kerkum & S.M. Wiersma, 1990. Chironomiden in zand- en slibbodems in de IJssel. Report no. 90.068x of the Inst. Inl. Water Managem. Wastew. Treatm., 18 pp.

Van Urk, G. & A. bij de Vaate, 1990. Ecological studies in the lower Rhine in The Netherlands. Limnologie aktuell 1: 131–145. G. Fischer Verlag, Stuttgart.

Van Urk, G., F.C.M. Kerkum & A. bij de Vaate, 1990. Caddis flies of the Lower Rhine. Proc. Symposium on Trichoptera, ms. 5 pp.

Van Urk, G. & H. Smit, 1989. The lower Rhine geomorphological changes. In G.E. Petts (ed.), Historical change of large alluvial rivers: western Europe. John Wiley: 167–182.

Van Vuuren, W.E., 1989. Cadmiumcalamiteiten op de Maas: een triest record in november 1988. H_2O 22: 23–25, 35.

Verhey, C.J., 1961. De vissen en de visvangst. In C.J. Verhey (ed.), De Biesbosch land van het levende water. Thieme, Zutphen: 139–164.

Wibaut-Isebree Moens, N.L., 1965. Onderzoek van grindgaten langs de Maas. Report RIVON, 44 pp.

Wiegerinck, J.A.M. & M.J. Heesen, 1987. Visserijkundige waarnemingen in het Haringvliet en Hollands Diep in de jaren 1976 t/m 1986. Report no. 31 of the Ministry of Agriculture and Fisheries, 42 pp.

Willemsen, J., 1983. Ketelmeer, IJsselmeer en Markermeer: waterkwaliteit, voedselketens en visstand. In G.P. Hekstra & W. Joenje (eds), Rijnwater in Nederland, Oecologische Kring, Arnhem: 33–51.

Wilson, R.S. & S.E. Wilson, 1985a. A reconnaissance of the River Rhine using Chironomidae pupal exuviae (Insecta: Diptera). Mem. Amer. Ent. Soc. 34: 361–385.

Wilson, R.S. & S.E. Wilson, 1985b. A survey of the distribution of Chironomidae (Diptera, Insecta) of the River Rhine by sampling exuviae. Hydrobiol. Bull. 18: 119–132.

Wolff, W.J., 1978. The degradation of ecosystems in the Rhine. In M.W. Holdgate & M.J. Woodman (eds), The breakdown and restoration of ecosystems. Plenum Press, New York: 169–187.

WOM, 1977. De Oude Maas als groene rivier. Report Working group 'Oude Maas', 117 pp.

Zonneveld, I.S., 1960. De Brabantse Biesbosch, a study of soil and vegetation of a fresh water tidal delta. Ph.D. Thesis, Agricultural University of Wageningen. Versl. Landbouwk. Onderz. 65.20, 3 vols., Pudoc, Wageningen.

Hydrobiologia **265**: 129–153, 1993.
E. P. H. Best & J. P. Bakker (eds), Netherlands-Wetlands.

An ecosystems approach to base-rich freshwater wetlands, with special reference to fenlands

Geert van Wirdum
DLO Institute for Forestry and Nature Research, P.O. Box 23, NL-6700 AA Wageningen, The Netherlands

Key words: fen vegetation, succession, water cycle, base state, nutrient state, conservation

Abstract

A survey of base-rich wetlands in The Netherlands is presented. The main area of their occurrence is the low-lying Holocene part of the country, until some thousand years ago a large and coherent wetland landscape: the Holland wetland. The development of various parts of the Holland wetland into marshes, fens and bogs can be understood from hydrological relations in mire basins, as recognized in the distinction of primary, secondary and tertiary mire basin stages. Presently, the remnants of the Holland wetland are separate base-rich wetlands. The succession of their vegetation reflects various abiotic conditions and human influences. Three main developmental periods are distinguished as regards these factors. The first, geological period of mire development is seen as a post-glacial relaxation, with the inertia due to the considerable mass of wetland as a stabilizing factor. Biological "grazing" influences, as an aspect of utilization by humans, converted base-rich wetlands to whole new types in the second, historical period. Presently, mass and harvesting have decreased in importance, and actual successions in terrestrializing turbaries seem to reflect rapidly changing environmental conditions. Human control could well become the most important factor in the future development of wetland nature. The present value of open fen vegetation strongly depends on the continuation of the historical harvesting. The development of wooded fen may help to increase the mass of wetland in the future. Best results in terms of biodiversity are expected when their base state is maintained through water management.

The vegetation and hydrology of floating fens in terrestrializing turbaries is treated in some more detail. Various lines and phases in the succession are distinguished. Open fen vegetation at base-rich, yet nutrient-poor sites is very rich in species threatened elsewhere. The fast acidification of certain such fens is attributed to hydrological and management factors. This acidification is illustrated in the profile of a floating raft sample. At the scale of these small fens, the elemental structure comprising base-rich fen, transitional fen and bog vegetation, is not as stable as it was in the large Holland wetland. A critical role seems to be played by the supply of bases with the water influx. The changing base state is supposed to change the nutrient cycling to such an extent that it would be correct to call this trophic excitation of the ecosystem, rather than just eutrophication. Eutrophication indicates a quantitative reaction to an increased nutrient supply, the internal system being unaltered. The drainage of fens, resulting in an increased productivity of the vegetation, provides another example of excitation, to the effect that the functional system is dramatically changed internally.

1. Introduction

Base-rich freshwater wetlands cover a considerable area in The Netherlands (Best *et al.*, 1993; Wolff, 1993). They represent internationally important natural values (Rooth, 1989), relying on bases supplied with groundwater or surface water, or retained in the local substrate. The basic ions, such as calcium and magnesium, are often associated with (hydro)-carbonate and with non-acidic conditions. This chapter covers the main types of base-rich wetland in the lower part of The Netherlands, exclusive of estuarine and riverine marshes, with their original relations. The main points to be treated in more detail are the requisite base supply in connection to the water cycle, and the control of the nutrient state. The focus is fen mires in terrestrializing turbaries and their vegetation, the main field of application nature protection, and the approach systems ecology. I have tried to emphasize some lines of thought not always apparent from published case studies, to which I refer for primary data. More research is evidently needed to see whether all the viewpoints given here will stand up to close scrutiny. Lakes are treated in detail by De Haan *et al.* (1993), and marshes in recent embankments by Joenje & Verhoeven (1993). As various aspects of Dutch fenlands have been dealt with in quite some detail by Den Held *et al.* (1992), Koerselman & Verhoeven (1992), Van Wirdum *et al.* (1992), Vermeer & Joosten (1992) and Wiegers (1992), there is no separate chapter on them in this volume. Some special attention is paid to them in the present chapter, however.

Base-rich freshwater wetlands: between the rivers and estuaries and the bogs

Base-rich freshwater wetlands during the origin of the Dutch landscape occurred in-between riverine ecosystems at one side, and base-poor wetlands at the other. Marshes are currently more clearly associated with riverine systems in The Netherlands. They are representative of the primary stage of terrestrialization of open water basins (Fig. 1). Only when the hydrodynamic environment has sufficiently calmed down, organic muds and peats, as in lakes and fens, can accumulate. As the inflow of base-rich water decreases, the wetland ecosystems ultimately become relying on water from the atmosphere, in the long run leaching any retained bases. In the absence of human influences they usually become poor in nutrients as well (Cruickshank, 1972), and many characteristic species which abound in such environments even seem to be unable to take full profit of supplementary nutrients under the prevailing acid conditions (Verhoeven *et al.*, 1988). This marks the transition to the bog stage of wetlands (Casparie, 1993; Schouwenaars, 1993). Mires, exclusive of estuarine and riverine marshes and of bogs, thus form the hub of the cases studied here: fenland with lakes, and marshes in recent embankments. They predominantly occur in the low-lying Holocene part of The Netherlands, once a large wetland composed of various types of mire (Fig. 2). The name "Holland" seems to refer to this ancient large mire landscape. The root "hol" (also "hel") indicates swampy or miry entities and is found in quite some toponyms in The Netherlands (Huisman, 1953). Many present wetlands can be considered transformed fragments of that old Holland mire.

Large base-rich wetlands are typical of receiving, rather than shedding land systems

Dune slacks (Van Dijk & Grootjans, 1993) and stream valleys (Higler & Verdonschot, 1993) also comprise important base-rich wetlands, but these are part of now largely base-poor land systems: the dunes and the higher, Pleistocene part of the country, respectively. These land systems are involved in natural leaching processes, possibly accelerated by "acid rain". In view of water and element cycling they represent shedding landscapes at the national scale. Several still base-rich wetlands which obtain their bases from sources within these landscapes indeed exhibit strong acidification trends (Van Dam & Buskens, 1993). Others for the time being escape acidification since they receive base-rich groundwater (Groot-

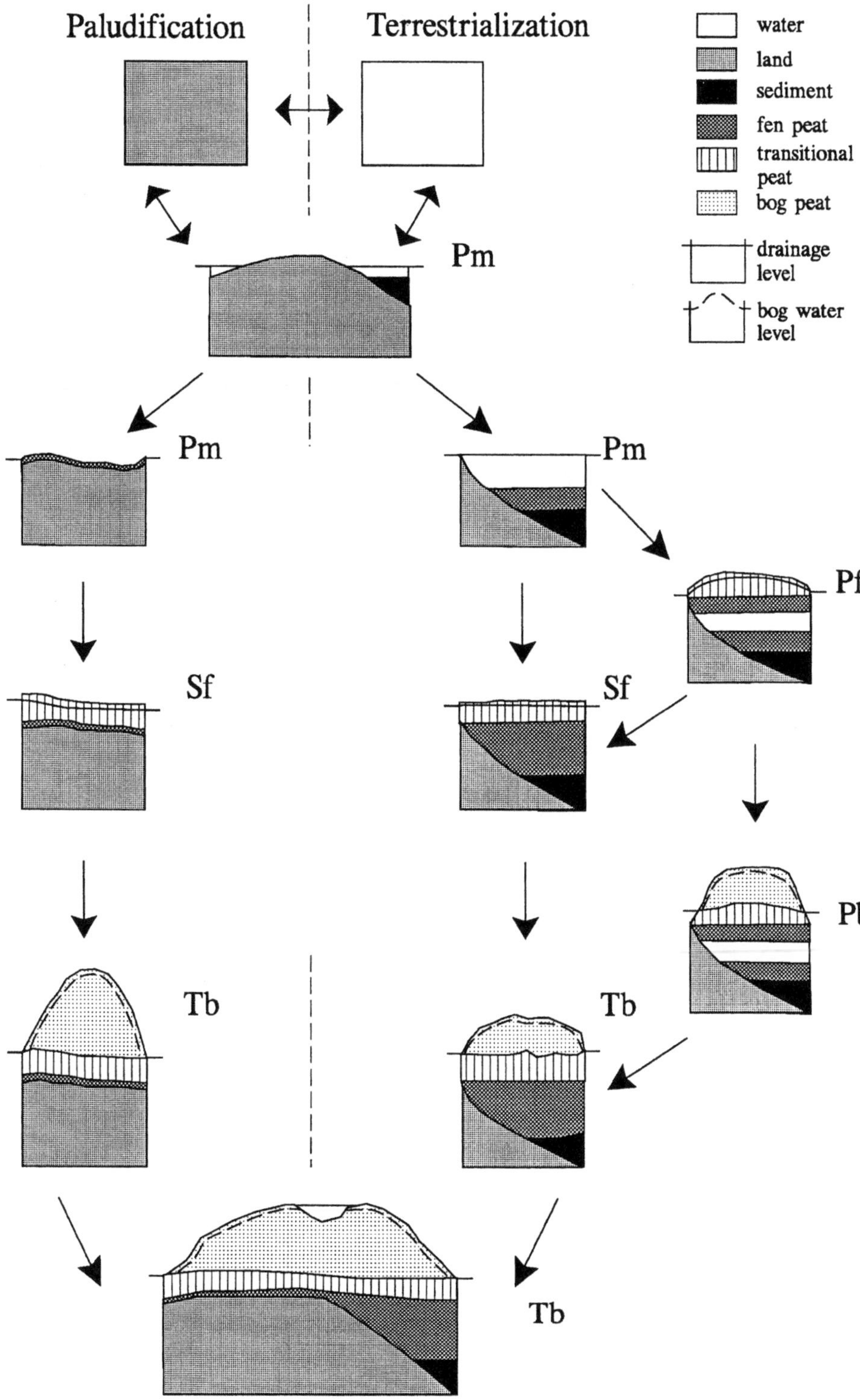

Fig. 1. Scheme of paludification and terrestrialization. P, S, T, m, f, b: primary, secondary, and tertiary basin stage, marsh, fen, and bog, respectively. After Van Wirdum (1979a).

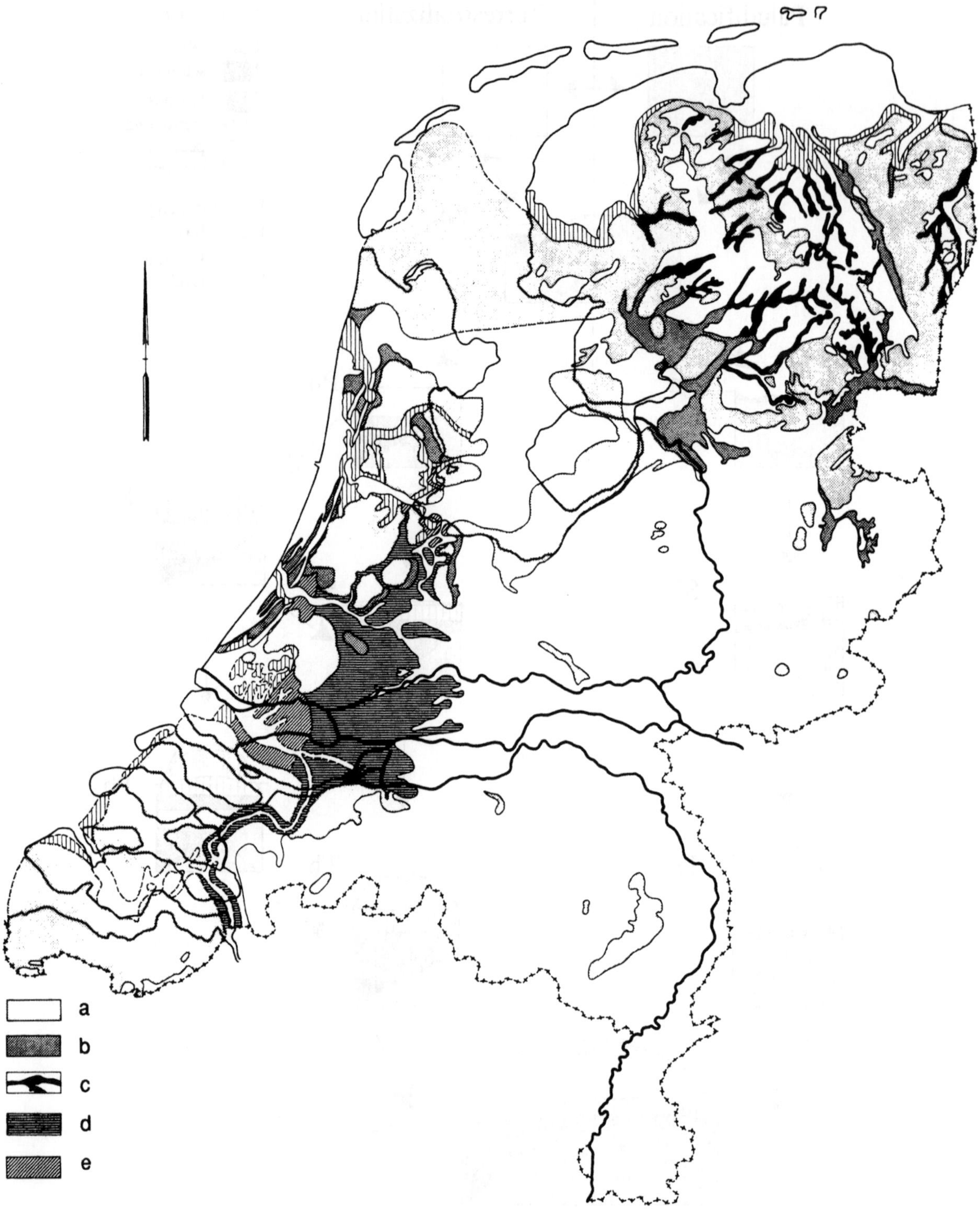
a
b
c
d
e
f
g
0 10 20 30 50 km

jans, 1985; Everts & de Vries, 1991). The base-rich wetlands considered here rather represent receiving land systems. Together with water as a carrier, receiving land systems usually receive both bases and nutrients, the latter especially in densely populated catchment areas.

Base-rich is not synonymous to nutrient-rich

"Base-rich" should not be confused with "eutrophic". Richness in bases usually provides an environment in which nutrients are physiologically well available to plants, so promoting productivity (*e.g.*, Etherington, 1975). However, in permanently water-logged sites wetness and low-redox conditions depress decomposition, and thus control concentrations of free nutrients to levels too low to support a wealthy growth of eutraphentous vegetation (Armstrong, 1975). Furthermore, base-rich sites can be meso- or oligotrophic due to the absence or exhaustion of a source of nutrients, as in case of a long continued vegetation harvesting (Koerselman & Verhoeven, 1992). The exception rather than the rule among the Dutch base-rich wetlands, such meso-oligotrophic sites are exceptionally rich in rare plant species (Van Wirdum, 1991). They have therefore attracted much attention from the point of view of nature conservation.

Present-day wetland protection: another human factor in ecology

The position and dynamics of base-rich wetlands have changed a lot in the history of the Dutch landscape. This is substantiated by treating the relevant periods as templates for wetland development: the geological, the historical, and the present templates. The basic idea is the association in each template of characteristic values of changeability (*vs.* stability) in time, and disjunction (*vs.* connectivity) in space. The association is probably determined by few master factors which changed as the templates replaced each other. Reviewing the development of fenlands in the present template, and the efforts to improve this template, it is to be expected that future wetland development will inevitably reflect human concepts and understanding of ecology as well as the side effects of land-use and the influences of environmental pollution. In the end, political factors will possibly even re-define ecology. In the applied field of nature protection, it is important to recognise the decisive role of human plan, without depriving nature of all it needs to be an artist.

2. A survey of base-rich wetlands in The Netherlands

Mire basin stages: hydrological relations in the Holland wetland

The ecology of the development from marsh to bog, and its relation to shedding and receiving land systems can be described by the application of the primary, secondary, and tertiary types of mire basins according to Moore & Bellamy (1973). A tentative characterization of these types is presented in Table 1. The types essentially form a sere, the mire sere (paludisere, Fig. 1).

The classification into primary, secondary, and tertiary mire basins is based on water storage characteristics. It does not equally cover other aspects of ecology, such as those usually addressed with the concepts marsh, fen, transitional fen, and bog, based on vegetation and local hydrology. The basin stages are used as a main entry here since they reflect regional rather than local hydrological relations, and since they facili-

Fig. 2. Approximate distribution of peat types ca. 1st century AD, indicating the character of the mire landscape in the Mid-Subatlantic period. Botanical composition of peats, nutrient state and base state of mires: a: Sphagnum and Ericaceae, oligotrophic, base-poor; b: Phragmites, Carex and Amblystegiaceae, mesotrophic, base-rich; c: Carr species, meso-eutrophic, base-rich, formed in brook valleys; d: Phragmites and forest species, formed in eutrophic, base-rich riverine mires; e: Various types with silt and clay, formed in eutrophic, base-rich freshwater-tidal mires; f: Phragmites, Scirpus, formed in eutrophic, base-rich brackish mires; g: Various types of peat locally present. After Stiboka (1965).

Table 1. Characteristics of primary, secondary and tertiary mire basins (tentative).

Basin stage	Primary	Secondary	Tertiary
Ecosystems	Marshes, lakes	Fens	Bogs
Source of water	Large rivers	Small rivers, groundwater	Precipitation
Water storage [1,2]	+ + +	+	–
Dominant cycle	Sediments	Solutes	Water
Sedimentation [1]	+ + +	(+)	–
Peat accumulation [1]	(+)	+ +	+ + +
Trees [1]	+ +	+	–
Herbs [1]	+ +	+ +	+
Mosses [1]	(+)	+ +	+ + +
Animals [1]	+ + +	+ +	+
Plant genera and families	*Phragmites, Scirpus, Typha*	*Carex, Amblystegiaceae*	*Sphagnum, Ericaceae*
Main (former) human use	Hunting, fishing, water regulation	Crop production	Fuel production
Main external threats	Change of sediment transport, "normalization" of rivers, soil pollution	Change of water supply, water pollution	Climate change, atmospheric pollution

[1] Importance decreases from + + + to – ; (+) locally only.
[2] The capacity temporarily to store water at high floods or rainfall, rather than the volume of water comprised.

tate an understanding of the spatial arrangement of mires (Fig. 3). Typical relationships between basin storage and locally defined ecosystems are given in Fig. 1 and Table 1. Moore & Bellamy used the term mire for actively accumulating peatlands, as do Stanek & Worley (1983) and Gore (1983). This term is extended here, so as to include sedimentation environments and degrading peatlands, which are all genetically related. Marshes in sedimentation environments are quite characteristic of primary mire basins. Fen mires typically abound in basins at their secondary, rather than primary stage. The floating fens discussed in this chapter form an exception due to

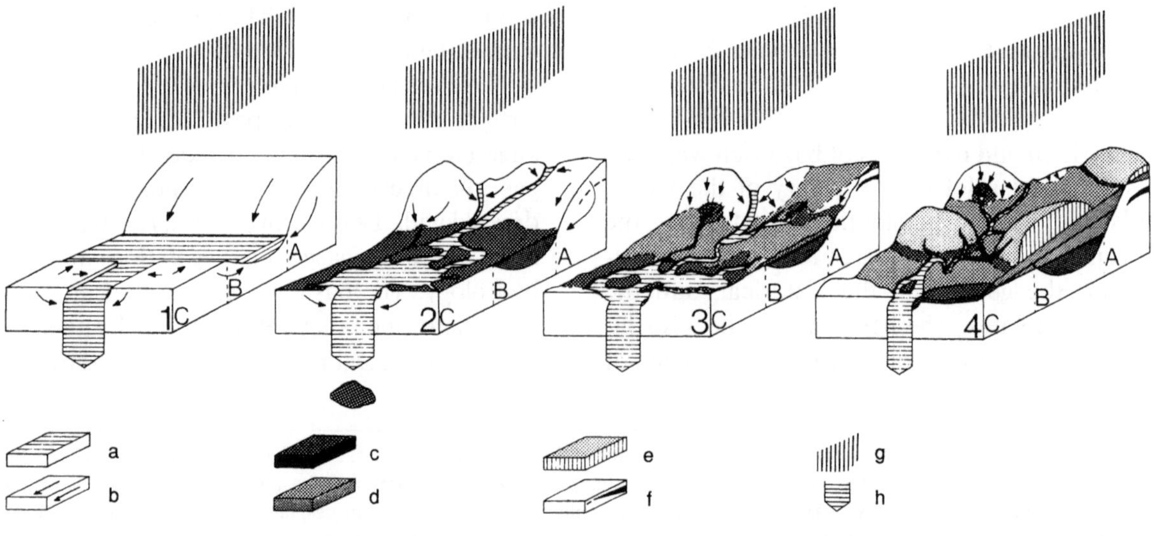

Fig. 3. Terrestrialization and paludification in the land system. A: shedding, B, C: receiving land segments; 1–4: phases in wetland development; a–h: open water, land with flow direction of water, primary mire with sediment, secondary mire with fen peat, tertiary mire with bog peat, impermeable layer, precipitation, and water outflow, respectively. From: Van Wirdum (1979a).

the removal of peat by man (see below), followed by a renewed progressive terrestrialization. Apparently the biocoenoses in parts of a basin may lead the development to a next stage. Strictly speaking they form local basins nested within the larger one. Similarly, while the tertiary stage is indicated by bog ecosystems, bog pools and degrading bog surfaces represent environments more akin to primary basin conditions.

The suggested interaction between adjoining wetland types in different stages of development is reflected by the composition of the large postglacial Holland wetland (Figs 2, 3). The "building block" of such a landscape, with shedding and receiving elements, is shown in Fig. 4. This building-block structure can be recognized at various scale levels, down to the hummock-hollow pattern so characteristic of many fens and bogs. The building blocks make the landscape coherent, but at the same time sensitive to a wide spread of initially local disturbances (Van Wirdum 1979b, 1991).

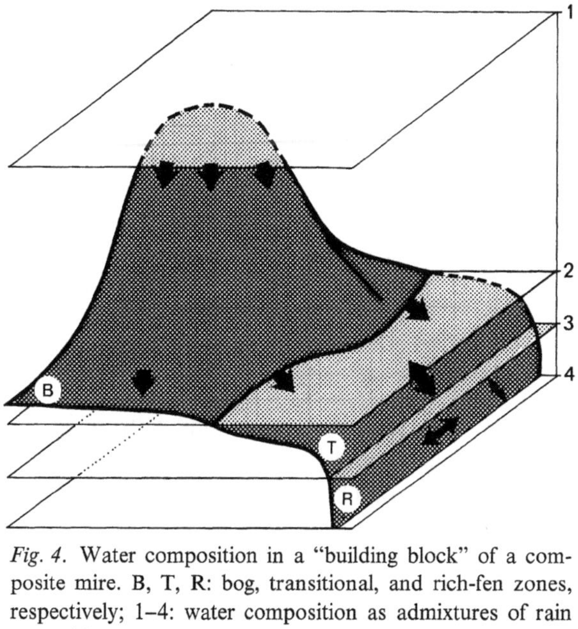

Fig. 4. Water composition in a "building block" of a composite mire. B, T, R: bog, transitional, and rich-fen zones, respectively; 1–4: water composition as admixtures of rain water (1) and groundwater (4), 2: 80% rain water, 3: 60% rain water. Typically, the building block is stable when the drainage level is concurrent with the water composition. From: Van Wirdum (1979b).

The primary basin stage: the filling of a geomorphological basin

A wetland gradually filling in a water basin is in the primary basin stage. Typically, sediment transportation with flowing water is an important process in primary mires. The water is usually supplied by rivers. In case of constant water inputs to the relevant landscape section, the ongoing sedimentation urges an increased outflow. This may cause paludification of a downstream land segment (Fig. 3). Typically, ecosystems in primary mire basins have a diverse community structure. They are not as rich in plant species as those in secondary mire basins, but more animal groups and species are represented, and their total mass is substantial.

Marshes and lakes are typical of primary mire basins. Considerable areas of salt marsh are still present in The Netherlands (Bakker *et al.*, 1993). Due to the intensive land-use and the improvement of rivers for navigation, only small areas of freshwater marsh have been left along the banks of the large rivers and in their cut-off meanders. The public awareness of the natural values of such marshes has led to a policy of rendering present farmland on the floodplains of the rivers back to nature, and of increasing the influence of the river water in such areas. Back in 1421, a hazardous flooding turned farmland polders southwest of Rotterdam into a freshwater tidal marsh area, "De Biesbosch" (Zonneveld, 1959). Since 1970, dams for flood protection in the so-called Delta area have strongly diminished tidal effects in De Biesbosch (De Booijs, 1982). The area now develops into a floodplain marsh. However, along the neighbouring Oude Maas river a narrow, but even older band of marsh with well-developed freshwater tidal biocoenoses still stands the expansion of the Rotterdam urban-industrial environment (Werkgroep Oude Maas, 1977). Joenje & Verhoeven (1993) treat the special case of marshes on mud flats originating from recent embankments in former estuaries and in the lake IJsselmeer area.

Most lakes (De Haan *et al.*, 1993) do not show any clear tendency of mire succession. Slowly filling in with organic mud they represent a somewhat deviating type of primary basins. Most terrestrializing turbaries (this chapter) in The Netherlands geomorphologically belong to the primary type as well, even though a floating raft may now entirely cover the original body of open water.

Just like some floodplain marshes, many of the remaining terrestrializing turbaries are maintained as they are part of primary basins with a key role in the water management of The Netherlands. They are used to temporarily store surplus and supply water. Meanwhile, the attractive scenery of broads and half-open fenland is very important for water recreation. Presently, its importance for nature protection is mainly based on the communities of floating fen. They can survive here in the absence of a rising water level since they float. Thus, their maintenance is less demanding with respect to the surrounding lands than in case of a secondary mire basin proper.

The secondary basin stage: raising the level of the original basin

When the water has become caught in the mazes of the sediments and peats, and peat accumulation continues above the original drainage level, the secondary basin stage is initiated. In this stage part of the water input from surrounding landscape sections is incorporated in the new peat. The mire's drainage level is raised and the mire area will extend to a new equilibrium value. The source of water inputs is rivers and groundwater. Water mostly seeps through the peat, not allowing of any substantial transportation of sediments. Solute transport appears to be a dominant factor. The main secondary wetland type is fens. Fens are characterized by a permanently water-logged peat soil not supporting a rich animal life. Scrub and carr are common features in the natural vegetation of fens in secondary basins. Open fen vegetation is often due to the reclamation of such wooded fen by man. However, in the Holland wetland large areas of open fen were formerly maintained by marine or slightly brackish influences. In larger wetland areas, a mosaic of primary and secondary basins is often present. Fenlands in stream valleys thus constitute extremely diverse landscapes. The fen vegetation in terrestrializing turbaries is formed in floating fens in primary wetland basins (see above), but apart from the abundance of reeds it is more akin to the vegetation usually found in secondary mire basins. Later in this chapter due attention will be paid to the base supply to these peculiar fens. The undisturbed parts of the ancient fenland polders, many dune slacks, and fens in stream valleys basically belong to the secondary type. Many of these are now degrading due to drainage.

Fenlands in secondary basins have mostly been used by humans for agriculture and for the cutting of fire wood. In the historical period, dairy farming became the most important way of utilization. Under this land-use, however, the peat is gradually oxidized. By definition, this implies a consumptive and possibly even terminal type of land-use.

The tertiary basin stage: bog peat accumulating above the regional drainage level

When the capacity to raise the drainage level, or the driving force to maintain a water inflow are exhausted, the basin exclusively relies on rain water inputs. This is the tertiary stage characterized by bogs. Differences between types of bog especially have to do with climatically determined differences in the water cycle. Although the transition towards this stage is of great relevance to the present discussion of base-rich wetlands, the stage itself is treated in the case studies on bogs (Casparie, 1993; Schouwenaars, 1993). Bog peats have a very high carbon content; they constitute considerable carbon stores (Gorham 1990). Their utilization, especially for fuel, has caused the disappearance of most bogs in The Netherlands. In the lower part of The Netherlands they have been substituted by polders, lakes and fenlands. Land-use almost inverted the water level relations in the building blocks of the former Holland mire: river valleys with silted-up floodplains now have their water levels well above those in fenlands and polders.

Mass, utilization and human control as factors in the ecology of base-rich wetlands

It is a pity that no fens and bogs have remained reasonably undisturbed during the last thousand years in The Netherlands. However, a large area of wetland converted through land-use during many centuries acquired a peculiar and rich nature, some of which has been preserved. Today, entirely new fenlands are growing-in abandoned turbaries, which replaced former wetlands. After all, there is not just one main line of succession from primary to tertiary mire basins, or from marsh to bog. Many derived, regressive, or resumed states line up in a succession of human influences. Indeed most went through three major developmental periods providing "templates" for their shaping: the geological, historical, and present templates of mire development (Table 2; Van Wirdum *et al.*, 1992). The application time of each template strongly decreases from geological to present, and so does the inertia the wetland devices owe to it. The historical and present templates moreover undid substantial earlier influences.

Autogenic succession: the relaxation of eco-devices

Palaeo-ecological studies have revealed trends which are often referred to as an "autogenic" succession from open water or marsh via fen to bog (see, *e.g.*, Overbeck, 1975). Such autogenic succession was thought to be repeated after re-creation of an early stage. According to this view, the early stages are instable: relaxation would lead to the "natural equilibrium" or climax stage of bog, concurrent with the tertiary basin stage. Various opinions are held as regards the relative role of vegetation, hydrology, and climate in this process (for a critical review and case studies, see Grosse-Brauckmann, 1974, 1976; Tallis, 1983). Van Wirdum *et al.* (1992) conclude that a whole "eco-device" (Van Leeuwen, 1982; Van Wirdum, 1986), inclusive of water, substratum, and fauna and man, rather than just the vegetation determines the course of relaxation at the given geographic location. Most ecologists presently seem to agree that undisturbed relaxation is the exception rather than the rule in nature, and that relatively small differences in abiotic conditions or human influences may have a strong influence upon the would-be climax and upon the successional process. Conditions often change long before an equilibrium is established. Actual successions in such instances consist of a track of successive segments representing different lines in the idealized scheme.

Present successions are different from the geological record

The various lines of succession observed in recent Dutch wetlands (Segal, 1966; Den Held *et al.*, 1992; Van Wirdum *et al.*, 1992) are different from the geological record: the present wetlands seem to be much more changeable than the ancient ones. Apparently wetlands differ as regards their inertia with respect to a changing environment. The mass and extent of sediments and peat incorporated in a wetland increase the momentum of ongoing successional processes. This would explain climatically determined trends in the geological record (Casparie, 1993). Most present wetlands are relatively small and functionally isolated from each other. Their succession has little momentum and main trends in it may change at

Table 2. The importance of mass, harvesting and human control in the geological, historical and present templates of mire development.

Factor	Template: Geological	Historical	Present
Mass (nature)	Dominant	Decreasing	To increase
Harvesting (Man as grazer)	Increasing	Dominant	Not to decrease
Human control (plan)	Absent	Increasing	Decisive

the mercy of (mainly human) impacts in the environment. Whole new trends have been generated typical of the present human impacts. They are held characteristic of nature-on-decline. Such trends were described by Succow (1988) for mires in eastern Germany.

The massive Holland wetland from 5000 BP to 1000 BP: the geological period

The low-lying part of The Netherlands (Fig. 2), with extensions along the valleys of rivers and rivulets, once represented the large delta of the rivers Rhine, Meuse, and Scheldt. The delta came into existence when the post-glacial rise of the sea level about 5000 BP (Atlantic, Neolithic period) had slowed down far enough to allow of the formation of beach bars with a lagoonal area behind (Pons, 1992). Subsequently, the coastline shifted according to phases of transgression and regression of the sea, but it was stable enough to promote the formation of a large mire landscape, some 7 m below the present sea level. A general pattern of base-rich and base-poor mires developed, composed of building blocks as shown in Fig. 4. Due to the precipitation surplus characteristic of the climate, the most remote interfluvial parts developed into bogs, whereas swamps and marshes were found along the river branches. In between, base-rich fens were abundant.

Even in the prehistoric period it is difficult to imagine the base-rich freshwater wetlands as wetlands without human interference (see Ellenberg, 1978). The delta was colonized by men by the Neolithic period, especially along the streams and the higher Pleistocene areas (Bloemers *et al.*, 1981). Although the geological and climatological conditions will have restricted human utilization and its influence on the vegetation, such human influence may have been substantial in large areas, except for the bogs. In Roman times the local inhabitants no doubt were a major ecological factor, even on many of the bogs (Borger, 1992; Casparie, 1972). The combined influence of early land-use and transgressions of the sea resulted in an increased area of base-rich mires at the cost of bogs. The various parts of the deltaic area, however, were still strongly coherent, and geological factors can in general be traced at many places at the same time: they ruled the hierarchy. The mass of the coherent Holland wetland provided a basis for the local development of derived and regressive types, and for a consequent resumption of the relaxation process.

Utilization of separate wetlands from 1000 BP to 100 BP: the historical period

Roughly since the eleventh century, an increased growth of the Dutch population, the building of embankments and substantial hydrotechnical works, agriculture and grazing on the mires' surface, and the use of peat as a resource for fuel and, near the coast, salt, turned the hierarchy bottom up. Humans now largely determined, not just influenced, the development of nature. They created an artificial separation between distinct land units.

In particular, excavation for fuel and horticultural peats caused an almost complete set-back of the development and a re-shaping of an initial state for mire development according to the prevailing social and topographic conditions. Many large lakes (broads) were thus formed, but also extensive areas consisting of a multitude of smaller, rectangular dredging lakes (*petgaten*). Such *petgaten* are the fundamental terrestrializing entities in the fen mires studied later in this chapter. Where the peat was not cut or dredged away, it provided a surface for agriculture. This was facilitated by drainage. As a consequence, the land surface was lowered by the combined influence of subsidence and oxidation of peat (Borger, 1992). These wetlands are no longer actively accumulating peatlands.

The historical template should not just be regarded destructive to the then existing wetlands. They underwent a development which led to a large extension of base-rich and not eutrophicated ecosystems. The growing human population drew biomass, for food, fuel, and construction, from them. At several places, this resulted in a long-term harvesting regime, which gave rise to the development of specific ecosystems with a low nutrient state (Bakker, 1979). Land-use determined by biological functions typical of humans thus emerged as the dominant factor of the historical template (see below). The scale and composition of the landscape and the velocity of change are typical of the human biological "program". In terms of mass, the human biomass outgrew the geological mass in importance.

The historical template gave shape to the wetlands involved until ca 1850–1940. From that time onwards the freshly started terrestrialization in *petgaten*, and the succession of the vegetation in wet grasslands and lakes was deflected due to influences typical of the present template. Depending on the state reached when this deflection occurred, the historical template is still reflected in the present wildlife communities, however. This is most significant in base-rich but nutrient-poor fens with an annual mowing regime.

Polders and ditches: wetlands converted in the historical period

The ancient fenland polders constitute a most important example of wetlands converted in the historical period. These base-rich wetlands typically consisted of poorly drained pastures and hay fields with a dense network of ditches. The grassland vegetation used to be very rich in species until ca 1950 (De Vries, 1929; Van Leeuwen, 1954). The polders used to attract large populations of certain bird species thus specialising as "meadow birds" (Beintema, 1986). Note that, in spite of considerable subsidence of the peatland surface, this very type represents a transformation of the old mire surface without any sudden elimination of the peat mass and of the inertia inherited from the geological template. An increased use of fertilisers and improved drainage have now transformed most of the grassland into species-poorer types hardly justifying the wetlands' label. Only some meadow birds could stand this change, and may even have been favoured by it, as long as human activities were not physically disturbing (Beintema & Müskens, 1987). The conservational value of the grasslands now heavily rests upon the populations of these birds (Wolff, 1993).

The ditches in between the grassland parcels definitely represent wetland. The total length of ditches in The Netherlands is estimated at ca 3–4 10^5 km (Higler, 1979). As to the vegetation of ditches and ditch borders, a clear similarity exists to the types found in larger fenlands. A comprehensive survey was made by De Lange (1972). Melman (1991) has substantiated evidence for good management of ditch-banks as a contribution to the survival of fenland nature in The Netherlands. As an anecdote, it can be added that polder ditches are so common in the Holocene part of The Netherlands, that their wildlife in many present researchers raised the first ecological curiosity. In this respect the rich macrofauna contributed considerably to the popularity of the ditch habitat. The public awareness of these values has now diminished as a result of strong urbanization.

Human control of wetlands in the present period

The present template is especially characterized by a variety of polluting agents and rapidly changing land use. The technological capacity of humans has taken over the dominant role of geological and biological factors. In general this has resulted in a substantial loss of species diversity in wetlands. No time was given, until now, for evolution, or settlement of fit species. The changeability of the present template is too high for most of nature to cope with. Moreover, the human technological "program" had a world-wide influence, as shown by the almost synchronous degradation of geologically and biologically determined ecosystem functions. This observation in turn led to specific measures aimed at the protection and conservation of nature by purposeful management, which in itself marks the impact of man changing his mind. Most natural values of fenlands presently rely on management continuing the harvesting process as practised in the historical period. There have been new developments of nature too. As a by-effect of hydro-technical works newly embanked areas went through initial phases of great interest to wildlife. Some of the resulting marshes were preserved as nature reserves (Joenje & Verhoeven, 1993), and further new wetlands are planned (Nature Policy Plan, 1989).

The most obvious factor of the present template is the increase of human control. It has become clear that this control extends to the global scale, and should include targets with respect to nature and environment. Now and in the near future, the momentum of natural processes will be derived from the mass of human artifacts, inclusive of laws, regulations and thoughts, *i.e.*, the human will. Even if a fiasco results, the main factor will have been (failure of) human control!

Functions and management of wetlands in the future

With some exaggeration the above treatment of the post-glacial development of wetlands in the lower part of The Netherlands can be summarized as follows:

The new climate, the geology of the continental fringe, and the gradual re-entering of flora and fauna determined the first period of relaxation, the formation of the Holland wetland. Although various important thresholds were passed and caused acceleration or retardation in the process, nothing was new enough to change things as drastically as did the changing technological capacity of humans. First their biological "grazing" influences intensified so as to become dominant. The scale and changeability of the wetlands were no longer determined primarily by geological characteristics and the various adapted organisms, but rather by human ways of life. Recently, however, this role has been taken over by man-made machines and rules. This now renders ecology a social and political item, and the reverse: the future of our wetlands depends on future human factors concurrent with the irreversibly developing universe. The role humans see for wetlands in the future may well lead their further ecological development.

Base-rich freshwater wetlands, as much as many other, have long been regarded wastelands, challenging improvement for dairy farming (Balátová-Tuláčková, 1976; De Vries, 1948), wood, reed and fish production (Rodewald-Rudescu, 1974). Recreation and water quality control are more recently developing fields of use. The most significant issue here, however, is wildlife conservation, with landscape and environment as connected notions. With little natural left, the Dutch have decided to improve their wetlands, and to "build" new ones. The social and economical relevance of this policy has revitalized broad discussions about nature and the human role. Some advocate the abandonment of traditional open fen in favour of carr development. Others remind of the enormous increase of carr already observed, and the near absence of anything akin the species-rich variants of swampy alder woodland known from the past. The latter require a flushing with base-rich water which is no longer provided for in the present mire basins when ditch management is stopped and fluctuations of the water level are prevented. So why not provide for incidental, controlled flooding artificially, still others call, and they try to design sufficiently large areas that can be made free of human interests intolerant of flooding. All in all, a real control dilemma is started. In this dilemma the disappearance of many species of base-rich fens, especially the harvested ones, is risked, long before there is any clear indication that enough "new nature" can be generated to accommodate a similar natural diversity without any human controlled harvesting. This is not merely a problem of management of the existing eco-devices. It also involves the selection and development of most opportune ones, and the control of the environment.

Control of the environment has to take precedence of connectivity increase

As regards human control, the role of mass and momentum is reminiscent of a typical navigation problem. Often the destination of a vessel will not be straight ahead of the current compass. Special circumstances may even require the destination to be changed. Similarly, the future state of a mire is not necessarily the end of the presently indicated succession. A navigator will even intentionally change the apparent direction from the goal set when currents and storms are anticipated. In nature policy the currents that have already caused the breakdown of the formerly coherent Dutch wetland fleet have to be calculated by the captains of nature protection in order safely to wheel the remaining cockle shells over the torrential ocean of the anthropogenic landscape. This seems to be quite an effort compared to the easy floating of large mires over the relatively calm seas of the Atlantic and Subboreal periods, as observed by palaeo-ecologists. Based on their lack of momentum the present fragmental wetlands have quite different reaction coefficients. It is noteworthy, in this respect, that the peat excavations now terrestrializing, though occurring in aggregations, only measure some $400 \times 30 \times 3$ m^3 (see below)! Once it is recognized that hydroseral succession is not just an automatism driven by the vegetation, it follows that the spatial discontinuity in the present wetland landscape inspires a diversity of simultaneous and asynchronous successions. Only through this very discontinuity some of the wetlands escaped most of the obtrusive influences of the present template. It seems imperative, therefore, that any further increase of connectivity in the landscape is preceded by an improved control of pollution in the ecological field, the ocean of the navigation problem. Especially in the flat lowlands of The Netherlands the absence of isolation can facilitate an almost simultaneous collapse of nature in a great area. The most hostile factor in this country presently is the eutrophication of water and atmosphere by intensive agriculture.

3. Vegetation and hydrology of terrestrializing turbaries

Methods to relate vegetation and hydrology on an ecological basis

Various methods have been developed or adapted in The Netherlands to measure hydrological factors and to relate them to the mire vegetation, especially in the 1970s (see Van Wirdum, 1991). This started a field of research now called ecohydrology (*e.g.*, Grootjans, 1985; Koerselman, 1989a, b; Van Diggelen *et al.*, 1991; Everts & De Vries, 1991; Kemmers, 1990; Schot, 1991; Wassen, 1990). In order to relate the results of water sampling to the observed characteristics of the vegetation, two methods of interpretation are used, which are shortly explained here.

Interpretation of water analyses in view of the water cycle

The water cycle quantitatively describes the flow of water between the seas, the atmosphere, and the land (rivers and aquifers). During this cycle the chemical composition of the water adapts to these main compartments (Gibbs, 1970). A "typical" analysis can be assigned to each extreme type (Van Wirdum, 1991; Van Wirdum *et al.*, 1992). The extremes are called lithotrophic (deriving its characteristics from the lithosphere), atmotrophic (from the atmosphere), and thalassotrophic (oceanic), respectively. A weighted coefficient of correlation is used as a similarity index, enabling an arrangement of analyses in the litho-atmo-thalasso field. Actual analyses tend to form swarms in this field, which can be named after the most similar extreme type, or after types connected by the swarm. A litho-atmocline thus shows a series of water analyses connecting lithotrophic and atmotrophic end members. It is the typical sequence found in the paludisere (Moore & Bellamy, 1973). A thalasso-atmocline and, recently, various swarms reflecting the influence of polluted Rhine water, are also quite common in The Netherlands. For a detailed description of this method, graphing procedures, and the reference analyses, see Van Wirdum (1991).

Interpretation of nutrient state and base state by floristic analysis

Due to the indicative significance of species, it is possible to derive an indication of the "apparent" environmental conditions by application of indicator lists (Kruijne *et al.*, 1967; Ellenberg, 1974).

Such lists should be applied with caution. The information in them is to a great extent based on the expert judgment of vegetation ecologists, and reflects the main geographical and ecological areas of interest of the experts. Moreover, the factors for which the indications have been scaled are poorly defined, and for most species experimental evidence is lacking. It has been shown (J. Wiertz *et al.*, 1992) that the apparent behaviour of several species differs from the general Central-European pattern when they grow in the Dutch dunes, or in woodlands. On the other hand, Van Wirdum & Van Dam (1984) found that the estimations by Kruijne *et al.* (1967) for Dutch grasslands, and those by Londo (1988) for The Netherlands in general, on the whole were consistent with the Ellenberg indications (1974). Van Wirdum (1991) compared various literature sources, inclusive of the Finnish indicator system (Eurola *et al.*, 1984). What in the Finnish system is defined as a response to the trophic state of a site, in other sources appeared to be listed rather as an alkalinity response. Van Wirdum also discusses the possible effects of estimations made by association analysis, i.e., the use of indicated site values in order to derive a response pattern for species not yet considered in the lists under construction. He suggests a different procedure to cope with these effects. That procedure is followed here in order to present a general survey of the fen environment.

Usually, a site is supposed to be environmentally homogeneous, and the average indicator score for each factor is held representative. Especially in mires with a hummock-hollow pattern an association of quite different micro-milieus exists. The indicator information is therefore summarized by computing the frequency distribution of the indicator scores for linked factors, *e.g.*, for base state and nutrient state. In order to be able to test differences between such distributions, very simple indicator scales are used (Van Wirdum, 1991). Only the extreme values for each factor are distinguished, reserving a third class for all intermediate or indeterminate indications. For two factors, this procedure already yields nine possible states of micro-sites. The naming of base state values is chosen according to the fairly strict relation with the water supply: acidic is atmotrophic; alkalinic, in freshwater environments, is lithotrophic. The intermediate base state is indicated as circum-neutral. As regards the nutrient state, oligo-, meso-, and eutrophic are used as names.

The results of application of this method are shown in radial diagrams for an appropriate selection of the combined states (Fig. 5). In such diagrams the vegetation of a site is shown in an environmental ring connecting different combinations of base state and nutrient state, considered poles of attraction. The centre is connected to each pole by a radial axis. The summed scores for each combined state are scaled in order to map the largest one at a distance of one radius length from the centre, *e.g.*, at the relevant pole. The other scores are plotted to the same scale and the points on the axes so obtained connected. The patch thus formed has a characteristic polarity suggestive of the environmental constitution of the vegetation's site. In many cases, a base pattern exists, which is similar for almost all sites compared: the normal pattern. This normal pattern may well hide the differences caused especially by the deviating micro-milieus in the sites, and by less common species. This problem is overcome by normalization in advance of the plotting: the scores for individual sites, groups or types are divided, for each factor combination, by the score of the sum total of all groups. Figure 5 shows radial diagrams for the types of fen vegetation described below. Five polar axes were selected for illustration in these diagrams. One axis just aggregates all factor combinations with eutrophic and mesotrophic nutrient states, with the exception of the combination litho-mesotrophic. The latter and the three oligotrophic combinations have been assigned to separate axes. They reflect the special preference of mire species which are threatened by eutrophication. Special care is given to the number and arrangement of the poles in order to exclude the occurrence of isolated scores, *i.e.*, significant scores for a particular combination, surrounded by zero scores for both neighbour poles. Due to the con-

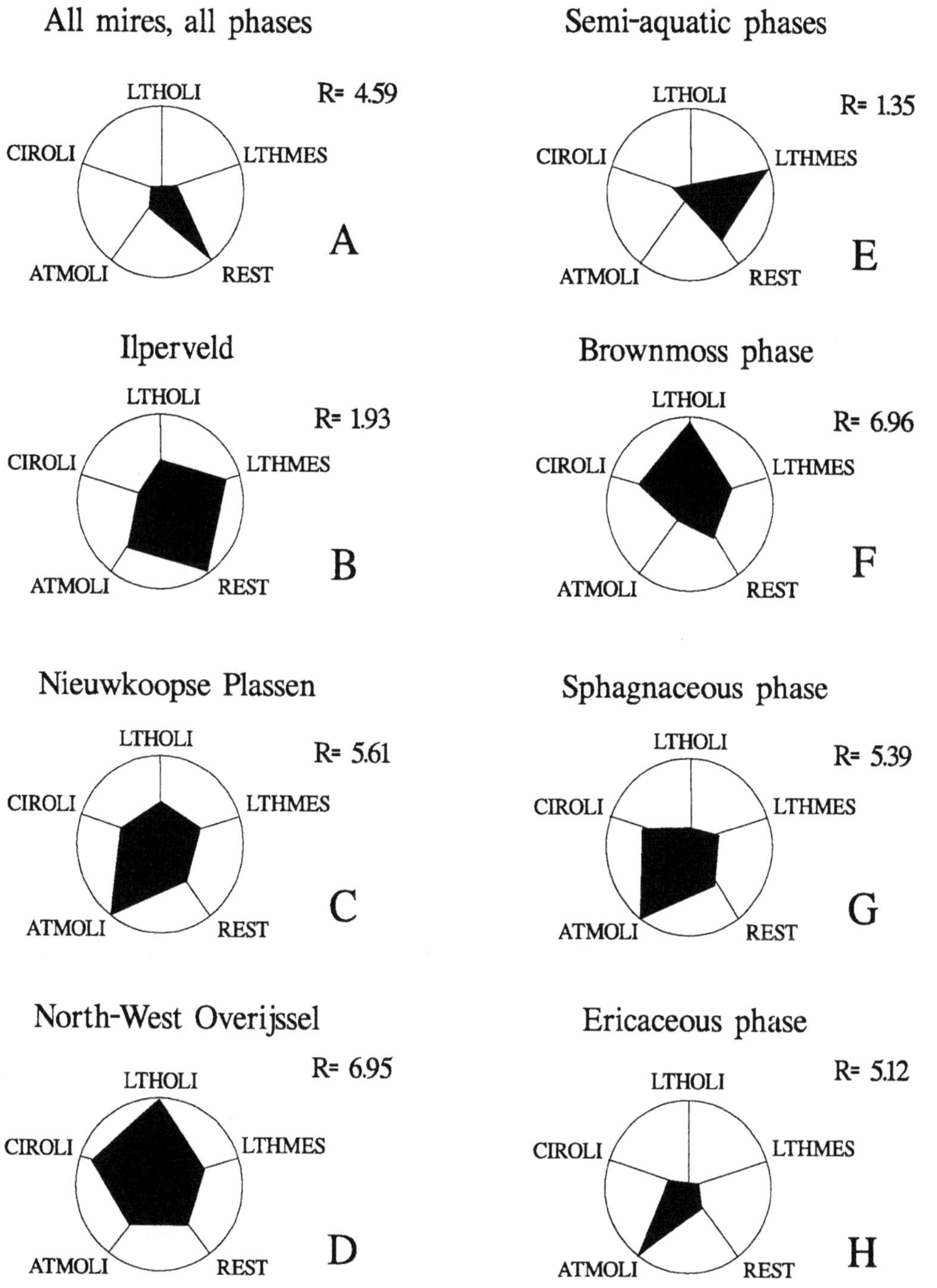

Fig. 5. Relative importance of various combinations of base and nutrient state in seres and phases of terrestrializing turbaries. The axes represent atmo-oligo- (ATMOLI), circumneutral-oligo- (CIROLI), litho-oligo- (LTHOLI), litho-mesotrophic (LTHMES), and all remaining (REST) combinations, respectively. R: frequency per relevé of red-list species. A: normal pattern, used for the normalization of all other figures; B–D: bulrush, reed, and quagfen sere, respectively; E–H: semi-aquatic, brownmoss, sphagnaceous, and ericaceous phase, respectively. According to data in Den Held *et al.* (1992) and Van Wirdum *et al.* (1992) for the Ilperveld, Nieuwkoopse Plassen and North-West Overijssel mire areas.

struction of the diagrams, isolated scores would be invisible in them.

Vegetation and succession of terrestrializing turbaries

The vegetation of terrestrializing fen mires, their hydrology, nutrient dynamics, and conservation, have been dealt with in quite some detail by Den Held *et al.* (1992), Koerselman & Verhoeven (1992), Van Wirdum *et al.* (1992), Vermeer & Joosten (1992), and Wiegers (1992). In the present discussion, I try to add some understanding at the ecosystems level, and to further substantiate the importance of state transitions. Since *petgaten* and *kragges* are very particular and important structures to these fen mires, not treated in any detail in the general literature, I also describe these entities (see also Van Wirdum, 1991).

The petgat, basic entity of terrestrializing turbaries

The largest areas with natural fen vegetation are presently found in former turbaries. The turbaries are located mainly in regions covered with bog during the geological period. During the historical period the reclamation of surrounding land had seriously decreased the area of land available for the storage of water at high floods, so urging the flood waters into the remaining mires. This influence, together with early land-use on the bog surface, the post-glacial sea level rise, and the continuing effect of eustatic sinking of the lower part of The Netherlands, caused a drowning of the bogs. Sometimes clay was deposited on top of the peat. The drowned-bog peat could not be easily drained, and the deeper layers had to be dredged from below the water level. This was done in so-called *petgaten*, some 30 m wide and often several hundreds of meters long. The reasons for this peculiar form were both hydrological and backed by traditional ownership and land-use. The land-use started from linear structures in the landscape: the border of the already inhabited, slightly higher sand regions, levees along rivers, or roads and canals. As the population grew, the ownership became defined according to narrow parcels of land from each owner's home base down into the "wilderness".

When the peat working started, access was maintained along the sides of these parcels, where standing baulks of peat were left and used for drying the dredge and cutting it to turves. As the excavations proceeded, they extended over areas of fen peat and peats covered with clay. In case such peatland had been used as pasture, the parcel boundaries often were trenches, which were filled in with the top sods of the land first in order to create such baulks. Working in *petgaten* had the additional advantage of enabling a lowering of the water table with small wind pumps. The peat working thus resulted in large complexes of narrow *petgaten* (Fig. 6). In several cases, the baulks were too narrow to stand wave erosion. This led to the formation of large broads, similar to the Norfolk Broads in England (Lambert *et al.*, 1960). As demanded by socio-economic factors, in some cases the peat was removed down to the very mineral bottom, whereas in other cases some peat was left. Accordingly, the bottom of a *petgat* may consist of sand, clay, or peat. The depth varies between ca 1 m and ca 4 m. A *petgat* may have an open connection to a larger network of surface water, or it may be surrounded by land at all sides. The walls and the bottom of a *petgat* are hydrologically very important parts inherited from the geological template of landscape develop-

Fig. 6. Aerial view of an area of terrestrializing *petgaten*. Photography: the author.

ment. Their resistance to a throughflow of water quantitatively determines the water balance and water sources of the basins relaxing from the dredging of peat and subject to primary mire development.

The first vegetational phases of terrestrialization

Although the formation of a new, however small, basin for mire development resembles the large disturbance by the post-glacial rise of the sea level in the geological period, the environmental template was quite different in the historical period. First of all, a large relative rise of the water level occurred at once, and the original vegetated land surface was removed. *Petgaten* have very steep sides, not easily allowing of the establishment of helophytic vegetation. In some cases layers of remaining peat or disposed sods of upper peat rose to the water surface, and formed a substrate for the germination of helophytes. In some shallow *petgaten*, *Phragmites* and *Scirpus* could develop from remaining rhizomes. This especially occurred where such species had already settled after earlier flooding with sea or river water. In most cases, however, a vegetation of aquatic macrophytes was formed: the aquatic phase of terrestrialization. According to Van Wirdum *et al.* (1992) the succession towards helophytic vegetation is often discontinuous with the aquatic phase, the helophytic vegetation developing from the sides of the *petgat* or on floating mud layers, and replacing the aquatic one. The rhizomes of the helophytes, especially those of *Phragmites*, *Typha*, *Scirpus*, *Menyanthes* and *Equisetum*, form a floating raft (*kragge*) just below the water level: the semi-aquatic phase of terrestrialization. The *kragge* is then colonized by *Carex* species, mosses and hepatics, and some herbs. The average level of the *kragge* in this brownmoss phase is just about the water level, with a pattern of hummocks and hollows. In slightly calcareous freshwater environments *Scorpidium scorpioides* may become a dominant species in the moss layer. This phase may last for some tens of years. It is characteristic of the historical template that the semi-aquatic and brownmoss phase vegetation used to be harvested for reed and stable litter (in autumn and winter) and for fodder (in summer), respectively. The harvesting regime prevented a settlement of shrub and tree species, and improved the milieu for many moss and herb species. Hence, the relevant types of vegetation are exceptionally rich in species. This type of rich-fen in small quagmires is called quagfen (Van Wirdum, 1991).

Den Held *et al.* (1992) and Van Wirdum *et al.* (1992) distinguish between three seres of terrestrialization: the brackish, or bulrush sere (named after *Scirpus lacustris ssp. tabernaemontani*), the freshwater, or quagfen sere, and the intermediate reed sere (named after the dominant *Phragmites australis*). While *kragges*, and so quagfen, occur in all seres, the semi-aquatic and brownmoss phases, with very weak *kragges*, last much longer in the relatively calcareous freshwater environment of the quagfen sere. Hence, this sere is presented as the typical case, and given the general name. *Scirpus* and *Phragmites* are also present in the quagfen sere, but not as dominant species. Their place is taken by *Carex diandra* and *Carex lasiocarpa*. Figure 5 summarizes the environmental indications derived from the floristic composition (scaled frequency of species) of the seres and phases, respectively. The diagram also shows the frequency of Dutch red-list species per relevé for each sere and phase. The brownmoss and sphagnaceous phases (see below), especially in the quagfen sere, contribute most to the natural value of terrestrializing turbaries. This is due to a high contribution of rare species characteristic of a litho-oligotrophic, *i.e.*, base-rich but nutrient-poor, micro-milieu. The indications are backed by water analyses presented by Van Wirdum *et al.* (1992) and authors referred to there.

The general scheme of seres and phases of terrestrialization is shown in Fig. 7. When the harvesting of the vegetation is stopped, scrub or carr usually develops, except in brackish environments. Sallows (*Salix cinerea-aurita agg.*) and alder (*Alnus glutinosa*) already settle on flotants and semi-aquatic phase *kragges*. However, such transitions have not been described to any great detail in The Netherlands. Possibly, the formation of flotants, and then scrub, in *Stratiotes*

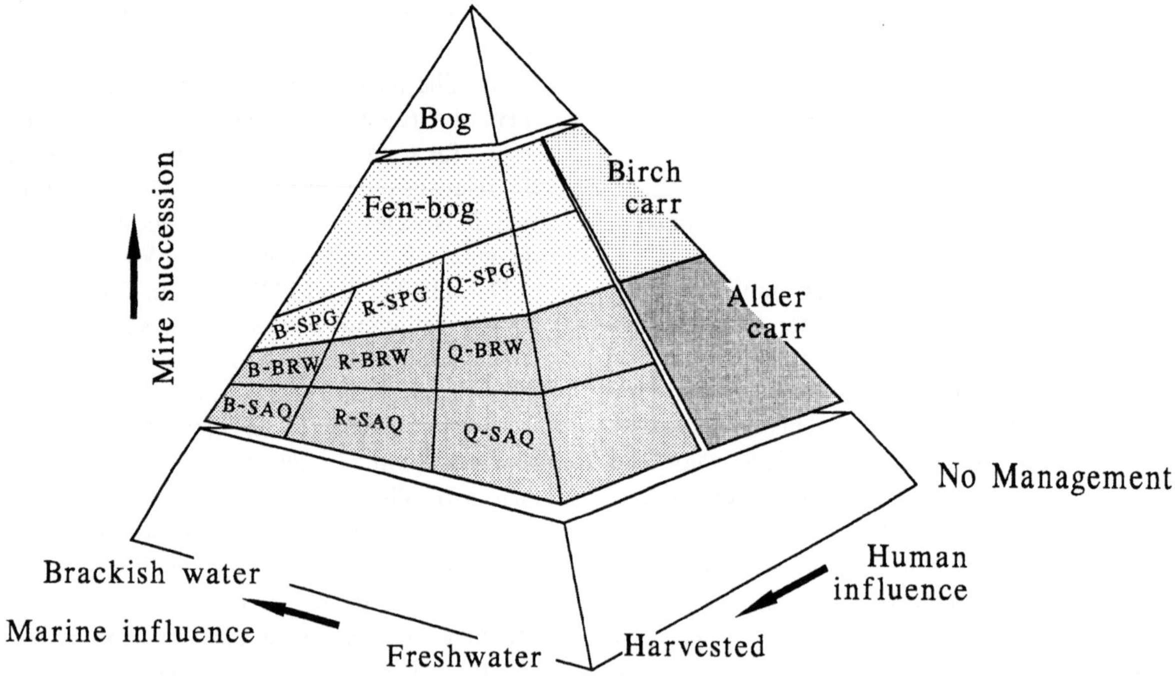

Fig. 7. Types of vegetation in terrestrializing turbaries, as they depend on harvesting, water salinity, and atmotrophiation during mire succession. B, R, Q: bulrush, reed, and quagfen sere, respectively; SAQ, BRW, SPG: semi-aquatic, brownmoss, and sphagnaceous phase, respectively. Highlighted: base-rich fens.

aloides vegetation, an example of the aquatic phase, provides one of the few examples of continuity between aquatic and later phases of terrestrialization. In such cases the typical moss layer of the brownmoss phase will only develop in its full composition and extent if the shrubs are removed and a harvesting regime installed within some ten or twenty years.

The kragge, result and regulator of terrestrialization; further vegetational phases

The development of a *kragge* also interferes with the local hydrology (Van Wirdum *et al.*, 1992; Van Wirdum, 1991). The water in the surface layer of the *kragge* is no longer mixed with the water underneath and gradually approaches rain water in chemical composition, especially in the hummocks. This process proceeds faster in environments with brackish groundwater than in freshwater *petgaten*. Van Wirdum *et al.* (1992) ascribe this to the greater density of brackish water, favouring the early development of an upper layer of freshwater, and to the relative similarity of the ionic compositions of rain and brackish water. *Sphagnum* species (bog mosses) seem to prefer this ionic composition with sodium, rather than calcium, as the dominant cation. Once the environment at the *kragge* surface has become suitable for *Sphagnum*, populations of species of this genus can easily expand and progressively acidify the milieu. This marks the transition into the sphagnaceous phase of terrestrialization.

In case of an inflow of surface water from surrounding ditches or canals, as is often the case, the thickening *kragge* provides a weak barrier in the main flow channel above the *petgat* bottom. Van Wirdum (1991) gives clear evidence of the resulting zoned hydrologic and vegetation pattern (Fig. 8). Near the entrance an often eutraphentous type of vegetation, characteristic of neutral to alkalinic milieu, is found, grading into a typically oligo-mesotraphentous type, also indicative of a neutral to alkalinic base state, and finally replaced by an acidophilous bog vegetation. The natural increase of the importance of rain water

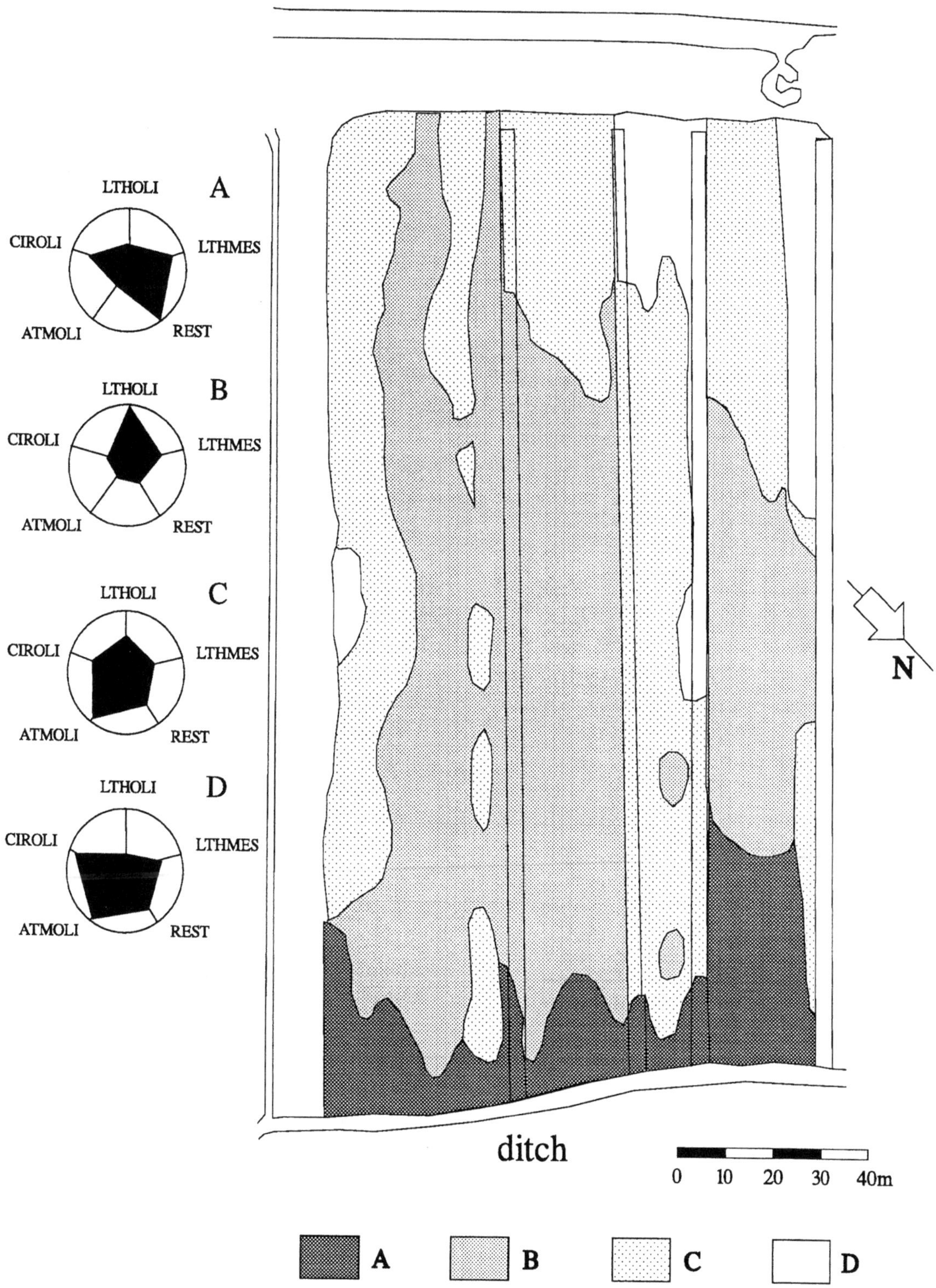

Fig. 8. Zonation of the vegetation from a ditch (below) into more isolated mire. A–C: eutrophic reed swamp, brownmoss fen with a relative abundance of litho-oligotrophic micro-milieus, and atmo-oligotrophic sphagnaceous fen, respectively. Zone D represents a slightly drier litterfen type. Radar diagrams as in Fig. 5. After: Van Wirdum (1991).

in the local water balance has been termed atmotrophiation (Van Wirdum *et al.*, 1992). An ongoing atmotrophiation may lead to initial bog stages, termed the fen-bog, also ericaceous or dwarfshrub phase, with *Erica tetralix* as a characteristic species.

The *kragge* typically reaches a thickness of 30–70 cm in The Netherlands. The thicker ones often result from a long-term reed harvest, promoted by the irrigation of reed beds, primarily for frost protection. The irrigation retards atmotrophiation and it favours the development of roots and rhizomes near the *kragge* surface. As a side effect such thick *kragges* no longer follow the water level as easily as before. When irrigation stops, the upper peat may run dry and start to oxidize, promoting a less valuable, somewhat ruderal type of vegetation.

Figure 9a–c shows a *kragge* section. The section was cut in a quagfen which has shown a fast atmotrophiation during the last twenty years. The original *Phragmites* and *Equisetum* rhizomes can be recognized in the lower part, as well as the *Scorpidium* layer (Fig. 9b,c), and the *Sphagnum* layer. The fast atmotrophiation in this case is well documented. It was largely caused by a combination of disturbing measures marking a changing harvesting management, and by the filling-in of supply ditches. The disturbance is reflected in the profile by a layer of *Carex elata* rhizomes and shoots, which will have promoted the fast formation of a rain water layer at the top of the *kragge*.

Terrestrializing turbaries and their hydrology, base and nutrient economy

Petgat hydrology, the main environmental variate

The pattern of vegetation described above resembles the "building block" shown already in Fig. 4, and characteristic of the Holland mire in the geologic period. However, there is a very important difference as regards the hydrology. In most cases, the supply of water to quagfens is from the eutrophicated surface water system, and the driving force behind this is a leakage of water towards deeper aquifers, rather than the occurrence of temporarily high levels in the surface water system. In the geological period high levels frequently must have caused a flooding of valley mires with base-rich, but relatively nutrient-poor waters in a zone several hundreds to thousands of metres wide. Similarly, in the historical period large areas of developing quagfen were flooded since it was often impossible to dispose of excess water by means of wind pumps and sluices only, especially during storms. In the end the presently constant water levels will result in a relatively sharp boundary between a narrow eutrophic zone near the canals, and acidic, and partially mineralizing peatland at greater distance. The intermediate, mesotrophic and circumneutral-alkalinic zone is mostly narrow and often even too narrow to allow of a full development of the characteristic vegetation. By consequence the rich-fen vegetation so characteristic of the quagfens in the historical period are threatened with an almost complete disappearance, irrespective of the harvesting management.

The cause of this threat is found in the water balance of the *petgaten* (Koerselman, 1989a; Van Wirdum, 1991). The large-scale drainage of lands for agriculture in The Netherlands, and the maintenance of low water levels in many deep polders, together with a still increasing production of tap water from deeper aquifers, ever increases the driving force for the leakage of phreatic water from the mires. This is compensated for by the supply of surface water, mostly derived from the Rhine and Meuse systems and now heavily polluted. The compensation is based on a strategy of keeping the water levels constant with the help of motor pumps. As a result, the transmissivity of the flow channel in the terrestrializing *petgaten* determines the water levels in the mire area. The *kragge* structure, in turn, determines the degree of development of an atmotrophiated top layer. Paradoxically, as the leakage is stronger, the width of the zone influenced by base-rich surface water is extended. In a case study, with considerable leakage (2 m a^{-1}), a zone of 10–30 m width appeared eutrophic. The mesotrophic rich-fen zone reached from 30–160 m from the surface water system, and further down into the mire

a

b

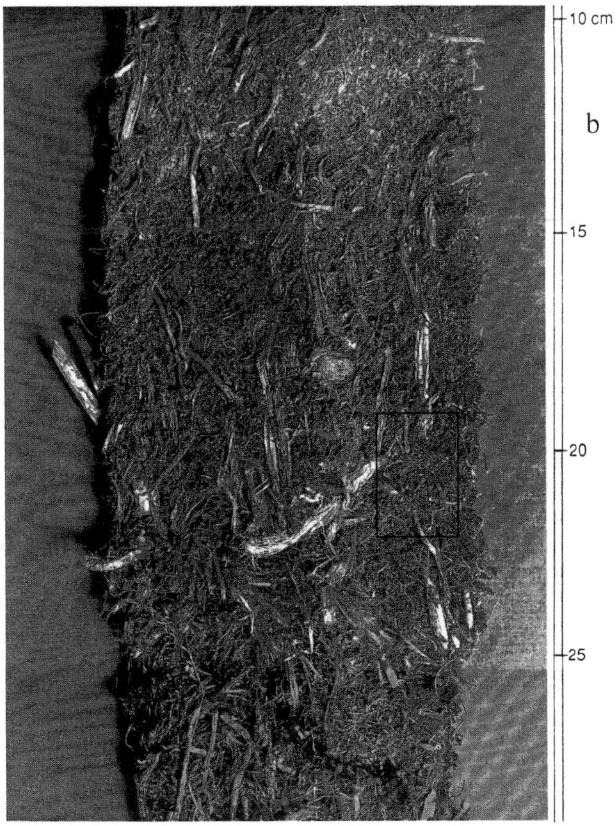

c

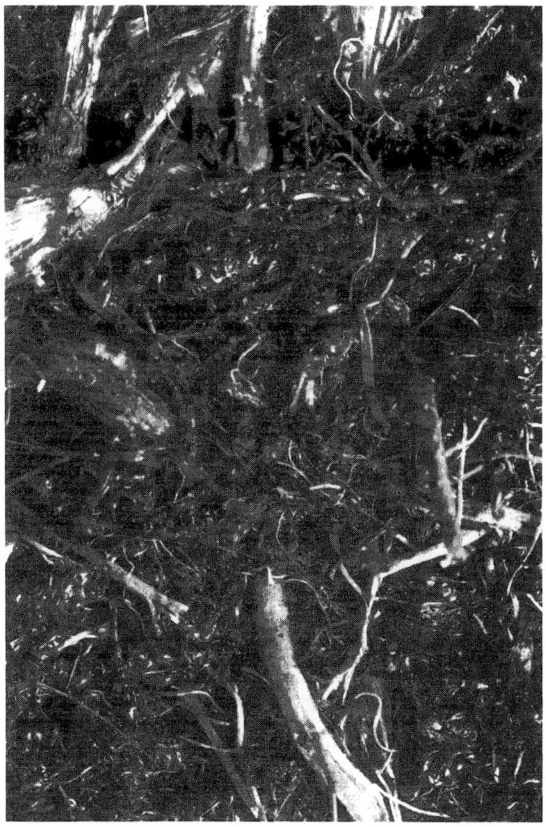

Fig. 9. Kragge section from quagfen in the sphagnaceous phase (a), detail of brownmoss horizon (b), with remains of *Scorpidium scorpioides* (c). Photography: J. van Osch (TFDL-DLO).

acidic vegetation was found (Fig. 8). Presently, nature management pays much attention to the maintenance of ditches and canals in the fenland areas, and to the control of eutrophication in the surface water system. The surface water system is also very important in fenlands with upwelling groundwater. In case the water surplus can not be disposed of, a dynamic equilibrium will be reached with no seepage on average. The top of the *kragge* in this equilibrium situation will again become saturated with the precipitation surplus water.

In the geological period, with natural drainage only, the central bog areas will have fed part of their precipitation surplus to the intermediate mire zones. These zones probably were subject to a regular variation between rain water and surface water influences, which will have determined the local mosaic of site types.

The base state of the peat as a nodal parameter between hydrology and nutrient economy

So far the various sources of water in floating fens in terrestrializing turbaries have been emphasized. Often a changing water supply is not immediately shown by a vegetational change. Apparently, even the relatively small mass of peat in the *kragge* is capable of a substantial exchange of bases. Recent investigations to be published elsewhere have indeed shown that the surface layer of a base-rich brownmoss phase *kragge* holds enough bases to buffer the water composition for several years when it becomes exposed to rain water supply only. In recent modelling approaches (Van Wirdum, 1991; Van Wirdum *et al.*, 1992) it is consequently suggested to interface hydrological models to ecological ones by using the base state as a nodal parameter. The base state can be estimated with reasonable accuracy from the water balance, the peat mass, and the composition of water supplied to the relevant sites. Moreover, a direct check is possible by analysis of soil and water samples. Within each site, a characteristic difference in height of the water table with respect to the *kragge* surface, and the distance from the surface water system should be taken into account in distribution functions in order to estimate the frequency distribution of types of sub-sites (base state, wetness). Within the sub-sites, the individual plants probably rely on the nutrient state, rather than on the base state. It has appeared (Kemmers, 1990) that especially a peat saturated with calcium provides binding surfaces for phosphorus, thus stabilizing the concentration of this element in the interstitial water at a low (meso-oligotrophic) level. The normal atmotrophiation does enhance the availability of phosphorus, but the more acidophilous vegetation of the sphagnaceous phase seems incapable of using the surplus (Verhoeven *et al.*, 1988; Koerselman & Verhoeven, 1992).

All in all, quite some effort is still needed to interface models of hydrology and community ecology, each with their proper scales and reaction coefficients. Nodal parameters and distribution functions provide a means for such interfacing at the eco-device level.

The transition to a different system state: excitation versus (internal) eutrophication

On several occasions the replacement of the vegetation by a more eutraphentous type, still in a circum-neutral environment, has been suggested in the field. A similar difference is clearly seen in space, for instance by comparing the quagfen vegetation and water chemistry in the Vechtplassen area to the vegetation in North-West Overijssel, and also by comparison of quagfens within the Vechtplassen area (Koerselman, 1989b). The possible occurrence of such differences has been attributed by Van Wirdum (1981) to a difference in the ionic composition of the water, especially associated with the relative abundance of calcium. Several mechanisms have been suggested by various authors in order to explain this. Most emphasis is laid on the possible formation of phosphate precipitates (Koerselman & Verhoeven, 1992), but this has never been proven in the field. For the time being the availability of binding surfaces for phosphate, statistically shown by Kemmers (1990) is held a more probable explanation. This is to be incorporated in the "ecological" part of models.

The exchange of bases between the water and the peat can be described in hydrological models. It is imperative to be aware of the fact that the driving force behind this exchange is the relative presence of ions in the water and on the soil complex, rather than the absolute concentrations. Polluted Rhine water, as supplied to many quagfens in the Vechtplassen and Overijssel regions, has similar, or even slightly higher concentrations of calcium than typical lithotrophic water. The concentrations of sodium and potassium, and of sulphate, are much higher, however. This means that sodium from the water may be bound on the exchange complex, calcium going into the solution. When this occurs, the binding capacity for phosphates would decrease, and the availability of phosphates to fast growing plants and to microbiota increased, so increasing the apparent nutrient state of the site. Such apparent increase of the nutrient state, without any substantial eutrophication in the water source is usually called "internal eutrophication" in The Netherlands. At the systems level this term is confusing.

In systems theory, things internal to the system are the things that do not really change in the long run, although they may show dynamic fluctuations. Eutrophication, as described for many aquatic and agricultural systems, indicates that the same, unchanged system is supplied with more nutrients, and that it is capable of a response by higher production of biomass. A good example is the supply of fertilizer to farmland in order to raise production. This can be quantitatively described in an ecosystem model. The "internal eutrophication" suggested above rather indicates a real state change of the ecosystem. Other processes, usually not provided for in existing models, become decisive. A correct name for such system changes would be (trophic) excitation, in the same sense as excitation is known from physical chemistry. Excitation raises the system to a higher energy state, basically changing its behaviour, aggregation and stability.

Trophic excitation is a very common and important process when human influences become apparent in nature. Another example is the increased availability of nitrogen and phosphorus as a result of drainage of fenlands. Such trophic excitation has been used to the benefit of the Dutch for many centuries, as shown in this chapter. Trophic excitation indeed changed the ecosystem type of many wetlands once accumulating substantial amounts of peat. They were rendered peat consuming, rather than peat producing systems.

Acknowledgements

Dr Adriana H. Prins, dr Bryan D. Wheeler, dr Elly P.H. Best and dr Jan P. Bakker critically read the manuscript and suggested improvements. The Wienand Staring Centre kindly permitted the use of Fig. 2.

References

Armstrong, W., 1975. Waterlogged soils. In J.R. Etherington, Environment and plant ecology. Wiley, London: 181–218.

Bakker, J.P., J. de Leeuw, K.S. Dijkema, P.C. Leendertse, H.H.T. Prins & J. Rozema, 1993. Salt marshes along the coast of The Netherlands. Hydrobiologia 265: 73–95.

Bakker, P.A., 1979. Vegetation science and nature conservation. In M.J.A. Werger (ed.), The study of vegetation. Dr W. Junk Publishers, The Hague: 247–288.

Balátová-Tuláčková, E., 1976. Rieder- und Sumpfwiesen der Ordnung Magnocaricetalia in der Záhorie-Tiefebene und dem nördlich angrenzenden Gebiete. Veda, Bratislava, 258 pp.

Beintema, A.J., 1986. Man-made polders in The Netherlands: a traditional habitat for shorebirds. Colonial Waterbirds 9: 196–202.

Beintema, A.J. & G.J.D.M. Müskens, 1987. Nesting success of birds breeding in Dutch agricultural grasslands. J. of Applied Ecol. 24: 743–758.

Best, E.P.H., J.T.A. Verhoeven & W.J. Wolff, 1993. The ecology of The Netherlands wetlands: characteristics, threats, prospects and perspectives for ecological research. Hydrobiologia 265: 305–319.

Bloemers, J.H.F., L.P. Louwe Kooijmans & H. Sarfatij, 1981. Verleden land. Meulenhoff, Amsterdam, 192 pp.

Borger, G.J., 1992. Draining-digging-dredging; the creation of a new landscape in the peat areas of the low countries. In J.T.A. Verhoeven (ed.), Fens and bogs in The Netherlands. Kluwer Academic Publishers, Dordrecht: 131–171.

Casparie, W.A., 1972. Bog development in southeastern Drenthe. Dr W. Junk Publishers, The Hague, 271 pp. + figs.

Casparie, W.A., 1993. The Bourtanger Moor: endurance and vulnerability of a raised bog system. Hydrobiologia 265: 203–215.

Cruickshank, J.G., 1972. Soil geography. David & Charles, Newton Abbot, 256 pp.

De Boois, H., 1982. Veranderingen in het milieu en de vegetatie in De Biesbosch door de afsluiting van het Haringvliet. With summary in English. Thesis, Agricultural University, Wageningen, 55 + 155 pp.

De Haan, H., L. van Liere, Sj.P. Klapwijk & E. van Donk, 1993. Structure and function of fen lakes in relation to water table management in The Netherlands. Hydrobiologia 265: 155–177.

De Lange, L., 1972. An ecological study of ditch vegetation in The Netherlands. Thesis, University of Amsterdam, 112 pp.

De Vries, D.M., 1929. Het plantendek van de Krimpenerwaard III. Over de samenstelling van het Crempensch Molinietum coeruleae en Agrostidetum caninae. De Bussy, Amsterdam, 403 pp.

De Vries, D.M., 1948. Method and survey of the characterization of Dutch grasslands. Vegetatio 1: 51–57.

Den Held, A.J., M. Schmitz & G. van Wirdum, 1992. Types of terrestrializing fen vegetation in The Netherlands. In J.T.A. Verhoeven (ed.), Fens and bogs in The Netherlands. Kluwer Academic Publishers, Dordrecht: 237–321.

Ellenberg, H., 1974. Zeigerwerte der Gefässpflanzen Mitteleuropas. Scripta Geobotanica 9, Göttingen, 97 pp.

Ellenberg, H., 1978. Vegetation Mitteleuropas mit den Alpen. Eugen Ulmer, Stuttgart, 981 pp.

Etherington, J.R., 1975. Environment and plant ecology. Wiley, London, 347 pp.

Eurola, S., S. Hicks & E. Kaakinen, 1984. Key to Finnish mire types. In P.D. Moore (ed.), European mires. Academic Press, London: 11–119.

Everts, F.H. & N.P.J. de Vries, 1991. De vegetatieontwikkeling van beekdalsystemen (Eng. Summ.). Historische Uitgeverij, Groningen, 223 pp.

Gibbs, R.J., 1970. Mechanisms controlling world water chemistry. Science 170: 1088–1090.

Gore, A.J.P., 1983. Introduction. In A.J.P. Gore (1983), Mires: swamp, bog, fen and moor; General Studies. Elsevier, Amsterdam: 1–34.

Gorham, E., 1990. An ecologist's guide to the problems of the 21st century. The American Biology Teacher 52(8): 480–483.

Grootjans, A.P., 1985. Changes of groundwater regime in wet meadows. Thesis, University of Groningen, 146 pp.

Grosse-Brauckmann, G., 1974. Zum Verlauf der Verlandung bei einem eutrophen Flachsee; I. Heutige Vegetationszonierung. Flora 163: 179–229.

Grosse-Brauckmann, G., 1976. Zum Verlauf der Verlandung bei einem eutrophen Flachsee; II. Die Sukzessionen, ihr Ablauf und ihre Bedingungen. Flora 165: 415–455.

Higler, L.W.G., 1979. Sloten. In Rijksinstituut voor Natuurbeheer, Levensgemeenschappen. Pudoc, Wageningen: 57–63.

Higler, L.W.G. & P.F.M. Verdonschot, 1993. Stream valleys as wetlands. Hydrobiologia 265: 265–279.

Huisman, J.A., 1953. De hel-namen in Nederland. Rede. Wolters, Groningen, 18 pp.

Joenje, W. & B. Verhoeven, 1993. Wetlands of recent Dutch embankments. Hydrobiologia 265: 179–193.

Kemmers, R.H., 1990. De stikstof- en fosforhuishouding van mesotrofe standplaatsen in relatie tot mogelijkheden van aanvoer van gebiedsvreemd water (Eng. abstract). In B. Beltman, W. Koerselman, A.M. Kooijman & A.F.M. Meuleman (eds), Workshop Interne Eutrofiëring. The Utrecht Plant Ecology News Report 10: 7–22.

Koerselman, W., 1989a. Groundwater and surface water hydrology of a small groundwater-fed fen. Wetlands Ecol. and Management 1(1): 31–43.

Koerselman, W., 1989b. Hydrology and nutrient budgets of fens in an agricultural landscape. Thesis, University of Utrecht, 164 pp.

Koerselman, W. & J.T.A. Verhoeven, 1992. Nutrient dynamics in mires of various trophic status: nutrient inputs and outputs and the internal nutrient cycle. In J.T.A. Verhoeven (ed.), Fens and bogs in The Netherlands. Kluwer Academic Publishers, Dordrecht: 397–432.

Kruijne, A.A., D.M. de Vries & H. Mooi, 1967. Bijdrage tot de oecologie van de Nederlandse graslandplanten (Eng. summ.: Contribution to the ecology of the Dutch grassland plants). Pudoc, Wageningen, 65 pp.

Lambert, J.M., J.N. Jennings, C.T. Smiths, C. Green & J.N. Hutchinson, 1960. The making of The Broads. Research Series 3, The Royal Geographic Society, London, 153 pp.

Londo, G., 1988. Nederlandse freatofyten (Summary: Dutch freatophytes). Pudoc, Wageningen, 108 pp.

Melman, T.C.P., 1991. Slootkanten in het veenweidegebied (Eng. Summ.). Thesis, University of Leiden, 338 pp.

Moore, P.D. & D.J. Bellamy, 1973. Peatlands. Elek Science, London, 221 pp.

Nature Policy Plan, 1989. Natuurbeleidsplan. SDU, 's-Gravenhage, 179 pp.

Overbeck, F., 1975. Botanisch-geologische Moorkunde. Wachholtz, Neumünster, 719 pp.

Pons, L.J., 1992. Holocene peat formation in the lower parts of The Netherlands. In J.T.A. Verhoeven (ed.), Fens and bogs in The Netherlands. Kluwer Academic Publishers, Dordrecht: 7–79.

Rodewald-Rudescu, L., 1974. Das Schilfrohr. Schweizerbart'sche Verlagsbuchhandlung, Stuttgart, 302 pp.

Rooth, J., 1989. Lijst van internationaal belangrijke wetlands in Nederland. In A.L. Spaans (ed.), Wetlands en watervogels. Pudoc, Wageningen: 95–109.

Schot, P.P., 1991. Solute transport by groundwater flow to wetland ecosystems. Thesis, University of Utrecht, 134 pp.

Schouwenaars, J.M., 1993. Hydrological differences between bogs and bog-relicts and consequences for bog restoration. Hydrobiologia 265: 217–224.

Segal, S., 1966. Ecological studies of peat-bog vegetation in the North-Western part of the province of Overijssel (The Netherlands). Wentia 15: 109–141.

Stanek, W. & I.A. Worley, 1983. A terminology of virgin peat and peatlands. In C.H. Fuchsman & S.A. Spigarelli (eds.), Proc. Int. Symp. on Peat Utilization. Bemidji State University, Bemidji: 75–102.

Stiboka, 1965. De bodem van Nederland. Stichting voor Bodemkartering, Wageningen, 292 pp.

Succow, M., 1988. Landschaftsökologische Moorkunde. Gustav Fisher, Jena, 340 pp.

Tallis, J.H., 1983. Changes in wetland communities. In A.J.P. Gore (ed.), Mires: swamp, bog, fen and moor. Elsevier, Amsterdam. Ecosystems of the world 4A: 311–347.

Van Dam, H. & R.F.M. Buskens, 1993. Ecology and management of moorland pools: balancing acidification and eutrophication. Hydrobiologia 265: 225–264.

Van Diggelen, R., A.P. Grootjans, R.H. Kemmers, A.M. Kooijman, M. Succow, N.P.J. de Vries & G. van Wirdum, 1991. Hydro-ecological analysis of the fen system Lieper Posse, eastern Germany. Journal of Vegetation Science 2: 465–476.

Van Dijk, H.W.J. & A.P. Grootjans, 1993. Wet dune slacks: decline and new opportunities. Hydrobiologia 265: 281–304.

Van Leeuwen, C.G., 1954. Een verdwijnende levensgemeenschap: het blauwgrasland. Natuur en Landschap 7: 84–93.

Van Leeuwen, C.G., 1982. From ecosystem to ecodevice. In S.P. Tjallingii & A.A. de Veer (eds), Perspectives in Landscape Ecology. Pudoc, Wageningen: 29–34.

Van Wirdum, G., 1979a. Veen, venen en moerassen. In Rijksinstituut voor Natuurbeheer, Levensgemeenschappen. Pudoc, Wageningen: 99–114.

Van Wirdum, G., 1979b. Dynamic aspects of trophic gradients in a mire complex. CHO-TNO, The Hague, Proc. and Inf. 25: 66–82

Van Wirdum, G., 1981. Linking-up the natec subsystem in models for the water management. CHO-TNO, The Hague, Proc. and Inf. 27: 108–128.

Van Wirdum, G., 1986. Water-related impacts on nature protection sites. CHO-TNO, The Hague. Proc. and Inf. 34: 27–57.

Van Wirdum, G., 1991. Vegetation and hydrology of floating rich-fens. Datawyse, Maastricht, 310 pp.

Van Wirdum, G., A.J. den Held & M. Schmitz, 1992. Terrestrializing fen vegetation in former turbaries in The Netherlands. In J.T.A. Verhoeven (ed.), Fens and bogs in The Netherlands. Kluwer Academic Publishers, Dordrecht: 323–360.

Van Wirdum, G. & D. van Dam, 1984. Standplaats en plant; bewerking ecologische indicatiewaardenlijsten. Studiecommissie Waterbeheer Natuur, Bos en Landschap, Utrecht, 116 pp.

Verhoeven, J.T.A., A.M. Kooijman & G. van Wirdum, 1988. Mineralization of N and P along a trophic gradient in a freshwater mire. Biogeochemistry 6: 31–43.

Vermeer, J.G. & J.H.J. Joosten, 1992. Conservation and management of bog and fen reserves in The Netherlands. In J.T.A. Verhoeven (ed.), Fens and bogs in The Netherlands. Kluwer Academic Publishers, Dordrecht: 433–478.

Wassen, M.J., 1990. Water flow as a major landscape ecological factor in fen development. Thesis, University of Utrecht, 199 pp.

Werkgroep Oude Maas, 1977. De Oude Maas als groene rivier. Raad van Advies voor Natuurbehoud en Natuurbeheer van Rijnmond, Rotterdam, 117 pp., appendices.

Wiegers, J., 1992. Carr vegetation: plant communities and succession of the dominant tree species. In J.T.A. Verhoeven (ed.), Fens and bogs in The Netherlands. Kluwer Academic Publishers, Dordrecht: 361–395.

Wiertz, J., J. van Dijk & J.B. Latour, 1992. MOVE: vegetatiemodule; de kans op voorkomen van ca. 700 plantesoorten als functie van vocht, pH, nutriënten en zout. RIN-rapport 92/24, Instituut voor Bos- en Natuuronderzoek, Wageningen, 137 pp.

Wolff, W.J., 1993. Netherlands-Wetlands. Hydrobiologia 265: 1–14.

Zonneveld, I.S., 1959. De Brabantse Biesbosch, een studie van bodem en vegetatie van een zoetwatergetijdendelta (a study of soil and vegetation of a freshwater tidal delta). Pudoc, Wageningen, 396 pp., appendices, summary 210 pp.

Hydrobiologia **265**: 155–177, 1993.
E. P. H. Best & J. P. Bakker (eds), Netherlands-Wetlands.

The structure and function of fen lakes in relation to water table management in The Netherlands

H. de Haan[1], L. van Liere[2], Sj.P. Klapwijk[3] & E. van Donk[4]
[1]*Limnological Institute, Vijverhof Laboratory, Rijksstraatweg 6, 3631 AC Nieuwersluis, The Netherlands;* [2]*National Institute of Public Health and Environmental Protection, P.O. Box 1, 3720 BA Bilthoven, The Netherlands;* [3]*Waterboard of Rijnland, P.O. Box 156, 2300 AD Leiden, The Netherlands;* [4]*Agricultural University Wageningen, Department of Nature Conservation, P.O. Box 8080, 6700 DD Wageningen, The Netherlands*

Key words: fen lakes, hydrology, eutrophication, productivity, element cycling, recovery

Abstract

Most relatively large (> 250 ha) and shallow (< 2 m), Dutch lakes originate from large-scale peat mining. Originally, their water tables were regulated by natural processes such as evapotransporation, precipitation, drainage and seepage (up and downwards). These lakes used to be oligo to mesotrophic, and the phytoplankton consisted mainly of diatoms and green algae. However, 30 years ago filamentous cyanobacteria were also present.

Owing to the intensification of agriculture in the surroundings of the lakes, their management developed more and more towards a man-made reservoir function. This management resulted in inflow of eutrophic water from the agricultural areas in winter. In summer the lakes were increasingly used as a water supply. This supply is compensated by water imported mainly from the River Rhine.

The impacts of this reservoir function on the limnology of the Dutch fen lakes are discussed with the Tjeukemeer, the Loosdrechtse Plassen and the Reeuwijkse Plassen as examples. All these lakes have become typical eutrophic waters with high turbidity, dominance of filamentous cyanobacteria and bream and complete loss of submerged macrophytes. Finally, methods and results of manipulation procedures to restore these waters are discussed.

Introduction

The wetlands of The Netherlands have constantly changed during our era. Before the first settlers arrived, some 5000 years ago, these changes were governed by climatological and geohydrological processes. But thereafter mankind started to affect the genesis of the environment by the construction of dikes and reclamation of land. Until the introduction of the wind mill by the end of the 16th century, the anthropological impacts on landscape and hydrology of The Netherlands were relatively small. However, the application of modern hydrotechnology allowed large-scale management of wetlands. Consequently, nowadays most of the larger (> 250 ha) Dutch lakes are subjected to a man-made hydrology resulting in a reservoir function. In this chapter the general limnological responses of some typical cases to this artificial hydrology are described.

Generally most larger Dutch lakes occur in areas with peaty soils (Fig. 1). Among these lakes

Fig. 1. Map of The Netherlands showing the location of fen areas.

intensively studied are the Tjeukemeer in the northern Netherlands and the Loosdrechtse Plassen and the Reeuwijkse Plassen in the midwestern part of the country.

In the Middle Ages excavation of the peat lands and bogs began. Later this activity occurred on an 'industrial' scale leaving small and shallow (< 2 m) water bodies separated by narrow strips of land. Ultimately these strips were too weak to resist wind-caused turbulence and larger lakes developed.

Recently the peat lands have been increasingly exploited giving these areas a completely artificial waterflow. This water management is mainly governed by agricultural needs. Consequently many fen lakes have become part of a so-called boezem system. A boezem consists of an interconnected system of ditches, canals, ponds and lakes. Shortages of water can be offset by inflow of mainly Rhine water into the boezem. Water surplus in surrounding low-lying areas (polders) is pumped into the boezem waters and ultimately drained into the North Sea. Thus the water chemistry of many fen lakes fluctuates seasonally between that of the Rhine and that of the drainage water from their catchment areas.

Water management of Tjeukemeer, Loosdrechtse Plassen and Reeuwijkse Plassen

Tjeukemeer

The Tjeukemeer, located in Friesland in the northern Netherlands (Fig. 2) is a shallow ($\bar{z}$ = 1.5 m) boezem lake with a surface of 2150 ha. Originally, the water of Friesland was governed by natural climatological and geohydrological processes. Assuming that the mean annual precipitation (770 mm) and evapotranspiration (500 mm) (see Colenbrander, 1989) and the episodic downward seepage (< 1 mm d^{-1}) (see Broers, 1987) in Friesland have not changed much, the high water table of the Friesian lake area could rise quickly during prolonged precipitation. Consequently, the low-lying peaty land, used as hay land, was frequently flooded in wet seasons. In order to improve agriculture by lowering the water table of south-western Friesland, pumps were constructed near Lemmer (1920) and Staveren (1967). Additionally, surplus water in northern Friesland was led into the North Sea via locks at Harlingen and Lauwersoog (Fig. 2b). However, in dry seasons evaporation losses could not be offset by inflow until the creation of the IJsselmeer (1932) by damming of the Zuiderzee. At present the water table of Friesland is completely manageable.

The Tjeukemeer has a season-dependent water regime. In summer, water from the IJsselmeer (Fig. 2b) is allowed in. In winter, superfluous water from the surrounding peaty areas is pumped into the lake (Fig. 2c). The water brought in from the IJsselmeer originates largely from the Rhine with a relatively high chloride (> 250 mg l^{-1})

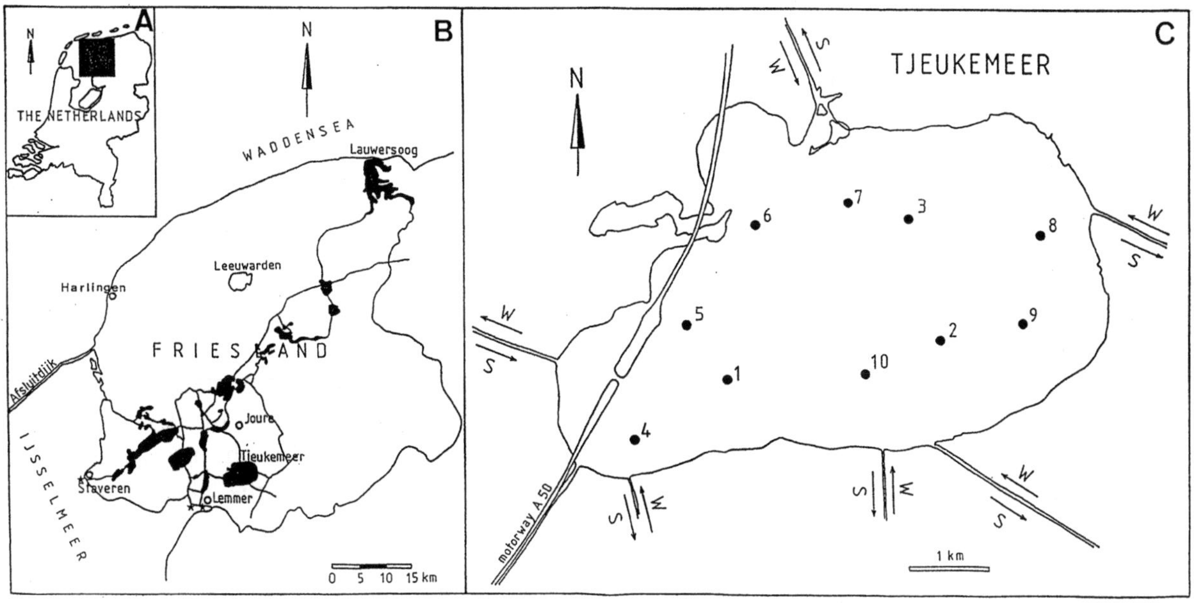

Fig. 2. Location of Tjeukemeer and its sampling stations. The arrows in panel (C) indicate the water flow direction in winter (W) and summer (S). After De Haan (1982).

concentration. Thus the presence of Rhine water in the lake can be traced by measuring chloride concentrations.

The hydrology of the Tjeukemeer as indicated by the mean annual chloride concentration between 1970 and 1986 is shown in Fig. 3. Comparing this concentration with those of the Rhine and the IJsselmeer clearly demonstrates the hydrological interrelation of the Rhine and the Tjeukemeer. The interdependence of the climate and the water management is so strong that the mean annual water composition of the lake can be predicted either from annual precipitation or the annual amount of IJsselmeer water let in at Lemmer (Fig. 4; De Haan & Moed, 1985; De Haan & Voerman, 1988). A relatively small inflow of IJsselmeer water means a relatively large contribution of water from the peaty catchment to the

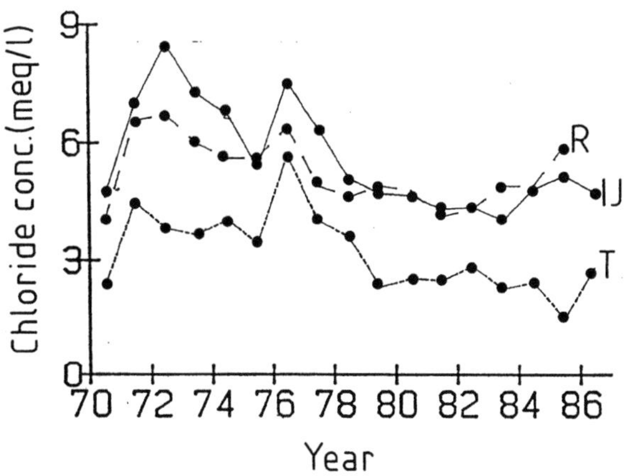

Fig. 3. Mean annual chloride concentrations in the Rhine (R), Tjeukemeer (T) and IJsselmeer (IJ) during 1970–1987. After De Haan & Voerman (1988).

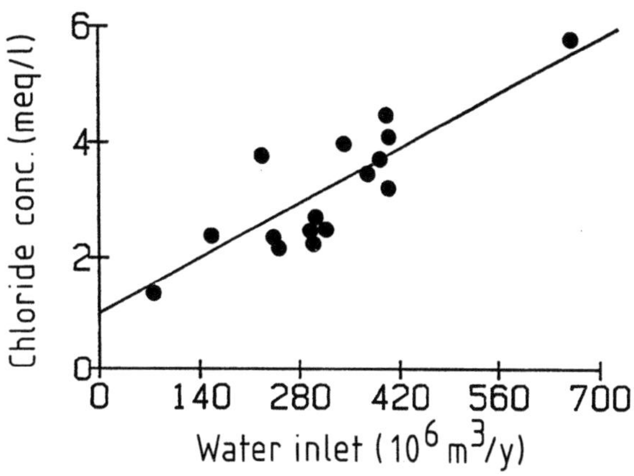

Fig. 4. Correlation between the mean annual volume water inlet at Lemmer and the mean annual chloride concentration in Tjeukemeer. After De Haan & Voerman (1988).

Tjeukemeer water composition. A large inflow of the latter water type is reflected by high concentrations of humic substances. On average, Tjeukemeer water contains 23% of water from the IJsselmeer. The water residence time in the lake is about three months.

A water balance was calculated for the south-western lake area of Friesland (Table 1). The typical water management in summer and winter is clearly visible. The change of sign in the storage reflects the reversed water flow in the different seasons. The inflows of IJsselmeer water and polder water are the main components of the water balance. The large storage illustrates the reservoir function of the boezem lakes of Friesland. About 65% of the water transported from the peaty area is flushed through the Tjeukemeer. For water imported in the area this proportion is ca. 25% (Van Huet *et al.*, 1987).

Loosdrechtse Plassen

The Loosdrechtse Plassen system is part of the Vechtplassen area between Amsterdam and Utrecht in the central Netherlands (Fig. 5). Their water surface area is 1450 ha and their average depth about 1.7 m. In this section most attention is paid to lake Loosdrecht.

The hydrology of Loosdrecht is determined by the man-made water management in the Vechtplassen area (Engelen, 1986). The reclamation of several surrounding polders had a major impact on the hydrology of the lake. The reclamation forced most of the originally upwelling groundwater to by-pass the lake basin directly into polders such as Bethune (Fig. 5). Also direct seepage of lake water to the polders occurred. To maintain the water level in summer, 4–6 × 10^6 m^3 Vecht water had to be supplied. A different water regime occurred after 1932 when the Community of Amsterdam started to use water from the polder Bethune as a drinking water source. For this purpose 17–25 × 10^6 m^3 polder water yearly flowed via Loosdrecht into Loenderveen. However, the quality of the lake water had decreased so much by 1959 that the polder water was directly transferred via a separate canal into a deepened and isolated basin of Loenderveen. Another process diminishing the ground water flow to Loosdrecht is the drinking water pumping station about 10 km north-east of Loosdrecht. In 1988 this station extracted 16 × 10^6 m^3 of ground water (Van Liere *et al.*, 1989a). In 1984 the inlet of water from the Vecht to maintain the water level, was replaced by dephosphorized water from the Amsterdam Rhine Canal (Fig. 5). This manipulation was carried out to decrease the external phosphorus load of the Loosdrechtse Plassen. The main ground water components in the area around the Loosdrechtse Plassen as treated above are summarized in Fig. 6.

Table 1. Water balances of the south-western fen lakes of Friesland estimated after Van Huet (unpublished results), of the Loosdrechtse Plassen after Kal (1986) and of the Reeuwijkse Plassen after Hooykaas *et al.* (1989). Amounts are in 10^6 m^3.

Area and Period	In			Out			Storage
SW Friesland	Precipitation	IJsselmeer	Polders	Evaporation	Polders	Pump at Lemmer	
January 1986	3.7	0.0	16.3	0.0	0.0	60.4	− 40.4
July 1986	2.2	57.4	1.9	4.1	5.9	0.0	51.5
Loosdrechtse Plassen	Precipitation	Vecht inlet	Natural inflow*	Evaporation	Infiltration*	Discharge	
April–September 1983	6.0	6.0	4.3	6.2	6.0	6.1	− 2.0
October 1983–March 1984	6.8	0.0	12.8	1.7	3.7	13.3	− 0.6
Reeuwijkse Plassen	Precipitation	Boezem inlet**	Effluents	Evaporation	Infiltration	Discharge	
January 1986	1.8	0.16	0.003	0.06	0.1	1.7	0.103
July 1986	0.7	0.71	0.005	1.53	0.1	0.0	− 0.215

* Including Bethune polder and/or exchange with Kievitsbuurt;

** Including Willens polder.

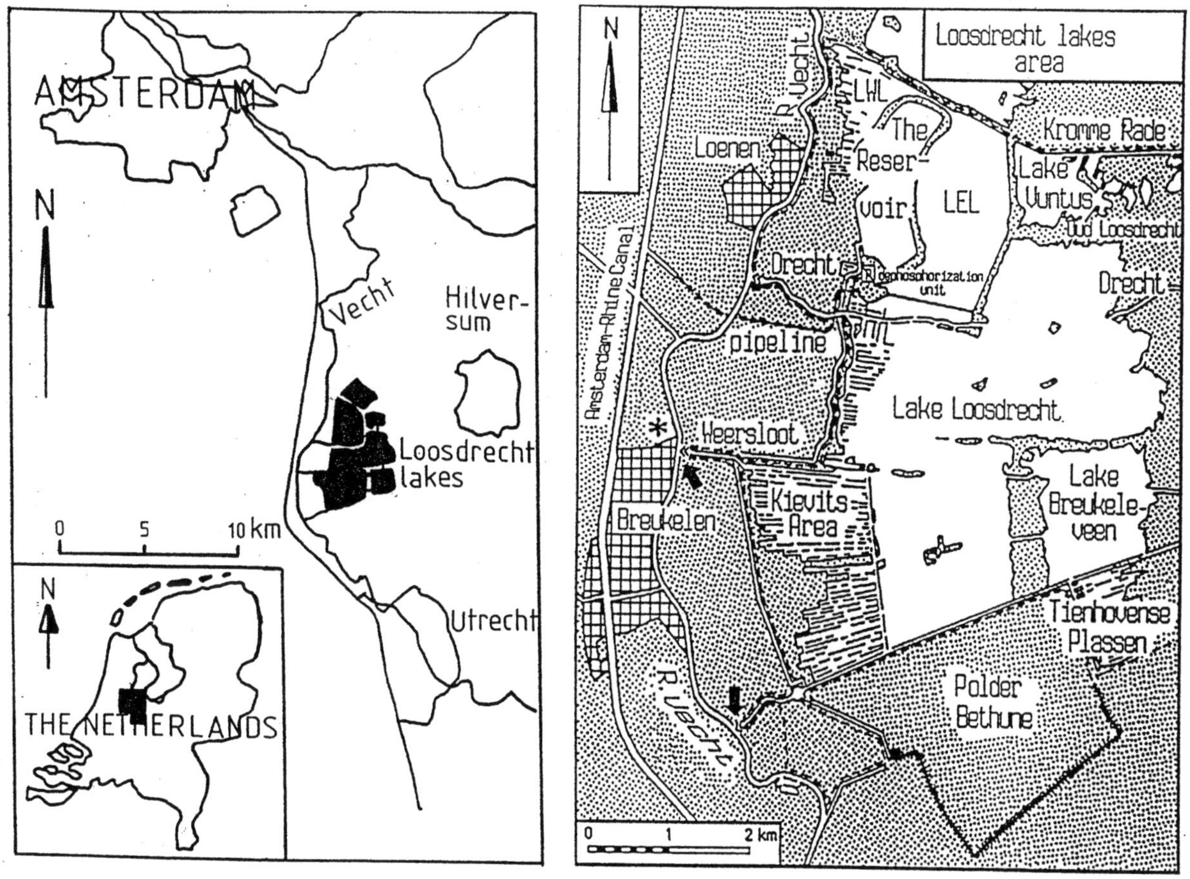

Fig. 5. The location of the Loosdrechtse Plassen. Arrows indicate inlet sluices for Vecht water supply. LWL (Lake Western Loenderveen) and LEL (Lake Eastern Loenderveen) are hydrologically separated from the rest of the system. Star: Limnological Institute. After Van Liere *et al.* (1990b).

A typical water balance of the area, before the Amsterdam Rhine Canal served as a dephosphorized water supply, is given in Table 1. The roles of the external water source and the increasing downward seepage (infiltration) in summer are obvious in this balance.

Reeuwijkse Plassen

The Reeuwijkse Plassen are situated in the area of the Rijnland Water Authority (Fig. 7b). Not all Rijnland lakes are fen lakes. The Braassemer, for instance, has a clay bottom and the bottom of the Westeinder is a mixture of sand and peat. However, the Reeuwijkse Plassen (Fig. 7c) are characteristic fen lakes.

The surface area of the Reeuwijkse Plassen is 674 ha and they have an average depth of 1.9 m. They are the remnants of a much larger area of fen lakes originating from peat mining. Due to their very low level, about 2 m below main sea level, the Reeuwijkse Plassen are separated from the Rijnland boezem, but they still depend on it for their water supply in the case of water shortages. From the surrounding polderland, water drains into the lakes at times of water surplus. In this section most attention is paid to lakes Elfhoeven and Nieuwenbroek since they have been studied most intensively.

The water table management of the Reeuwijkse Plassen depends on that of the Reeuwijk polder (Fig. 8). At the points indicated water can be let into the polder. Surplus water is pumped from the

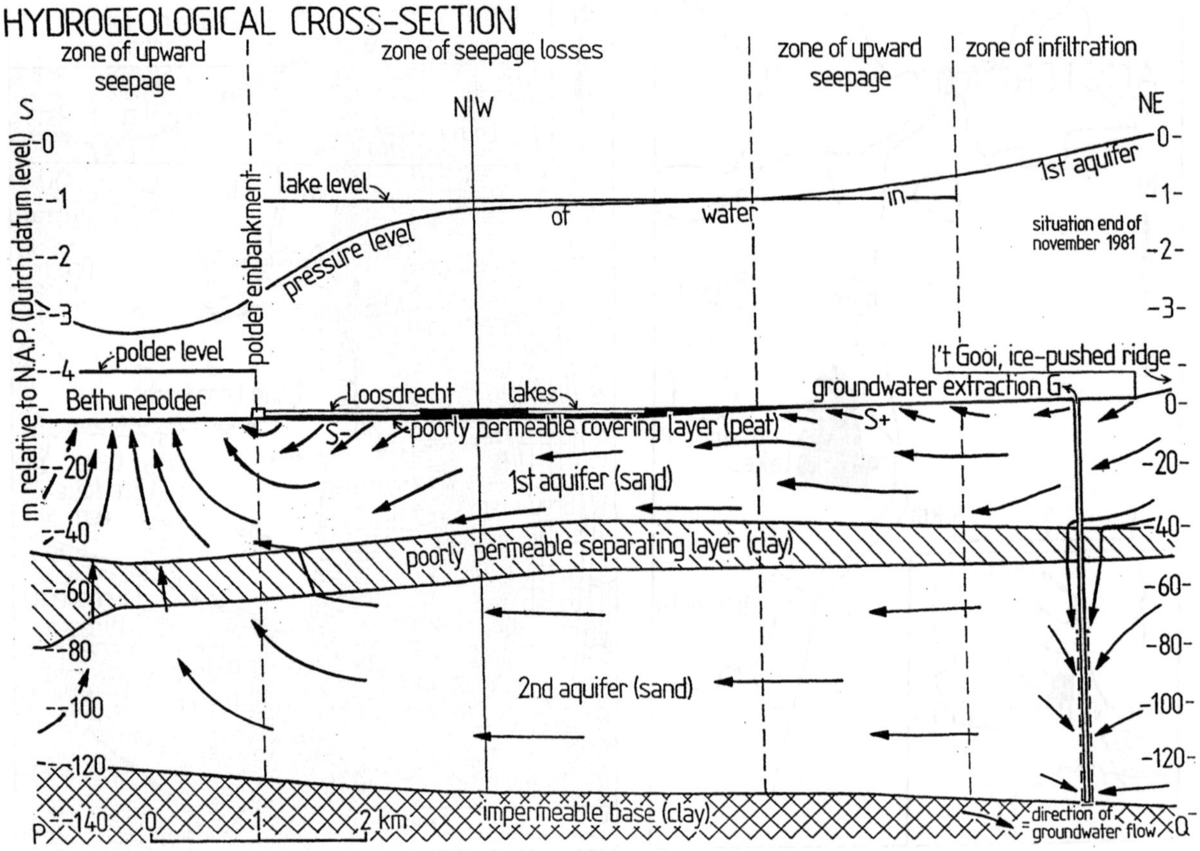

Fig. 6. Hydrological cross-section along an axis through Loosdrecht from Drecht to Polder Bethune (see also Fig. 5; from Kal *et al.* (1984). Reproduced with permission of Schweitzerbart'sche Verlagsbuchhandlung, Stuttgart, Germany.

polder into the boezem at Gouda and Bodegraven. The water level of the boezem is kept constant at 0.6 m below mean sea level. Water needs of the boezem of Rijnland are offset by inlet of water from the Hollandse IJssel, a branch of the Rhine. Excess water in the boezem can be pumped indirectly into the North Sea (Hooykaas *et al.*, 1989).

The peaty bottom of the Reeuwijkse Plassen is porous. Consequently, there is a constant downward seepage of water. The seepage through the bottom of the lakes is estimated at 1.3 × 10^6 m^3 y^{-1}. However, between 1983 and 1987 the net annual precipitation on the lakes varied between 11.9 and 13.5 × 10^6 m^3. Therefore, the lakes have a water surplus which is pumped into the boezem (Hooykaas *et al.*, 1989). Yet, in relatively dry seasons boezem water is transferred into the lakes in amounts varying from 0.8 to 2.5 × 10^6 m^3 y^{-1} in the period 1983–1987.

A typical water balance for the Reeuwijkse Plassen is presented in Table 1. Natural precipitation and evaporation dominate this balance.

Water chemistry and the impact of management

The chemical composition of water and sediments during the last few decades is described. Four groups of substances are distinguished: major ions, nutrients, metals (essential and heavy), and organic micropollutants.

Major ions

The oldest chemical data on fen lakes concern the Loosdrechtse Plassen. These data are from the

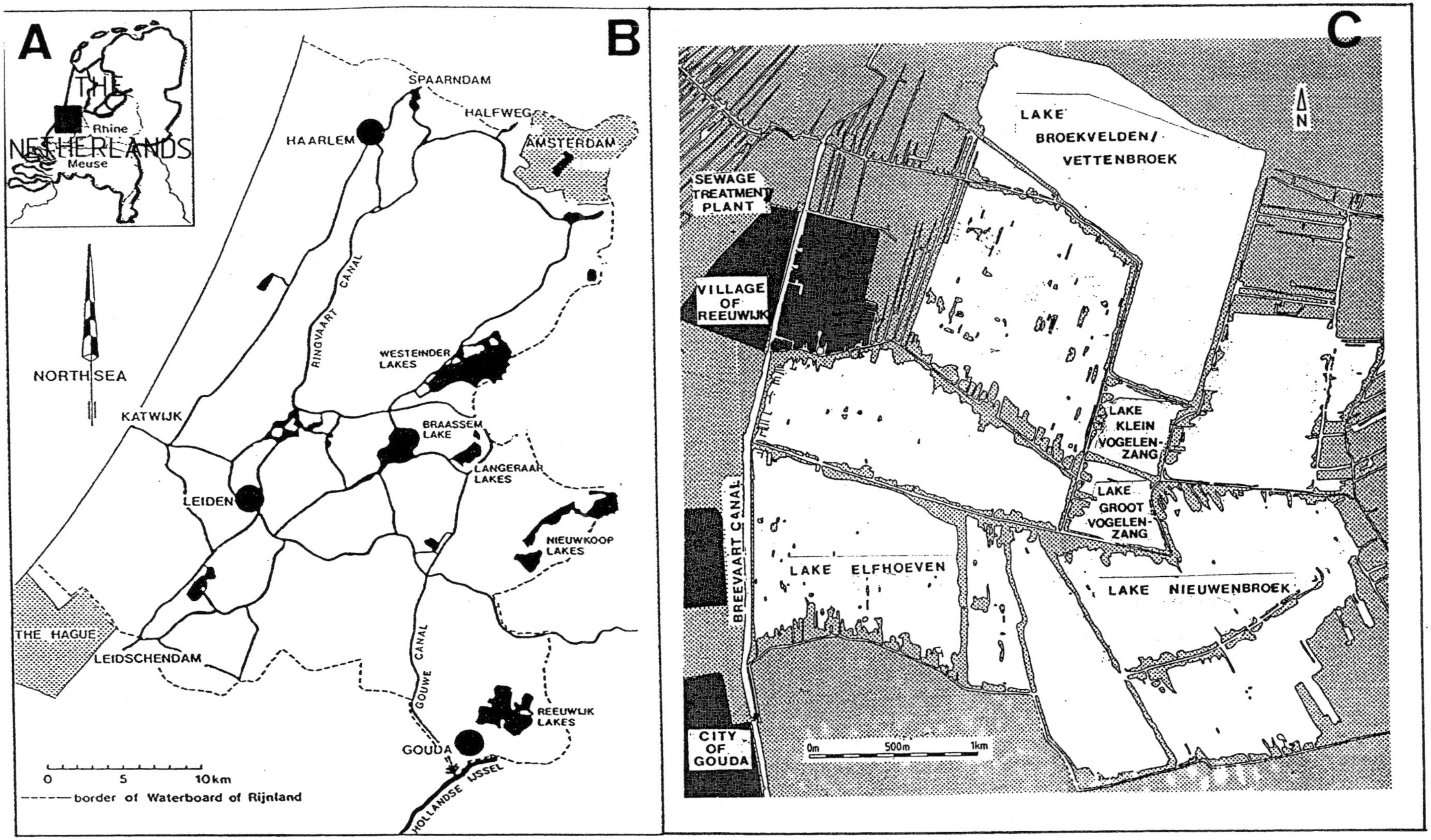

Fig. 7. The location (A) and the area (B) of the Rijnland Waterboard (inset) and of the Reeuwijkse Plassen (C).

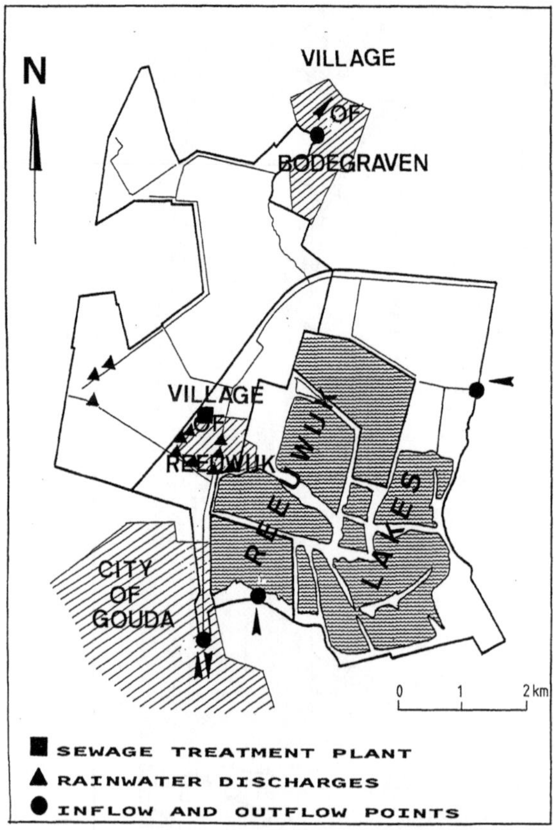

Fig. 8. The polder Reeuwijk with the location of the sewage treatment plant, sewage discharges, inflow and outflow locations.

period 1920–1930 and enable us to gain insight into the impacts of the inflow of water to the water quality of the lakes (Table 2). Most striking is the increase in the chloride content of the Vecht which is related to that of the Rhine. The high chloride content of the Amsterdam Rhine Canal is influenced also by that of the Rhine. Substitution of the presently used water supply from the Amsterdam Rhine Canal by that from the seepage water would help restore the earlier, lower, concentrations of bicarbonate and chloride. The same trend of increasing chloride concentrations has been found in the boezem of Rijnland since 1941 (Klapwijk, 1988).

The chemistry of Tjeukemeer has not been studied before 1960, so that no long-term changes in the ionic composition can be substantiated. The salinity of the autochthonous water in this area is also lower than that of the allochthonous water. The ionic composition fluctuates seasonally. The concentrations of calcium, magnesium, sodium, chloride and sulphate are usually highest in summer and are in the ranges of 2–3, 1–1.5, 1.5–6 and 1–1.5 meq l^{-1}, respectively.

Nutrients

The oldest information available on nutrient (phosphorus and nitrogen) concentrations in fen lakes also concerns the Loosdrechtse Plassen. The data listed in Table 3 demonstrate that Loosdrecht has been seriously eutrophicated during the last 60 years.

Reintroduction of seepage water as a water supply will not reduce the external nutrient loads.

Table 2. Mean concentration (mg l^{-1}) of major ions in Loosdrecht, Vecht, Rhine, and seepage water from the glacial ice-pushed ridge east of the lake (see Fig. 6). Data are derived from Barendregt *et al.* (1989) and Van Liere *et al.* (1989a).

Water	Element						
	Na^+	K^+	Ca^{2+}	Mg^{2+}	HCO_3^-	SO_4^{2-}	Cl^-
Loosdrecht 1920–1930	19	4	40	5	92	34	39
Loosdrecht 1985	30	4	48	5	153	19	46
Vecht 1930	–	–	82	–	182	29	44
Vecht 1985	46	11	76	10	171	69	124
Rhine 1930	–	–	66	–	165	47	61
Rhine 1985	78	7	84	–	63	82	205
Amsterdam Rhine Canal	78	7	84	12	171	74	138
Seepage water	32	4	48	5	104	38	50

Table 3. Mean concentrations (mg/l^{-1}) of nutrient species in Loosdrecht and three types of inlet water. Data are derived from Van Liere *et al.* (1989a).

Water	Nutrient Species			
	PO_4–P	NH_3–N	NO_2–N	NO_3–N
Loosdrecht 1920–1930	0.00	0.12	0.01	0.56
Loosdrecht 1985	0.02	0.14	0.03	3.10
Vecht	2.95	0.29	0.64	19.2
Amsterdam Rhine Canal	0.76	0.62	0.32	18.0
Seepage water	0.29	0.44	–	4.34

It is not known to what extent the high nutrient concentrations in the seepage water are due to polluted drainage water. The recent mean annual concentrations of total phosphorus and nitrogen in Loosdrecht are around 0.15 and 1.0 mg l^{-1}, respectively.

Since 1941 two to four fold higher phosphate concentrations have been observed in the boezem of Rijnland. Inorganic nitrogen concentrations almost doubled (Klapwijk, 1988). This increase has led to mean annual total phosphorus and nitrogen concentrations in *e.g.*. Elfhoeven of ca. 0.13 and 2.0 mg l^{-1}, respectively (Van der Vlugt & Klapwijk, 1987). The total phosphorus concentration in Tjeukemeer fluctuated since 1968 around 0.2 mg l^{-1}, the total nitrogen concentration around 4 mg l^{-1} (De Haan & Moed, 1985). The relatively high total nitrogen concentration in Tjeukemeer indicates a relatively large impact of agriculture.

The concentrations of total phosphorus and nitrogen in fen lakes are closely connected to the water balance and the quality of the different water sources. In Tjeukemeer this relation is reflected in significant ($0.654 < r < 0.812$; $0.001 < P < 0.01$) correlations between these concentrations and the climate dependent net inflow of water into the boezem in south-western Friesland (Fig. 9). The concentrations in the lake are highest in wet seasons, i.e., periods with a negative net inflow. The loads from the peaty polders are then higher than when high evaporation is offset by inlet of IJsselmeer water.

A similar relation exists between the total phosphorus concentration of the lake water and the amount of Vecht water supplied to the Loosdrechtse Plassen (Van Liere *et al.*, 1989b). After 1984, when the inlet of Vecht water was replaced by dephosphorized Amsterdam Rhine Canal water, this relationship changed of course.

Table 4. External phosphorus load (g P m^{-2} y^{-1}) to the Loosdrechtse Plassen and the contribution (%) of the main sources after Van Liere *et al.* (1989a)

Year	Source						Load
	Polder Bethune	River Vecht	Amsterdam Rhine Canal	Hinterland	Discharge Recreation	Deposition	
1931–44	20	40	–	11	20	9	0.56
1947–57	16	56	–	7	16	5	1.12
1959–70	6	55	–	8	24	6	0.98
1970–75	7	67	–	11	12	2	1.23
1976	1	91	–	3	4	1	4.03
1977–83	9	66	–	11	11	3	1.10
1984–85	23	11	3	31	23	9	0.35
1986–87	19	38	4	21	11	6	0.47

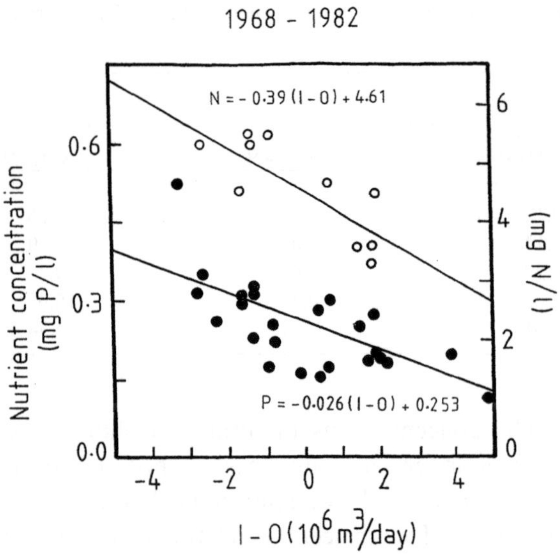

Fig. 9. Correlations between the mean half-yearly concentrations of total P and total N in Tjeukemeer and the daily inflow minus daily outflow (I-O) of water in south western Friesland during 1968–1982. From De Haan & Moed (1985). Reproduced with permission of Internat. Ass. Water Pollut. Res. Contr., London, UK.

The man-made inflow of allochthonous water thus mainly determined the eutrophication of the fen lakes in The Netherlands. This conclusion is illustrated by the increasing external phosphorus load on the Loosdrechtse Plassen, between 1932 and 1983 (Table 4). These data demonstrate also the effectiveness of the dephosphorization measures taken in 1984. For Tjeukemeer no historical load data are available. Recent data of Van Huet *et al.* (1987) estimate the external load at about 2 g P m^{-2} y^{-1}. In abnormally wet years like 1985 this load may increase to 2.9 g P m^{-2} y^{-1}. Contributions of deposition, effluents, etc, do not contribute substantially to this load.

From the phosphate balance of the Reeuwijk polder it appears that the external water supply is not the largest component of the external phosphorus load. In 1986 the Rijnland Water Authority started phosphate removal at the sewage treatment plant of Reeuwijk (see Fig. 8). Consequently, the external load decreased from more than 2 to about 1.2 g P m^{-2} y^{-1}. However, different water bodies within the area may have different gross loads, *e.g.* Elfhoeven and Nieuwenbroek have gross loads of 1.6 and 0.5 g P m^{-2} y^{-1}, respectively (Klapwijk, 1988).

Not only open water but also sedimentary phosphorus concentrations will have changed owing to the eutrophication of the fen lakes. Unfortunately, lack of historical data means that these changes have not been documented. Until recently no measurements of sedimentary phosphorus were made. From these measurements it appears that the horizontal distribution of sedimentary phosphorus is connected to the morphometry and hydrology of the system.

Van Huet & De Haan (1992) found a relationship between the sedimentary phosphorus content and water flow in the Tjeukemeer. At places with relatively low current, i.e., at canal in and out flow points, higher contents were observed than in the canals. In the centre of the lake also, relatively low contents occur, but they are influenced by wind-induced resuspension and tranport rather than water current. In the Loosdrechtse Plassen the morphometry and hydrology also appear to determine the horizontal distribution of sedimentary phosphorus. The highest contents occur at sheltered places and around the inlet of phosphorus-rich Vecht water (Boers *et al.*, 1984).

All the three fen lakes discussed here exhibit phosphorus contents in the top layer (0–5 cm) of the sediment of the same order of magnitude (0.3–1.1 mg P g^{-1} dry weight) (Boers *et al.*, 1984; Klapwijk, 1988; Van Huet & De Haan (1992). This rather constant content may be caused by a similarity in the properties of the peaty sediments. But, it may also indicate that the phosphorus dynamics of the toplayer of the sediment has been determined by the recent eutrophication due to more or less equal external phosphorus loads. The nitrogen content of the top layer of the Reeuwijkse Plassen sediments varies around 4 mg N g^{-1} dry weight. For the other fen lakes no systematic sedimentary nitrogen data are available.

Metals

Only recently data on the metal chemistry of the fen lakes in The Netherlands are available. The metal contents of the sediments in the Loosdrechtse Plassen were studied by Kühnel (1985). The aquatic concentrations of some essential metals in relation to the water management of Tjeukemeer were investigated by De Haan *et al.* (1990). Incidentally aquatic metal concentrations have been determined in the Rijnland lakes. On the other hand Siebers (1985) investigated the horizontal distribution of sedimentary metals.

The gradients in the Vecht and the Loosdrechtse Plassen indicate that inlet of the river water has polluted the lakes. The lead and copper concentrations decrease gradually eastwards into the lakes but iron seems to be of autochthonous origin (Fig. 10). Apparently, the metals adsorb to suspended matter at mixing of both water types and are flocculated. An exception is cadmium which behaves differently from zinc, lead, copper and nickel. Cadmium (data not shown) accumulates in the south corner of the lakes causing concentrations exceeding the standard (10 mg g^{-1} dry weight) for using this sediment as fertilizing mud. In the area with lower sedimentary metal concentrations, diffuse sources contribute mostly to these concentrations.

The aquatic metal concentrations in the Reeuwijkse Plassen have been analysed incidentally.

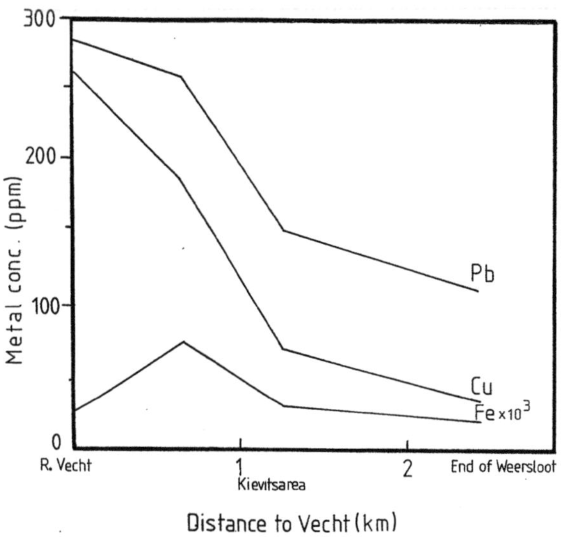

Fig. 10. Metal concentrations in the sediment of Loosdrecht as a function of the distance from the Vecht. After Kühnel (1985).

These surveys suggested higher aquatic copper (6 $\mu g\ l^{-1}$) and little lower zinc (5 $\mu g\ l^{-1}$) concentrations relative to those in Tjeukemeer being < 2 $\mu g\ l^{-1}$ and 8 $\mu g\ l^{-1}$, respectively. Among 16 metals in sediments in Elfhoeven, the concentrations of nickel, lead and zinc are comparable to those in the open Loosdrechtse Plassen far east from the Vecht. The copper concentrations are ca. 3 times higher in Elfhoeven. In this lake arsenic, lead and zinc and, to a lesser extent, manganese, cobalt, copper and aluminium exhibit irregular horizontal distributions in the top layer of the sediments. Additional analysis of wind-induced particle transport may relate these distributions to the complex morphometry of the lake.

The aquatic concentrations of iron, manganese, copper, cobalt, zinc and aluminium in Tjeukemeer were studied in relation to their occurrence in the main water sources of the lake. The impact of wet deposition on the metal loads was included in this study. High loads of iron and manganese occur in relatively wet years like 1985 when large volumes of polder water are flushed through the lake (Table 5). In contrast, copper and zinc loads are likely to be higher in relatively dry years when a high net inflow of IJsselmeer water occurs. The inputs of copper and zinc by wet deposition and water flow are almost equal in wet years.

The results listed in Table 5 suggest climate-dependent behaviour of the metal concentrations in Tjeukemeer. As expected the concentrations of

Table 5. Annual metal loads (tonnes) to Tjeukemeer by water inflow and mean wet deposition between 1984 and 1986. The net inflow of IJsselmeer water ($10^6\ m^3$) at Lemmer is also given. For more details see De Haan *et al.* (1990).

Year	Load						Net inlet
	Fe	Mn	Cu	Zn	Co	Al	
1984	5.36	0.48	0.04	0.21	–	–	42
1985	99.98	6.81	0.04	0.29	0.02	15.48	−35
1986	24.92	1.23	0.15	0.48	–	–	67
Mean deposition load	2.04	0.12	0.05	0.28	–	2.03	

e.g.. iron and chloride as a tracer of IJsselmeer water, are significantly inversely correlated (De Haan & De Boer, 1986). However, principal component analysis indicated stronger relations between copper, zinc, cobalt and aluminium and wind speed than the hydrological tracer, chloride. Manganese and iron, however, were strongly related to the UV light absorbance at 250 nm, i.e., the concentration of humic substances, indicating their polder origin. Thus in Tjeukemeer, wind-induced resuspension can contribute more to the aquatic metal dynamics than the water management.

Organic micropollutants

Virtually nothing is known about the micropollutants in fen lakes. The analysis of micropollutants in Nieuwenhoek was recently started by the Water Authority Board of Rijnland. A total chlorinated hydrocarbon pesticide concentration of $49\,\mu g\,l^{-1}$, a total polychlorinated biphenyls (PCB) concentration $< 1\,\mu g\,l^{-1}$ and a polyaromatic carbohydrates (PAC) concentration of 4.11 mg g^{-1} of dry weight in the sediment were found.

In peaty lake bottoms in the neighbourhood of Tjeukemeer PCB's have been determined (Claassen & De Jongh, 1988). The concentrations were about 60% below the state limit ($50\,\mu g\,g^{-1}$ of dry weight) set for soil sanitation. The sediments of the Vecht are heavily polluted with PCB's, PAC's, hexachlorbenzene and dichlorphenyldichlorethane (Van Rooy *et al.*, 1989). Therefore, horizontal distribution of these substances analogous to those observed for metals in the sediments of the Loosdrechtse Plassen may be expected.

Biotic structure

In this section the plant and animal communities are described in relation to the changes in water management of the fen lakes.

Planktonic organisms

The phytoplankton of the Loosdrechtse Plassen has been studied more than 60 years. The main historical changes are summarized by Van Liere *et al.* (1989a). The oldest observations are from 1922 by Heymann (1922) who noted that, even at a wind force of 6 Beaufort, the lake bottom was visible. Initially, the increased external nutrient loading was reflected in increased macrophytic densities. The regular occurrence of the cyanobacterium *Microcystis aeruginosa*, reflected the changing nutrient status. In 1954 the phytoplankton was rich in species but nearly half of the summer planktonic biomass consisted of the cyanobacteria *Microcystis aeruginosa* and *Gloeotrichia echinulata*. In the subsequent decades macrophytes decreased and in 1980 the submerged vegetation had virtually disappeared. At present the phytoplankton is dominated by the cyanobacteria *Oscillatoria redekei* and *O. limnetica* and the prochlorophyte *Prochlorothrix hollandica*. The corresponding dramatic changes in transparency, chlorophyll-*a* concentration, and sestonic dry weight are summarized in Table 6. The values of the seston dry weight include dead algae and detritus.

The first phytoplankton records of the Tjeukemeer date from 1955–1960 (P. Leentvaar, pers. comm.; Moed & Hoogveld, 1982) and indicate dominance of *Melosira* spp., *Scenedesmus* spp. and *Aphanizomenon flos aquae*. The mean summer phytoplankton biomass (chlorophyll-*a*) dramatically increased between 1968 and 1972 (Table 6), largely due to an increasing cyanobacteria population. From 1972 onwards the phytoplankton consisted in early spring of diatoms like *Asterionella formosa, Diatoma elongatum, Melosira* spp. and the centric diatoms. These organisms were succeeded in early summer by green algae, mainly *Scenedesmus* spp., and the cyanobacteria *O. agardhii, O. redekei, O. limnetica* and *Lyngbya* spp., of which *O. agardhii* ultimately dominated in summer. The biomass of this 'monoculture' of *O. agardhii* may reach values of 17 mg l^{-1} (Table 6).

The phytoplankton in the Reeuwijkse Plassen was studied for the first time in 1960 (P. Leentvaar, pers. comm.; Klapwijk, 1981). Cyanobacteria were most abundant. Nieuwenbroek, particularly, is nowadays almost completely domi-

Table 6. Summer means of transparency, chlorophyll–*a* concentration, seston and zooplankton biomass in Loosdrechtse Plassen, Tjeukemeer, Elfhoeven and Nieuwenbroek.

Lake and period	Transparency (m)	Cholophyll–*a* ($\mu g/l^{-1}$)	Seston (mg/l^{-1})	Zooplankton (mg/l^{-1})	Reference
Loosdrecht					
1955	>1.7	4	0.15	0.15	Van Liere *et al.* (1986b)
Loosdrecht					
1986	0.3–0.5	130	29	0.6	Van Liere *et al.* (1986b)
Tjeukemeer					
1969	–	35	–	–	De Haan & Moed (1985)
1972–1973	–				
1982	0.3	100	17	1.6	Gulati (1975)
Elfhoeven		130	–	–	De Haan & Moed (1985)
1983–1985	0.6	80	18.4	5.3*	Van der Vlugt & Klapwijk (1987)
Nieuwenbroek					
1983–1987	0.4	111	26.0	1.6*	Van der Vlugt & Klapwijk (1987)

* Wet weight

nated by *Lyngbya* spp. Elfhoeven has a substantial population of green algae (Van den Hove, 1988). Despite its higher gross phosphorus load, Elfhoeven has a lower seston biomass (Table 6). *O. redekei* predominated before 1970 and was succeeded by *L. limnetica*. The seston dry weight in summer was about 25 mg l^{-1} (Table 6).

The three fen lakes exhibit comparable biological consequences of their eutrophication regardless their geographical distribution (Table 6). They have similar phytoplankton biomass in terms of chlorophyll-*a* and are dominated by filamentous cyanobacteria. However, each lake is dominated by a different species. The reasons for this differing dominance are not known. Difficulties in determining differences in the morphology of the filaments involved, may have led to problems in nomenclature.

The zooplankton population in the Loosdrechtse Plassen in 1955 was characterized by *Ceriodaphnia pulchella, Diaphanosoma brachyurum* and *Eudiaptomus gracilis* (Geelen, 1955). Their densities remained low and varied around 5 ind l^{-1}. Thirty years later the zooplankton density had increased 8 times, while the biomass was 4 times higher (Table 6) (Gulati, 1984; Van Liere *et al.*, 1989a). Large species had disappeared and were replaced by small species. The density of *Chydorus sphaericus* had increased 4-fold, of *Daphnia* spp. 13 and of *Bosmina* spp. 15-fold. *Cyclops* spp., including nauplii, have also increased significantly. The rotifer density increased by one order of magnitude and consisted mainly of *Keratella cochlearis*. New species as *Anuraeopsis fissa* and *Pompholyx sulcata* appeared. All these species indicate eutrophic conditions. Thus, the zooplankton composition in time reflects the increased eutrophication of the Loosdrechtse Plassen.

The zooplankton biomass in Tjeukemeer was of the same order of magnitude as in the Loosdrechtse Plassen (Table 6) in the period 1969–1971 (Vijverberg & Richter, 1982; Gulati, 1975). The contribution of cladocerans and copepods to this biomass was 80 and 20%, respectively. Eight cladoceran species are common in Tjeukemeer of which *Leptodora kindtii* is the largest. However, *Daphnia hyalina* contributes mainly to the cladoceran biomass while the number of the smaller species of *Bosmina* and *Chydorus* may greatly exceed that of *Daphnia*. The copepod population was dominated (66%) by *Acanthocyclops robustus*. In a later study Lammens *et al.* (1985) confirmed the *Daphnia* data of Vijverberg & Richter (1982).

The zooplankton biomass in Nieuwenbroek consists substantially (6–17%) of rotifers. In Elfhoeven, this proportion does not exceed 4% (Van der Vlugt & Klapwijk, 1987).

Benthic invertebrates

The benthic invertebrates have been studied most intensively in Tjeukemeer. This study includes only the last two decades (Beattie, 1982; Lammens *et al.*, 1985), so the impact of the changes in water management on the benthos can not be determined. The benthic invertebrates were dominated by chironomids with relatively low density (50–1200 individuals/m^{-2}) and biomass (0.1–2.5 g fresh weight/m^{-2}). A main cause of this low abundance is connected with the large wind influence on the open and shallow lake. Even at low wind speeds (> 5 m s^{-1}) the top layer is resuspended by turbulence. This instability of the substrate inhibits the development of the chironomids which is reflected in the decreasing biomass with increasing distance from the lake shore (Fig. 11). In smaller, more sheltered lakes higher densities may occur (Beattie, 1982). In Tjeukemeer 7 species were common. Smaller species such as *Cladotanytarsus* had the highest densities, but the larger *Einfeldia* dominated the biomass. The largest chironomid *Chironomus*, although occurring in low densities, constituted about 20% of the biomass.

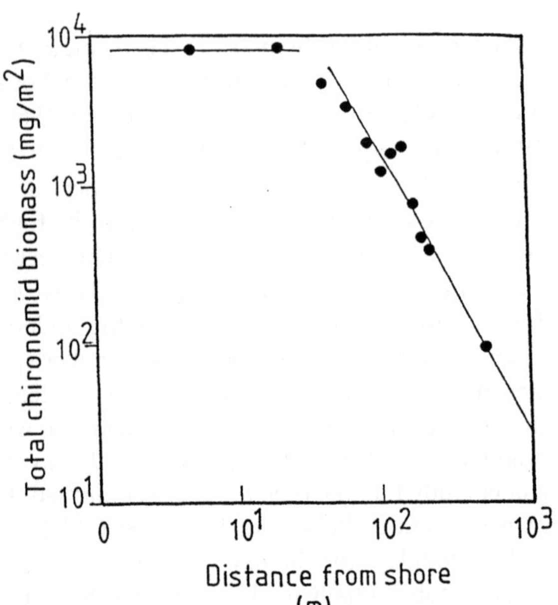

Fig. 11. Chironomid biomass as function of distance from the shore in Tjeukemeer. After Beattie (1982).

Macrophytes

Increased eutrophication leads to lower macrophytic diversity and density and higher turbidity due to increased algal (cyanobacterial) growth. For the Loosdrechtse Plassen this process is qualitatively rather well documented (Van Liere *et al.*, 1989a). For the Reeuwijkse Plassen the decline and subsequent disappearance of several submerged macrophytes has also been qualitatively described. For the Tjeukemeer nothing on the aquatic macrophytes had been reported before 1965. In 1965 *Potamogeton* beds covered about 10% of the lake surface (Chambers, 1971). These beds disappeared in 1972 and this disappearance coincided with a drastic increase in the cyanobacterial biomass (Moed & Hoogveld, 1982).

Around 1930 the Loosdrecht water was clear and low densities of mainly *Nitella* and *Potamogeton* occurred. Inlet of polluted Vecht water, drainage of domestic waste water and flushing the lake with polder Bethune water, led to increased nutrient loading. Consequently, macrophytic vegetation increased. Large parts of the lake bottoms were covered with *Nitellopsis obtusa*. In the littoral zone and between the banks of the Kievits Area (see Fig. 5) *Potamogeton, Stratiotes* and *Myriophyllum* became abundant. In 1961 the characeans and other submerged aquatic macrophytes were nearly absent. By 1980 the submerged vegetation had virtually disappeared (Best *et al.*, 1984).

In the Reeuwijkse Plassen, like in Tjeukemeer, the late 1950s/early 1960s were the turning point for the abundance of submerged macrophytes. Nowadays emergent species, indicative of nutrient richness, such as *Nymphea* and *Nuphar*, are abundant in the Reeuwijkse Plassen. Also *Phragmites* and *Typha* are present but less abundant than 80 years ago. Less than 1% of the area of Tjeukemeer is covered by emergent macrophytes.

Fish

The fen lakes have largely been stripped from submerged macrophytic vegetations since they were eutrophicated. Thus, Tjeukemeer was dominated by pikeperch (*Stizostedion lucioperca*) and bream (*Abramis brama*) (Lammens, 1986). Although 27 fish species have been observed only 8 species are quantitatively important (Van Densen & Vijverberg, 1983).

A special relation exists between the man-made water regime and the fish community structure of Tjeukemeer. The smelt (*Osmerus eperlanus*) biomass depends indirectly on the amount of IJsselmeer water transferred to the Friesian boezem in spring, because smelt larvae developed at spawning grounds near the inlet sluice at Lemmer, (see Fig. 2b) are then transported into the lakes. This hydrologically steered 0+ smelt density can easily vary 4-fold (Van Densen & Vijverberg, 1982). Unbalanced predator-prey ratios are therefore damped by artificial smelt imports from the IJsselmeer.

The Loosdrechtse Plassen were dominated also by bream while pikeperch was the main predator (Van Densen *et al.*, 1986). In Klein Vogelenzang, a small part of the Reeuwijkse Plassen, 15 fish species with an estimated biomass of 150 kg ha^{-1} were observed (Raat, 1989). The fish community was also dominated by bream (100 kg ha^{-1}), while only 10% of the total biomass consisted of pikeperch, pike (*Esox lucius L.*) and perch (*Perca fluviatilus* L.).

Birds and mammals

Too few data on numbers of birds breeding on and around fen lakes are available to establish general relationships between the water management-caused eutrophication and changing bird communities. Additionally, the available data are semiquantitative and concern only short periods. However, there is evidence that the little bittern (*Ixobrychus minutus*) became extinct in two of the lake systems probably because of eutrophication. Regular yearly counts of this species from 1940 indicate gradual decrease in numbers after 1955 when the eutrophication-caused turning point of the submerged vegetation also occurred in the Rijnland lakes. The species also decreased in numbers in the Loosdrechtse Plassen. The extinction can be connected with eutrophication-induced decreased breeding habitat (reed and cat-tail beds) and decreased density of prey-organisms (small fish, insects). Other observations of decreased numbers of *e.g.* reed warblers, common terns and black terns are not supported by quantitative data.

As far as we know systematic studies on mammals dependent on the Dutch fen ecosystems have not been made. Recently much attention has been paid to the otter (*Lutra lutra*) as an indicator of water quality. In this context historical data have been gathered which indicate that around 1930 the otter was common in fen lakes. Even in the 1960s Van Wijngaarden & Van de Peppel (1964) noted otters in the northern part of the Reeuwijkse Plassen. Otters may have occurred until 1985 in the Loosdrechtse Plassen and Reeuwijkse Plassen, but they were certainly extinct in 1987. The probably complete extinction of otters in The Netherlands has been connected with accumulation of organic micropollutants (Broekhuizen, 1986). Other contributing factors are disappearance of the ecological infrastructure required for exchange between different populations, traffic accidents, and decreased breeding habitats.

Functional system characteristics

It is clear from the previous sections that there is little quantitative information available to describe the biological impacts of the increasing eutrophication, linked to the hydrological manipulations of the lakes. The recent structural developments have of course consequences for the functioning of the systems. However, studies on ecosystem functioning of peat-land lakes were only initiated about two decades ago with the food-chain study of Tjeukemeer. Therefore, long-term changes in the functioning of Dutch fen lakes in relation to their anthropogenic management are not documented. Nevertheless the investigations made indicate the resultant functioning of fen lakes suffering from prolonged external nutrient loadings. In this context productivity, element cycling, and trophic interactions will be discussed.

Production at various trophic levels

The production of the different trophic levels of the food chain in Tjeukemeer and Loosdrecht are summarized in Fig. 12. The primary production of both lakes is almost equal (about 400 g C m^{-2} y^{-1}), provided 1 g C equals 2 g dry weight. This high level of primary production is characteristic of eutrophic lakes (Wetzel, 1975). Assuming a nutrient controlled production, it is not surprizing that the primary production in both systems is the same, since their external phosphorus loads were almost the same until 1984. However, other factors than phosphorus such as light climate and nitrogen may contribute to the regulation of the phytoplankton growth.

Despite the similarities in primary production, the composition of the phytoplankton community responsible for this production differs in both systems. This different composition may mean different food quality although all dominant phytoplankton species are filamentous species. It is uncertain whether the difference in secondary production reflects a different food quality because the accuracy of these production estimates is fairly low. Therefore, it can be concluded that both lakes exhibit similar production features. The production data indicate that prolonged and comparable external nutrient loading in different parts of the country has turned the fen lakes, originally being mesotrophic, into highly productive systems with a rather simply structured food chain (low species diversity). The food chains of the Reeuwijkse Plassen have been hardly studied.

So far there is no unambiguous evidence for phosphorus limitation in the Loosdrechtse Plassen. Even 5 years after the inlet of dephosphorized water, algal bioassays in laboratory enclosures do not clearly indicate phosphorus-limited growth. For Tjeukemeer a multiple addition bioassay (water enrichment with all essential elements except the one tested) indicated nitrogen limitation of phytoplankton growth (De Haan *et al.*, 1982). However, field data of Golterman (1983) collected in tubes with open ends pushed into the sediments indicated light limitation.

Klapwijk (1988) compared five different methods to assess growth limiting factors in the Reeuwijkse Plassen: (1) bioassays with the natural phytoplankton community; (2) bioassays with test algae (*Scenedesmus quadricauda*); (3) nutrient uptake experiments; (4) nutrient ratios and (5) statistical relationships between chlorophyll-*a* and possible limiting factors. The different methods did not give the same results. Generally, both types of bioassay and nitrogen-carbon ratios indicated nitrogen limitation. The other methods primarily predicted phosphorus limitation. Although the reason for these results is not always obvious, Klapwijk argues that the conditions of the biosassay with the natural phytoplankton resemble most the *in situ* situation and give most reliable indications. Based on this approach Nieuwenbroek with the lowest phosphorus concentrations but with the highest sestonic concentration (see Table 6) is closer to phosphorus limitation than Elfhoeven with the highest phosphorus concentrations and gross phosphorus load.

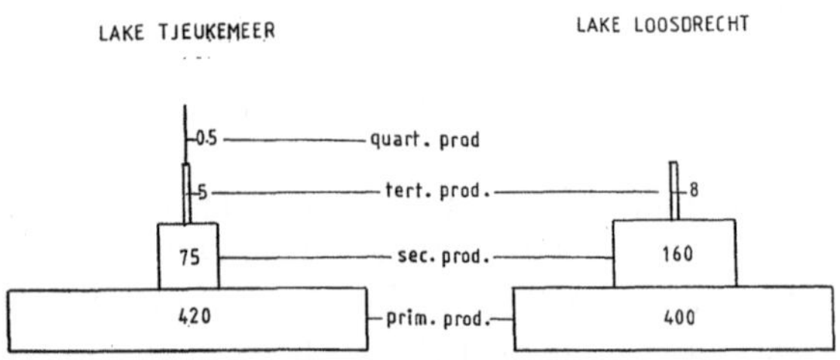

Fig. 12. Schematic representation of the production (g C m^{-2} y^{-1}) at the different trophic levels of the food chains in Tjeukemeer and Loosdrecht. After Van Densen & Vijverberg (1983) and Gezondheidsraad (1989).

Element cycling

In the previous paragraphs production aspects of the carbon balance of Tjeukemeer and Loosdrecht have been presented. The mineralization of organic matter in these shallow fen lakes has been less intensively studied. Nevertheless, available studies particularly on Loosdrecht reveal a rather detailed picture of the carbon and phosphorus dynamics.

Lakes are open systems. They fix inorganic carbon from the atmosphere and receive varying amounts of allochthonous organic matter. Lakes with peaty catchments may have large inputs of humic substances. Tjeukemeer, *e.g.*., may receive annual loads of organic matter varying between 35 and 160 g C m^{-2}. About 75% of the carbon fixed is mineralized by aerobic and anaerobic heterotrophic bacteria. Mineralization by animals, mainly herbivores and detrivores is low (Gons, 1983). It is not known where most mineralization occurs, i.e. in the water column or in the sediment. The water column and the sediment of shallow fen lakes are frequently mixed by wind-induced turbulence. In deeper lakes mineralization occurs mainly in the sediments.

The carbon cycle for Loosdrecht (Fig. 13) was calculated (Gezondheidsraad, 1989). Phytoplankton was responsible for the entire primary production which is characteristic for the eutrophic fen lakes. Only 43% of carbon fixed was used for secondary production, and more than 50% was subjected directly to mineralization via detritus. About 20% of the detritus originating from phytoplankton was consumed by benthic invertebrates. About 25% of the carbon fixed

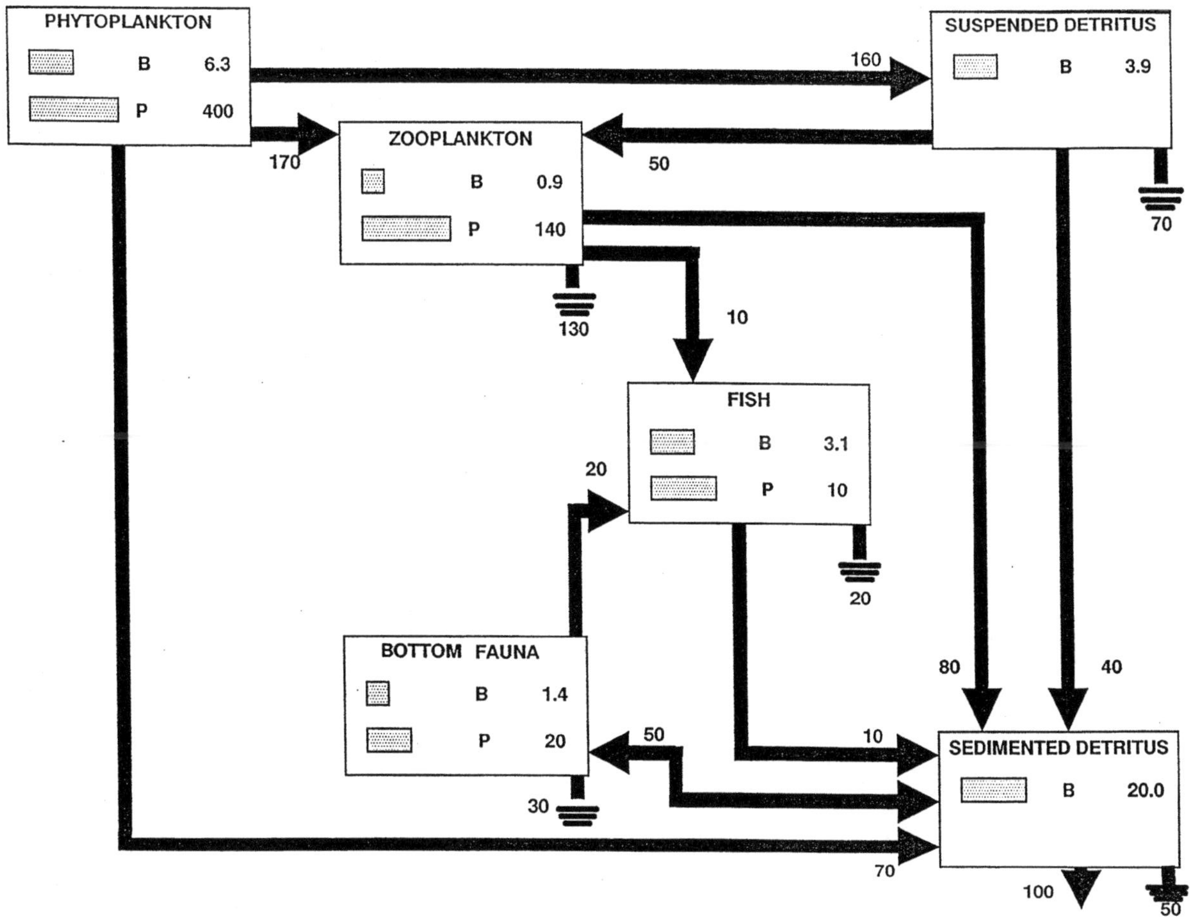

Fig. 13. Carbon cycle of Loosdrecht. Large rectangles are compartments; arrows flows; B, biomass in g C m^{-2}; P, production in g C m^{-2} y^{-1}. Primary production is given as a net rate. Respiration and mineralization of detrital matter are indicated by a 'sink' sign. The outward arrow from the bottom detritus compartment is the estimated net accumulation of refractive material. The rectangles in front of biomass (B) and production (P) indicate quantities on a logarithmic scale. Data concern the years 1984–1986.

appeared to escape mineralization possibly contributing to sediment formation (humification).

The total phosphorus flows through the main aquatic compartments of the food chain of Loosdrecht were estimated (Fig. 14) from carbon: phosphorus ratios and the carbon cycle (Van Liere *et al.*, 1989b, 1990a). Despite dephosphorization of the water supply, external phosphorus loading and phosphorus release from the sediment are still contributing to the phosphorus cycling. Grazing of phytoplankton by zooplankton is important for regeneration of phosphorus available to algal growth. Fish contribute considerably to the phosphorus pool. Harvesting fish from systems like eutrophic Loosdrecht may reduce the total phosphorus pool considerably. Aquatic plants do not contribute significantly to the phosphorus cycling. The within lake phosphorus cycling in the water column can retard the effects of restoration measures, like the reduction of external phosphorus loads.

The phosphorus cycling within the water column was recently extended to and linked to the cycling of carbon in the upper 2 cm of the sediment (Janse & Aldenberg, 1990). No big changes in chlorophyll and phosphorus levels were indicated for the period 1982–1987 applying this model. However, decreased phosphorus: carbon ratios of seston were predicted after 1984, as were observed *in situ*. This trend indicates increasing phosphorus deficiency in phytoplankton coinciding with the reduction in external phosphorus loading in 1984. Simulations further suggested that the system adapted within two years to the reduced external load. However, the reliability of these predictions is disputable.

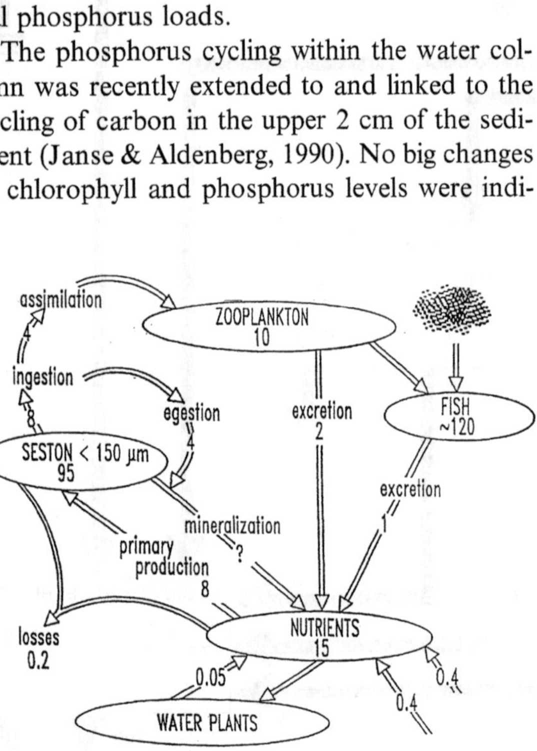

Fig. 14. Phosphorus cycle of lake Loosdrecht. Circles represent compartments; arrows flows. Concentrations are in μg P l^{-1}; flows in μg P l^{-1} d^{-1}. Data concern means for the period April–September during 1984–1987. From Van Liere *et al.* (1989b); fish data from E.H.R.R. Lammens (pers. comm.). Reproduced with permission of the publisher.

Trophic interactions

Interactions between the different structural components of the aquatic community with respect to nutrient loading of the Dutch fen lakes are now considered. We try to relate the increased nutrient concentrations to trophic interactions, ultimately resulting in the present food chain.

At excess external nutrient loading compared to internal loading, epiphytic algae and subsequently phytoplankton profit by increased production. The macrophytic growth is inhibited by increased seston concentrations, decreased transparency and stimulated epiphytic growth. This process can reinforce itself by release of nutrients due to breakdown of primary producers. But increased epiphytic growth will cease as the substratum decreases due to loss of the submerged vegetation. Ultimately the contribution of macrophytes and associated algae in the littoral zone to the increased total production of the lake will become practically zero.

The presence of macrophytes is a prerequisite for the occurrence of fish species like pike, tench, rudd, and crucian carp. When this vegetation disappears these species lose their habitat and roach, perch, ruff, bream, white bream and eel become abundant. In the eutrophic, turbid and 'open' Dutch lakes high biomass and the occurrence of older bream coincide with high pikeperch density and low biomass of smaller fish species, such as roach, perch, ruff and smelt. This relationship suggests interactions between these fishes like competition and predator-prey relations. Zooplankton plays an important role here because of its large contribution to the secondary production and its role as main food source of all young fish.

Fishes usually prefer larger crustacea such as *Daphnia hyalina* and *Leptodora kindtii* as their food and thus high densities of 0 + fish drastically affect the composition of the zooplankton community. For example the development of 0 + fish was very good in Tjeukemeer in 1980 and 1981 and the zooplankton was dominated by small species. In the two subsequent years the situation was reversed. The lack of large zooplankton (*D. hyalina*) has consequences for the feeding of bream which shifts to chironomids. If the zooplankton were continuously dominated by small species then the bream population would decrease until it could be sustained by the chironomids.

The increased turbidity due to eutrophication seems to favour pikeperch and bream. Bream is an efficient filterfeeder and is rather independent of water transparency. These conditions also favour pikeperch at the expense of the clear water predator pike. Pikeperch selectively eat on slim species such as roach and perch and, therefore, the population of the broad bodied bream increases. Bream is also favoured by a relatively high growth rate and the higher availability of zooplankton due to the selective predation of roach and perch by pikeperch. Thus enormous bream and pikeperch populations developed in the turbid fen lakes.

Recovery from perturbations (and resilience)

Generally, ecosystems are assumed to possess feed-back mechanisms to neutralize impacts of stress as caused for instance by eutrophication. Consequently, eutrophicated ecosystems are resistent against management measures meant to reduce water turbidity. In this context long periods, i.e., several years, between the start of the restoration and visible improvement are not surprising (De Kruijf *et al.*, 1988). The stress neutralizing mechanisms of the ecosystem can be both abiotic and biotic of nature. For instance, the abiotic adsorption of phosphate to sediments maintains an internal phosphorus cycle which is governed by biotic processes such as mineralization, grazing and primary production. Usually, prolonged stress leads to ecosystems with reduced stability and diversity (retrogression).

Retrogression of a stressed ecosystem proceeds via different stages with various durations. It is not unlikely that the impact of stress like eutrophication may sometimes be irreversible. Retrogression then has passed a point-of-no-return and the ecosystem may develop towards a new stage (De Kruijf *et al.*, 1988). Nevertheless, restoration measures should be directed towards less eutrophied ecosystems. Restoration to a derived state necessitates profound knowledge of the ecosystem concerned.

Restoration of eutrophic aquatic ecosystems is increasingly applied in The Netherlands. This mostly comprises measures to decrease the number of cyanobacteria in favour of submerged and floating-leaved macrophytes. Measures applied are (1) nutrient manipulation (reduction of external phosphorus and nitrogen loading); (2) hydrological manipulation (flushing with water relatively poor in nutrients); (3) sediment manipulation (dredging); (4) chemomanipulation (immobilization of nutrients) and (5) biomanipulation (changing the size of a biotic compartment).

The different measures are based on two different assumptions. The first one is that decreasing nutrient loads cause lower algal biomass, a shift from cyanobacteria to green algae and increased growth of submerged aquatic macrophytes. Applications of this approach are: (1) reduction of external nutrient loads; and (2) reduction of internal phosphorus loads by dredging and chemomanipulation.

The second assumption is that decreased piscivorous populations cause increases in planktivorous biomass, reductions in crustacean herbivorous biomass, and increases in phytoplankton biomass. Conversely, sufficiently large populations of predatory fish limit the biomass of planktivores, favour large-bodied crustacean zooplankton, which in their turn reduce the seston increasing the underwater light intensity.

In the restoration of fen lakes in The Netherlands, the external phosphorus load of the Loosdrechtse Plassen has been reduced since 1984 from > 1 g P m^{-2} y^{-1} to 0.35 g P m^{-2} y^{-1}. No conclusive trends in the concentrations of total phosphorus and chlorophyll-*a* in Loosdrecht were visible within 5 years. This retarded recovery may be due to the phosphorus accumulated in the biomass and the sediment, which sustains an internal phosporus cycle between phytoplankton and zooplankton. Additionally, lower phosphorus concentrations may not be reflected in lower cyanobacterial biomass, since these organisms use phosphorus more efficiently at phosphorus than at light limitation. Thus phosphorus-limited cyanobacteria reach higher biomass in the presence of the same phosphorus concentration (Fig. 15). Therefore, restoration to the stage at which green algae and diatoms dominate the phytoplankton community, may not be possible along the reverse of the same pathways leading to the eutrophication effects. Additional measures like the ones mentioned earlier and a reduction in the loading of external nitrogen compounds could accelerate this restoration (Van Liere *et al.*, 1989b).

To date dredging has not been used successfully for the restoration of Dutch fen lakes.

Hydrological manipulation may result in decreased numbers of cyanobacteria and increased Secchi depths as demonstrated in Veluwemeer, where flushing in autumn and winter with phosphorus-poor water removed the 'winter inoculation' of *Oscillatoria agardhii* (Hosper & Meijer, 1986). Consequently, the mean summer concentration of chlorophyll-*a* decreased from about 250 $\mu g\,l^{-1}$ in the 1970s to about 100 $\mu g\,l^{-1}$ in the 1980s.

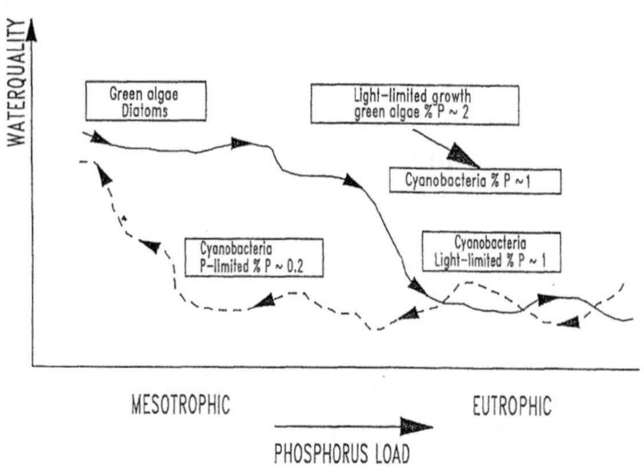

Fig. 15. Relative changes in phytoplankton composition of lakes during increased and decreased phosphorus loading. From Van Liere *et al.* (1989b). Reproduced with permission of the publisher.

The reinstatement of the original water sources and balance would probably be the most effective restoration measure for the Loosdrechtse Plassen, because the upward seepage of ground water would occur again which is nowadays mainly directed towards the deep surrounding polders. However, seepage water may presently be polluted also.

Chemomanipulation has only recently been applied to one of the Reeuwijkse Plassen, Groot Vogelenzang (Boers & Wijesooriya, 1989). Here $FeCl_3$ (100 g Fe m^{-2}) was added in autumn in 1989. Water transparancy increased dramatically from 40 to 80 cm (Quaak *et al.*, 1990). However, chemomanipulation will probably have less favourable effects in the larger (> 250 ha) peat land lakes where wind-induced resuspension of sediments frequently occur.

Reduction of external nitrogen loads has not yet been applied, mainly because of technical difficulties in managing nitrogen loads. Moreover, dominance of another, nitrogen-fixing, cyanobacterium can be expected at lower external nitrogen loads when phosphorus concentrations are still high (Zevenboom & Mur, 1980), which is also undesirable.

Biomanipulation has a rather short history in The Netherlands (Gulati & Van Donk, 1989). So far it has been limited largely to small lakes of which Zwemlust (A = 1.5 ha, z = 1.5 m) was a successful example of biomanipulation, but this lake has a sandy bottom.

In Breukeleveen (A = 150 ha), one of the Loosdrechtse Plassen, a fish management experiment was carried out in the expectation that it would yield similar effects to Zwemlust. The reasoning behind this experiment was to stimulate the development of filter-feeding crustaceans, which in turn should reduce the phytoplankton biomass. For this purpose 89 kg planktivorous and benthivorous fish per hectare were harvested in March 1989 leaving 22–43 kg ha^{-1}. Subsequently, predatory fish (pike, pikeperch) were returned to the lake while 0 + pike (70,000) and daphnids (32×10^6) were introduced. Sink nets with infra-red sensors were placed in the in and outlets of the lake to prevent import of fish but allowing free boat passage. No significant changes in the phytoplankton population occurred the first summer, while the fish standing stock appeared to be equal to that before the manipulation due to high recruitment. This manipulation apparently was not successful on a short term.

In Klein Vogelenzang (Reeuwijkse Plassen) a biomanipulation project was started in spring 1989. Bream were harvested and 0 + pike were introduced. As a result the zooplankton biomass increased enormously and transparency increased from 0.2 m in July–August to 1.25 m in October (Van der Vlugt *et al.*, 1990).

Conclusions and recommendations

Despite all the restoration efforts it is not yet clear at which external nutrient loads the manipulated systems will actually visibly function as restored systems. Especially, more knowledge is needed on the effects of the various restoration measures in fen lakes exposed to wind and wave action. In the years to come knowledge of how to manage recovered ecosystems may be extended. In addition to further reduction of external nutrient loads periodical harvesting of aquatic macrophytes, improving the habitat of piscivores, and periodical and selective harvesting of cyprinids may be necessary (Grimm & Van Donk, 1989).

References

Barendregt, A., M.J. Wassen & J.T. De Smidt, 1989. Regeneratie en behoud van ecosystemen door integraal waterbeheer. In Van Liere L., R.M.M. Roijackers & P.J.T. Verstraelen (eds), Integraal waterbeheer in het Goois/Utrechts stuwwallen- en plassengebied. Commissie voor Hydrologisch Onderzoek TNO, The Hague. CHO/TNO Rapporten en Nota's No. 22: 45–58.

Beattie, D.M., 1982. Chironomid populations in the Tjeukemeer. Ph.D. Thesis, University of Leiden, 150 pp.

Best, E.P.H., D. De Vries & A. Reins, 1984. The macrophytes in the Loosdrecht lakes: A story of their decline in the course of eutrophication. Verh. int. Ver. Limnol. 22: 868–875.

Boers, P.C.M., J.W. Th. Bongers, A.G. Wisselo & Th.E. Cappenberg, 1984. Loosdrecht Lakes Restoration Project: Sediment phosphorus distribution and release from the sediments. Verh. int. Ver. Limnol. 22: 842–847.

Boers, P.C.M. & W.A.D.D. Wijsooriya, 1989. Fosfaatfixatie met ijzer: een methode om de interne fosfaatbelasting te verminderen? H_2O 22: 501–504.

Broekhuizen, S., 1986. Een otter Lutra lutra (L., 1758) uit het natuurreservaat de Rottige Meenthe met niersteen, levertumor en PCB's. Lutra 29: 298–302.

Broers, H.P., 1987. Grondwaterstromingsstelsels in Zuid Friesland. Province of Friesland, Leeuwarden, 78 pp.

Chambers, M.R., 1971. Studies on the littoral fauna of the Tjeukemeer. Ph.D. Thesis, University of Liverpool, 281 pp.

Claassen, T.H.L. & A.W.J.J. De Jongh, 1988. De otter als normsteller voor kwaliteit van het oppervlaktewater. H_2O 21: 432–436.

Colenbrander, H.J., 1989. Water in the Netherlands with annex selection of current research topics. TNO Committee on Hydrological Research, The Hague. Proceedings and information TNO Committee on Hydrological Research 37, 96 pp.

De Haan, H., 1982. Physico-chemical environment in Tjeukemeer with special reference to speciation of algal nutrients. Hydrobiologia 95: 205–222.

De Haan, H. & J.R. Moed, 1985. Phosphorus, nitrogen and chlorophyll-*a* concentrations in a typical Dutch polder lake, Tjeukemeer, in relation to its water regime between 1968 and 1982. Wat. Sci. Technol. 17: 733–743.

De Haan, H. & T. De Boer, 1986. Geochemical aspects of aqueous iron, phosphorus and dissolved organic carbon in the humic Lake Tjeukemeer, The Netherlands. Freshwat. Biol. 16: 661–672.

De Haan, H. & J. Voerman, 1988. The chemistry of a typical Dutch reservoir, the Tjeukemeer, in relation to its water management between 1970 and 1986. Freshwat. Biol. 20: 1–10.

De Haan, H., J.B.W. Wanders & J.R. Moed, 1982. Multiple addition bioassay of Tjeukemeer water. Hydrobiologia 88: 233–244.

De Haan, H., J. Voerman, T. De Boer, J.R. Moed, J. Schrotenboer & H.L. Hoogveld, 1990. Trace metal chemistry of a Dutch reservoir, the Tjeukemeer. Freshwat. Biol. 24: 391–400.

De Kruijf, H.A.M., E. Jagtman & H. De Haan, 1988. Omgaan met Stofstromen in Oppervlaktewateren. Waterkwaliteitsverbetering door ecologisch beheer. In Bijlsma A. (ed.), NIBI, Utrecht: 57–82.

Engelen, G.B., 1986. Interaction of hydrological systems and eutrophication of Loosdrecht lakes. Hydrobiol. Bull. 20: 17–25.

Geelen, J.F.M., 1955. The plankton in the lakes in 1954.

Amsterdam Municipal Waterworks. Report no. 38. Gezondheidsraad, 1989. Ecological standards for water management. Description of the parameters. Third advice (in Dutch). Gezondheidsraad, The Hague, 424 pp.

Golterman, H.L., 1983. Algal bioassays and algal growth controlling factors in eutrophic shallow lakes. Hydrobiologia 100: 59–64.

Gons, H.J., 1983. De koolstofkringloop in drie Nederlandse meren. In Parma S., H.M. Van Emden & J. Castelein (eds), Oecologie van meren en plassen. PUDOC, Wageningen: 85–119.

Grimm, M.P. & E. Van Donk, 1989. Mogelijkheden voor de toepassing van actief biologisch beheer in de Vechtplassen. In Van Liere L., R.M.M. Roijackers & P.J.T. Verstraelen (eds), Integraal waterbeheer in het Goois/Utrechts stuwwallen- en plassengebied. Commissie voor Hydrologisch Onderzoek TNO, The Hague. CHO/TNO Rapporten en Nota's. No. 22: 167–180.

Gulati, R.D., 1975. A study on the role of herbivorous zooplankton community as primary consumers of phytoplankton in Dutch lakes. Verh. int. Ver. Limnol. 19: 1202–1210.

Gulati, R.D., 1984. The zooplankton and its grazing as measures of trophy in the Loosdrecht lakes. Verh. int. Ver. Limnol. 22: 863–867.

Gulati, R.D. & E. Van Donk, 1989. Biomanipulation in The Netherlands: applications in fresh water ecosystems and estuarine water– an introduction. Hydrobiol. Bull. 23: 1–4.

Heymann, J.A., 1922. Onderzoek aan de Loosdrechtse plassen op maandag 12 Juni 1922. Amsterdam Municipal Waterworks. Report No. 24.

Hooykaas, L.J., M. Donze & Sj.P. Klapwijk, 1989. Fosfaatbalans van de polder Reeuwijk en de Reeuwijkse Plassen. Vakgroep Gezondheidstechniek en Waterbeheersing. Technical University of Delft, Mededeling No. 25.

Hosper, S.H. & M.-L. Meijer, 1986. Control of phosphorus loading and flushing as restoration method for Lake Veluwe. Hydrobiol. Bull. 20: 183–194.

Janse, J.H. & T. Aldenberg, 1990. Modelling the eutrophication of the shallow Loosdrecht lakes. Verh. int. Ver. Limnol. 24: 751–757.

Kal, B.F.M., 1986. Monthly mass balances for compartments of the Loosdrecht Lakes system: approach and preliminary results. Hydrobiol. Bull. 20: 27–39.

Kal, B.F.M., G.B. Engelen & Th.E. Cappenberg, 1984. Loosdrecht Lakes Restoration Project: Hydrology and physicochemical characteristics of the lakes. Verh. int. Ver. Limnol. 22: 835–841.

Klapwijk, Sj.P., 1981. Limnological research on the effects of phosphate removal in Rijnland (in Dutch with an English summary). H_2O 14: 472–483.

Klapwijk, Sj.P., 1988. Eutrophication of surface waters in the Dutch polder landscape. Ph.D. Thesis, Hoogheemraadschap van Rijnland, Leiden, 227 pp.

Kühnel, V., 1985. Heavy metals in the Loosdrecht lakes sediment (in Dutch). Institute of Earth Sciences, Free University of Amsterdam, WQL Report 1985–5, 36 pp.

Lammens, E.H.R.R., 1986. Interactions between fishes and the structure of fish communities in Dutch shallow, eutrophic lakes. Ph.D.Thesis, Agricultural University, Wageningen, 99 pp.

Lammens, E.H.R.R., H.W. De Nie, J. Vijverberg & W.L.T. Van Densen, 1985. Resource partitioning and niche shifts of bream (*Abramis brama*) and eel (*Anguilla anguilla*) mediated by predation of smelt (*Osmerus eperlanus*) on *Daphnia hyalina*. Can. J. Fish. Aquat. Sci. 42: 1342–1351.

Moed, J.R. & H.L. Hoogveld, 1982. The algal periodicity in Tjeukemeer during 1968–1978. Hydrobiologia 95: 223–234.

Quaak, M.P., P.C.M. Boers, Sj.P. Klapwijk & J.C. Van der Vlugt, 1990. First results of phosphate fixation using ironchloride in lake Groot Vogelenzang look very promising. H_2O 23: 114–117.

Raat, A.J.P., 1989. De visstand in Klein Vogelenzang in januari 1989. OVB, Nieuwegein, Project RW/OVB: 89–01.

Siebers, H.H., 1985. Patterns and variability of phosphate and heavy metals in sediments of two shallow lakes. Ph.D. Thesis, Technical University of Delft, 172 pp.

Van den Hove, L., 1988. Resultaten planktononderzoek 1983 t/m 1986 in de Reeuwijkse Plassen. Rapport Hoogheemraadschap van Rijnland. Hoofdafdeling Waterbeheer, Afdeling Waterhuishouding, 12 pp.

Van Densen, W.L.T. & J. Vijverberg, 1982. The relations between 0 + fish density, zooplankton size and the vulnerability of pikeperch, *Stizostedion lucioperca*, to angling in the Frisian lakes. Hydrobiologia 95: 321–336.

Van Densen, W.L.T. & J. Vijverberg, 1983. De rol van de vis in het voedselweb van het Tjeukemeer. In Parma S., H.M. Van Emden & J. Castelein (eds), Oecologie van meren en plassen. PUDOC, Wageningen: 57–84.

Van Densen, W.L.T., C. Dijkers & R. Veerman, 1986. Fish community of the Loosdrecht lakes and the perspective for biomanipulation. Hydrobiol. Bull. 20: 147–163.

Van der Vlugt, J.C. & Sj.P. Klapwijk, 1987. Waterkwaliteitsonderzoek in de Reeuwijkse plassen 1983–1985. H_2O 20: 86–91.

Van der Vlugt, J.C., Sj.P. Klapwijk & A.J.P. Raat, 1990. Visstandbeheer en waterkwaliteit. Eerste resultaten van een experiment in Klein Vogelenzang (Reeuwijkse Plassen). H_2O 23: 118–121.

Van Donk, E. & C. Collé, 1988. Schistosome dermatitis, a possible complication of food-web manipulation in swimming waters (in Dutch with an English summary). H_2O 24: 696–699.

Van Huet, H.J.W.J., H. De Haan & T.H.L. Claassen, 1987. Fosfaateutrofiëringsonderzoek in het merengebied van zuid-west Friesland. H_2O 20: 131–135.

Van Huet, H.J.W.J. & H. De Haan, 1992. Horizontal and vertical distributrion and speciation of phosphorus in sediments of interconnected and eutrophic polder lakes in SW Friesland, The Netherlands. Arch. Hydrobiol. 123: 361–378.

Van Liere, L., L. Breebaart, W. Kats & J.J. Buyse, 1989a. De waterkwaliteit in het Loosdrechtse plassengebied. In Van

Liere L., R.M.M. Roijackers & P.J.T. Verstraelen (eds), Integraal waterbeheer in het Goois/Utrechts stuwwallen-en plassengebied. Commissie voor Hydrologisch Onderzoek TNO, The Hague, CHO/TNO Rapporten en Nota's No. 22: 265–279.

Van Liere, L., R.D. Gulati, F.G. Wortelboer & E.H.R.R. Lammens, 1989b. From food to fish and from fish to food (in Dutch). Landschap 6: 33–46.

Van Liere,L., R.D. Gulati, F.G. Wortelboer & E.H.R.R. Lammens, 1990a. Phosphorus dynamics following restoration measures in the Loosdrecht lakes (The Netherlands). Hydrobiologia 191: 87–95.

Van Liere, L., O.F.R. Van Tongeren, L. Breebaart & W. Kats, 1990b. Trends in chlorophyll and total phosphorus in Loosdrecht lakes, The Netherlands. Verh. int. Ver. Limnol. 24: 707–710.

Van Rooy, P.T.J.C., P. Vertegaal, D. De Smit, J.T. De Smidt & P.J.T. Verstraelen, 1989. De restauratie van de Vecht: een studie naar de ontwikkelingsmogelijkheden. In Van Liere L., R.M.M. Roijackers & P.J.T. Verstraelen (eds), Integraal waterbeheer in het Goois/Utrechts stuwwallen-en plassengebied. Commissie voor Hydrologisch Onderzoek TNO, The Hague, CHO/TNO Rapporten en Nota's No. 22: 181–195.

Van Wijngaarden, A. & J.B. Van de Peppel, 1964. De otter, *Lutra lutra* (L.) in Nederland. RIVON, Zeist, Rapport, 132 pp.

Vijverberg, J. & A.F. Richter, 1982. Popultion dynamics and production of *Daphnia hyalina* Leydig and *Daphnia cucullata* Sars in Tjeukemeer. Hydrobiologia 95: 235–260.

Wetzel, R.G., 1975. Limnology. Saunders, Philadelphia, 743 pp.

Zevenboom, W. & L.R. Mur, 1980. N_2-fixing cyanobacteria, why they do not become dominant in Dutch hypertrophic lakes. In Barica J. & L.R. Mur (eds), Hypertrophic ecosystems. Dr W. Junk Publishers, The Hague: 123–130.

Hydrobiologia **265**: 179–193, 1993.
E. P. H. Best & J. P. Bakker (eds), Netherlands-Wetlands.

Wetlands of recent Dutch embankments

W. Joenje[1] & B. Verhoeven[2]
[1]*Agricultural University, Bornsesteeg 69, 6709 PD Wageningen, The Netherlands*
[2]*Waterkant 83, 8307 AZ Ens, The Netherlands*

Key words: wetland, polder, drainage, desalination, colonization, vegetation-succession, groundwater regimes, meso-series

Abstract

Wetlands of new embankments are characterized with respect to soil and hydrological regime and the processes of desalination and vegetation succession. Soil- and height gradients, the lack of local drainage and the generally low nutrient content provide conditions for *dynamic mesoseries* (groundwatertable low in summer, saturated in winter) and locally for *stable mesoseries* (relatively high in summer, saturated in winter).

Relations between vegetation and herbivores (geese, ducks, cattle) are discussed.

Introduction

The skill of coastal engineering has evolved in The Netherlands over at least a millennium. The dikes and dams, sluices and mills which protect against flooding of the land drained from sea and swamp, are backbone structures in the Dutch lowlands, which indeed, are predominantly man-made.

Following previous storm floods the dikes have been repaired and often increased in height. However, the storm surge of 1953 triggered a new approach: on some places long stretches of dikes were enforced and heightened, but elsewhere new high dams, huge sluices and sophisticated storm-surge barriers have been constructed in the estuaries of the southwest and the north. This shortening of the coastline eliminated the most valued parts of the natural coastal wetlands: the estuaries, their mudflats and shallows and gradients in salinity and nutrients.

However much the extinction of the tidal lands has been felt as a loss, parts of the newly formed landscapes have not been used as farmland and offer a rich variety of spontaneous, 'natural' developments. Gradually the outstanding value of these newly embanked wetlands has been formally recognized; substantial areas have received the status of nature reserve and proper nature-management schemes are being developed with increasing input of local and national experiences.

Stimulated by the Dutch examples and technology, neighbouring countries have also hardly left alone any estuary or free discharging river along the North Sea coasts. The related losses of natural habitats are partly counterbalanced by the development and careful management of new wetlands. Indeed, the whole chain of new habitats which has been created along the North Sea coasts presents a major ecological feature.

The knowledge of natural succession obtained in these new territories over several decades, and experience of habitat construction have promoted the recent ideas on, and applications of, "nature technique": the habitat construction *de novo* and the subsequent management of these natural successions according to formulated aims. Also outside the new embankments these ideas have now developed to become part of nature conservation strategies in a much wider sense.

In this chapter the soil- and hydrological processes in the new wetlands of recent Dutch embankments are summarized and related to the development of plant species and communities. The descriptions and conclusions often hold for other areas outside The Netherlands on the coasts of the North Sea.

The polder concept

A polder is considered to be an area whose hydrology has been made independent from the surrounding watertable. This definition usually is restricted to areas which have been reclaimed mainly for agricultural purposes. Within this view falls a wide variety of lowlands such as coastal polders, peat and swamp embankments and reclaimed lakes. However, in the context of this paper, stretches of water with controlled water table, such as dammed rivers and reservoirs and the holms and watermeadows belonging to them, can all be included under the heading "polder", although the polder status of such areas may be quite arbitrary.

The first human inhabitants in the Netherlands lived on high sites, like beach-ridges, creek- and river levees and outcrops of pleistocene deposits. Later, man-made mounds safeguard their dwellings from flooding. At the time the local population started to protect their grounds on the high saltmarsh against the sea by means of dikes, the first coastal polders came into being. During low tide the excess water was drained off through flapgates, but the rise of the sea level and the subsidence of the land increased the need for lifting or pumping devices.

The large and relatively elevated inland peat and peat-on-clay areas became also cultivated. Small streams as well as the main rivers, provided a certain drainage. But here too the drainage caused subsidence (by mineralization and compaction, enhanced by peat digging) and this increased pressure for the construction of pumping installations.

A special category of polders consists of reclaimed lakes (443 'droogmakerijen' (in Dutch), comprising 240 former natural lakes and 203 lakes originating from peat excavation), which cannot be drained without mills (Schultz 1987).

The excess water in polders and 'droogmakerijen' is often lifted to an intermediate reservoir system, (the 'boezem') from which it is sluiced or pumped into a river or the sea. Figure 1 shows the principal categories of land protected against flooding in the Netherlands.

Drainage was always a cumbersome activity, so in most polder-like areas wetlands were common. Nowadays modern pumping stations allow a deep and constant drainage which eliminates many wetland conditions. But in the newest coastal embankments large wetlands form part of the project. This text deals only with the latter wetlands and derives examples in particular from the Lake Grevelingen and the Lauwersmeer (Fig. 1).

For a description and ecological survey of the effects on the flora and vegetation of embankments and drainage in the southwest of the Netherlands, the reader is referred to Beeftink (1975).

Coastal embankments

The builders of the oldest coastal polders avoided the embankment of low lying areas: there the tidal movement hampered dike construction and drainage by gravity of such low patches was difficult. Wetlands formed by the closing of narrow creeks lost their natural value by deeper drainage or disappeared by landfill. However, these polders have a variety of wetlands originating from peat excavation for salt making and clay digging for dike repair.

Later on the closing of somewhat wider tidal gaps became feasible and sandy and muddy creek beds were transformed into valuable brackish or fresh wetland strips after embankment. In the second half of this century this development led to the enclosure of large sea arms and estuaries, dictated by the greater safety requirement against

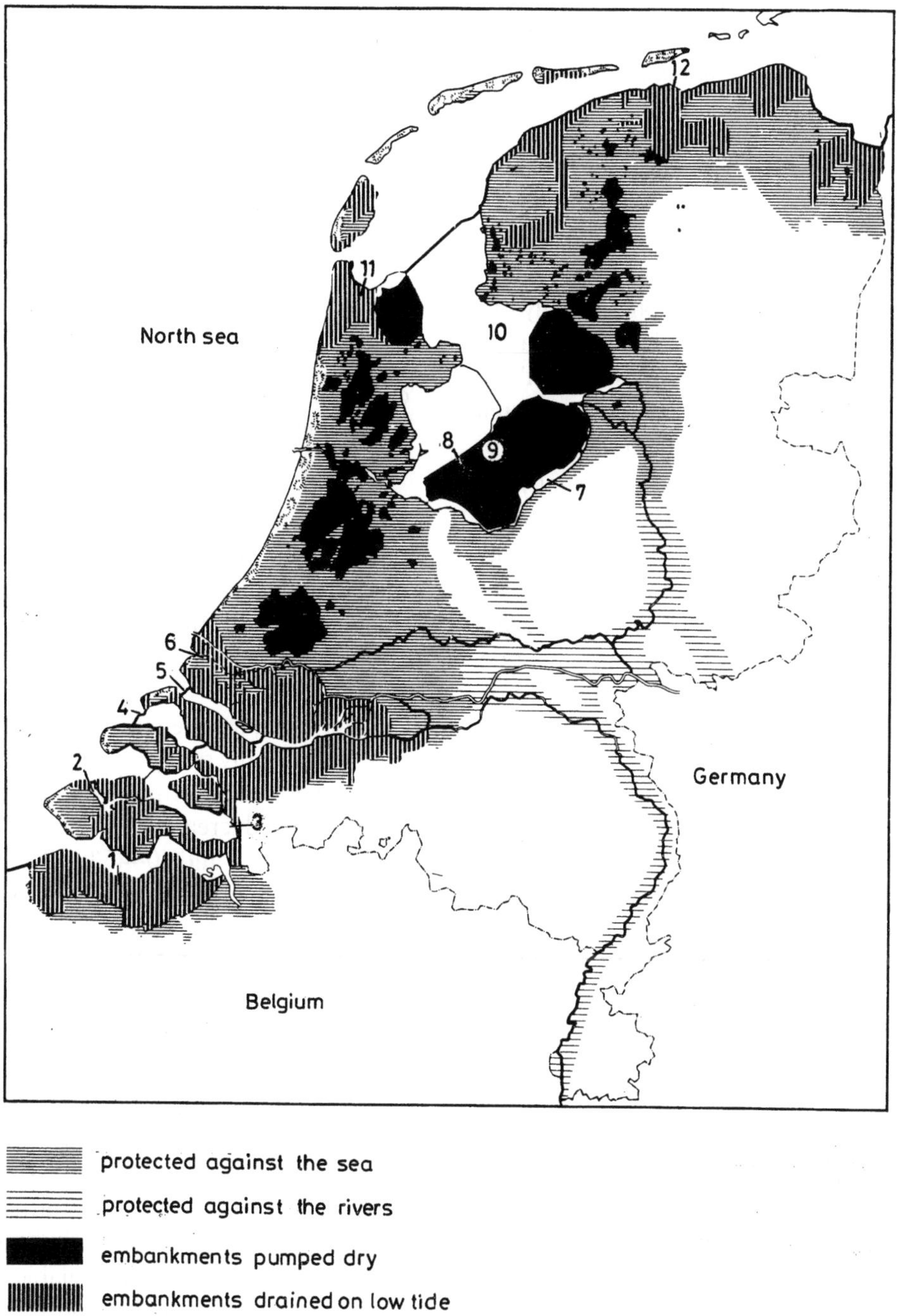

Fig. 1. Areas in The Netherlands protected against flooding (after Schultz, 1987). (Key: 1. Braakman; 2. Veerse Meer; 3. Markiezaatsmeer; 4. Grevelingen Meer; 5. Haringvliet; 6. Brielse Meer; 7. Border Lakes; 8. Oostvaarders Plassen; 9. Flevo polder; 10. IJsselmeer: 11. Amstelmeer; 12. Lauwersmeer.)

stormfloods and the improvement of the drainage of the hinterland, rather than by the wish to gain new land.

In these recently embanked areas the surface of the land is often equalled by the water surface and it is in these new polders where large areas of wetlands are formed. Unlike in older polders, the water quality is only partially determined by the precipitation and the drainage water from the polder itself, because the bulk of the water received in its reservoir lake stems from far larger catchment areas in the predominantly agricultural hinterland.

The abiotic environment

Generally the terrestrial part of the new embankment comprises at least two morphologically different areas: vegetated land of higher elevation (former saltmarshes, in Dutch 'schorren' or 'kwelders') and the void or scarcely vegetated lower areas (tidal flats, in Dutch 'slikken' or 'wadden'). The higher marshes are usually reclaimed for agricultural use; they are too dry to become a wetland. The lower flats however are subjected to the natural fluctuations governed by the new hydrological regime (target-polderlevels, precipitation and drainage) replacing the tide. They often remain significant as wetlands, provided that their hydrology meets a wetland definition and the land use does not interfere with natural developments, as does military use, recreation or agriculture.

The embanked waters may remain saline (Grevelingen), change to brackish (Veerse Meer, Braakman) or fresh conditions (Lauwersmeer, Amstelmeer, Haringvliet, Brielse Meer, Markiezaatsmeer and the IJsselmeer with its Borderlakes and – within the Flevopolder – the Oostvaardersplassen (Fig.1)). Summer and winter water levels may be different. In some embankments (Lauwersmeer, Braakman) the contents of the lake are replaced several times during winter by inflow from the hinterland; in others the level is maintained through supply by precipitation and drainage from the surrounding land and adjusted by in- and outlets (Grevelingen). In the saline Veerse Meer the unusual conditions of a high summer- and a low winterwatertable are now being reconsidered.

The sediments, generally rich in lime vary from rather fine grained sands to clay. Almost all soil properties, be it physical or chemical, are correlated with the amount of lutum (particles $< 2\,\mu m$), at least within a certain geographical area and

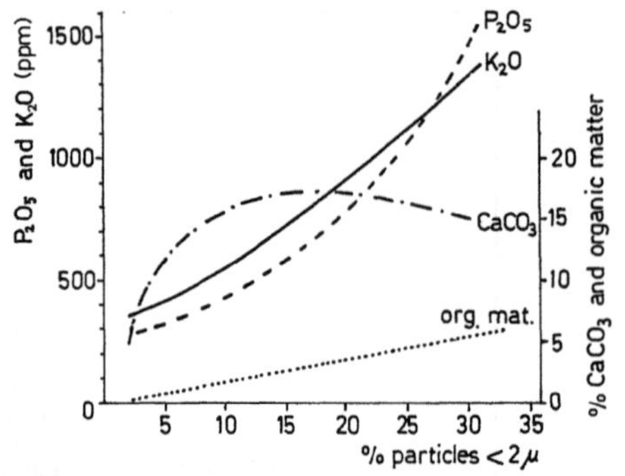

Fig. 2. The relationship between the amount of P_2O_5, K_2O, $CaCO_3$, organic matter and the lutum content in marine coastal sediments (Verhoeven, 1952).

environment (Fig. 2). A fresh silt shows a higher pore volume than an older deposit; a vegetated mud contains more organic matter per g of lutum, than a bare mud. In samples with the same lutum content the lime content in Zeeland (southwest) is as a rule higher than in the Wadden Sea area.

After drainage the clayey sediments which are no longer or only rarely inundated, dry and shrink, resulting in crack formation and subsidence. The pore volume decreases from 60–80% to 60–65% in the saltmarsh and 40–45% in the older polders. Although the total pore space diminishes, the number of wide pores increases; hence the aeration improves and in a complex of chemical (oxidation-)reactions the soil 'ripens.' This process of initial soil formation is well documented from the IJsselmeer polders (Zuur, 1961).

Percolating precipitation water washes the salt downwards. The leaching rate depends mainly upon the amount of water passing through the soil, although the soil moisture characteristics play a part too. The precipitation of only one average winter, as far as available for downward passage, may amount to 300 mm (for a sandy soil) and will be sufficient for the complete leaching of salt from the rhizosphere. However, in conditions of impeded drainage (high ground watertable, impervious layers) this may take several decades.

Capillary rise of groundwater during the summer (up to 150 mm in wetland conditions) may resalinize the surface layers. More complicated salinity patterns in the riparian zone may develop, either in relation to seepage of deeper groundwaters, or, in contrast, to an influx of lake water (fresh or saline) into the border zone (Fig. 3 a, b). The patterns of salt and water movements provide the predominant factors for the development of the vegetation.

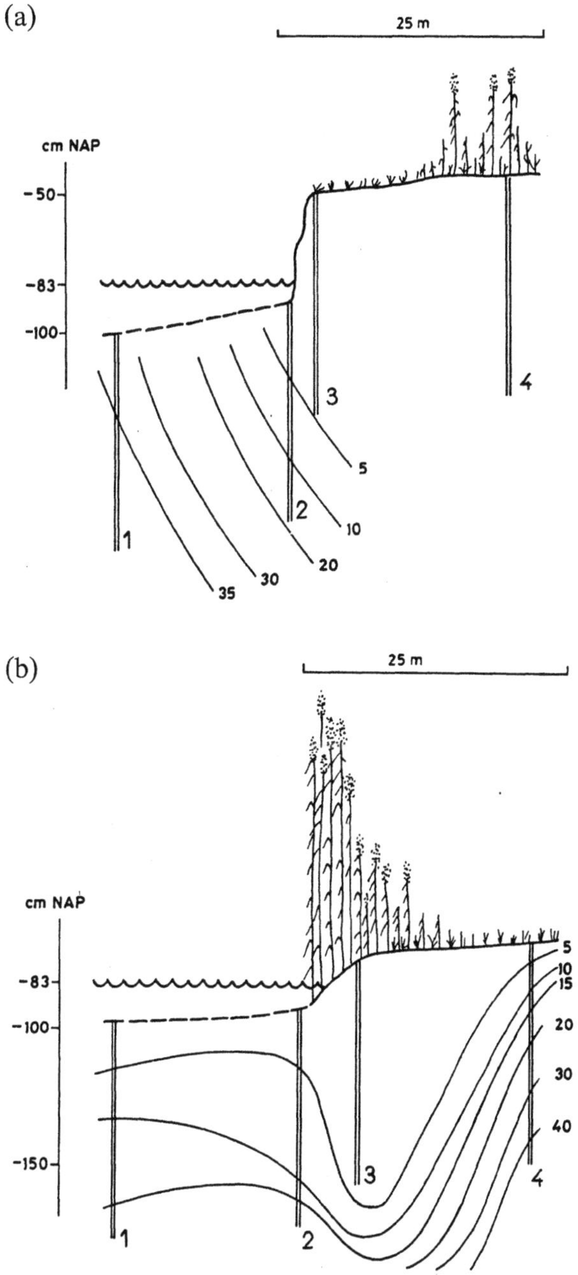

Fig. 3. Isohalines (g l^{-1} soil moisture) in the borderzone of the (fresh) reservoir in the Lauwerszeepolder, 10 years after the enclosure (after Pot, 1983); (a) a high and well-drained sandflat (b) a low, silty sandflat with impeded drainage.

Excluding the sands poor in lutum, the young marine sediments are moderately rich in phosphorus and potassium, although the K content tends to decrease with desalination. Part of the organic nitrogen, especially from decaying organisms, is readily mineralized. The NH_3 and NO_3 produced is lost in the early years by leaching and denitrification or accumulated in new plant and animal biomass (Bakker, 1958; Van Schreven, 1965).

Below the coarser sandflats a freshwater lens can be formed within a few years, as is described for the Braakman (Smit & Visser 1987) and the Lauwersmeer (Fig. 3a) and a fresh seepage border zone is maintained (Fig. 4a, b). In siltier sands the summer evapotranspiration causes a concave groundwater table far below the level of the lake and as a consequence the lake water seeps into the border area; a possible result of this process is depicted in Fig. 3b. This process may be reversed in winter. A remarkable phenomenon is reported from the Stampersplaat (Grevelingen), where a freshwater lens was formed to a depth of 10 m in 8 years, which extends under the lake (Fig. 4b), in this way enabling reed (*Phragmites australis*) to grow in the saline border zone (Visser *et al.*, 1987).

Succession on embanked tidal flats

The natural succession of a desalinating former tidal landscape has been documented on several occasions, especially in the Netherlands (a.o. Feekes 1936, 1943; Beeftink 1971, Joenje 1978). The process is described in some detail using data from the Lauwerszee-polder (Joenje, 1978).

Abiotic conditions and changes

The estuary of the Lauwerszee (9100 ha) was embanked in May 1969 and the tidal channels changed into a reservoir lake of some 2,100 ha, with a summer level at NAP (Dutch Ordnance Level) −0.83 m, enlarged by 200 ha at the winter level of NAP −0.93 m. The average depth of

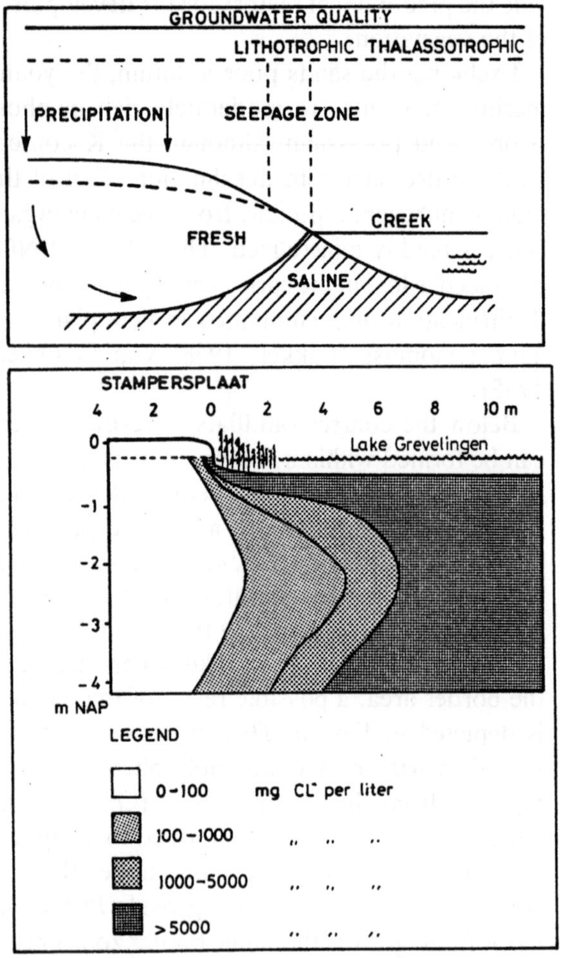

Fig. 4. (a) Hydrological processes in the borderzone along the western creek of the Braakman (after Smit & Visser, 1987). (b) Isohalines in the border of the Stampersplaat (Grevelingen) and reed vegetation, indicating desalinized subsoil *(loc. cit.)*.

this lake (the Lauwersmeer) is 2.30 m, with large shallows and some deep areas down to 10 m. About 300,000 ha of the mainland discharges on the Lauwersmeer. The water residence time in winter is only a few days, but in dry summer periods this may amount to several months. The saline waters have been replaced by eutrophic freshwater within a few months after enclosure.

The land surface comprises some higher saltmarshes and vegetated sedimentation fields (Dutch: landaanwinningsvakken) and a large surface of former tidal flats, 2100 ha of which are lower than 0.9 m above the average reservoir level. Without means of local drainage this area remains a true wetland, of which the lower parts are occasionally flooded by the lake. The level can rise half a meter and more, especially in stormy periods, when the sea levels at low tide remain high and prevent the discharge of water through the sluices (Fig. 5). Areas below NAP −1.20 m remain under water permanently and areas above NAP are almost never inundated, wheras the soils at NAP −0.50 m are flooded several times a year.

After enclosure the banks of the lake are attacked by waves at only one height level. Much sand has been transported into the deeper parts of the lake and consequently in many places the shoreline retreats. The soil type of the drained flats varies from fine sands relatively poor in lutum (between <3% to 3–5%) and silty sands (5–12% lutum). The grain sizes of the sand fraction decrease with increasing lutum content. Sands predominate in the centre and the North; in the other directions the lutum content rises, reflecting the declining tidal influence.

In sandflats the groundwatertable follows the reservoir level providing that (a) the sands (poor in lutum) have a good permeability, (b) the subsoil is permeable and (c) the flats have no silty 'shoulders' along their borders. In the less permeable sands the groundwater in summer may descend far below the polder level, or rise to the soil surface, depending on precipitation and evapo-transpiration. The lack of lateral groundwater movements and the low pore volume of these soils cause quick saturation, after which the precipitation flows as surface run-off along the height gradient into the lake. This means a less effective desalination.

In the lower areas the poor desalination is even counteracted by the capillary rise of saline groundwater during periods of high evaporation. This wetland condition with (ground-) waterlevels up to the soil surface during longer periods of the year, is illustrated in Fig. 6. The hydrological regime is called dynamic *mesoseries* (after Drost & Visser, 1981). Sometimes the groundwater levels only descend a few dm, *e.g.* under the influence of lateral supply of groundwater from higher areas; this situation, the stable *mesoseries*, offers

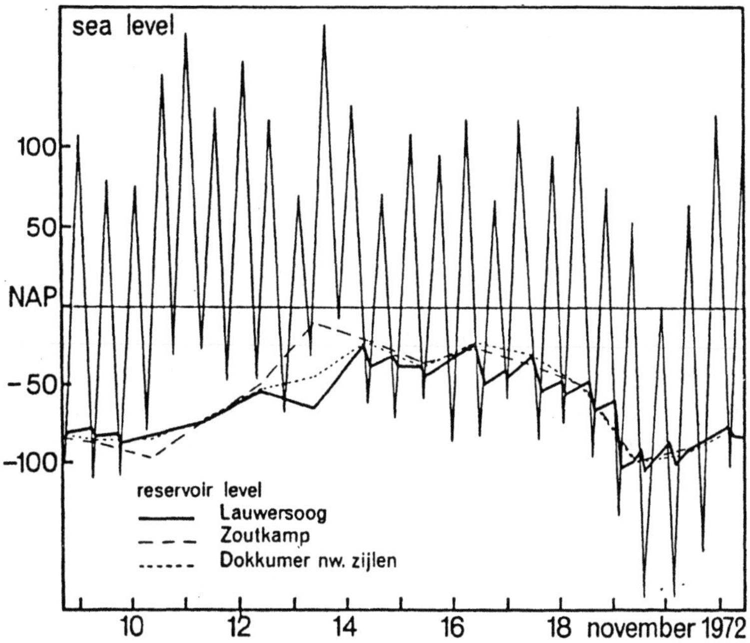

Fig. 5. The effect of a storm, 13 November 1972, on the water levels in the Lauwersmeer (Joenje, 1978).

a potential habitat for many interesting vegetation types and plant species from environments related to dune slacks.

The lime content of the soil varies from 2–3% in the sands to over 10% in the silty sands. As indicated earlier (Fig. 2), many other properties are correlated with the lutum content.

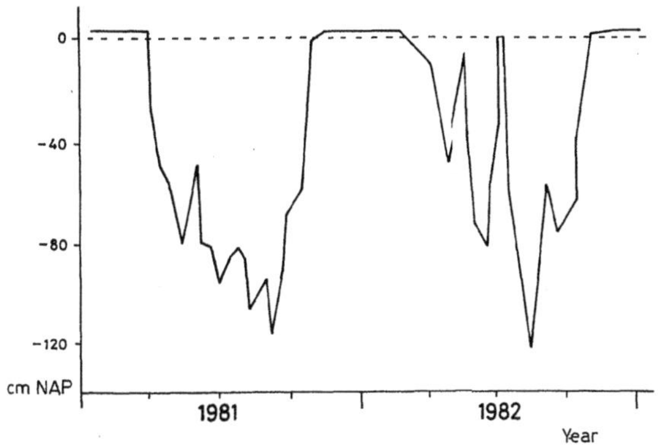

Fig. 6. The fluctuations of the groundwater level on a sandflat in the Lauwerszeepolder over a period of two years (dynamic *mesoseries*) (RIJP, 1985).

Pioneer plant species

The emerging new land presents a bare and salty environment. Only gradually, although in a few places rather quickly, the salt concentrations fall below 2 to 5 g per liter soil moisture (Fig. 7), at which levels several plant species from fresh environments can survive. Lethal salt concentrations may return occasionally, because of evapotranspiration. Plant colonization in the early years

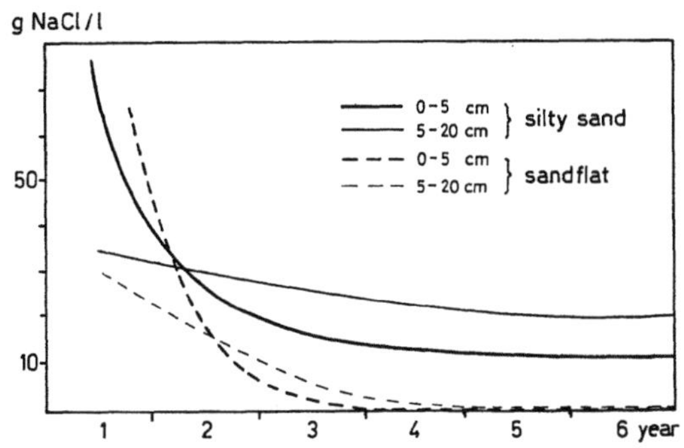

Fig. 7. Desalinizing upper soil layers of sand and silty sand in the Lauwerszeepolder during 6 years following the embankment.

is almost exclusively by halophytes and halotolerant species and the duration of this first phase is strongly dependent upon the rate of desalination (most data from Joenje, 1978a).

At the time of the enclosure, in May, the seeds of these plant species growing on the saltmarshes of the former estuary, were already present in the soil, transported there by the tidal currents and in densities correlated (again) with the lutum content of the soil. The sandiest area in the centre and the North remained void the first year and with densities of the pioneer plants of the annual species of *Salicornia brachystachya* and *S. dolychostachya*, *Suaeda maritima* and *Atriplex prostrata* increasing in the direction of the surrounding marshes (Fig. 8).

This sparse pioneer vegetation only reached densities of about 4000 plants per ha, but they thrived well and consolidated their early establishment by seeding the empty space successfully. Their seed dispersal was effected in the sheets of water blown across the saturated surface of the practically bare flats.

In the first year following embankment the hydrology and height level of the drained, saline surfaces barely influenced the colonization of the halophyte species, however this changed markedly in the following years.The populations of the annual halophytes demonstrated an exponential increase during the first four years, until they reached very high densities which caused competitive interactions between individuals. Exponential increase of biomass was also stopped by competition with newly arriving species (Fig. 9). Several species of halophytes, notably *Spergularia salina*, *Aster tripolium* and several species of *Puccinellia (P.maritima, P.cappillaris, P. distans ssp. distans, P. d. ssp. borealis*) had to start from

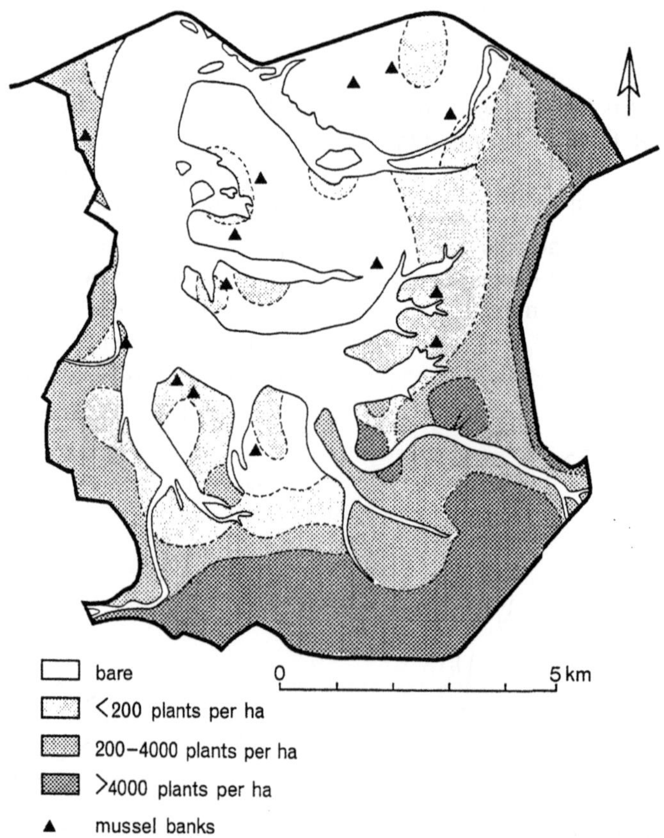

Fig. 8. The density of the pioneer vegetation the first year of the Lauwerszeepolder.

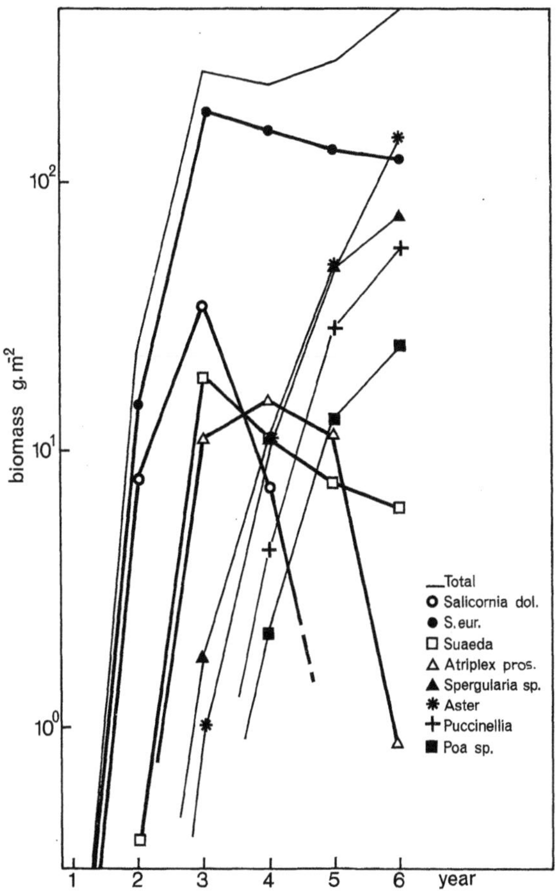

Fig. 9. The rise of the principal colonizing species in the first 6 years of the Lauwerszeepolder.

a few individuals (lack of a seed bank) or had to migrate in from the saltmarshes. Arriving much later, these species were equally able to increase their numbers exponentially, although the area was already densely occupied by the first invaders, notably the *Salicornia spp.* (Fig. 9).

The colonization by salt-tolerant species producing the characteristic patterns of seedling aggregates around the parent plants, was followed by the increasing dominance of perennial species (notably *Aster tripolium, Puccinellia spp., Spergularia maritima, Poa trivialis*). After 7 years the annual halophytes began to give way to a natural grassland vegetation, which in following years often became dominated by *Agrostis stolonifera* and *Phragmites australis*.

The foregoing description of the colonization as summarized in Fig. 9, parallels the qualitative descriptions given by Feekes for the Wieringermeer-polder (Feekes, 1936) and for the sandflats along the friesian coast of the IJsselmeer (Feekes, 1943). The colonization of the sandier Middelplaten in the Lake Veere (Beeftink *et al.* 1971) shows essentially the same pattern of 'come and go' with partly the same species (Fig. 10).

Apparently, where desalination proceeds over a period of several years, an orderly series of species waves occurs. By contrast, the plant col-

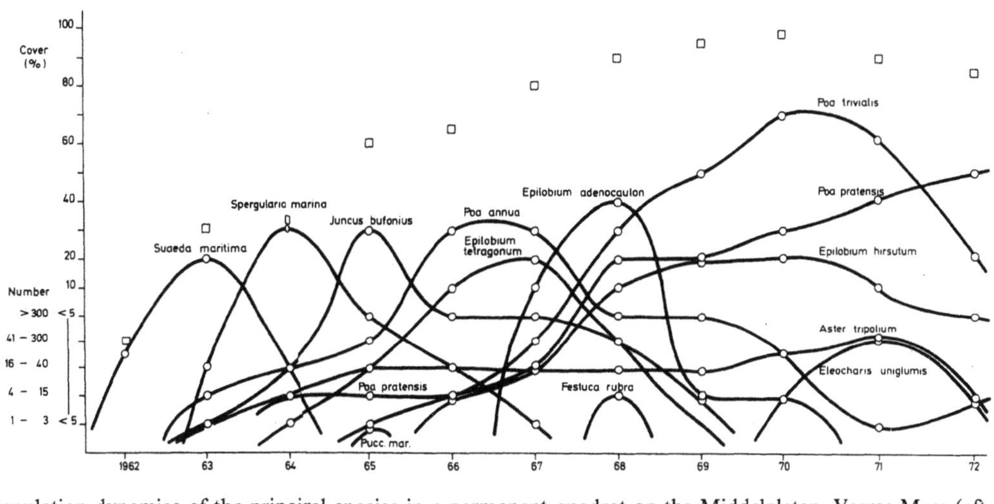

Fig. 10. Population dynamics of the principal species in a permanent quadrat on the Middelplaten, Veerse Meer (after Beeftink *et al.*, 1971).

onization of more quickly desalinizing musselbanks and former sedimentation fields, all silty and well-drained, proceeded capriciously and quickly. On isolated musselbanks 38 species of vascular plants were already present in the first year; apart from halophytes, common weeds from the surrounding mainland predominated, either with hard-scaled seeds (seeds of *Rumex, Polygonum, Chenopodium* a.o. apparently survived the marine conditions, being deposited before the enclosure), or arriving by air in the first year (species of *Chamaerion, Epilobium, Tussilago, Sonchus, Cirsium and Senecio*).

Within a few years perennial ruderal species came to dominate and the resulting dense and highly productive vegetation consisted of *Epilobium hirsutum, Chamaerion angustifolium, Urtica dioica, Cirsium arvense* and several grass species. This succession series is well-known from similar areas after embankment, *e.g.* in the southwest of the Netherlands (Slikken van Flakkee, Markiezaatsmeer) and in the new polders of the German Bight. The musselbanks and other small areas rapidly colonized by a glycophyte vegetation formed the centres from which later desalinizing places were effectively colonized.

Fauna

As with the plants, several animal species display a widespread, temporary dominance and some of them developed huge populations or 'plagues,' a most typical feature of early stages of succession. The vole population in the former saltmarsh increased and threatened the crop on the adjacent mainland, but also attracted many birds of prey, notably Short-eared Owl, species of Harriers and Buzzards. Several waves of Chironomid midges, a plague of noctuid caterpillars and mass occurrence of several species of beetles and spiders are reported (Meijer 1980). Most significant and striking however, were the high numbers of geese and ducks (Barnacle Goose, Wigeon, Teal etc.), attracted by the annual halophyte vegetation. The waterfowl harvested almost all the ears from *Salicornia* by stripping them off, even anticipating the crop in their arrival date (Fig. 11; Van Eerden, 1981).

Salicornia plants on the desalinizing flats contained less salt than in marine conditions, where they normally are not grazed.

The in winter saturated surface offered ideal conditions for Teal to sieve the seeds of *Salicornia* and *Spergularia spp.* from the silty puddles on a large scale. In later years Grey-lag Geese were observed destroying the soggy mat of *Agrostis stolonifera*, by trampling and digging for rhizomes and for the tubers of *Triglochin palustris*.

Vegetation succession on the sandflats

Whereas the annual halophytes occupied both the higher and lower flats, the subsequent invaders were each confined to one of the three main habitat types: the lower flats (silty, rather persistently saline and waterlogged and/or flooded in winter), the flats out of reach of most inundations by the

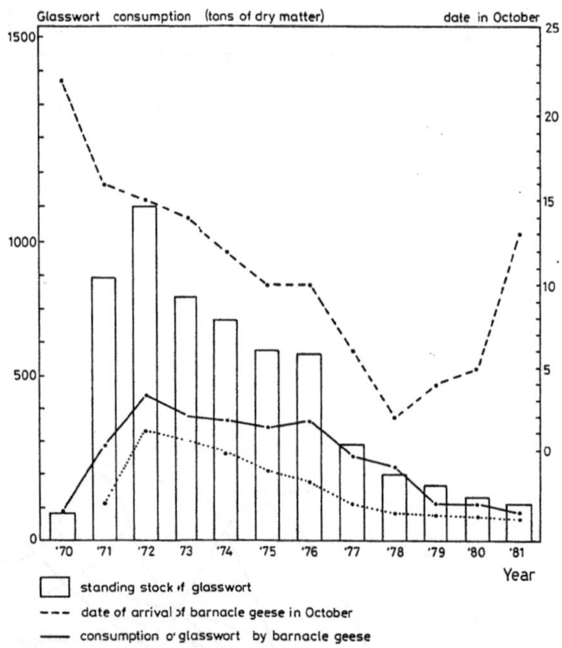

Fig. 11. Consumption of *Salicornia* seeds by Barnacle Goose and Wigeon in the Lauwerszeepolder, together with the dates of arrival of the geese, over a period of 12 years (after Van Eerden, 1985).

lake (still brackish with impeded drainage) and the higher sandflats (quickly desalinized and with a lower water table).

The silty sands after 8 years harboured a saltmarsh-like vegetation with species as *Puccinellia maritima, Triglochin maritima, Juncus gerardii, Carex extensa, Plantago maritima*; even species such as *Glaux maritima, Halimione portulacoides* and *Limonium vulgare* were occasionally found. After 6 to 8 years the less salt-tolerant species like *Agrostis stolonifera* and *Alopecurus geniculatus* gradually gained importance (Table 1).

The sandflats had attracted several species from dune slack-type environments, such as *Carex oederi pulchella, C. distans, Centaurium pulchellum, C. littorale, Odontites verna serotina, Parnassia palustris* and *Schoenus nigricans* (Bruinenberg *et al.*, 1980) (Table 1).

The higher situated and well-drained sands showed an early development of *Bryophytes (Bryum bicolor, Funaria hygrometrica* a.o., viz. Joenje & During, 1977), followed by many euchorous phanerogams from fresh habitats like *Epilobium spp., Erigeron canadense, Senecio vulgaris* etc., but also Reed and Willows were among the successful invaders (Table 1). *Calamagrostis epigejos, Hippophae rhamnoides, Festuca rubra, Carex arenaria* readily colonized this nutrient-poor and in some periods very dry habitat (*xeroseries*). This dry area occupying a few tens of ha, is stricktly speaking not a wetland, but forms the upper part of the characteristic soil- and height gradient of the type '*limes divergens*' (Van Leeuwen, 1966), a prerequisite milieu for botanical developments towards stable communities with rare and less common species.

Successional trends

The successional trends, arising from the establishments described, are towards *Junco-Molinion*-type vegetation on the high sands and *Lolio-Cynosuretum*-type vegetation on the siltier and moist or flooded areas, as well as towards many transitions. The development is illustrated in the

Table 1. Groups of plant species characterizing the three main habitat types (*xero-*, dynamic *meso-*, and stable *mesoseries*) of the flats in the Lauwerszeepolder after 8 years; silty sands reflect saline conditions.

Dry sandflat (*xeroseries*)	Sandflat (*d-mesoseries*)	Silty sands (*s-mesoseries*)
Cerastium fontanum	*Agrostis stolonifera*	*Agrostis stolonifera*
Cirsium arvense	*Alopecurus geniculatus*	*Armeria maritima*
C. vulgare	*Calamagrostis epigejos*	*Artemisia maritima*
Erigeron canadense	*Carex arenaria*	*Carex extensa*
Epilobium adnatum	*C. distans*	*Glaux maritima*
E. parviflorum	*C. serotina pulchella*	*Halimione portulacoides*
E. adenocaulon	*C. pulchellum*	*Limonium vulgare*
Phragmites australis	*Hippophae rhamnoides*	*Juncus gerardii*
Senecio vulgaris	*Juncus alpino-articulatus* ssp. *atricapillus*	*Plantago maritima*
S. arvensis	*Juncus maritimus*	*Spergularia media*
Taraxacum sp.	*Odontites verna* ssp. *serotina*	*Triglochin maritima*
Juncus bufonius	*Parapholis strigosa*	
Salix alba	*Parnassia palustris*	
S. viminalis	*Plantago coronopus*	
	Rubus caesius	
	Sagina nodosa	
	S. maritima	
	Salix arenaria	
	Schoenus nigricans	
	Senecio jacobaea	

Table 2. Plant species found after 8 years, indicating a trend towards *Junco-Molinion-* (nutrient-poor sands) or *Lolio-Potentillion-* (water meadow) type communities.

Species of *Lolio-Potentillion*	Species of *Junco-Molinion*
Agrostos stolonifera	*Agrostis capillaris*
Alopecurus geniculatus	*Agrostis canina*
Angelica silvestris	*Anthoxanthum odoratum*
Anthriscus silvestris	*Calamagrostis canescens*
Bidens tripartitus	*Cerastium glomeratum*
Cardamine pratensis	*Cirsium palustre*
Cicuta virosa	*Eriophorum angustifolium*
Eupatorium cannabinum	*Festuca ovina* ssp. *tenuifolia*
Lycopus europaeus	*Hieracium pilosella*
Lythrum salicaria	*Holcus lanatus*
Lychnis flos-cuculi	*Hypochaeris radicata*
Mentha aquatica	*Juncus articulatus*
Phalaris arundinacea	*J. effusus*
Potentilla anserina	*J. conglomeratus*
Rumex acetosa	*Luzula campestris*
R. crispus	*Poa pratensis*
Sium erectum	*Rumex acetosella*
S. latifolium	
Solanum dulcamara	
Stachys palustris	
Valeriana dioica	

species lists, given in Table 2. After a decade the uppermost soil layer of the high sands has already lost part of its Calcium carbonate by leaching, a process which continues together with the lowering of the pH and the accumulation of humus and nutrients in the upper soil layer (Joenje, 1978a). Decalcification processes in marine sediments are discussed by Verhoeven (1962).

On the silty sands, by contrast, the succession shows an extensive spread of *Agrostis stolonifera*, followed by a gradually thickening stand of Reed, often together with *Salix spp.* (notably *Salix repens* and in higher places *S. alba, S, viminalis* a.o.). Comparing this case with some older areas (*e.g.* the sandflats on the friesian westcoast of the IJsselmeer (Slager & Smit, 1987)), it can be concluded, that the *syntaxa* mentioned above will only gradually become saturated with their 'characteristic' species. The number of species could rise considerably, provided that the biomass of the dominants mentioned is removed periodically by some kind of management or 'nature exploitation', be it grazing, mowing or even turf cutting or an occasional controlled burn. This effect of the removal of the biomass, well-known in Dutch nature management practice and often discussed scientifically (e.g. Grime, 1979), makes it a crucial measure in most of the management schemes applied in the new polder nature reserves. Under the next heading the various activities applied in different polders will be considered, together with the prevailing ecosystem characteristics.

Perspectives for maintenance and management

With respect to the general parameters of an ecosystem – the biomass-, energy- and nutrient flows – the new wetlands cannot be taken as isolated areas. Indeed the low lying grounds are often directly influenced or inundated by nutrient-rich adjacent waters linking the system to an often large agricultural hinterland. The waters (reservoirs) then tend to eutrophicate (Berger, 1985). The lower and mostly silty grounds may also receive (excess) nutrients from inundation waters. They may remain saline for a period of several decades because of impeded drainage.

As long as there is no surface drainage system, the low grounds are characterized by a groundwater-table of the *hygroseries-* or stable *mesoseries*-type (Fig. 12). This environment is to be regarded as a proper wetland condition. The higher and drier (sandier) grounds, well out of reach of inundations, have the groundwater regime of the *xeroseries*.

The two environments mentioned are mostly linked by characteristic soil- and height transitions, either steep or gradual, reflecting the tidal origin of the landscape and also presenting a most valuable abiotic base for natural succession towards stable and species-rich communities. It follows, that developments in the area and the value as a nature area have to be judged within this landscape- and ecological context.

The lower area, base- and nutrient-rich, and the higher sands where leaching prevails, follow different successional series. The environmental dynamics in the lower zone (deposits of debris,

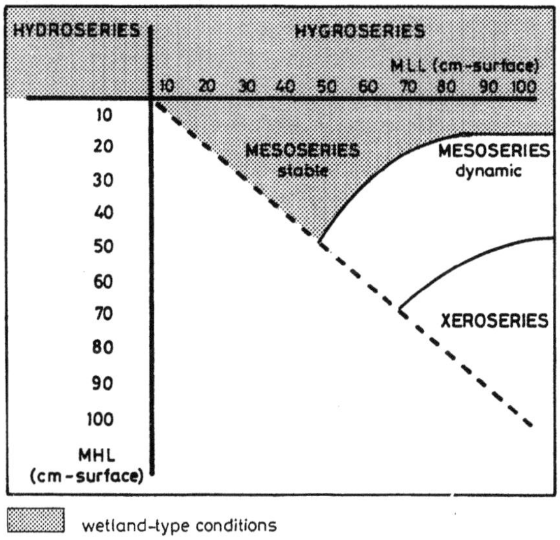

Fig. 12. The relation between (*hygro-*, *meso-* and *xero-)series* and the average shallowest groundwater level (after Smit & Visser, 1987). The hatched area represents wetland condition (MHL: mean highest level; MLL: mean lowest level).

silt, icesheets) and the changes in (ground)water levels sustain a considerable annual production by a coarse vegetation of Reed and Willows, the debris of which is accumulating. This development is found in the Lauwerszeepolder, on the Wadden-islands, along the Amstelmeer, along the borderlakes of the IJsselmeer-embankments, as well as in the Delta region.

An older example is found on the sandflats along the friesian coast of the IJsselmeer. In Fig. 13 a transect of the Piamer Kooiwaard, a sandflat, is depicted. Here, after some 50 years a 'laissez-faire' management has produced a *Phragmites, Scirpus maritimus* and *Salix* wilderness poor in species. A distinct peat deposit is found on the intermediate level of the height gradient, which is above most inundations. Here the influence of precipitation prevails and acid conditions develop, in which *Eriophorum angustifolium* is found and *Sphagnum* increases. Since the start of a mowing regime and after the removal of the willowtrees, a species-rich meadow vegetation has developed along the height gradient, gradually replacing the *Phragmition* communities. By now 180 species of vascular plants with *e.g. Carex tumidicarpa, C. lasiocarpa, C. diandra, Dactylorhiza majalis* and *Pedicularis palustris*, and, slightly higher, *Rhinanthus angustifolius, Lychnis flos-cuculi*, are reported (Slager & Smit, 1987; viz. also Feekes, 1943). Still a halophyte vegetation persists in some depressions, with *Puccinellia maritima, P. distans, Juncus gerardii, Triglochin maritima, Glaux maritima* and *Parapholis strigosa* (compare Tables 1 and 2). This resilient saline

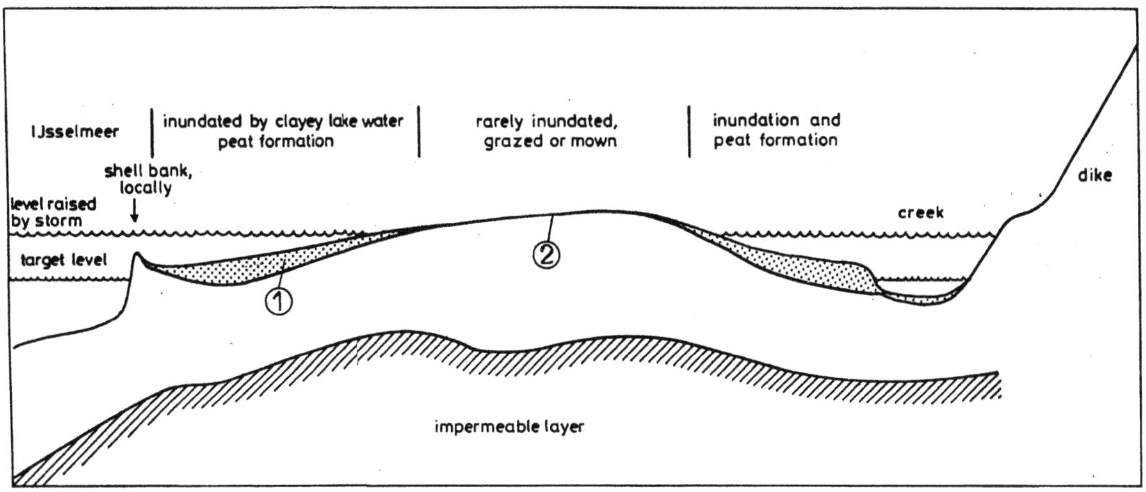

Fig. 13. The hydrological zones in the Piamer Kooiwaard (after Slager & Smit, 1987). (1) Vegetation of *Phragmites, Scirpus maritimus* and Willows. (2) Grassland locally rich in species.

element resembles that described by Westhoff (1969) for permanent quadrats in the marshes of the former Zuiderzee.

A very interesting example of these successional transitions is reported from Denmark (Gravesen, 1972) in the area Tipperne of the Ringkøping fjord, as a result of the partial removal of the marine influences. Furthermore, the brackish to fresh conditions along the coasts of the Baltic create numerous locations with quite similar vegetation types (Gillner, 1960, Wallentinus, 1973).

The land use or nature management most often applied, is extensive grazing by cattle and horses; sheep are vulnerable in these areas since they tend to become infested by the liver fluke. The large areas even make it a profitable use (Lauwerszeepolder, Grevelingen). It not only restores and maintains the ecological conditions for smaller and less competitive plant species (open sward, locally saline), but it also keeps the vegetation accessible as a feeding ground for geese and other waterfowl and wintering waders, as well as providing the breeding biotopes for meadowbirds (Van Eerden, 1985; Joenje, 1981).

Experimental management of these new landscapes has been tried out not only in the Netherlands, but also in Great Britain, Germany and Denmark. Indeed, the results are difficult to compare in view of the various activities and the (phyto) geographical differences.

Threats

Finally, one factor common to all these new or still young nature areas has to be stressed as the most crucial for retaining their value as natural wetland communities. This is the groundwater regime, the (stable) *mesoseries*, in relation to the level of the surface (reservoir) waters; preferably, though not necessarily, occasional inundations should occur in the winter half of the year.

Whether it is recreation, military training or agriculture, almost every activity on these areas involves drainage or other alterations to water regimes. Thus we may conclude that in order to develop nature conservation interests in a wetland perspective,

- the management of the water level fluctuations, preferably lower summer and higher winter levels, has to be controlled and consistent and
- no surface drainage system should be imposed.

References

Bakker, D., 1958. Over de betekenis van stikstof voor de natuurlijke vegetatie der IJsselmeerpolders. Van Zee tot Land 26: 53–66.

Beeftink, W.G., 1975. The ecological significance of embankment and drainage with respect to the vegetation of the southwest Netherlands. J. Ecol. 63: 423–458.

Beeftink, W.G., M.C. Daane & W. de Munck, 1971. Tien jaaroecologische verkenningen langs het Veerse Meer. Natuur enLandschap 25: 50–63.

Berger, C., 1985. Hydrobiologische aspecten van het Lauwersmeer. In: RIJP, Vijftien jaar afgesloten Lauwerszee. Lelystad: 117–125.

Bruinenberg, J., W. Joenje & T. Wierenga, 1980. Hapaxanth species of coastal beach plains colonizing embanked sandflats. Acta Bot. Neerl. 29: 497–508.

Drost, H.J. & J. Visser, 1981. Het grondwaterregime als structurerende factor voor de begroeiing in afgesloten estuaria met een toepassing in het Grevelingenbekken. Flevobericht 163: 201–207. Rijksdienst voor de IJsselmeerpolders, Lelystad.

Feekes, W., 1936. De ontwikkeling van de natuurlijke vegetatie in de Wieringermeerpolder, de eerste groote droogmakerij van de Zuiderzee. PhD. Thesis, Wageningen.

Feekes, W., 1943. De Piamer Kooiwaard en Makkumerwaard. Ned. Kruidk. Arch. 53: 288–331.

Gilllner, V., 1960. Vegetations- und Standortsuntersuchungen in den Strandwiesen der Schwedischen Westkuste. Acta Phytogeographica Suec. 43: 1–98.

Gravesen, P., 1972. Plant communities of saltmarsh origin at Tipperne, Western Jutland. Bot. Tydsskr. 67: 1–32.

Grime, J.P., 1979. Plant strategies and vegetation processes. John Wiley & Sons, Chichester.

Joenje, W., 1978. Plant colonization and succession on embanked sandflats. PhD. Thesis, Groningen.

Joenje, W., 1982. Nature in new Wadden-polders: conservation by exploitation. In: Polders of the World. ILRI, Wageningen.

Joenje, W. & H.J. During, 1977. Colonisation of a desalinating Waddenpolder by bryophytes. Vegetatio 35: 177–185.

Meijer, J., 1980. Colonization of the Lauwerszeepolder by some elements of the arthropod fauna. PhD. Thesis, VU Amsterdam.

Pot, R., 1983. Ontziltingsgeschiedenis van enkele oevers in de

Lauwerszeepolder. Report Dept. Plant Ecology, State Univ. Groningen.

Schultz, F. 1987, In: J.J. Vonk & R.T. de Boer, 1989. Inleiding tot de inrichting van het Landelijke gebied. Pudoc, Wageningen.

Slager, H. & G.F.J. Smit, 1987. De Waarden langs de friese IJsselmeerkust: samenhang tussen bodem, hydrologie en vegetatie. Landschap 4: 264–273.

Smit, G.F.J. & J. Visser, 1987. De samenhang tussen bodem, hydrologie en vegetatie in het Veerse Meer en de Braakman. Landschap 4: 274–288.

Van der Meijden, R., E.J. Weeda, F.A.C.B. Adema & G.J. de Joncheere, 1988. Flora van Nederland. Wolters-Noordhoff, Groningen.

Van Eerden, M.R., 1985. Aspecten van de ontwikkeling van de Vogelbevolking. In: RIJP, Vijftien jaar afgesloten Lauwerszee. Resultaten van onderzoek en ervaringen met inrichting en beheer. Flevobericht 247: 145–172. Rijksdienst voor de IJsselmeerpolders, Lelystad.

Van Schreven, D.A., 1965. Stikstofomzettingen in jonge IJsselmeerpolder gronden. Van Zee tot Land 41, Tjeenk Willink, Zwolle.Verhoeven, B., 1952. Embankment and cultivation of marine foreland. Soil Science 74: 91–95.

Verhoeven, B., 1952. Embankment and cultivation of marine foreland. Soil Science 74: 91–95.

Verhoeven, B., 1963. On the calciumcarbonate content of young marine sediments. Bull. Intern. Inst. Land Reclamation and Improvement 4: 6–27.

Visser, J., G.J. Rook & D.J. Fluijt, 1985. Ontziltings- en verziltingsprocessen op de oevers van het Grevelingenmeer. Landschap 2: 226–234.

Wallentinus, H.G., 1973. Above ground primary production of a *Juncetum gerardii* on a Baltic seashore meadow. Oikos 24: 200–219.

Westhoff, V., 1969. Langjährige Beobachtungen an Aussüszungs- Dauerprobeflächen beweideter und unbeweideter Vegetation an der ehemahligen Zuiderzee. In R. Tüxen (ed.) Experimentelle Pflanzensoziologie. Dr W. Junk Publishers, The Hague: 246–255.

Zuur, A.J., 1961. Initiele bodemvorming bij mariene gronden. Med. van de Landbouwhogesch. en de Opzoekingsstations van de Staat te Gent 24: 7–33.

Hydrobiologia **265**: 195–201, 1993.
E. P. H. Best & J. P. Bakker (eds), Netherlands-Wetlands.

Nutrient-poor freshwater wetlands

L.W.G. Higler
DLO Institute for Forestry and Nature Research, P.O. Box 46, 3956 ZR Leersum, The Netherlands

Key words: nutrient-poor waters

Abstract

Until the introduction of fertilizers in the 19th century nutrient-poor conditions prevailed in the larger part of The Netherlands. Freshwater wetlands consequently were of nutrient-poor nature. During the first half of the 20th century many surface waters gradually eutrophied, but most waters in the Pleistocene area and even some polders in the Holocene area kept their nutrient-poor conditions. Only during the last decades virtually all waters became eutrophied or polluted. The main sources of nutrients are agriculture, precipitation and water from the large rivers.

The lack of nutrients

The areas with originally nutrient-poor waters are depicted in Fig. 1 and cover over 50% of The Netherlands.

The trophic status of surface waters in the beginning of the 19th century was dominated by factors such as the composition of the soil and subsoil, the amount and composition of precipitation and the influence of the sea and rivers. Sandy soils prevail in the eastern and southern part of The Netherlands, predominantly from Pleistocene origin. Clay, silt and peat soils occur largely in the northern and western part, the Holocene district. The precipitation surplus of 200 to 400 mm a year is inherent to the atlantic climate. This causes a downward water flow in the higher parts of the country, the Pleistocene district and the dunes, causing leaching of minerals and nutrients to deeper soil levels. Surface waters in these areas, therefore, used to be poor in minerals and nutrients. This phenomenon occurred in stream valleys, actively growing raised bogs and moorland pools, largely in the Pleistocene district, and in the wet dune slacks along the western and northern coast. It did not occur in the rest of the country where the surface waters must have been rich in minerals and nutrients, because the rivers and the sea had brought fertile silt.

Social developments up to the twentieth century and their consequences for the environment

The larger part of the country was very nutrient-poor and hostile to human settlement until the Middle Ages. From this time on the environment gradually changed in the Pleistocene district by the development of cattle and sheep farming. The grazing by cattle and sheep caused a further oligotrophication in the forests, but the dung fertilized the areas near the villages around the stables where the animals were kept at night (Bieleman, 1987). In this way gradients between nutrient-poor and nutrient-rich areas came into existence. Small nuclei of fertilized tillages became surrounded by far larger areas of former heathlands and later also aeolian drift sands (Schroevers,

1980, 1986). In the same period the human population started on one hand draining and, on the other hand reclaiming peat bogs, and using human excrements for fertilization purposes, especially around cities. The turf excavation from the peat bogs for burning purposes simultaneously was removal of organic material from one site, creating nutrient-poor stagnant waters, and increase in organic matter at other sites. The first reclamation of lakes on sea silt started in the 15th century and, subsequently, international trade introduced minerals and nutrients from England, Scotland and the Baltic countries in the 17th and 18th centuries (Schroevers, 1980, 1986). The introduction of artificial fertilizers (phosphate) in the 19th century, finally, initiated a vast intensification in and expansion of agriculture.

Oligotrophy in the first half of the twentieth century

The possiblities for nutrient-poor surface waters did not differ very much from those on the map of Fig. 1 in the beginning of the century. The general character of many surface waters basically had not changed, although the application of fertilizers and pollution from a growing human population undoubtedly had raised their trophic status. Hundreds of nutrient-poor and mildly nutrient-rich moorland pools, remnants of nutrient-poor peat bogs and numerous small streams fed by the peat bogs and by groundwater were present in the Pleistocene district, which was still an area with poor soils, heathlands and small farms. Most streams were still naturally shaped and unpolluted.

A chain of marshes and lakes, resulting from peat excavation and partially fed by streams from the higher grounds stretched from northeast to southwest bordering the Pleistocene. These lakes and broads varied from mildly nutrient-rich to nutrient-rich and in the southwest some of them were brackish. There were places where the formation of peat, fed by rainwater, had started again, thus forming nutrient-poor conditions. The stretch of marshes and lakes was interrupted by the Zuiderzee which was in open connection with the Wadden Sea. The former peat bogs in the Holocene district had been reclaimed now, but in the heart of some of the polders nutrient-poor conditions still existed (De Vries, 1929; C.G. van Leeuwen, pers. comm.). The eutrophication was visible near the boundaries of the polders, where the reservoirs with nutrient-rich water, farms and villages formed the eutrophying component of a gradient via mildly nutrient-rich to nutrient-poor.

In the dune areas there were wet valleys, small and some larger lakes, most of which containing fresh water (Cl^- content $< 300 \text{ mg l}^{-1}$), low in nitrate and phosphate, but especially south of Bergen (roughly northwest of Amsterdam) with higher Ca-contents. Many small rivulets drained nutrient-poor groundwater from the dunes into the polders.

In summary the surface waters in The Netherlands showed a rich variety in different types, where the lack of nutrients was a normal condition, especially in the Pleistocene district and where pollution was a rare and local phenomenon.

The eutrophication of originally nutrient-poor systems

The main factors determining the trophic status of systems as described earlier have changed in such a way that conditions for nutrient-poor water hardly can be found nowadays. The composition of the soil has changed by enormous amounts of artificial fertilizers. The Netherlands' position with respect to the use of artificial fertilizer as compared to that in the rest of Europe is depicted in Fig. 2 (Schroevers, 1980). The use of liquid manure from fattening farms has increased considerably since 1965, the year in which the map was drawn. The total number of livestock in 1989 is 112 million animals (Agricultural Research Institute, 1990). It is subdivided over 13.7 million pigs, 4.8 million cows, 1.4 million sheep, 90 million chickens, 713,000 ducks and 900,000 turkeys. The food for most of these animals comes from other countries, the manure is staying within

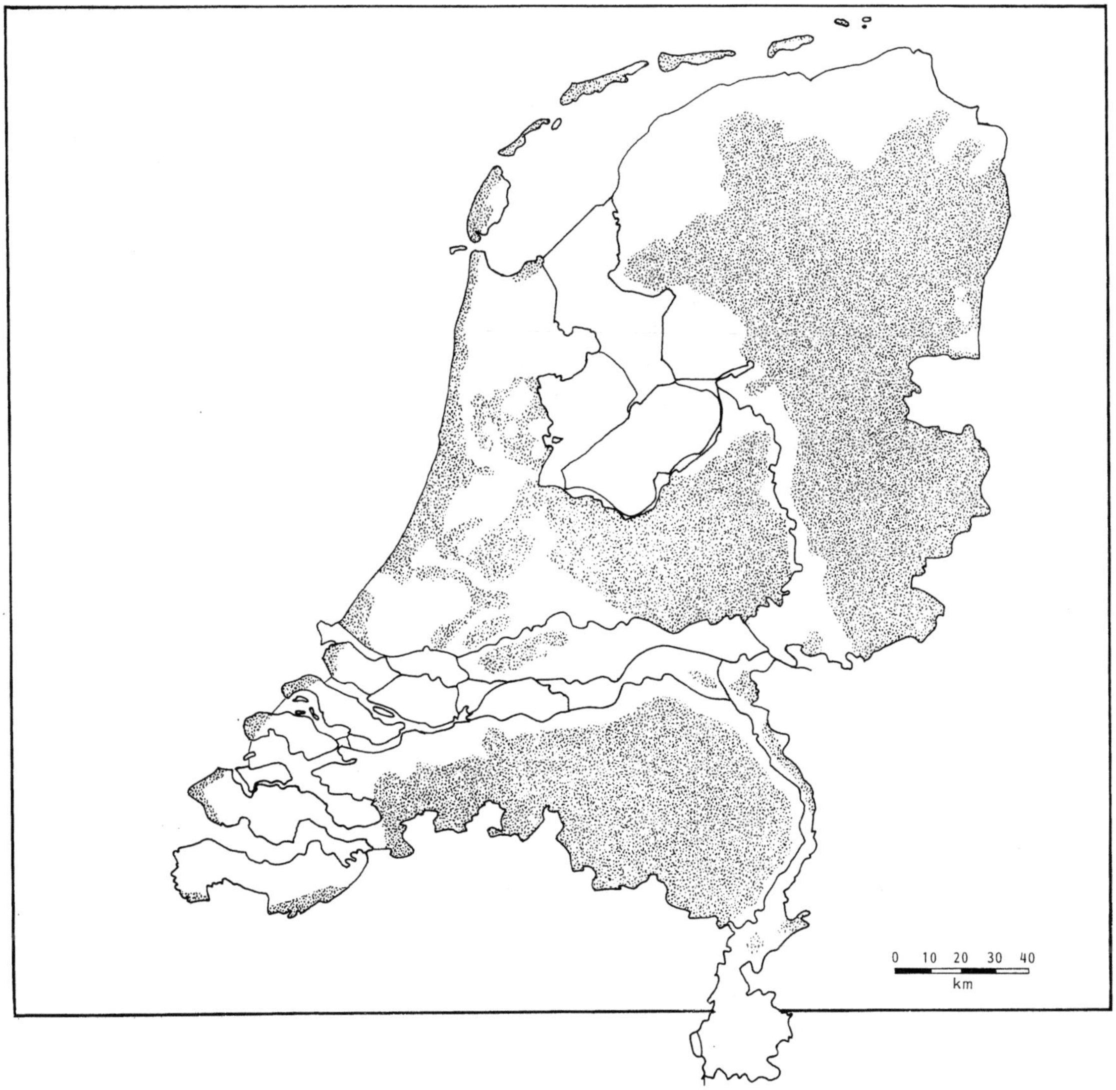

Fig. 1. Areas with potentially nutrient-poor wetlands in The Netherlands.

our boundaries and causes an overfertilization of especially the nutrient-poor sandy soils. By surface run-off and leaching to the groundwater many surface waters and seepage areas are being eutrophied in this way.

Precipitation nowadays has not only acidified, but also contains large amounts of nitrogen in different chemical combinations, resulting in N-loads of some tens to 100 kg $ha^{-1} yr^{-1}$ at present. This was about 7 kg ha^{-1} yr^{-1} on average in 1938 (Leeflang, 1938). The total load of ammonia, produced in The Netherlands, is 260.10^6 kg a year, most of which comes from the abovementioned livestock. Merely this eutrophying effect of the precipitation is sufficient to raise the trophic status of most nutrient-poor waters in the country. The paradoxal situation now exists, that the former nutrient-poor area, where the highest concentrations of fattening farms are to be found, is the most eutrophied part of the country.

The input of nutrients has, apart from the already mentioned sources, been increased by effluents of purification plants and many other direct and indirect sources. In 1980 there was an

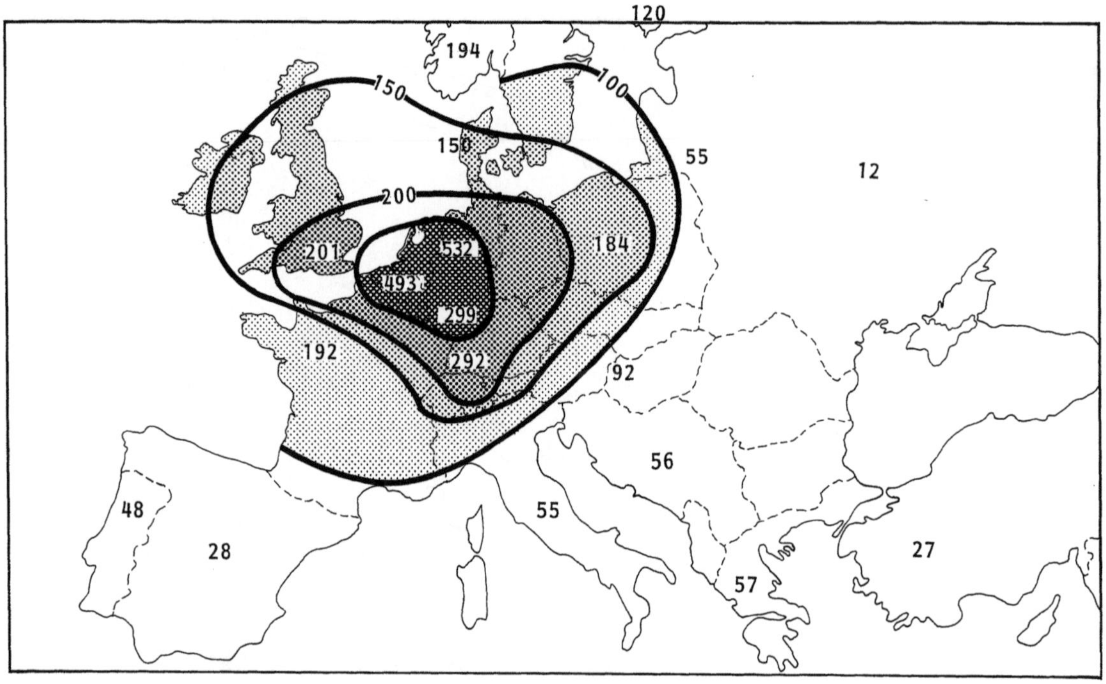

Fig. 2. Mean quantities of artificial fertilizer per unit arable land in Europe in 1965, expressed as the total of nitrogen, phosphate and potassium in kg ha^{-1} (Schroevers, 1980).

accumulation of 132.10^6 kg P of which 14.10^6 kg ended up in surface waters (Olsthoorn, 1984) and recent estimations show that tens of thousands of hectares of agricultural land are potentially saturated with phosphates (Breeuwsma & Schoumans, 1987). In contrast with former ideas about the retention capacity of soils for phosphates, there is proof of leaching out to groundwater and seepage water feeding streams and ditches (Breeuwsma *et al.*, 1989). Each year 300.10^6 kg N is reaching the groundwater and several pumping stations for drinking water had to be closed because of overstep of the international standard of 50 mg NO_3 l^{-1}. The increase in NO_3 contents in streams by the feeding groundwater has been observed already in the late 1960s (Higler & Repko, 1981).

Finally, the influence of the river Rhine and to a lesser extent of the river Meuse has to be mentioned. Figure 3 shows the areas where water from these rivers is introduced. The Netherlands, which was basically a wet country, is drying out nowadays by the use of groundwater for drinking water purposes and by the lowering of the groundwater table for agricultural purposes (Leuven & Bles, 1989). In the larger part of the country the groundwater table is lowered with at least 20–35 cm, near pumping stations sometimes over a meter. In dry periods of extremely dry years practically all surface waters in the country can be supplied with allochthonous water (Third Water Policy Plan, 1989). This results in a uniform water quality with a generally higher nutrient load than originally present in the majority of Dutch surface waters (Roelofs, 1989).

Threats and perspectives

Despite the presence of many nutrient-poor wetlands until the beginning of the century, and for the larger part even until the 1950s the present situation is such that even the last remnants are disappearing rapidly. Prognoses for the future confirm the negative picture that has been given here. A study group of the Utrecht University

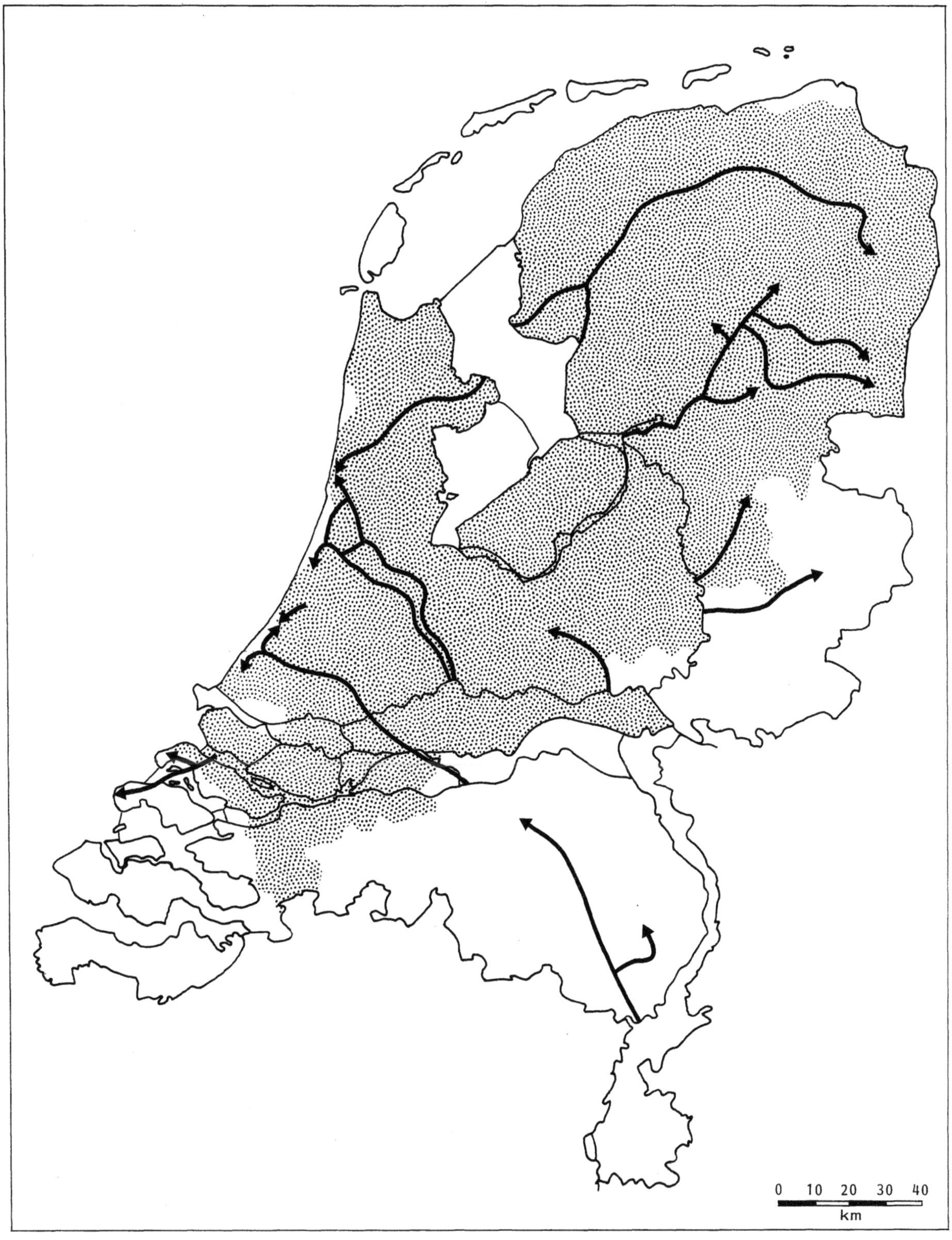

Fig. 3. Areas in which water from the river Rhine is or can be introduced.

studied developments of nutrient-poor environments on Pleistocene soils in the period 1900–2025 and the results of this study point to a further deterioration of the groundwater and surface water (Study Group on Future Planning of Oligotrophic Environments, 1983). Its statement that it may be possible to protect the remaining 25% of nutrient-poor areas by the instalment of buffer areas deserves more attention than hitherto is given. The group describes an optimistic scenario that must be worked out in the near future to preserve at least some examples of nutrient-poor wetlands. This scenario contains sweeping changes in governmental policy, in general mentality with respect to the environment and in agricultural aims. Proposals with respect to concentration of new buildings in existing agglomerations, desisting from the building of new roads and scaling down in agriculture are hesitatingly propagated nowadays.

The main threats for the remaining nutrient-poor wetlands are the nutrient contents of precipitation, runoff from fertilized land, groundwater and allochthonous water (predominantly from the river Rhine). Sources and seepage areas are for their nutrient status dependent on groundwater composition which will worsen definitely. Small watercourses in the Pleistocene part of the country receive water from surface runoff, interflow and base flow. Meinardi & Van der Valk (1989) investigated the N-contents in streams and small watercourses in the Pleistocene sand area and calculated the total N-load in 1990 and 2010 on the basis of the present status of fertilization. The results show that winter concentrations are higher than summer concentrations in many areas and that in general a further increase in N-content is to be expected. In a very few cases the 2010 situation is similar to that of 1990. The authors postulate an increasing eutrophication by phosphates, either from drainage of the phosphate-saturated areas, or from surface runoff. A total stop in fertilization will not influence the increasing release of phosphates. Some small stagnant water bodies in the Pleistocene district and the dunes which are isolated from groundwater or receiving nutrient-poor groundwater have a little chance to remain nutrient-poor. This also holds for the few remnants of raised bogs in the Pleistocene district. Both water types, however, keep receiving more or less eutrophied precipitation. There is little chance that nutrient-poor freshwater wetlands will subsist.

References

Agricultural Research Institute, 1990. Landbouwcijfers 1990. Agricultural Research Institute, The Hague, 249 pp.

Bieleman, J., 1987. Boeren op het Drentse zand 1600–1910: een nieuwe visie op de oude landbouw. Studia Historica: part XV. Hes, Utrecht, 834 pp.

Breeuwsma, A. & O.F. Schoumans, 1987. Forcasting phosphate leaching from soils on a regional scale. In W. van Duyvenboden & H.G. van Waegeningh (eds), Proc. Int. Conf. Noordwijk aan Zee. Vulnerability of soil and groundwater to pollutants: 973–981.

Breeuwsma, A., J.G.A. Reyerink, D.J. Brus, H. van het Loo & O.F. Schoumans, 1989. Fosfaatbelasting van bodem-, grond- en oppervlaktewater in het stroomgebied van de Schuitebeek. The Winand Staring Centre, Wageningen. Report, 91 pp.

De Vries, D.M., 1929. Het plantendek van de Krimpenerwaard III. Over de samenstelling van het Cempensch Molinietum coeruleae en Agrostidetum caninae. Een phytostatische bijdrage tot de associatiewetenschap. Kruidkundig Arch. Afl. 2: 1–401.

Higler, L.W.G. & F.F. Repko, 1981. The effects of pollution in the drainage area of a Dutch lowland stream on fish and macro-invertebrates. Verh. Internat. Verein. Limnol. 21: 1077–1082.

Leeflang, K.W.H., 1938. De chemische samenstelling van den neerslag in Nederland. Chemisch Weekblad 35: 658–664.

Leuven, R.S.E.W. & F.J.J. Bles (eds), 1989. Verdroging in Nederland. Oorzaken, omvang en oplossingen. Stichting Natuur en Milieu, Utrecht, 188 pp.

Meinardi, C.R. & J.P. Van der Valk, 1989. Het stikstofgehalte in beken en waterlopen van het Nederlandse zandgebied. National Institute of Public Health and Environmental Protection. Report nr. 728472016, 49 pp.

Olsthoorn, C.S.M., 1984. De fosfaatbelasting van het Nederlandse oppervlaktewater. Chemisch Magazine. Bijvoegsel bij Chemisch Weekblad 14: 213–216.

Roelofs, J.G.M. (ed.), 1989. Aanvoer van gebiedsvreemd water. Omvang en effecten op ecosystemen. Department of Aquatic Ecology and Biogeology, University of Nijmegen, 146 pp.

Schroevers, P.J., 1980. Een kwestie van schaal. Veranderingen in flora en fauna weerspiegelen economische geschiedenis. Phase 1: 30-39.

Schroevers, P.J., 1986. Eutrophisation et paysage. Lecture in the Paris colloqium 'Protection de l'environnement et aménagement des deux fleuves européens le Rhône et le Rhin, en France et aux Pays-Bas'. Unpublished report. Research Institute for Nature Management, Leersum, 12 pp.

Study Group on Future Planning of Oligotrophic Environments, 1983. Lange-termijnontwikkelingen van voedselarme milieus en grondwater van de pleistocene zandgronden. Een verkenning van de periode 1900–2025. State University of Utrecht, 289 pp.

Third Water Policy Plan, 1989. Derde Nota Waterhuishouding. SDU, The Hague, 297 pp.

Van der Maarel, E. & P.L. Dauvellier, 1978. Naar een globaal ecologisch model. SDU, The Hague, 2 volumes.

Hydrobiologia **265**: 203–215, 1993.
E. P. H. Best & J. P. Bakker (eds), Netherlands-Wetlands.

The Bourtanger Moor: endurance and vulnerability of a raised bog system

W.A. Casparie
Biological-Archaeological Institute, University of Groningen, Poststraat 6, 9712 ER Groningen, The Netherlands

Key words: climatic factors, hydrology, anthropogenic influences, drainage

Abstract

From the study of a raised bog complex in the former Bourtanger Moor in the northeastern Netherlands, which was formed between ca. 4500 BC and 1700 AD, it is clear that raised bogs are both very robust and vulnerable systems, which can develop within fairly broad climatological limits, but that they are also extremely sensitive to the effects of drainage caused by reclamation carried out by man. By means of good management, aimed primarily at restoring the typical ombrogenous hydrological system, in all likelihood it will be possible to develop the growth of raised bog in nature reserves suitable for this purpose, namely peat-bog relicts.

Introduction

Vulnerability of raised bogs

Raised bogs are known to be vulnerable systems. Of the raised bogs of The Netherlands, for example, none has survived the onslaughts of the last 500 years or so. The typical open raised-bog landscape no longer exists in The Netherlands. Most of the raised bogs have disappeared as a result of erosion or peat-digging following artificial drainage. This is also the case with the Bourtanger Moor.

The original raised-bog plant communities have become extinct. Whatever has remained in the form of remnants of raised bogs cannot develop by itself into a living raised bog. This indicates the great vulnerability of raised bogs.

The former Bourtanger Moor

The Bourtanger Moor is situated in the Northeastern part of The Netherlands. In former times it constituted the natural border with Germany over a considerable distance. Since the 17th century AD more than 98% of it has been dug away.

The actual study area is situated in the Hunze valley, an ice marginal valley that originated in the Saalian glacial, in which the western branch of the Bourtanger Moor developed (Fig. 1). The peat stratigraphy of this area is well-known on the basis of extensive paleobotanical research (Van Zeist, 1955; Casparie, 1972; Casparie & Streefkerk, 1992). Follow-up studies of the hydrology have been carried out to obtain insight into the possibilities for the regeneration of raised bog in nature reserves (Streefkerk & Oosterlee, 1984; Streefkerk & Casparie, 1987, 1989). Our knowledge of raised bog formation and the mechanisms

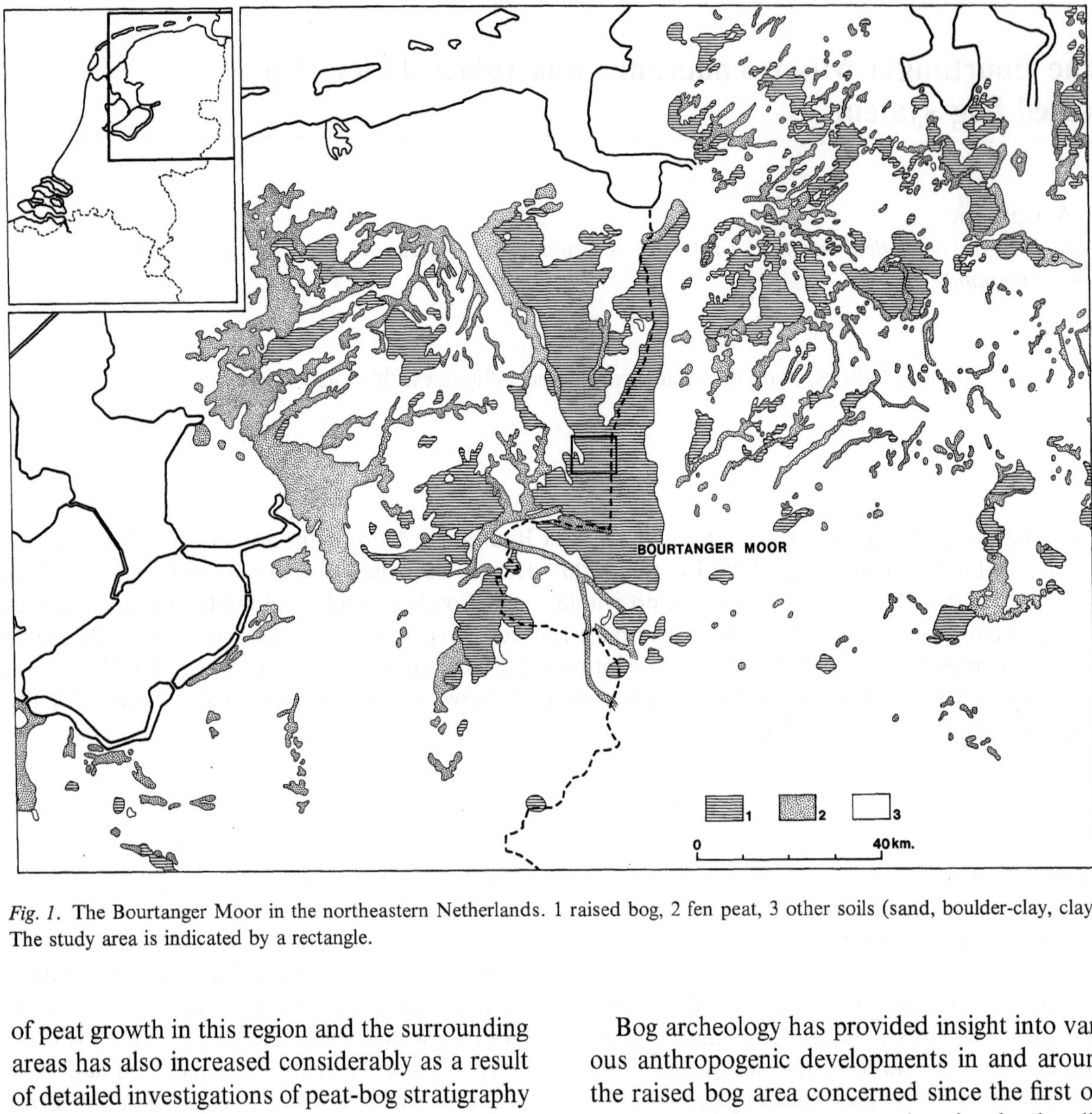

Fig. 1. The Bourtanger Moor in the northeastern Netherlands. 1 raised bog, 2 fen peat, 3 other soils (sand, boulder-clay, clay). The study area is indicated by a rectangle.

of peat growth in this region and the surrounding areas has also increased considerably as a result of detailed investigations of peat-bog stratigraphy (Dupont, 1985, 1986; Van Geel, 1978; Van Geel & Dallmeyer, 1986; Van Geel *et al.*, 1989; Van der Straaten, 1981; Teunissen, 1970).

To the west of the study area there are Pleistocene boulder-clay and cover-sand deposits, constituting the Drenthe Plateau. The eastern rim of this plateau is called the Hondsrug. Since ca. 3000 BC these higher soils have been intensively occupied by the people of the successive agrarian cultures of the Neolithic, Bronze Age, Iron Age and historical periods. The bog itself was not occupied.

Bog archeology has provided insight into various anthropogenic developments in and around the raised bog area concerned since the first occurrence of human occupation in the locality (Casparie, 1980, 1982, 1984a, 1984b, 1986).

The aim of this chapter

The bog development in the Bourtanger Moor has been described on the basis of the peat stratigraphy (Casparie & Streefkerk, 1992). The time course of peat growth indicates that raised bogs are robust systems. An ombrotrophic raised-bog system develops here, that is able to maintain itself for about 6000 years. A deposit of *Sphagnum* peat of a thickness of many metres formed between ca. 4500 BC and ca. 1700 AD. The reclamation activities of man terminate the ombrotrophic peat formation in the 18th century AD, involving drainage. Various physico-chemical and anthropogenic factors that influence the raised bog formation are discussed.

The possibilities for the regeneration of raised bogs in several Dutch nature reserves are suggested. These possibilities are based on data from paleo-ecological and hydrological studies, carried out by, among others, Streefkerk & Casparie (1989). Practical experience with respect to the regeneration of raised bog is described by Schouwenaars (1992).

The recorded peat-bog development

The minerotrophic peat development until ca. 4500 BC is described briefly in this section. Subsequently, the ombrotrophic peat formation is discussed with reference to three 'moments in the process of raised bog formation': ca. 4500 BC, ca. 2000 BC, and ca. 1000 AD. These three points in time correspond to the beginning of the formation of the raised bog, the inception of the formation of poorly humified *Sphagnum* peat in the otherwise highly humified *Sphagnum* peat-bog system, and the formation of the poorly humified *Sphagnum* peat, respectively. The direct data concerning the last of these developments are relatively scanty. Therefore the general picture outlined here is based on extrapolation of data obtained from various places in this region of raised bog, the extent of which is often very local. The three above-mentioned moments in the formation of raised bog are indicated in the highly schematic west-east cross section through the peat in the study area (Fig. 2).

The formation of raised bog is accompanied by the development of distinct spatial structures. These are to be regarded as limiting conditions with respect to the raised bog hydrology in particular. This formation process will be outlined briefly.

Peat-bog development until ca. 4500 BC

Since ca. 11000 BC the Hunze valley, in which this part of the Bourtanger Moor develops, has been so damp that here a Late Glacial Hypnaceae peat has been able to develop initially. This stage was followed by fen peat development, in which *Betula* is an important component from ca. 8300 BC onwards. Subsequently, after ca. 5000 BC *Alnus* plays an important role in the peat formation. The fen peat is rather mesotrophic than eutrophic, fed by the water contained in this valley.

Pinus spreads repeatedly over the bog surface after periods of drying out, *e.g.* around 7000 BC and 5300 BC.

Ca. 4500 BC: the start of the ombrotrophic peat growth

The highest groundwater level in the bog was attained around 4500 BC. The influence of precipitation on the peat-forming environment, *Alnus*-carr, increased. After a period of slight drying out there is an opportunity for *Pinus sylvestris* to become established on a wide scale on the surface of the fen peat. In this pine forest on an acidic peat soil, that extended over an area of about 10 km^2, a vegetation became established including various Ericaceae, *Polytrichum strictum*, and in the wettest places *Scheuchzeria palustris, Carex limosa* and also *Sphagnum rubellum*. These plant communities overgrew the pine forest in the course of 400 years. A highly humified *Sphagnum rubellum* peat developed in this process. The peat-bog soil remained waterlogged and the effects of occasional drying out were only superficial.

Ca. 2000 BC: the highly humified Sphagnum peat growth

The ombrotrophic peat growth could take place because of favourable climatological and hydrological conditions. A deposit of highly humified *Sphagnum rubellum* peat, 1 m thick, was formed, to which *Eriophorum vaginatum, Calluna vulgaris*

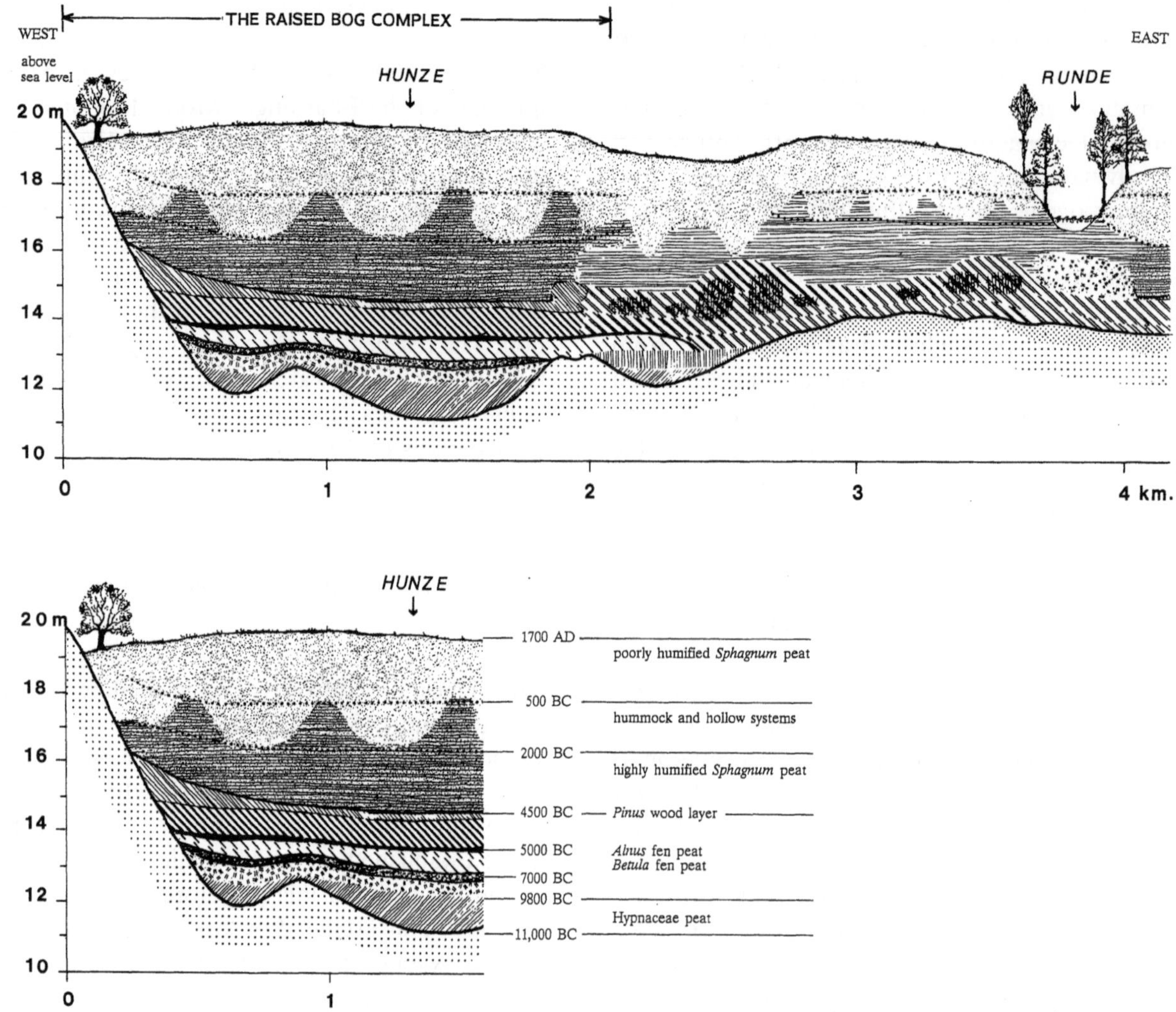

Fig. 2. Highly schematic cross section through the peat deposits in the study area. The raised bog complex under discussion concerns the western half of the deposits.

and other Ericaceae, and *Empetrum nigrum* contributed considerably. Distinct patterns of somewhat wetter and slightly less wet peat are discernible in the peat. A hummock-hollow pattern is present over considerable areas of the peat surface: somewhat drier hummocks with much *Eriophorum vaginatum* and *Calluna*, and wetter hollows, which play an important role in the water supply of the bog. After ca. 2000 BC these hollows change and the formation of poorly humified *Sphagnum* peat occurs, which is described in more detail below.

Ca. 1000 AD: the poorly humified Sphagnum peat growth

The ombrogenous peat deposit had attained a thickness of about 4 m. It showed a distinct zonation, brought about by a climatic change which involved cooler and more humid weather since ca. 2000 BC. Three deposits of ombrogenous *Sphagnum* peat can be distinguished: highly humified *Sphagnum* peat (from 4500–2000 BC), poorly humified *Sphagnum* peat (from ca. 500 BC), and in between the transition from highly humified to poorly humified *Sphagnum* peat (between ca. 2000–500 BC), showing a clear pattern of highly humified hummocks and poorly humified peat in the hollows.

The peat formation can be outlined as followed. On the bog surface of ca. 2000 BC, which had became wetter, many hollows become inundated. In these hollows *Sphagnum cuspidatum* became well-established, resulting in the formation of poorly humified *Sphagnum* peat. At the same time two new species of *Sphagnum* immigrated here: *S. papillosum*, that became established in the wet hollows and expanded there abundantly, and *S. imbricatum*, that became established on the sides of the hummocks and ultimately overgrew the highly humified *S. rubellum* hummocks. This process continued until ca. 500 BC. These two immigrant peat moss species are by far the most important constituents of the poorly humified *Sphagnum* peat.

Spatial structures

The upward growth of the ombrogenous peat deposit requires an appropriate water supply. During the process of formation of the raised bog a number of spatial structures develop, such as hummock-hollow systems, the domed complexes, bog ponds, contact zones and bog streams. Together these structures ensure (a) a water surplus on the bog surface which feeds the ombrogenous peat formation (*acrotelm*), (b) a sufficient water supply in the bog which permits the system to survive dry periods without rainfall (*catotelm*), and (c) the maintenance of oligotrophy. The bog should not dry out or become too wet, requiring a regulated water discharge (Casparie & Streefkerk, 1992).

In summary, the peat-forming environment continued to be oligotrophic and very wet, on the whole during a period of more than 6000 years of ombrotrophic peat development (between 4500 BC and 1700 AD). The environment for peat formation remained essentially unchanged despite any water excess or deficit in the peat-bog vegetation, and despite the supply of any nutrients, either in the form of aerosols or via wet or dry deposition.

Factors influencing raised bog growth

A large number of factors, both of a physico-chemical as well as of an anthropogenic nature, influence the formation of raised bog. Some of these factors have a negative influence on the ombrogenous peat formation, in some cases posing a direct threat to it. Three groups of influences can be distinguished: 1. The climatological development, notably the influence of precipitation and temperature; 2. External and internal processes that influence the actual formation of ombrogenous peat, notably the development of spatial structures and the hydrological development; 3. Interference by man, both in the form of processes causing eutrophication as well as activities like trampling, digging and reclamation.

The effect of climate on the formation of raised bogs

The amount of precipitation and the mean temperature in the months of June and July are especially important. The formation of raised bogs in the lowland plain of Northwestern Germany and in The Netherlands occurs only where the mean annual precipitation is $\geq$ 700 mm and the mean temperature in June–July is $\leq$ 16–17 °C. In regions with lower precipitation or higher temperatures periods of drought occur regularly of such duration that peat-forming *Sphagna* are unable to become firmly established, and consequently no raised bogs can develop.

Presently, the Bourtanger Moor has a mean annual precipitation of 775 mm; the mean July temperature is about 16 °C, and, therefore, the climatological conditions for raised bog formation are excellent. This was not always the case in the period from 4500 BC until the 18th century AD, when periods with lower precipitation alternated with periods in which the climate was considerably wetter. Since 4500 BC the mean July temperature has probably decreased by about 2 °C.

Under conditions of increasing precipitation the hollows become inundated. This threatens especially the growth of *Sphagnum rubellum*, that did indeed decline around 2000 BC. *S. papillosum* can flourish in very wet situations. Consequently the latter species strongly expanded in the hollows. Moreover, the transport of hummock material into the hollows increases with increasing precipitation. The hummocks are usually the most acidic parts of the bog surface. However, the hollows now acidify also, which stimulates the growth of *Sphagnum papillosum* in the hollows and that of *S. imbricatum* on the sides of the hummocks. The peat-bog growth becomes more exuberant after 2000 BC.

Under conditions of decreasing precipitation the growth of *Sphagna* is checked. There is less evaporation from these bog mosses, and thus the water remains available for a longer period and enables the bog to survive. If the water level decreases further than 10–20 cm below the bog surface, the formation of raised bog may come to a standstill. Not only do the *Sphagna* become less luxuriant, but also there are now good opportunities for the expansion of higher plants, that with their roots can reach the water at a deeper level. This poses a direct threat to the formation of ombrogenous peat. The hummock vegetation with *Calluna* and *Eriophorum vaginatum* expands over the sides of the hummocks into the hollows. Although the evaporation from this vegetation is lower than from bog mosses, aeration of the peat deposit is higher.

Raised bog has a number of survival mechanisms to prevent its deterioration due to drought. These mechanisms are mainly, though not exclusively, of a hydrological nature. Yet if *Molinia caerulea* and *Betula pubescens* are able to become established on the bog surface, then the survival mechanisms of the raised bog are inactivated.

Raised bog growth mechanisms

There are several mechanisms in the development of raised bogs that clearly contribute towards maintaining the system. The most important one concerns the survival of periods of drought, resistance to inundation and the prevention of eutrophication, within the still relatively broad climatological limits within which raised bog formation is possible. The spatial structures play an important role in this context.

Under conditions of moderate drought, particularly in periods without any rainfall, the bog respiration (German: *Mooratmung*) plays an important role. The bog surface subsides somewhat, allowing the water level to remain sufficiently high to permit the *Sphagna* to survive. The acrotelm conditions are thus maintained. The water supply in the bog ponds (Dutch: *meerstallen*) guarantees a sufficient water supply in the growth season. The peripheral drainage via the hollows of the domed complexes permits maximal use of the water for the process of peat formation. The water supply in the contact zones maintains the waterlogged state of the bog complex, i.e. the catotelm, together with the great resistance of the highly humified peat.

The bog burst (German: *Moorausbruch*) signifies a rapid discharge of a great excess of nutrients. This kind of erosion can be called the 'self-cleaning capacity' of the raised bog. The erosion gulleys thus formed, that may attain a depth of some 2 m, become wet again so rapidly that the raised bog formation is stimulated to take place under distinctly more oligotrophic conditions. With this supplementary discharge mechanism also any extensive inundation of the bog surface is prevented.

The influence of prehistoric man on the peat-bog formation

Since ca. 3000 BC the Hondsrug, the whole region immediately to the west of the study area, has been intensively occupied by agrarian communities. The presence of settlements, cemeteries, field complexes and other anthropogenic traces such as iron-smelting ovens has been established in this region. The prehistoric occupation was accompanied by intensive deforestation, often of hundreds of hectares in a relatively short time (Casparie, 1992), by the production and intensive use of agricultural land right up to the edge of the bog. The wood was used for many purposes, such as building material, *e.g.* for a wooden trackway (Fig. 3), as fuel, *e.g.* for supplementary fertilization of the soil, heating, the preparation of food, cremations, charcoal production and for the smelting of bronze and iron. This last activity is well-known from this bog region bog-iron ore being present in abundance.

The whole region of the higher soils was in use by humans at 1500 BC, as became clear from pollen diagrams. The remnants of woodland must have been of minimal extent. These remaining areas would have been predominantly secondary forest, because the original forest had disappeared altogether, probably already by 2400 BC. Soil dust, smoke and ash from the prehistoric activities regularly sedimented as aerosols in the raised bog complex situated downwind (Van Geel *et al.*, 1989). From 1350 BC onwards bog-iron ore has been dug up in this bog region (Fig. 4), sometimes from a depth of 5 m. Between 800 and 500 BC an attempt was made to reclaim an area of bog for agricultural purposes with the bog surface being interfered with by means of shovels, by burning and by fertilizing. This attempt was fruitless, because the bog was still too wet. Around 200 BC willow beds were present on the mineral soil immediately next to the bog margin, for the production of willow wands, that could be used, for example, for a bog trackway (Fig. 5). At this time the other higher soils west of the raised bog area largely consisted of vast arable field complexes, the Celtic Fields (Brongers, 1976).

The deforestation and the agricultural use of the soil on the Hondsrug led to considerable changes in the soil composition. Wind-blown drift-sand occurred from 2400 BC onwards: initially not on a large scale but it does indicate changes in the hydrological regime. The surface drainage of this region occurred largely by gulleys, that opened out into the bog area. The water discharge, that must have changed substantially both in a quantitative (more peak discharges) and qualitative way (considerably enriched water), had a eutrophicating influence on the peat forming environment. Nevertheless the ombrogenous peat formation continued in an oligotrophic environment. The threats posed by anthropogenic causes have been sufficiently neutralized, as shown in the following section.

The Bourtanger Moor as a robust ecosystem

Ombrogenous peat growth and relevant physico-chemical factors

The large number of external and internal influences did not change the peat forming environment essentially over a period of about 6000 years. It remained a system of peat formation in a nutrient-poor environment, fed exclusively by precipitation. The interplay of climatological factors and spatial structures formed, as it were, a stable framework, within which the ombrotrophic conditions in the Bourtanger Moor were maintained for thousands of years. The climatological factors involved were the quantity of precipitation and the temperature, and their annual distribution. The spatial structures of the raised bog were sufficiently flexible to accommodate considerable fluctuations in the annual precipitation and possibly also in the temperature. There are no known incidences of inundation that disturbed the peat growth for a long period of time. The drainage capability of the raised bog could readily adjust to the temporary accumulation and rapid discharge of enormous water masses. The contact zones probably played a key role in this connection (Casparie & Streefkerk, 1992). Only very rarely this system malfunctions, that is if a bog burst can be regarded as such.

Periods of temporary drying-out are clearly visible in the peat deposits, notably in the vertical peat faces, in the form of deposits of more highly humified peat. These incidences of drying-out are partly due to climatological factors, partly due to hydrological factors and partly due to the activities of prehistoric man. In the vertical peat faces of the study area a number of climatologically wetter and drier phases after 2500 BC can be discerned. Wetter phases can be dated to 2000

BC, 800 BC, 500 BC (the bog burst) and 1300 AD. Dry phases, with marked expansion of hummock vegetation with respect to hollows and often with some charcoal identified in the peat, can be dated to 2200 BC, 1800 BC, 1200 BC, 850 BC, 200–400 AD, 800 AD and 1200 AD.

Significant hydrological developments occurred mostly during wet phases (floodings).

Ombrogenous peat growth and anthropogenic influences

The activities of prehistoric man in the bog, like the construction of wooden trackways, the digging-up of bog iron ore and the laying-out of a few fields, occurred on only a limited scale spatially. The (disturbing) influence on the raised bog formation was of local significance, and negligibly small with respect to the entire raised bog. The bog trackways actually stimulated the formation of raised bog locally. This is caused by the weight of the timber on the water-saturated soft peat surface (Casparie, 1984b).

The eutrophicating influence of prehistoric man is apparently only noticeable to a small extent. How can this be explained? In the first place it is possible that in prehistoric times the eutrophication pressure exerted on the raised bog was still very small. Although there is no clear quantitative information available with regard to the study area, this influence may not simply be underestimated. The upland soils have been intensively occupied.

The inorganic nutrients P and N are only minimally absorbed by *Sphagna*. They remain with the H^+-ions in the bog water mostly in the form of weak acids in solution and form a potential source of eutrophication. Maybe they leave the raised bog system via the superficial water discharge before they are able to cause any conspicuous eutrophication. The establishment of more demanding higher plants in the bog-forming environment is thus prevented. This is indicative of an efficient and relatively rapid discharge of the nutrients from the raised bog system, almost certainly within a year.

Fig. 4. The Bronze Age plank footpath XVII (Bou) near Bargeroosterveld, dendrochronologically dated at 1359 ± 9 yrs BC. The track, made of *Quercus* timber, was used for winning bog-iron ore.

More information is available on the influence of heavy metals. Van Geel *et al.* (1989) analyzed the content of heavy metals in a few raised bog profiles, in relation to increasing human influence on the regional vegetation. In their profile Engbertsdijksveen-IB there is a distinct correlation between the content of Pb (the only heavy metal analyzed here) and indicators of human influence, as recorded in the pollen diagram, between ca. 1000 BC and 1000 AD.

Fig. 3. The Neolithic wooden trackway XXI (Bou) at Nieuw-Dordrecht, C14-dated to about 2100 BC (uncalibrated). The trackway-surface layer is made of (cleft) trunks of *Quercus, Alnus, Betula* and *Tilia*.

Fig. 5. The Iron Age hurdle trackway XIV (Bou) near Emmerschans, C14-dated to 170 BC (uncalibrated). The hurdles were made of four-year-old *Salix* rods from willow coppices nearby the bog margin. The track was probably used for the winning of bog-iron ore.

Their profile Meerstalblok-K2 is from the Bourtanger Moor, about 10 km south of our study area. It has been analyzed for its content of a large number of heavy metals, for the period ca. 550–1050 AD. The same correlation appears to be present here too, although the profile analyzed is situated relatively far away from areas occupied in prehistoric times. The general picture presented by Van Geel *et al.* (1989) shows that the deposition of heavy metals from anthropogenic aerosols must have been considerable already in prehistoric times, and this would have had an influence on the peat-forming vegetation. This certainly applies to the bog formation in our study area, that was extremely exposed to the influx of soil dust out of the air.

The small influence of the relatively severe pressure of heavy metals on the ombrotrophic peat formation can be satisfactorily explained. *Sphagna* very readily accumulate heavy metals, that are bound especially in the peat-forming level right at the top of the raised bog. On the cell walls part of the heavy metals are exchanged for H^+-ions. According to Van Geel *et al.* (1989), this occurs predominantly by means of passive accumulation instead of metabolic activity. Moreover, part of this deposition probably remains present in the water and is carried away after some time in a dissolved state.

The influence of anthropogenic aerosols is reflected in local eutrophication of the peat-forming environment. After 3000 BC *Scheuchzeria palustris* established itself regularly in the wet hollows in the western half of this bog area, and was sometimes able to survive for a considerable period. However, the whole system in its entirety remained oligotrophic, very acidic and very wet.

Ombrotrophy and the hydrological regime

The ombrotrophy is maintained predominantly by the superficial discharge of water. This concerns both the quantity of water discharged and the level of the water discharge in the peat deposit. With regard to the water balance of a living raised bog, it is assumed that the mean evaporation rate is 550 mm/year (Streefkerk & Casparie, 1989). Under undisturbed conditions the vertical drainage is small: for a living raised bog 40 mm/year, but in more or less dried out peat-bog relics considerably higher, 80–200 mm/year (Schouwenaars, 1992). The latter situation can be excluded for the study area between 4500 BC and ca. 1700 AD. The vertical discharge in this period was not more than 40 mm/year (Streefkerk & Oosterlee, 1984). With a precipitation of 700 mm/year, the minimum value for raised bog formation, a superficial water discharge of 110 mm/year can be estimated. This amount of discharge is considered necessary on the one hand to keep the bog surface sufficiently wet for *Sphagnum* growth, and on the other hand to carry away in good time the excess of nutrients, that would otherwise threaten the raised bog formation. This discharge, that is necessary for maintaining the ombrotrophic system, is called the system-linked discharge (Casparie, 1984a; Streefkerk & Casparie, 1987, 1989; Casparie & Streefkerk, 1992; see also Schouwenaars, 1992). The mean precipitation of 775 mm/year, presently recorded in this region, signifies in this context the supply of an ample quantity of water to maintain sufficiently oligotrophic conditions in a raised bog system that is overloaded with nutrients, as has been present in the study area concerned since ca. 3000 BC. Thus on the one hand the precipitation feeds the ombrotrophic system. On the other hand, it ensures that an excess of nutrients is carried away in a relatively short time, possibly in one vegetation season.

Exploitation in the past and at present

Bog development and maintenance

Considering the many factors that influence the ombrogenous peat formation, raised bog systems are in fact robust systems. This can be ascribed especially to the broad climatological limits within which raised bog formation is possible, to the capability of raised bog formation to withstand a variety of threatening situations, to a number of specific physiological properties of the *Sphagna*, and to the development of spatial structures, that ensure the maintenance of the right hydrological conditions. The most important potential threat appears to be an excessive supply of nutrients. Examples of this are to be found in many drowned coastal raised bogs. In these bogs the supply of nutrient-rich groundwater to the oligotrophic environment or the occurrence of marine flooding hampered ombrotrophic peat growth and induced in many cases the forming of mesotrophic and even eutrophic fen peat. In inland raised bogs, on the other hand, the influence of anthropogenic eutrophication in the past has been limited. Such ombrogenous systems can continue to exist for thousands of years, provided that they are not drained. It is possible, however, that this situation presently no longer applies, especially in regions with high air pollution. This subject falls outside the scope of this chapter.

The vulnerability of raised bogs lies to a great extent in the specific water supply of the peat-forming *Sphagna*: a situation of water surplus, fed exclusively by rainwater, close to the bog surface, that has to be maintained for a considerable part of the year. Under natural conditions there is no potentially disastrous threat to the water supply, whatsoever.

Artificial drainage and consequent degradation

In fact it is only man and his drainage activities that pose a fatal threat to the raised bog system, of which the Bourtanger Moor is an example. The systematic drainage of the large raised bogs since the 16th century AD has been carried out ever since the discovery that the growth of raised bog can be stopped in this way. If for a period of 5–10 years the bog surface is drained over a large area by means of ditches more than 30 cm deep, then the acrotelm conditions will no longer recover in the peat-bog vegetation (Streefkerk & Casparie, 1989). All precipitation that reaches the bog surface will sink to a lower level in the peat deposit, and as a result the peat-forming *Sphagna* will be unable to use it sufficiently. The degradation of the bog will then follow automatically.

Chances for survival of raised bogs in The Netherlands

Only with the aid of drastic measures in hydrological management situations in raised bog relics can be developed in such a way that they permit the regeneration of raised bogs in an environment closely resembling that of the original acrotelm (Schouwenaars, 1993; Streefkerk & Casparie, 1989). The potential capacity of (drained) raised bog deposits to function as a base for renewed peat-bog growth is maintained very long, as a general rule. The climatological conditions, notably the quantity of precipitation, its annual distribution, and the mean temperature, can be regarded as limiting conditions which in The Netherlands are favourable. The quality of present-day precipitation, however, appears to be disadvantageous for the initiation of the raised bog formation.

To devote any closer attention to specific measures necessary for the restoration and management of bogland nature reserves would fall outside the scope of this chapter. The essence of such measures lies in the restoration of the hydrological regime under ombrotrophic conditions or the replication of part of that regime. With respect to the latter alternative, it will become clear only after many years whether this is a successful method.

Regeneration of the raised bog growth on the basis of hydrological intervention is being attempted out in a number of raised bog relics: in the Fochterloërveen, the Bargerveen, the Engbertsdijksveen and in De Grote Peel. One of the management problems in this connection is that inundation can easily cause undesirable eutrophication of an insufficiently acidic environment. Research, such as that being carried out by Schouwenaars (1993), can provide a good basis for establishing which measures need to be taken to ensure effective management. This chapter is intended to show that the actual process of raised bog formation offers ample opportunities for achieving this objective.

Acknowledgements

The drawings were made by L. van der Velde of the Provincial Museum of Drenthe at Assen, and by G. Delger and H.R. Roelink of the Biological-Archaeological Institute (BAI) of the University of Groningen. The photographs were prepared by Mr J. Schurer of the Central Photo Department of the University. Mrs S.M. van Gelder-Ottway of Haren translated the text into English. Mrs G. Entjes-Nieborg (BAI) typed the manuscript. The author is greatly indebted to these persons.

References

Brongers, J.A., 1976. Air photography and Celtic field research in the Netherlands. Nederlandse Oudheden 6: 1–147.

Casparie, W.A., 1972. Bog development in Southeastern Drenthe (the Netherlands). Vegetatio 25: 1–271.

Casparie, W.A., 1980. Veenvorming. In: Het veen, natuurlijk en menselijk moeras. Museumfonds (Prov. Museum van Drenthe, Assen) no. 3: 2–32.

Casparie, W.A., 1982. The Neolithic wooden trackway XXI (Bou) in the raised bog at Nieuw-Dordrecht (the Netherlands). Palaeohistoria 24: 115–164.

Casparie, W.A., 1984a. Water en veen; waterbalans, sleutel tot behoud. Natuur en Techniek 52/2: 102–121.

Casparie, W.A., 1984b. The three Bronze Age footpaths XVI (Bou), XVII (Bou) and XVIII (Bou) in the raised bog of Southeast Drenthe (the Netherlands). Palaeohistoria 26: 41–99.

Casparie, W.A., 1986. The two Iron Age wooden trackways XIV (Bou) and XV (Bou) in the raised bog of Southeast Drenthe (the Netherlands). Palaeohistoria 28: 169–210.

Casparie, W.A., 1992. Neolithic deforestation in the region of Emmen (The Netherlands). Paläoklimaforschung.

Casparie, W.A. & J.G. Streefkerk, 1992 (in press). Climatical, stratigraphic and paleo-ecological aspects of mire development. In J.T.A. Verhoeven (ed.), Fens and Bogs in the Netherlands, pp. 85–133.

Dupont, L.M., 1985. Temperature and rainfall variation in a raised bog ecosystem. Ph.D. Thesis, University of Amsterdam.

Dupont, L.M., 1986. Temperature and rainfall variation in the Holocene based on comparative palaeoecology and isotope geology of a hummock and a hollow (Bourtangerveen, The Netherlands). Rev. Palaeobot. Palynol. 48: 71–159.

Schouwenaars, J.M., 1993. Hydrological differences between bogs and bog-relicts and consequences for bog restoration. Hydrobiologia 265: 217–224.

Streefkerk, J.G. & W.A. Casparie, 1987. De hydrologie van hoogveen systemen. Staatsbosbeheer-rapport 19, 1987: 1–119.

Streefkerk, J.G. & W.A. Casparie, 1989. The hydrology of bog ecosystems. Guidelines for management. Report Dutch National Forestry Service, 1989: 1–125.

Streefkerk, J.G. & P.J. Oosterlee, 1984. Een beschouwing over hydrologische ingrepen in het hoogveenreservaat Bargerveen. Staatsbosbeheer, Afd. Waterhuishouding.

Teunissen, D., 1970. Verslag van het pollenanalytisch onderzoek van het profiel Meerstalblok. Internal report, University of Nijmegen.

Van Geel, B., 1978. A palaeoecological study of Holocene peat bog sections in Germany and the Netherlands. Rev. Palaeobot. Palynol. 25: 1–120.

Van Geel, B. & A.A. Dallmeyer, 1986. Eine Molinia-Torflage als Effekt eines Moorbrandes aus dem frühen Subboreal im Hochmoor Engbertsdijksveen (Niederlande). Abhandlungen Van Landesmus. f. Naturkunde Münster 48, H2/3: 471–480.

Van Geel, B., R. Bregman, P.C. van der Molen, L.M. Dupont & C. van Driel-Murray, 1989. Holocene raised bog deposits in the Netherlands as geochemical archives of prehistoric aerosols. Acta Bot. Neerl. 38(4): 467–476.

Van der Straaten, C., 1981. Deuterium in organic matter. Ph.D. Thesis, University of Groningen.

Van Zeist, W., 1955. Pollenanalytical investigations in the Northern Netherlands, with special reference to archaeology. Acta Bot. Neerl. 4: 1–81.

Hydrobiologia **265**: 217–224, 1993.
E. P. H. Best & J. P. Bakker (eds), Netherlands-Wetlands.

Hydrological differences between bogs and bog-relicts and consequences for bog restoration

J.M. Schouwenaars
Dept. of Hydrology, Soil Physics and Hydraulics, Wageningen Agricultural University, P.O. Box 9101, 6700 HB Wageningen, The Netherlands; (present address: Dept. of Physical Geography and Soil Science, University of Groningen, Kerklaan 30, 9751 NN Haren, The Netherlands)

Key words: bogs, relicts, water balance, restoration

Abstract

The hydrology of bog relicts differs from that in undisturbed bogs. The surface layers of these relicts mostly consist of moderately to strongly humified, secondary weathered peat as a result of drainage and peat cutting. The hydrophysical properties of these layers cause relatively high groundwater level fluctuations.

Deep drainage systems, both in the bog relicts and in their surroundings, may have increased the downward seepage. Reduction of these downward water losses may be crucial for the restoration of the required hydrological conditions in certain bog relicts (hydrological bufferzone as external water management option). The potential of internal hydrological modifications, where the increase in storage capacity near the surface is essential, should be emphasized in many bog relicts. Considerable reductions in water level fluctuations can be achieved *e.g.* when the open water within the area is enlarged and when this water is equally distributed over the area with small peat ridges in between. In general, attention should be given to both the internal and external options in studies on water management.

Introduction

Virtually all raised bogs in The Netherlands have disappeared as a result of peat cutting and cultivation. The total area of bog relicts is presently about 9,000 ha. The bogs are severely disturbed by drainage, peat cutting and fire. Small remnants of the original bog vegetation, dominated by *Sphagnum spec.*, have survived locally. The nowadays dominating species are *Molinia caerulea, Betula pubescens, Calluna vulgaris* and *Erica tetralix*.

The most important bog relicts in The Netherlands are presented in Fig. 1.

Attempts have been made since about 1970 to restore in several bog relicts environmental conditions suitable for the reestablishment of the former bog vegetation. The most important measure is to raise and to control the water level. This can be done by both internal and external water management. Internal measures involve the increase in water storage capacity near the surface to limit water level fluctuations. External measures involve the decrease in downward water

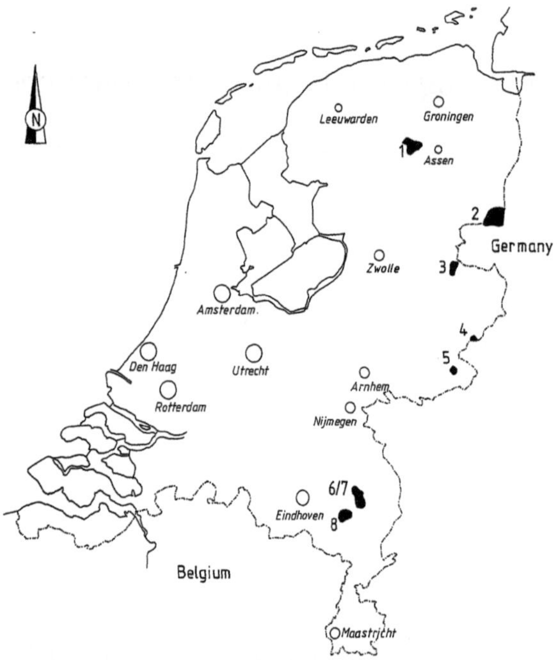

1 Fochteloorveen
2 Bargerveen
3 Engbertsdijkvenen
4 Haaksbergerveen
5 Korenburger Veen
6 Deurnese en Liesselse Peel
7 Mariapeel
8 Grote Peel

Fig. 1. Bog relicts in The Netherlands.
Total area (roughly):

1: 1,600 ha	5: 210 ha
2: 1,530 ha	6: 770 ha
3: 860 ha	7: 1,060 ha
4: 430 ha	8: 1,300 ha

losses to maintain a higher water level in summer, *e.g.* by the creation of hydrological bufferzones.

Ecological constraints for bog regeneration are not fully known. Many management practices are based upon the knowledge we have about required groundwater fluctuations. The groundwater reaches the surface during most of the winterperiod. In summer it should not be deeper than about 30–40 cm (Verry, 1988). The reconstruction of conditions which guarantee these limited fluctuations is possible in many bog relicts (Eggelsmann, 1988).

Here the consequences of the existing hydrological conditions in bog relicts for bog restoration will be discussed. Main attention is given to the quantitative hydrological aspects. Qualitative aspects with possible implications for the prefered water management measures will be discussed briefly.

The hydrology of raised bogs

Hydrophysical properties of Sphagnum peat

The upper decimeters of a young *Sphagnum* cover in a bog can be considered as an 'active layer' (Romanov, 1968). Ingram and Bragg (1984) defined them as the 'acrotelm'. It is in this layer that gas and water are exchanged with the atmosphere. The upper part mainly consists of green living stems of *Sphagnum*-mosses. They grow on top of dead moss parts which are only slightly decomposed. The lower boundary of the 'active layer' is formed by the permanently saturated, moderately decomposed peat layers.

A layer of *Sphagnum* moss consists of many large pores and is able to store an enormous amount of water. Romanov (1968) presented values for the bulk-density of the active layer ranging from 0.017 to 0.076 g cm^{-3} and for the volumetric content of the organic matter from 0.69 to 4.9 %. These upper layers obviously show more resemblance to water (solid matter content equals that of milk) than to soil. In the older and deeper layers the pore size is reduced due to decomposition and humification.

Due to the presence of many large pores in the upper layers, capillary rise is limited. Romanov (1968) observed a sharp decrease in capillary rise when the water table drops below 15–20 cm depth, which was confirmed by Schouwenaars (1990) in an experimental study in the Engbertsdijksvenen. In the latter study small weighable lysimeters were used to investigate evapotranspiration of *Sphagnum papillosum* and *Molinia caerulea*. The ratio between actual capillary water supply in a certain period and the needed capillary rise to fulfill the water requirements is strongly related to the depth of the watertable (Fig. 2).

From these lysimeter experiments values for the water storage coefficients in peat layers differing in degree of humification were obtained

Table 1. Water storage coefficients obtained from lysimeter experiments in the Engbertsdijksvenen (from Schouwenaars and Vink, 1990).

Type of peat	Depth (cm)	Storage coefficient[1]
Young living *Sphagnum* moss	0–15	0.23–0.34
Slightly humified *Sphagnum* peat (H2–3)[2]	10–30	0.11–0.17
Moderately humified *Sphagnum* peat (H3–4)	0–40	0.11–0.13
Strongly humified *Sphagnum* peat (H6–7) (with many living roots)[3]	0–35	0.14–0.33

[1] Regression coefficient obtained from linear regression.between change in water storage ($cm^3\ cm^{-2}$) and change in water depth (cm).
[2] Von Post-scale.
[3] In samples without roots, values varied from 0.05 to 0.10.

(Schouwenaars & Vink, 1990). Some results are summarized in Table 1.

The reversible processes of shrinkage and swelling in the upper bog layer (Ge: 'Mooratmung') reduce the risks for insufficient capillary water supply and inundation.

Evapotranspiration

Sphagnum mosses depend on capillary water supply to satisfy their transpiration demand, contrary to plants with a root system.

Dry summer periods longer than one week may result in a reduced evapotranspiration. In spring and autumn the potential evapotranspiration is lower than in summer and the water table usually is higher. Assimilation of the mosses is high in these periods, when compared to other plant species. Evapotranspiration from a *Sphagnum* vegetation is somewhat higher than from open water when water supply is not limited. This can be observed when water levels are about 5–10 cm below the tips of the *Sphagnum* stems (Fig. 2).

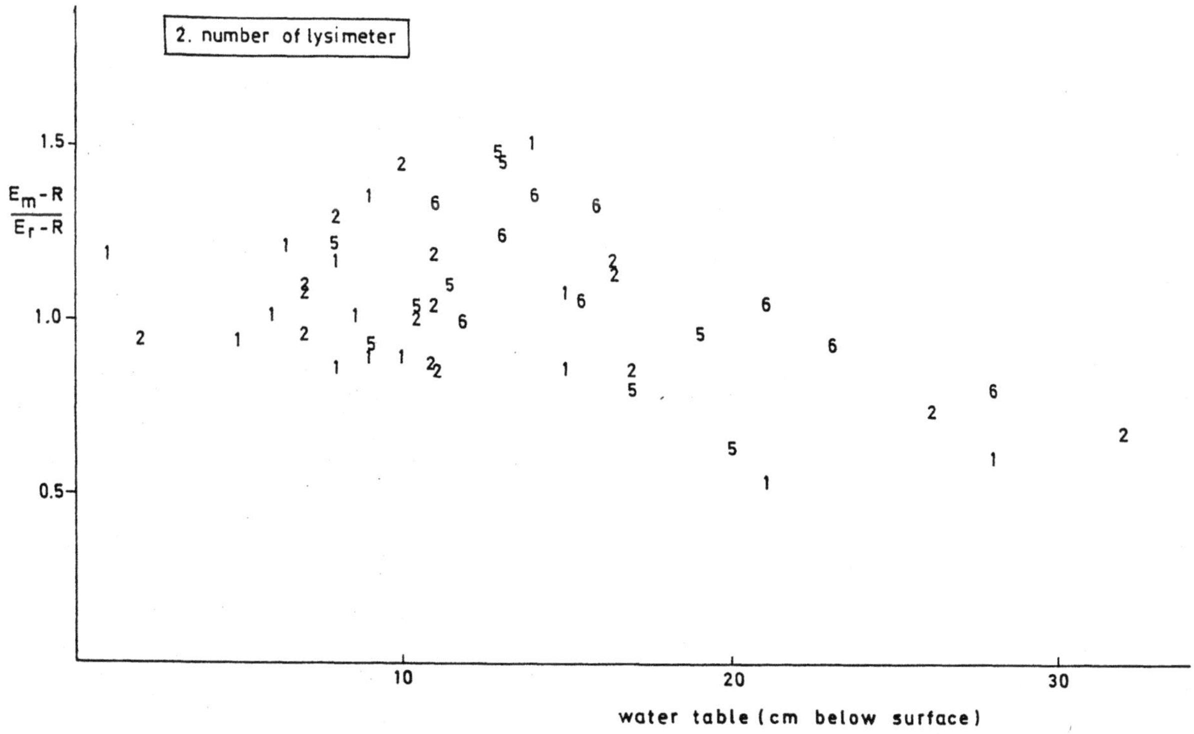

Fig. 2. The ratio between actual capillary water supply in a certain period (i.e. actual evapotranspiration (E_m) minus rainfall (R)) and the needed capillary rise to fulfill the water demand for 'reference' evapotranspiration (E_r) minus rainfall in relation to the depth of the water table in Sphagnum lysimeters (from Schouwenaars, 1990). Reference evapotranspiration was determined according to Makkink (Feddes, 1987).

When these tips are inundated evapotranspiration is reduced to the values of open water.

Downward seepage

Downward seepage from a bog is very limited. Eggelsmann (1960) found values of about 40 mm yr^{-1} for Königsmoor in Germany. This is caused by the very low hydraulic conductivity in the deeper strongly humified peat layers.

Superficial runoff

Runoff characteristics of bogs differing in latitude vary enormously because of different climatic conditions. In sub-arctic regions the freeze-thaw cycle determines the discharge pattern in time. In temperate regions high peak discharges can be observed after frost periods and under normal summer conditions no runoff from bogs is observed (Brandesten, 1987). Casparie & Streefkerk (1987) argued that a minimum yearly superficial discharge of 110 mm is needed to guarantee removal of nutrients from the bog system. More research is needed to show if this is a valid hypothesis.

The hydrology of bog relicts in The Netherlands

Hydrophysical properties and groundwater fluctuations

The hydrophysical properties of the upper bog layers are, partly irreversibly, changed after drainage. The *Sphagnum* mosses die off, heather and grass species (like *Erica tetralix, Calluna vulgaris* and *Molinia caerulea*) invade the site and these rooting plants are able to extract the water from deeper layers. The moss layer with its high water storage capacity, is replaced by a more compact slightly to moderately decomposed peat layer. Its water storage capacity is lower, resulting in larger groundwater fluctuations (Table 1).

These factors lead to a further lowering of the water table in summer and aeration will result in higher decomposition rates by (micro-)biological activity, ultimately reducing the potential for reestablishment of the former *Sphagnum* vegetation.

After removal of the upper peat layers the deeper, more humified, peat lies at the surface, of which the water storage capacity is low. Biological activity of roots and the presence of small peat cracks may, however, locally allow an increased water storage in the rootzone (Schouwenaars & Vink, 1990, Table 1). Limited water level fluctuations can be observed at sites with relatively much open water, equally distributed over the area. This water distribution pattern guarantees the infiltration of water into the peat remnants between the inundated sites. In strongly humified peat, infiltration already is severely limited at distances of more than 5 meter from an open water site (Schouwenaars, 1989).

Evapotranspiration

The actual and potential evapotranspiration of the dominating plant species in the Engbertsdijksvenen (Dutch bog-relict), *Molinia caerulea*, was examined, using lysimeters. Potential evapotranspiration of *Molinia* was closely related to its LAI-index development (Fig. 3; Schouwenaars (1990). This species proved capable to extract sufficient water from a peat soil to guarantee evapotranspiration at potential values, even after a dry period of one month with the water table at 70–80 cm depth.

Downward seepage

Peat excavation has resulted in a great variety in thickness of the peat layers in bog relicts. The hydraulic resistance of strongly humified peat is estimated at 3,500–4,000 days m^{-1} (Schouwenaars *et al.*, 1990). The annual downward seepage varies from 80 mm yr^{-1} to about 200 mm yr^{-1}, the average precipitation surplus in the region. In most part of the Engbertsdijksvenen the piezometric head in the underlying sandy aquifer

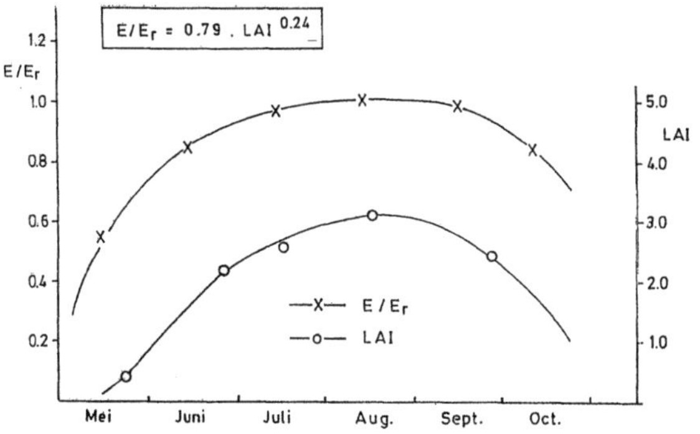

Fig. 3. Development of Leaf Area Index (LAI) and evapotranspiration ratio (E/E_r) for a *Molinia* vegetation during the growing season (obtained from lysimeter studies; Schouwenaars, 1990).
E: potential evapotranspiration (mm day^{-1});
E_r: *reference (Makkink) evapotranspiration (mm day*$^{-1}$*).*

always remains below the peat basis. Downward seepage at these sites is directly related to the ratio between water depth and thickness of the remaining peat layer (Fig. 4). At sites which receive no water from elsewhere, downward seepage cannot be larger than the precipitation surplus. As a consequence, also the average level of the water table will not be higher than the one resulting in a downward water loss equal to the average yearly precipitation surplus in the region.

Superficial runoff

Different peat excavation methods and subsequent management measures for water conservation have caused very site-specific runoff characteristics. The presence of many sites with open water within a catchment will generally reduce peak discharges because of their high storage capacity (a.o. Schouwenaars (1989)). Annual totals of runoff depend, of course, on weather conditions and downward seepage at a given site.

Water management and perspectives for *Sphagnum* regrowth

Experiences in the Netherlands with the restoration of bog relicts show that inundated sites should not be larger than 10–20 m in diameter to avoid wind-induced wave-movements. If these inundated sites are larger the establishment of floating mats of *Sphagnum* mosses is inhibited. Wind impact at an inundated site is reduced by the presence of many *Molinia* or *Eriophorum vaginatum* tussocks, which offer shelter and enlarge the area, suitable for *Sphagnum cuspidatum*-growth (Joosten, 1989).

The necessity to reconstruct the water storage capacity is emphasized by Eggelsmann (1987). This author describes the different techniques for water conservation in bog relicts in Lower Saxony (Germany). He recommends the creation of various dams parallel to the contours of the surface, which brings various small compartments about similar to the sawa-system, known from rice cultivation.

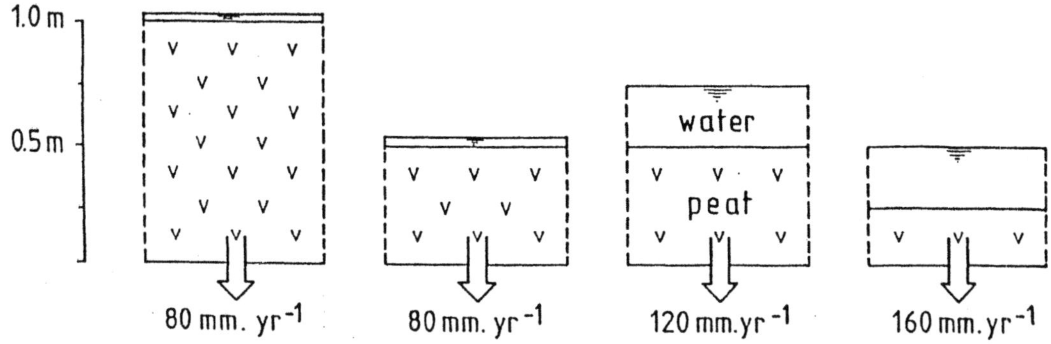

Fig. 4. Downward seepage for different sites in the Engberts-dijksvenen. (This schematic illustration shows that the downward flux is determined by the ratio of: the distance between the waterlevel and the peat basis and the thickness of the peat layer. When this ratio is 1, the downward seepage equals 80 mm yr^{-1}.)

In northwest Germany the hydrological guidelines for rewetting and renaturation of bog relicts are mainly based on studies of a.o. Eggelsmann (1988) and Blankenburg & Kuntze (1987). These authors concluded that strongly humified peat layers with a minimum thickness of 0.5 m are needed to guarantee seepage not to be higher than 50–60 mm yr^{-1}. They recommended the refilling of open drains reaching into the underlying aquifer, using strongly humified peat. In Lower Saxony some experiments are carried out which involve the construction of hummocks (ridges) and hollows (pools) (Eggelsmann, 1987).

Tüxen (1988) critizised the general character of the guidelines for renaturation activities and argues that for each area a detailed survey on ecological, pedological and hydrological characteristics is needed to define guidelines for its management.

Beets (1989) presented a method to assess the required water depth and open water fraction to guarantee limited water level fluctuations.

Internal water conservation measures, including the restoration of the water storage capacity near the surface have been proven to give satisfactory results in many bog relicts (a.o. Eggelsmann & Klose (1982), Joosten (1989)). *Sphagnum* regrowth was observed after water conservation measures in many bog relicts. Floating mats of vegetation have developed at permanently inundated sites, composed by *Sphagnum spec.* (especially *Sph. cuspidatum* and *Sph. papillosum*), *Andromeda polifolia*, and *Oxycoccus palustris*. The, in bog relicts common, smaller inundated sites which may fall dry after a long period without rainfall are often occupied by *Eriophorum vaginatum*. Its tufts offers the required shelter for wind and mostly a well developed *Sphagnum cuspidatum* cover can be found between this species.

The mosses dry out during dry periods, become yellowish-white and loose their assimilation capacity. They regain their carbon gaining activity in autumn, when air humidity is sufficient, even when the water table is still too low for shallow inundation.

Schouwenaars (1988a, b) evaluated various alternative water management options, using models. From these studies it appeared that internal water conservation options are crucial in areas where downward seepage is less than 100–150 mm yr^{-1}, the exact figure depending on the average annual precipitation surplus. When downward losses are higher, external measures to reduce these losses have to be considered (*e.g.* Poelman & Joosten, 1989).

Podschlod (1988) studied the importance of the presence of seeds and other diaspora in the upper peat layers. He concluded that the best perspectives for a rapid colonization of bare peat after rewetting were at sites where the former upper, slightly humified, layers are still present. Unfortunately, these layers have been removed completely in many bog relicts.

In areas where peat cutting is still going on, as is the case in many countries, it is recommendable to formulate hydrological constraints which have to be fulfilled when mining concessions come to an end. In this way equipment and practical experiences of the peat mining companies can be used. The constraints in these cases should guarantee, of course, the creation of suitable hydrological conditions for bog restoration. This approach is widely applied and accepted in peat mining concessions in Germany.

Conclusions and discussion

The hydrological characteristics of bog relicts differ from those of undisturbed bogs in two important aspects. In areas where the thickness of the remaining peat layers is less than 0.5 m, the downward water losses mostly are too high to guarantee a high water table during the summer period. In some of these areas the downward seepage may have been increased by drainage activities in adjacent agricultural lands.

Another aspect, relevant for all partly cut-away bogs, is the reduced water storage capacity near the surface. When humified peat lies at the surface, its soil physical properties bring about large fluctuations of the water level. When covered with grasses, like *Molinia caerulea*, the evapotranspiration during dry periods will be higher than for a *Sphagnum* vegetation and this further increases the lowering of the water table in summer.

The main objective of the water management measures for bog restoration presented in the previous section, is to create hydrological conditions which sufficiently compensate the adverse aspects mentioned above. When succesfull, the presence of growing *Sphagna* will gradually restore the water storage capacity near the surface and this restoring bog system will gradually regain the intrinsic regulation mechanisms. Artificial management measures to guarantee limited fluctuations will, therefore, only have a temporary function. Ultimately, a more self-regulating system, less depending on human interference will develop. For example, when after 50–100 years of development, a young living *Sphagnum* moss layer has developed at the top of more humified peat layers, its water storage properties and its evapotranspiration characteristics will cause reduced water level fluctuations. At a thickness of more than 30–40 cm the unsuited hydrophysical properties of the underlying humified peat will no longer affect the water level fluctuations.

It is questionable whether the restoration measures aimed at the reestablishment of different pioneer species like *Sphagnum cuspidatum, Sph. papillosum* and *Eriophorum vaginatum* in bog relicts will result in the reestablishment of the former ombrotrophic plant communities, and in the regeneration of the bog system. For instance, the changed input rates of nutrients from the atmosphere (*e.g.* nitrogen deposition) will affect the vegetation development in rewetted areas. From the present state of knowledge it has to be concluded that probably only the future, long term development of the different bog relicts will give sastisfactory answers to the many questions.

The knowledge on bog-systems indicate that they are robust with good perspectives for recovery (Casparie, 1993). Nevertheless, in particular the role of regulation mechanisms in the nutrient budget of (restoring) bog systems needs further examination. Therefore field experiments with both differences in water quantity (water levels, water balance) and water quality (nutrient deposition) are needed.

Acknowledgements

The author wishes to thank Dr W.A. Casparie for his suggestions and comments.

References

Beets, C.P., 1989. The relation between the area of open water in bog renmnants and storage capacity, with resulting guidelines for bog restoration. Proc. Symposium on Peatlands Ecosystems and Man – an impact assessment. Dundee, Scotland (in prep.).

Blankenburg, J. & H. Kuntze, 1987. Moorkundlich-hydrologische Voraussetzungen der Wiedervernassung von Hochmooren. Telma 17: 51–58.

Brandesten, C.O., 1987. Runoff from mire catchments. Trita-Kut series 1046. Royal Inst. of Technology, Dept. of Land improvement and Drainage, Stockholm, 89 pp.

Casparie, W.A., 1993. The Bourtanger Moor: endurance and vulnerability of a raised bog system. Hydrobiologia 265: 203–215.

Eggelsmann, R., 1960. Uber den unterirdischen Abfluss aus Mooren. Wasserwirtschaft 50: 149–154.

Eggelsmann, R., 1987. Okotechnische Aspekte der Hochmoorregeneration. Telma 17: 59–94.

Eggelsmann, R., 1988. Wiedervernassung von Hochmooren. Die Geowissenschaften 6: 317–322.

Eggelsmann, R. & E. Klose, 1982. Regenerationsversuch auf industriell abgetorftem Hochmoor in Lichtenmoor – erste hydrologische Ergebnisse. Telma 12: 189–205.

Feddes, R.A., 1987. Crop factors in relation to Makkink reference crop-evapotranspiration. In: Hooghart, J.C. (ed.), Evapotranspiration and Weather. TNO/CHO, The Hague, publication 39: 33–45.

Ingram, H.A.P. & O.M. Bragg, 1984. The diplotelmic mire: some hydrological consequences reviewed. Proc. 7-th Intern.Peat Congr. Dublin, vol 1: 220–234.

Joosten, H.J.H., 1989. Bog regeneration in the Netherlands; a review. Proc. Symposium on Peatland Ecosystems and Man – an impact assessment. Dundee, Scotland (in prep.).

Podschlod, P., 1988. Vegetationsentwicklung ehemaliger Torfabbaugebiete in Hochmooren des bayerischen Alpenvorlandes. Tuexenia 8: 31–53.

Poelman, A. & J.H.J. Joosten, 1989. On the identification of hydrological interaction zones for bog reserves. Symposium on Peatland ecosystems and Man – an impact assessment. Dundee, Scotland (in prep.).

Romanov, V.V., 1968. Hydrophysics of bogs. Ed. Monson Bindery Ltd. Jerusalem.

Schouwenaars, J.M., 1988a. The impact of water management upon groundwater fluctuations in a disturbed bog relict. Agricultural Water Management 14: 439–449.

Schouwenaars, J.M., 1988b. Hydrological Research in disturbed bogs and its role in decisions on water management in the Netherlands. Proc. of the Symposium on the hydrology of wetlands in temperate and cold regions, vol 1. Joensuu, Finland, 6–8 June Publications of the Academy of Finland 4 (1988): 170–177.

Schouwenaars, J.M., 1989. Hydrological characteristics of bog relicts in the Engbertsdijksvenen after peat cutting and rewetting. Proc. of the Symposium on Peatland ecosystems and Man – an impact assessment. Dundee, Scotland: 11–15 Sept. 1989 (in prep.).

Schouwenaars, J.M., 1990. A study on the evapotranspiration of Molinia caerulea and Sphagnum papillosum , using small weighable lysimeters. In J.M. Schouwenaars, 1990. Problem oriented studies on plant-soil-water relations. Ph.D. Thesis, Wageningen Agricultural University, 175 pp.

Schouwenaars, J.M., F. van Amerongen & M. Booltink, 1990. Hydraulic resistance of peat layers and downward seepage in bog relicts. International peat Journal 5 (in prep).

Schouwenaars, J.M. & J.P.M. Vink, 1990. Hydrophysical properties of peat relicts in a former bog and perspectives for Sphagnum regrowth. International Peat Journal 5 (in prep.).

Streefkerk, J.G. & W.A. Casparie, 1987. The hydrology of bog-systems (in Dutch). State Forestry Service, Utrecht, 120 pp.

Tüxen, J., 1988. Haben unsere Moore noch eine Zukunft? Telma 19: 333–344.

Verry, E.S., 1988. The hydrology of wetlands and man's influence on it. Proc. Symposium on the hydrology of wetlands in temperate and cold regions. Joensuu. Publ.Academy of Finland 5/1988: 41–61.

Hydrobiologia **265**: 225–263, 1993.
E. P. H. Best & J. P. Bakker (eds), Netherlands-Wetlands.

Ecology and management of moorland pools: balancing acidification and eutrophication

H. van Dam[1] & R.F.M. Buskens[2]
[1]*DLO Institute for Forestry and Nature Research, P.O. Box 46, 3956 ZR Leersum, The Netherlands;*
[2]*Catholic University of Nijmegen, Department of Ecology, Toernooiveld, 6525 EG Nijmegen, The Netherlands and Grontmij, P.O. Box 1265, 6502 BG Eindhoven, The Netherlands*

Key words: moorland pools, hydrology, chemistry, mass budgets, macrophytes, desmids, diatoms, cladocera, macrofauna, chironomids, fishes, amphibians, birds, primary production, decomposition, paleolimnology, historical ecology, acidification, eutrophication, guanotrophy, reclamation, afforestation, drainage, monitoring, management

Abstract

Moorland pools originally are shallow, often hydrologically isolated, soft-water bodies, with a low productivity. Some thousands of moorland pools originated from the late Pleistocene onwards in the heathland landscape in The Netherlands and adjacent areas, where soils have a poor buffering capacity. As the pools are largely fed by atmospheric precipitation, they are very vulnerable to changes in the environment, *e.g.* eutrophication and acidification.

Moorland pools are exposed to very high rates of wet atmospheric deposition: 44–50 mmol m^{-2} yr^{-1} sulphate and 84–103 mmol m^{-2} yr^{-1} ammonium. Mass budgets indicate that 20–70% of the input of sulphate by precipitation is reduced, 40–100% of the ammonium input escapes to the air or sediments, apparently due to nitrification, and 90–100% of incoming nitrate disappears. The production of alkalinity ranges from 12 to 52 meq m^{-2} yr^{-1}. The efficiency of these processes augments with pH-values of surface water increasing from 4.1 to 5.4. The accumulation of reduced sulphur compounds in the sediments is a threat in extremely dry summers, when desiccation causes oxydation of these compounds, resulting in very low pH-values (≤ 3.7).

Acidification by acid atmospheric deposition and eutrophication by agricultural acidification are the main threats to the moorland pool ecosystems and affect the species composition of assemblages of aquatic macrophytes, desmids, diatoms, macrofauna, fishes and amphibians, as has been shown by comparison of old and recent records on their distribution and paleolimnological methods. Afforestation exacerbates acidification and also reduces wind dynamics. Particularly the decrease of isoetids and desmids by both processes indicate the biological impoverishment of the pools.

Reductions of (potential) acid atmospheric deposition to less than 40 mmol m^{-2} yr and of ammonia to less than 30 mmol m^{-2} yr are necessary for recovery of the moorland pools. Methods for the addition of buffering material to a number of moorland pools, to counteract acidification until these deposition rates have lowered sufficiently, are given, as well as other methods for restoring the biological quality of moorland pools.

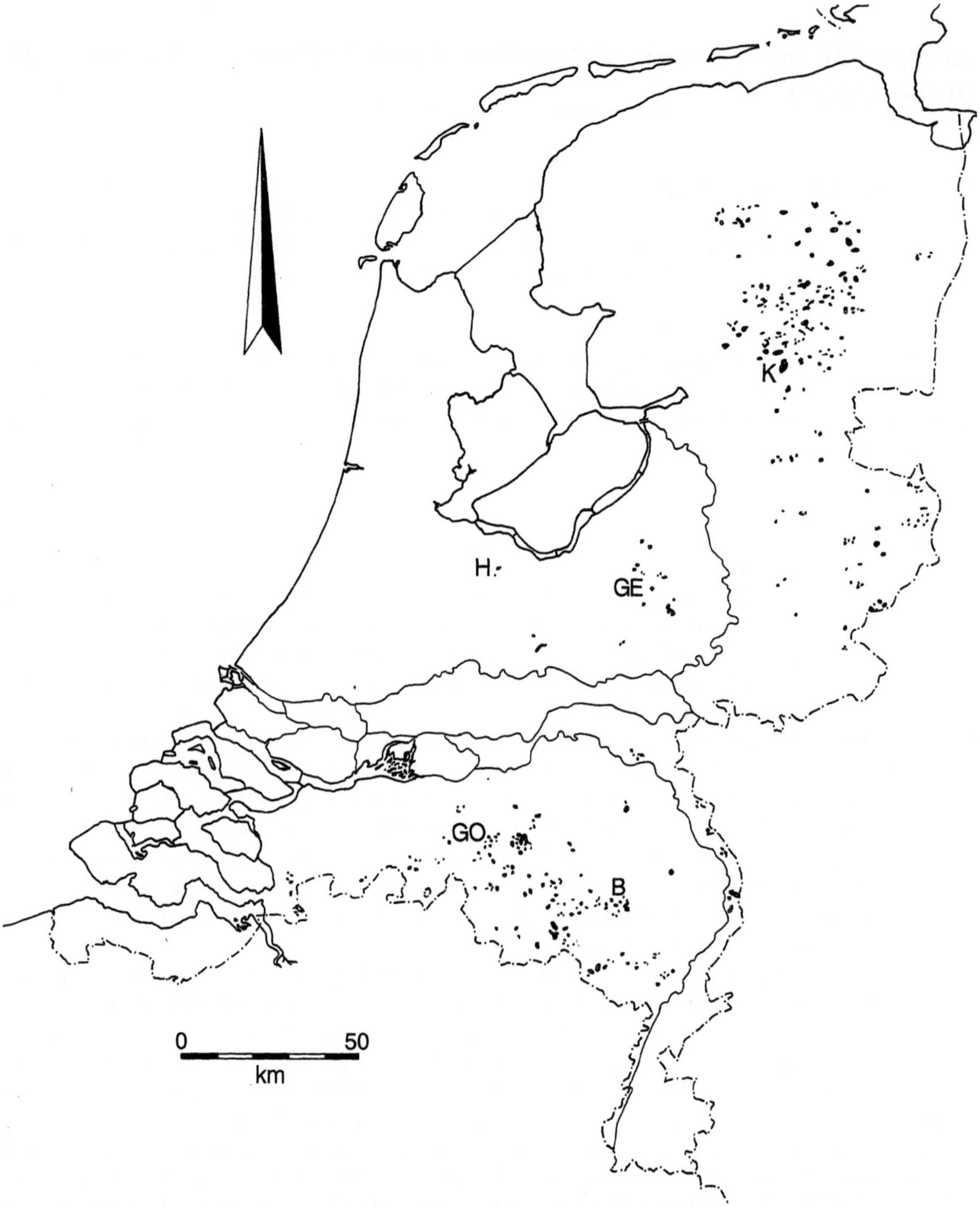

Fig. 1. Distribution of the 900 moorland pools, in which the littoral vegetation was surveyed in the years 1957–1959 by workers of the Foundation for the Investigation of Biocommunities. The moorland pools for case studies are indicated (K = Kliplo, GE = Gerritsfles, GO = Goorven, B = Beuven), as well as the location of the precipitation collector at Hilversum (H) (Leeflang, 1938).

Introduction

A moorland pool (Dutch: ven, fles, flaas; German: Heidetümpel; French: étang de bruyère) is an originally oligo- to mesotrophic and shallow soft-water body, situated in a sandy area, which is poor in lime and nutrients.

There are between 3,000 and 4,000 moorland pools present in the Pleistocene areas of The Netherlands, particularly in the northern province of Drenthe and the southern province of Noord-Brabant (Fig. 1)

The size of a typical moorland pool is between 0.5 and 1.5 ha, but many smaller and larger pools exist (Fig. 2). The maximum size is 75 ha. The maximum depth is rarely >3 m, the mean water depth at mean water level is typically between 0.5 and 1 m (Fig. 3). In the smallest pools the mean depth is only 1 dm at mean water level. The shore line development is between 1.1 and 4.0 (Van Dam, 1988). A layer of organic sediment covers at least a part of the mineral soil. The organic sediment is ≤ 0.05 m thick near the wind exposed north-eastern shore and up to 1.5 m thick near the wind sheltered south-western shore or the centre of the pool (Fig. 4).

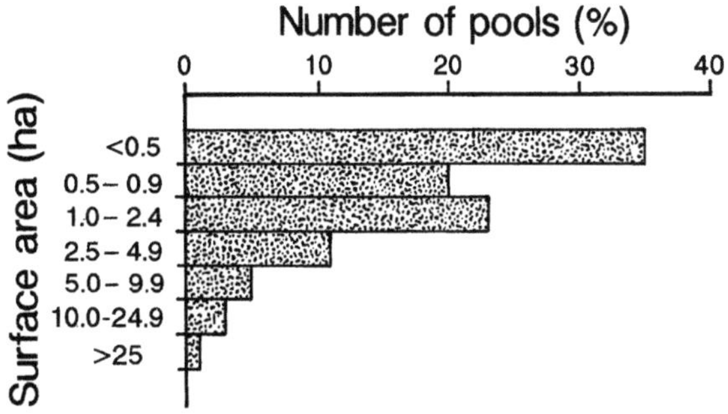

Fig. 2. Size distribution of 346 moorland pools in Noord-Brabant (after A.W.M. Mol, unpubl.).

After the formation of the pools the open water was colonized by aquatic macrophytes and in relatively deep pools peat was formed. In the shallowest pools the humic material was desiccated, pulverized and removed by the wind in extremely dry years. Many ponds gradually transformed into small bogs. The peat has been excavated since medieval times, for use as fuel and the bogs have been transformed into pools again. When the bulk of the peat was excavated the fresh sediments were sometimes removed and used as a fuel, after mixing with sods of heather (*Calluna vulgaris* (L.) Hull) and drying (Posthumus, 1911;

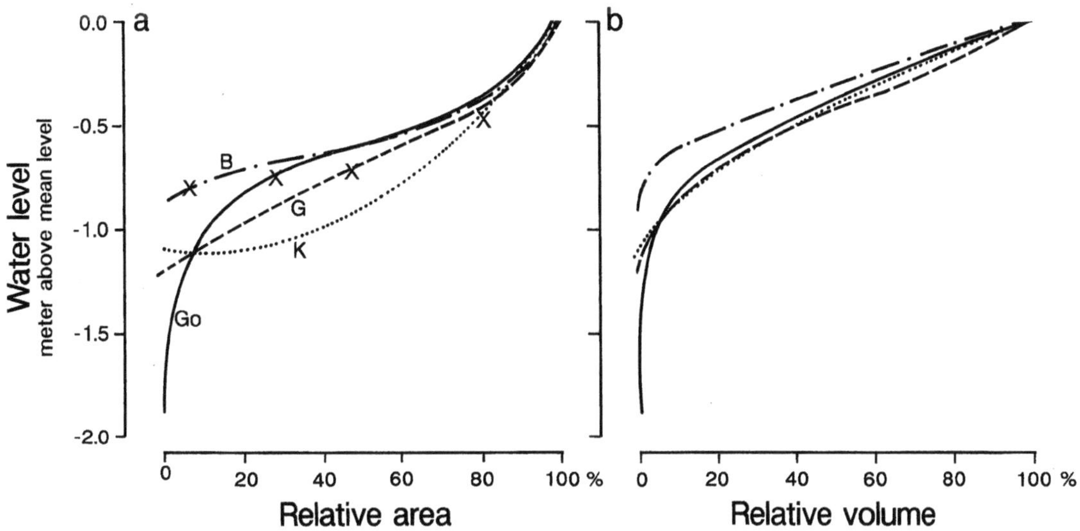

Fig. 3. Plots of water depth against relative area (a) and relative volume (b) at mean water level of the moorland pools Gerritsfles (G), Goorven (Go), Kliplo (K) and Beuven (B). The estimated lowest water level in the extremely dry summer of 1976 is indicated with ×. The relative area is the ratio between the actual surface area of the water and the surface area at mean water level (after Van Dam, 1988 and Buskens, 1990).

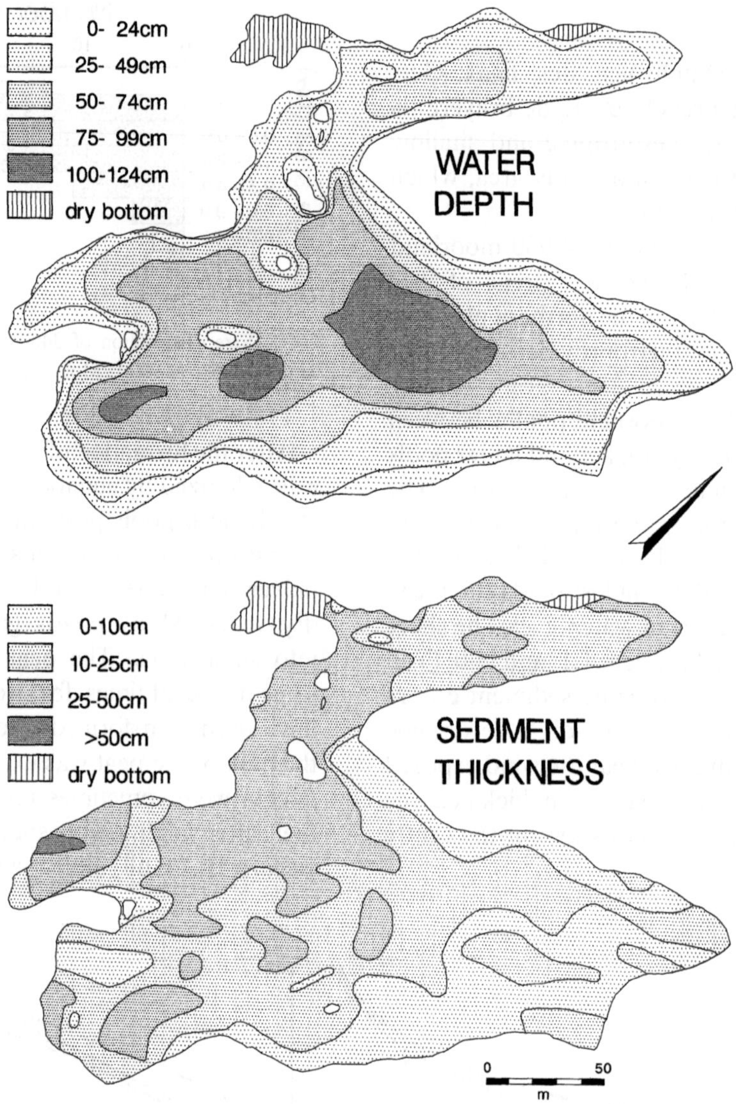

Fig. 4. Bathymetric map and map of sediment thickness of the pool Gerritsfles (Van Dam, 1987).

Van der Voo & Leentvaar, 1959; Westhoff *et al.*, 1973).

The pools with perched water tables have bottoms with an extremely high hydraulic resistance. They are oligotrophic because they are fed by precipitation only. The pools which are former river beds are often fed by groundwater and generally mesotrophic. These natural differences in trophic state were augmented by traditional small-scale activities as retting of flax, sheep washing, swimming, skating and inlet of nutrient-rich water for improvement of the fish status (Westhoff *et al.*, 1973; Arts *et al.*, 1988; Van Dam *et al.*, 1988a).

The natural and man-made variation among moorland pools allowed a rich variety of plants and animals typical for oligo- to mesotrophic environments to exist up to the first decades of the present century. Moorland pools are extremely important habitats for some macrophyte communities, particularly those belonging to the Littorelletea, and for algae, particularly desmids, which are most abundant in oligo- to mesotrophic soft-water bodies. As such waters are comparatively rare in The Netherlands and adjacent countries, moorland pools had a high value for nature conservation. However, this value has been strongly affected the last six decades by the changes invoked by modern agriculture: reclamation, drainage and eutrophication, and by the more recent phenomenon of acidification (Westhoff, 1979; Coesel & Smit, 1977; Coesel *et al.*, 1978; Leuven *et al.*, 1986c; Arts *et al.*, 1989).

It is the purpose of this paper to summarize the recent knowledge on hydrology, chemistry and biology of moorland pools, in relation to human influences on the pool ecosystems. This knowledge is required for a proper management of these valuable ecosystems for nature conservation (Nature Policy Plan). Changes in community structure will be emphasized, as these are much more sensitive to anthropogenic changes than ecosystem processes (Schindler, 1987). Examples will primarily be taken from a few well-known pool systems (Table 1). Goorven, Gerritsfles and Kliplo are pools in various stages of acidification. Beuven was eutrophied, but has been restored recently. Problems in these pools are representative of the main problems with management of moorland pools in general. In some cases, data from other pools are presented to provide a more general picture.

Origin and hydrology

Origin

According to their origin, the following types of moorland pools can be distinguished (*e.g.* Schimmel & Ter Hoeve, 1952; Broertjes, 1977; De Gans, 1982; Bakker *et al.*, 1986; Van Dam *et al.*, 1988b):

A. Depressions created by action of ice.

Pingo remnants, most abundant in the province of Drenthe and formed during the Weichsel ice age. The organic filling of the hollows started by the end of the Pleistocene era.

B. Depressions created by action of water.

B1. Remnants of snow meltwater valleys. Such pools are often oriented in rows, which may be oriented in another direction than the prevailing winds, which were from WSW to ENE.

B2. Former river beds. New river branches were formed when main branches of the wild river Meuse were obstructed by ice-floes. After melting of the ice the streams took another course and the new river branches remained as pools.

C. Depressions created by action of wind.

C1. Wind deflation hollows. These originated during the late Pleistocene or the early Holocene by deflation of cover sands and are shallow, generally less than 2 m deep. Typical examples have an elongated shape with the long axis in the prevailing wind direction. Beuven, and possibly also Goorven, belong to this type (Fig. 5).

C2. Hollows on a soil profile in coversand. By illuviation of dispersed humic compounds and subsequent precipitation in a hollow of a coversand area a perched water table was formed (Fig. 6). The pool Kliplo probably belongs to this type. Many of such pools originated about 3,000 y BP, particularly in Drenthe.

C3. Pools on placic horizons (iron pan). A placic horizon can be formed by precipitation of eluviated iron compounds at the interface of a waterlogged and a well drained soil. Such pools, with a perched water table, have been formed from Atlantic times (6,000–3,000 y BP) onwards until recently. During the first half of the 19th century the Gerritsfles site rapidly evolved from a wet heathland towards a pool, due to the large influx of iron with aeolian drift sand, which stopped around 1900, when the drift sand area was planted with Scots pines (*Pinus sylvestris* L.).

D. Artificial moorland pools.

Small pools have been created in the Pleistocene part of The Netherlands for various purposes, *e.g.* the use as a skating-rink or a fishpond, particularly in the present century.

Hydrology

A hydrological model (VENSIM) was developed for the simulation of water levels and budgets of chloride and water in isolated moorland pools by Houweling (Van Dobben *et al.*, 1993).

Table 1. Physiographical and historical data for selected moorland pools (Dickman *et al.*, 1987; Van Dam, 1988; Van Dam *et al.*, 1988b; Buskens, 1989).

	Goorven	Gerritsfles	Kliplo	Beuven
Surface area (ha)	2.4	6.8	0.6	58
Maximum depth (m)	1.9	1.2	1.1	0.8
Mean depth (m)	0.6	0.7	0.8	0.6
Shore line development	4.0	2.0	1.1	2.8
Hydrological type	Perched water table, isolated	Perched water table, isolated	Perched water table, isolated	Groundwater table, part of drainage system
Land-use in surroundings	Forest	Drift sand, heathland, shrubs, forest	Drift sand, forest	Heathland, forest
Age	Late Glacial	<200 y	Probably Subboreal	Late Glacial
Recent history				
1800–1850	Excavation of peat	Origin and extension	Excavation of peat	Probably stable situation
1850–1900	Eutrophication and alkalinization by inflow of agricultural drainage water	Eutrophication and alkalinization by washing of sheep	Eutrophication and alkalinization by washing of sheep and rearing of ducks	Probably stable situation
1900–1950	Acidification by hydrological isolation	Acidification by termination of sheep washing	Acidification by termination of sheep washing and rearing of ducks	Slight eutrophication by agricultural drainage water
1950–1985	Strong acidification by atmospheric deposition	Strong acidification by atmospheric deposition	Slight acidification by atmospheric deposition	Strong eutrophication and alkalinization by agricultural drainage water
1985–1990	Alkalinization by sulphate reduction and denitrification	Stabilization of acidification	Stabilization of acidification	Restoration and oligotrophication

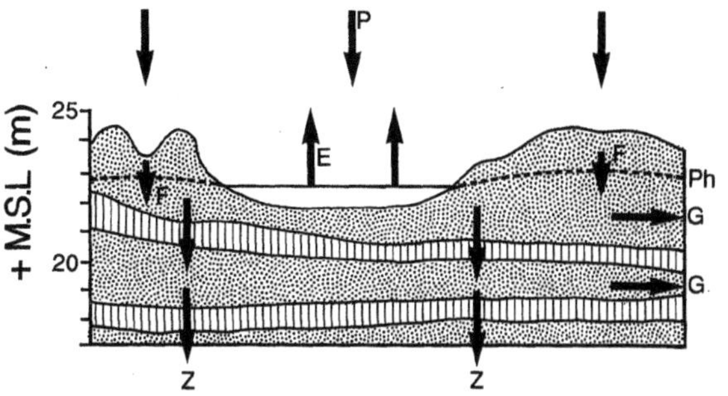

Fig. 5. Schematic cross-section of the moorland pool Beuven. The arrows indicate hydrological pathways. Pleistocene sand is dotted. Layers with humic compounds, loam and very fine sands with a high hydraulic resistance have been hatched. P = precipitation, E = evaporation, F = infiltration, Z = outseepage. G = direction of groundwater flow. Ph = phreatic level. MSL = mean sea level (after Buskens, 1989).

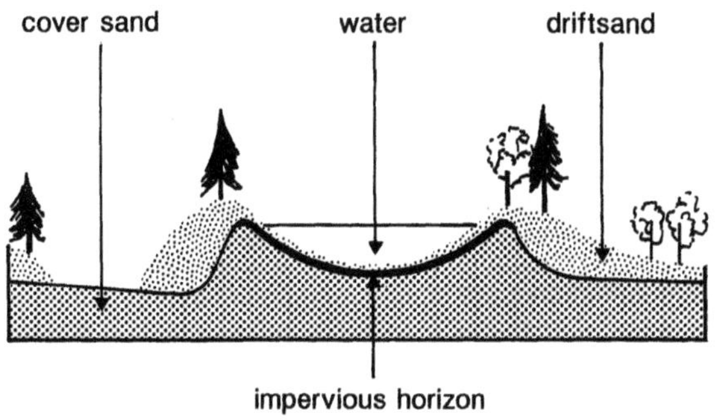

Fig. 6. Schematic cross-section of a moorland pool that developed on a soil profile in coversand by illuviation of dispersed humic compounds. Close to the pool the soil profile has disappeared by wind erosion, which was promoted by destruction of the original heatland podzol by overgrazing by sheep. Therefore the pool is presently located above the mean level of the environment (after Bakker *et al.*, 1986).

The model has been applied to the moorland pools Kliplo and Gerritsfles and seems to perform rather well. The correlation between simulated and observed pool water levels is 0.98 for both pools (Fig. 7). The highest water levels are observed in late winter or early spring and lowest levels in fall, particularly in years with relatively dry summers, *e.g.* 1982 and 1986.

The water level of Beuven can be regulated by in- and outlet of surface water, although the water

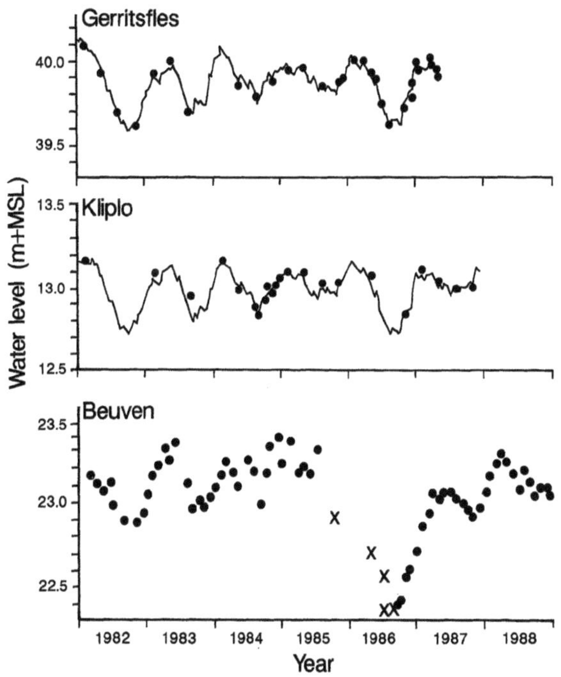

Fig. 7. Calculated (lines) and observed (dots) water levels of Gerritsfles and Kliplo. For Beuven observed water levels during the period of artificial drainage are indicated by ×.

levels follows roughly the same seasonal pattern as in Gerritsfles and Kliplo. Water levels were very low from fall 1985 to fall 1986, because the pool was artificially drained during this period for restoration purposes.

The water budgets, calculated with VENSIM for Gerritsfles and Kliplo, are presented in Table 2. The budget of Beuven was based on measurements of quantities of precipitation, surface water inflow and surface water outflow, calculations of outseepage as a function of water level and hydraulic resistance of the pool bottom. Inseepage was calculated as the balancing item (Buskens, 1989).

Gerritsfles and Kliplo are fed by precipitation exclusively. In Beuven the precipitation accounts for about 80% of the water supply. Evaporation accounts for ca. 60% of the water loss. The rest of the water leaves the pools by outflow either to groundwater or to surface water. In extremely dry years, *e.g.* 1921, 1947, 1959 and 1976 large parts of the bottom may be exposed (Fig. 3).

Table 2. Mean fluxes of water, selected ions and alkalinity and residence times in Gerritsfles, Kliplo and Beuven (Buskens, 1989; Van Dobben *et al.*, in prep.).

	Water (cm yr^{-1})	H^+	NH_4^+	K^+	Cl^-	NO_3^-	$SO_4^=$	o-PO_4	ANC* (meq m^{-2} yr^{-1})
		(……			mmol m^{-2} yr^{-1}			……)	
Gerritsfles (1982–1986)									
In: Bulk precipitation	80	19	84	4	74	40	44	0.7	− 14
Out: Outseepage	33	8	44	10	82	3	38	0.0	− 9
To air and sediment	47	9	33	− 3	− 5	35	8	0.6	− 12
Change in storage	1	2	8	− 3	− 3	1	− 2	0.0	6
Residence time (yr)	*0.9*	*1.3*	*1.1*	*5.7*	*2.5*	*0.2*	*1.8*	*0.2*	–
Kliplo (1982–1987)									
In: Bulk precipitation	86	18	89	4	78	43	44	0.4	− 17
Out: Outseepage	30	2	17	11	92	2	19	0.1	9
To air and sediment	53	15	62	− 6	13	42	24	0.4	− 24
Change in storage	3	1	10	− 1	− 1	0	1	− 0.1	− 2
Residence time (yr)	*1.0*	*0.3*	*0.6*	*9.3*	*4.0*	*0.1*	*1.4*	*0.6*	–
Beuven (1988)									
In: Bulk precipitation	88	15	103	6	52	44	50	0.6	− 3
Inseepage	11	8	2	3	23	23	24	0.0	− 49
Surface water inflow	9	0	3	29	66	2	44	0.0	75
Out: Outseepage	9	0	0	3	26	1	24	0.0	− 5
To air and sediment	65	23	109	− 3	2	77	86	0.8	− 52
Change in storage	− 3	− 3	− 2	− 1	− 6	− 11	− 108	− 0.2	86
Surface water outflow	37	3	1	39	119	0	116	0.0	− 7
Residence time (yr)	*0.5*	*0.2*	*0.0*	*1.5*	*1.2*	*0.1*	*1.4*	–	*0.1*

* Acid-neutralizing capacity = Σ (strong base cations) − Σ (strong acid anions) = $[Na^+] + [K^+] + [NH_4^+] + 2[Ca^{2+}] + 2[Mg^{2+}] - 2[SO_4^{2-}] - [NO_3^-] - [Cl^-]$ (Reuss & Johnson, 1986).

Chemistry

Atmospheric deposition

The chemical composition of surface water of moorland pools, particularly those with a perched water table, is determined by the composition of wet and dry deposition (Table 3).

Precipitation at Hilversum (Fig. 1) is dominated by chloride, as is usual in coastal areas, and also has a high concentration of excess sulphate (Table 3). The sulphate concentrations of the recently sampled sites are 10–17 times higher than the natural concentrations, due to anthropogenic impact; and those of ammonium 3–4 times higher than in 1870. These values, and also those for nitrate, are far higher than those presently found in remote areas of Europe (Hanssen *et al.*, 1990). The highest concentrations of sulphate, nitrate and ammonium are found at the Beuven site, which is in an area with extremely high ammonium deposition, due to intensive animal husbandry (Schuurkes *et al.*, 1988c; Erisman *et al.*, 1989). The potassium concentrations at Beuven are also extremely high.

For terrestrial ecosystems in The Netherlands the wet deposition has been calculated as 30% of the total potential acid deposition (Erisman *et al.*, 1989). Direct measurements of total deposition on aquatic surfaces are difficult to perform, but preliminary results indicate that total deposition on the open water surface of moorland pools is not very different from bulk deposition. Probably the total deposition nearshore is about twice as

Table 3. Average chemical composition (mmol m^{-3}), electrical conductivity and quantity of bulk precipitation near Hilversum and some moorland pools (after Leeflang, 1938; Van Dam, 1988; Buskens, 1989 and H.L.M. Kersten, unpubl.).

Location Period	Hilversum 1932–1937	Goorven 1982–1984	Gerritsfles 1982–1984	Kliplo 1982–1984	Beuven 1987–1988
Alkalinity	43	3	0	0	25
SO_4^{2-} (Excess SO_4^{2-})[a]	46 (40)	60 (57)	53 (47)	49 (45)	66 (62)
Cl^-	113	69	84	90	75
NO_3^-	2	48	50	49	63
Ortho-PO_4 (mmol m^{-3})	–	–	–	–	0.9
H^+	1	37	37	35	21
NH_4^+ (Excess NH_4^+)[b]	30 (–2)	108 (76)	94 (62)	87 (55)	133 (101)
K^+	–	5	4	4	50–68
Na^+	87[c]	55	68	73	53
Mg^+	41	7	8	9	10
Ca^{2+}	43	12	10	10	23
Mn^{2+}	0[d]	0.2	0.2	0.2	0.3
Al^{3+}	–	1	1	1	1
Fe^{3+}	0[d]	2	2	2	1
El. cond. (25 °C) (mS m^{-1})	3.1	5.8	4.4	4.4	–
Quantity (cm a^{-1})	72	85	86	89	72

[a] Excess SO_4^{2-} = [SO_4^{2-} in precipitation] – [SO_4^{2-} in seawater]/[Cl^- in seawater], * [Cl^- in precipitation], [b] Excess NH_4^+ = [NH_4^+ in recent precipitation] – [modelled NH_4^+ in precipitation in 1870] (Asman *et al.*, 1988), [c] including K^+, [d] below detection limit.

high as on the open water, because of the roughness of the helophytes (Buskens, 1989). For the present no reliable measurements of total deposition on moorland pools are present and consequently the atmospheric input of nitrogen and sulphur compounds is underestimated by an unknown amount.

Composition of pool water

The average chemical composition of poolwater in different periods is given in Table 4. Moorland pools are poor in solutes in comparison with other types of Dutch surface waters (Redeke, 1948). In the recent samples from Goorven, Gerritsfles and Beuven, sulphate is the most important anion, as in all acidified lakes and pools in The Netherlands (Leuven, 1988). In Goorven, sulphate has increased at the expense of the alkalinity (mostly bicarbonate). In the isolated pools Goorven, Gerritsfles and Kliplo, sodium is the dominant cation. The dominance of calcium in Beuven indicates the supply with surface and ground water.

Long-term changes of pH, alkalinity, sulphate and ortho-phosphate concentrations are presented in Fig. 8. These changes will be discussed below for some selected variables in relation to mass budgets.

Mass budgets

The mass budget for any substance in a moorland pool is

$$Q_p \cdot C_p + Q_{is} \cdot C_{is} + Q_{si} \cdot C_{si} - Q_{os} \cdot C_{os} - Q_{so} \cdot C_{so} - Q_s \cdot C_s - M_r = 0$$

where Q_p, Q_{is}, Q_{si}, Q_{os}, Q_{so} and Q_s are the volumes of precipitation, inseepage, surface water inflow, outseepage, surface water outflow and change in storage, and C_p etc. are the corresponding concentrations of the substance in these components. M_r is the amount of ions lost to air

Table 4. Average chemical composition (mmol m^{-3}) of surface water of some moorland pools (after Van Dam *et al.*, 1981; Buskens, 1989; H. van Dam, unpubl. and H.L.M. Kersten, unpubl.).

Pool	Goorven		Gerritsfles		Kliplo	Beuven	
Period	1919	1979–1989	1930	1979–1989	1979–1989	1983–1985	1987–1988
Number of measurements	1	42	1	42	42	21	31
SO_4^{2-} (Excess SO_4^{2-})[a]	104 (77)	246 (225)	198 (174)	120 (108)	58 (42)	307 (284)	375 (359)
Cl^-	536	407	468	243	313	440	313
Alkalinity	149	29	98	16	58	760	13
NO_3^-	0	4	0	8	5	47	8
Ortho-PO_4	0	0.2	0	0.2	0.3[c]	1.9	0.3
H^+	3	59	0	35	5	0	4
NH_4^+ (Excess NH_4^+)[b]	0 (−249)	181 (−8)	21 (−157)	94 (1)	50 (−61)	30 (−162)	7 (−127)
K^+	–	45	–	33	37	415	124
Na^+	–	303	–	194	248	374	246
Mg^{2+}	83	63	99	30	32	210	80
Ca^{2+}	83	68	98	35	31	319	267
Mn^{2+}	0	2	–	2	2[c]	2[d]	2
Al^{3+}	–	25	–	7	5[c]	9[e]	7
Fe^{3+}	<2	15	3	4	12[c]	13	2
CO_2	50	325	114	141	149	–	62
Conductivity at 25 °C (mS m^{-1})	9.7	13.1	10.0	7.5	5.9[c]	20.9	11.6
Colour (mg Pt l^{-1})	7	23	8	7	32	–	–
$KMnO_4$-consumption (mg l^{-1})	≥14	46	16	17	61[c]	–	–
pH (field) (−)	5.5	4.0	6.5	4.1	4.7	7.3	5.4

[a] Excess SO_4^{2-} = [SO_4^{2-} in pool water] − [Cl^- in pool water] * [SO_4^{2-} in seawater] / [Cl^- in seawater], [b] Excess NH_4^+ = [NH_4^+ in pool water] − [Cl^- in pool water] * [modelled NH_4^+ in precipitation in 1870] / [Cl^- in precipitation] (Asman *et al.*, 1988), [c] n = 35, [d] n = 12, [e] n = 4.

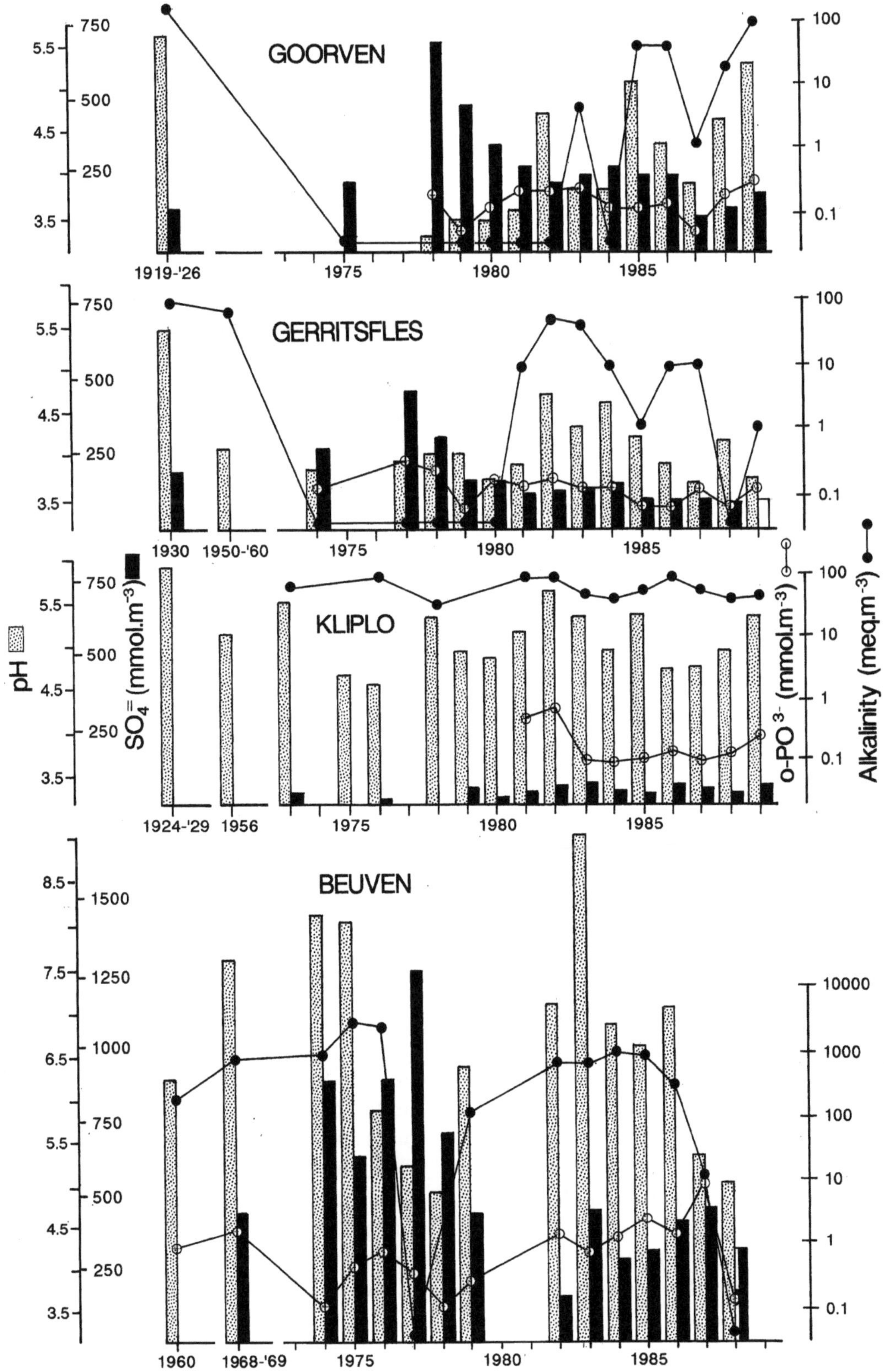

Fig. 8. Long-term changes in median values of selected physical and chemical variables.

or sediments. The water volumes and ion concentrations at the Kliplo, Gerritsfles and Beuven sites were calculated or measured. Subsequently M_r was calculated from the budget equation (Buskens, 1989; Van Dobben *et al.*, 1993). The results for selected ions are presented in Table 2.

Sulphur, pH and alkalinity

In Gerritsfles, Kliplo and Beuven respectively 17, 44 and 73% of the input of sulphate escapes to the air or sediment, apparently as a result of bacterial reduction of sulphate (Table 2). The efficiency of sulphate reduction correlates well with the differences in pH of the three sites (Table 4). Feijtel *et al.* (1989) calculated the net sulphate reduction in the sediment of Gerritsfles by diagenetic modelling of in situ sulphate concentration profiles to be 48 mmol m^{-2} yr^{-1}. E.C.L. Marnette (pers. comm.) measured a net sulphate reduction rate in Gerritsfles that is one order of magnitude higher than stated in Table 2. At least a part of this difference may be explained by the lack of reliable estimates of dry deposition of sulphur compounds (Schindler, 1988).

The reduction of sulphate goes along with the production of acid neutralizing capacity by the sediments, which is highest in Beuven and lowest in Kliplo.

In Goorven, the concentration of sulphate was lower in 1919 than in 1979–1989, but in Gerritsfles the concentration in 1930 was far higher than in 1979–1989 (Table 4). This is probably caused by low sulphate reduction rates in pools and lakes with sandy sediments as compared to pools and lakes with organic sediments (Schuurkes *et al.*, 1988b). Many moorland pools, including Gerritsfles, originally had a sandy bottom (*e.g.* Redeke & De Vos, 1932), which was buried under organic material due to decreased decomposition by acidification, as will be discussed later.

Concurrent with the increase in sulphate concentrations, pH and alkalinity in Goorven decreased between 1919–1926 and 1975–1981 (Table 4, Fig. 8). After the extremely dry summer of 1976, when 70% of the bottom of this pool was exposed to the atmosphere, the reduced sulphur compounds in the sediments oxidized and caused high concentrations of sulphate, and low pH-values after refilling (Vangenechten *et al.*, 1981; Van Dam, 1988). The sulphate concentration gradually declined, both by dilution with rain water and by sulphate reduction. The latter process also caused the increase of pH and alkalinity after 1981.

Basically, the changes in sulphate concentration, pH and alkalinity in Gerritsfles have the same pattern as in Goorven, but Gerritsfles recovered faster from the drought of 1976, because of the shorter residence time (Van Dam, 1988). There is no strong increase of pH after 1986, which is possibly due to cation exchange by the large masses of Sphagnum on the bottom of this pool (Clymo & Haward, 1982).

Also in Beuven a rise of sulphate and a drop of pH was observed after 1976, when the pool desiccated completely (Fig. 8). After the temporal draining of the pool for the restoration measures in 1985–1986, only a small increase of the sulphate concentration was observed, as the majority of the sulphur containing sediment was removed during the restoration. Nevertheless the decrease in sulphate concentration in 1988 caused an important change in the storage of sulphate in the pool water (Table 2).

In Kliplo the concentration of sulphate and the pH have been relatively constant since 1973 (Fig. 8), as only 20% of the bottom ran dry in 1976 (Van Dam, 1988).

In six Dutch moorland pools the average sulphate concentration increased from 107 mmol m^{-3} in the period before 1960 to 289 mmol m^{-3} in the period after 1960, in 21 pools the average alkalinity decreased from 563 to 129 meq m^{-3} over the same time span and the pH decreased from 5.7 to 5.0 in 86 pools. However, before 1960 the pH was evenly distibuted around the interval 5–6, which comprised 67% of the pools and after 1960 (mainly 1983–1984) the pH-distribution was distinctly bimodal with peaks in the intervals 3.5–4 and 6.5–9.5. Only 15% of the pools had a pH in the interval 5–6 (H.L.M. Kersten, unpubl.). This clearly demonstrates the acidification of isolated pools at the one hand and the alkalinization by eutrophication of non-isolated pools at the other hand. The decrease of sulphate and increase of pH between 1978 and 1983 has also been observed in other moorland pools in The Netherlands (Van Dam, 1988).

Nitrogen and potassium

In Gerritsfles, Kliplo and Beuven respectively 39, 70 and 100% of the ammonium input escapes to the air or sediments, apparently as a result of nitrification processes (Table 2). Most of the produced nitrate will be denitrified or taken up by water plants, as nitrate concentrations are low (Table 4). As the output of nitrate to the air or sediments is >88% denitrification must be very efficient.

Nitrification produces protons and therefore consumes acid neutralizing capacity. Denitrification produces acid neutralizing capacity. The net effect of nitrification and denitrification on alkalinity depends on the amounts of nitrate and ammonium that are transformed.

In an experiment where 138 mmol m^{-2} yr^{-1} ammonium, 82 mmol m^{-2} yr^{-1} sulphate and 10 mmol m^{-2} yr^{-1} nitrate (compare Table 3) were added to small moorland pools in a greenhouse, the initial pH decreased from 5.5 to 4.2 after two years, due to nitrification of ammonium. Denitrification was to small to compensate for the acidification by nitrification (Schuurkes *et al.*, 1987). However, the alkalinity increased after addition of nitrate to enclosures in an acid shallow lake in the coastal dunes, which is very similar to what happened in most moorland pools (Schuurkes *et al.*, 1988a).

Both in Goorven and Gerritsfles ammonium concentrations increased between 1919–1930 and 1979–1989 (Table 4). The mean ammonium concentrations of 15 Dutch moorland pools increased from 17 mmol m^{-3} before 1960 to 84 mmol m^{-3} after 1960 (H.L.M. Kersten, unpubl). The increase of the nitrate concentrations in Gerritsfles and Goorven in Table 4 is questionable, because of analytical problems in the past.

Particularly in Kliplo and Gerritsfles the sediments are an important source of potassium (Table 2). Therefore the potassium concentration in these pools is about three times as high as might be expected from the composition of precipitation (assuming that chloride is conservative). The average concentration of potassium in 188 acidified moorland pools in The Netherlands, sampled between 1983 and 1986 (M.M.J. Maenen, unpubl.) is about four times higher than expected on the basis of the potassium/chloride ratio in bulk precipitation at stations of the National Precipitation Chemistry Network in moorland pool districts in the same period. The elevated potassium concentrations are probably due to the high ammonium concentrations, because ammonium is very effective in replacing potassium on ion-exchange sites in soils (Sawhney, 1972). Schuurkes *et al.* (1987) measured elevated potassium concentrations in small artificial moorland pools that were fed by artificial precipitation.

Phosphorus and inorganic carbon

According to the mass balances (Table 2) over 85% of the input of ortho-phosphate to the three pools is stored in the sediments. Therefore the concentrations in surface water are low (Table 4). The sediments are rich in phosphates. In Beuven 5 mmol kg^{-1} dry weight and 0.68 mol m^{-2} were measured (Buskens, 1989). The release of phosphate from the sediments, and therefore also the concentration of phosphate in surface water, depends on the pH and the alkalinity of the overlying water, as is illustrated in Fig. 9. The low concentrations of ortho-phosphate at pH above 8.5 are caused by bloom of plankton algae. The low concentrations of phosphate at low pH-values are due to adsorption of phosphate by iron hydroxides and inhibition of decomposition at low alkalinity (Lijklema, 1980; Kok & Van de Laar, 1991).

Low concentrations of bicarbonate, (i.e. low alkalinity levels) are important to prevent internal eutrophication by release of phosphates from the

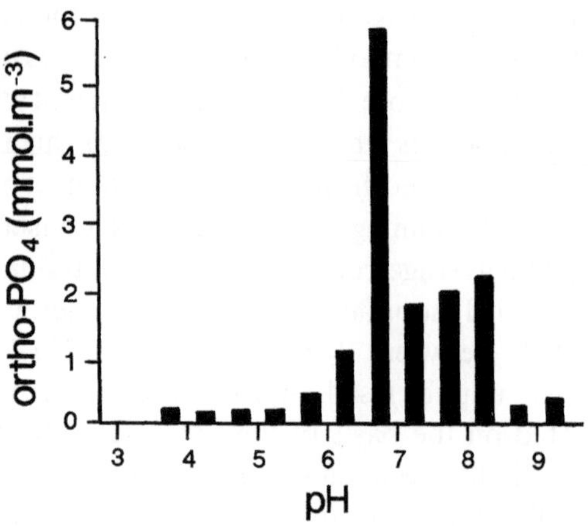

Fig. 9. Relation between mean concentrations of ortho-phosphate and pH in Beuven from 1975 to 1985 (Buskens, 1989).

sediments. Bicarbonate is also a source of carbon for many aquatic macrophytes, with the exception of *Juncus bulbosus* L. and the isoetid species, which use carbon dioxide. In poorly buffered non-acidified waters the concentration of carbon dioxide is usually between 6 and 40 mmol m^{-3} (Roelofs *et al.*, 1984). The carbon dioxide concentration in the Goorven sample from 1919 hardly exceeded these very low levels, but in the recent samples from this and the other pools elevated concentrations were measured (Table 4). These concentrations are in the range 66 to 870 mmol m^{-3}, which was found by Roelofs *et al.* (1984) in acidified moorland pools.

Macrophytes

Species composition

Historical and recent inventories of the most important aquatic macrophytes in the four pools are presented in Table 5, together with literature data on the ecology and geographic distribution of the species.

Most species are characteristic for oligo- to mesotrophic (pore water phosphate <20 mmol m^{-3}) soft water bodies (alkalinity predominantly $\leqslant 1$ meq l^{-1} and always <2 meq l^{-1}), with pH-values between 5.2 and 7.3. Many of these species have an atlantic, boreal or boreal-atlantic distribution. The flora of the four pools more or less represents the soft-water flora of The Netherlands. This flora is rare and relatively rich in species, due to the occurrence of both boreal and atlantic taxa (Arts & Den Hartog, 1990). Moorland pools are a very important habitat for these species within the country.

The combination of isoetid species has been described as the Isoeto-Lobelietum association, which is a part of the Littorellion alliance. The communities of atlantic plants are less homogeneous. The truly aquatic communities (with *Luronium natans, Myriophyllum alterniflorum*, etc.) have been accomodated in the Potamion graminei, while the amphiphyte communities are considered as associations of the Littorellion by Westhoff & Den Held (1969) and as associations of the Hydrocotyle-Baldellion and Eleocharition acicularis by Schaminée *et al.* (1990). The combinations of species from strongly acid waters are associations or (con)sociations of the Rhynchosporion alliance (Westhoff & Den Held, 1969) or impoverished associations of the Littorelletea class (Schaminée *et al.*, 1990). The groups of species from neutral to alkaline waters and acid-indifferent species include members of the Parvopotamion and Phragmition alliances.

Historical changes and management practices

The present vegetation of Goorven is quite uniform. In the open water *Nymphaea alba* is most conspicuous. *Juncus bulbosus* covered the water surface after the severe drought of 1976 until 1979 and decreased later on. In the period 1912–1959 this species was much less abundant. The enhanced levels of carbon dioxide in Goorven after 1976 stimulated the growth of this plant, which is able to take up carbon dioxide by the leaves (Roelofs *et al.*, 1984; Van Dam, 1987). Also in other moorland pools in The Netherlands, including Gerritsfles, *J. bulbosus* increased after the drought of 1976 (Sykora, 1979).

Goorven was a very shallow, acid pool with *Sphagnum* in the beginning of the 19th century,

Table 5. Predominant macrophyte species in different periods. Size of symbol is proportional to estimated abundance (smallest dot: rare species, largest dot: species covering almost whole bottom or forming a wide belt along the margins of the pool). Before the name of each species the optimal values for pH, alkalinity (eq m^{-3}) and phosphate concentration of sediment pore water (mmol m^{-3}) according to De Lyon & Roelofs (1986) and Arts *et al.* (1990a) are given. Values between parentheses are insignificant optima. ′ and ′′ indicate groups I (species intolerant to pH < 5) and II (species tolerant to pH < 5) respectively (Arts *et al.*, 1990). Geographical distribution (A = atlantic, B = boreal, W = wider) according to Arts & Den Hartog (1990) and Tutin *et al.* (1964–1980). Data prior to 1900 from paleolimnological analysis by Dickman *et al.* (1987) and Van Dam *et al.* (1988b). Data after 1900 from literature cited by Van Dam (1988) and Buskens (1989).

pH	Alk	PO_4	Distr.		Goorven				Gerritsfles				Kliplo				Beuven		
				From:	1800	1850	1912	1976	1800	1850	1916	1973	1800	1850	1924	1965	1934	1965	1986
				To:	1850	1900	1959	1984	1850	1900	1958	1984	1850	1900	1958	1984	1957	1985	1989
Species from strongly acid water																			
4.1	0.4	6	W	*Carex rostrata* Stokes			•	•			•	•			•	•	•	•	•
4.5	0.2	12	A	*Eleocharis multicaulis* Sm.			•	·	•		·	·			·	·	•	·	•
4.1	0.0	5	W	*Eriophorum angustifolium* L.			•	•		·	•	•			•	•			·
4.4	0.3	8	W	*Juncus bulbosus* L.	·	·	·	●	•	·	·	•		•		·	•	·	●
4.1	0.0	12	W	*Sphagnum* spec.	•	•	●	●	·	·	•	⬤	·	•	•	•			
4.4	0.1	11	W	*Utricularia minor* L.			·	·			·				•	·			·
Species from weakly acid water																			
–	–	–	W	*Agrostis canina* L.			·	·			•	•			•	·	·	·	•
5.5	0.5	7	W	*Hydrocotyle vulgaris* L.			·	•	·		·	•			·	•	•	•	•
5.6′′	0.3	6	A	*Hypericum elodes* L.			•	·									•	·	●
5.6	0.5	7	W	*Juncus effusus* L.			·	·			·	•			·	·	·	•	●
5.8′′	0.4	6	BA	*Littorella uniflora* (L.) Aschers.		·			·								⬤	•	●
5.1′′	0.2	–	B	*Lobelia dortmanna* L.		●			·	·	·						•		•
5.8′′	0.4	8	A	*Luronium natans* (L.) Raf.						·	·						•	·	•
5.2	0.4	11	W	*Potentilla palustris* (L.) Scop.			·				•	·			•	•		·	·
5.2′′[a]	–	–	B	*Sparganium angustifolium* Michx						·	•	·			●	·	•		
5.8′′	0.6	12	W	*Utricularia australis* R. Br.														·	•
Species from weakly acid to neutral water																			
–	–	–	W	*Callitriche* spec.														·	•
6.8′	0.8	4	A	*Echinodorus repens* (Lamk.) Kern & Reichgelt													•	·	•
7.3′	0.4	7	A	*Elatine hexandra* (Lapierre) D.C.													●	·	●
7.2′	1.1	11	W	*Eleocharis acicularis* (L.) Roemer & Schultes													•	●	•
7.3	(1.9)	11	W	*Eleocharis palustris* L.			·			·	•	•				·	•	•	•
6.3′	0.3	–	B	*Isoetes echinospora* Durieu													•	·	•
6.2′	0.4	12	W	*Lythrum portula* (L.) D.A. Webb														·	•
6.8′	0.7	13	BA	*Myriophyllum alterniflorum* D.C.			•		·										
6.3′	0.6	5	A	*Pilularia globulifera* L.								·					•		·
7.2′	1.4	18	W	*Potamogeton obtusifolius* Mert. & Koch															•
6.4′	0.7	7	A	*Potamogeton polygonifolius* Pourret		·	•												
7.1	1.3	(15)	W	*Potamogeton natans* L.			·	·		•	·	·			•	●		·	·
6.8′	1.0	6	W	*Ranunculus flammula* L.														·	•
6.9	1.2	13	W	*Riccia fluitans* L.														•	·
6.4′′	0.8	5	A	*Scirpus fluitans* L.			•				·								·
Species from neutral to alkaline water																			
7.6	2.8	(33)	W	*Elodea nutalli* (Planchon) St John														·	•
7.5	2.5	(30)	W	*Lemna minor* L.								·						•	
7.8	3.5	23	W	*Potamogeton berchtoldii* Fieb.															
8.0	3.0	47	W	+ *P. pusillus* L.													·	•	•
7.6	1.8	7	W	*Potamogeton gramineus* L.														·	•
pH-indifferent species																			
(6.3)	(1.6)	(12)	W	*Nymphaea alba* L.		·	●	•											
(7.1)	(2.3)	(14)	W	*Phragmites australis* (Cav.) Trin. ex Steud.			•	·							•	•	·	⬤	•
(7.1)	(1.8)	17	W	*Glyceria fluitans* (L.) R. Br.							•	·							

[a] Mean of pH-values at 14 sites in Western Europe between 1929 and 1983 (H. van Dam, unpublished results).

when the original bog vegetation had been removed for use as fuel (Table 5). Due to the inflow of buffered, slightly eutrophic agricultural drainage water, a rich assemblage of aquatic macrophytes, including isoetids, started to develop in the late 19th century. Many of the species from weakly acid to neutral water declined in the 20th century due to termination of the inflow of buffered water and the acidification by atmospheric deposition (Dickman *et al.*, 1987; Van Dam, 1988).

In the early 19th century the pool Gerritsfles was colonized by species from strongly acid waters (Table 5). In the late 19th and early 20th century also species from weakly acid and weakly acid to neutral water could develop due to the input of buffering compounds from sheep washing. After ca. 1930 many of these species disappeared due to the termination of sheep washing and the acidification by atmospheric deposition. The bottom of the open water zone was increasingly covered with *Sphagnum denticulatum* Brid. and places with a bare sandy bottom have almost disappeared (Van Dam, 1988; Van Dam *et al.*, 1988b). The decrease of *Glyceria fluitans* over the same period can be attributed to termination of the use of artificial fertilizer in a meadow close to the pool. The increase of *Juncus bulbosus* and *J. effusus* since 1950 are probably due to atmospheric deposition of acidifying substances and nutrients respectively.

In Kliplo, the original bog vegetation, which had developed since the Subboreal or early Subatlantic period, was excavated around 1850 and in the secondary pool *Sphagnum* and *Juncus bulbosus* became important taxa (Table 5). In the second half of the 19th century the nutrient concentration and consequently the pH of the pool increased and species from weakly acid waters, including the very rare *Sparganium angustifolium*, became important. The increase of nutrients was caused by sheep washing and possibly the use of the pool as a duck-decoy (Van Dam *et al.*, 1988b). During the last few decades the abundance of *S. angustifolium* decreased in favour of *Potamogeton natans* (Van Dam 1988).

Most of the water bottom of Beuven was covered with extensive swards of *Littorella uniflora* and other species of the Littorellion alliance before 1957 (Table 5). Due to eutrophication by agricultural drainage water the abundance of plankton algae increased and the area of Littorellion taxa decreased probably by light limitation. Nearshore, the area of stands of reed (*Phragmites australis*) greatly expanded. In 1985–1986 the majority of the reed was removed and the agricultural drainage water was diverted. South of the pool a reservoir was constructed where a part of the agricultural drainage water was allowed to enter. In the reservoir the concentration of nutrients decreased by sedimentation and microbial activity. Occasionally water from the reservoir was let into the pool to maintain the buffer capacity. Consequently the Littorellion alliance re-occupied large parts of the pool bottom by germination of seeds preserved in the bottom.

The changes in these four pools are similar to those in other moorland pools in The Netherlands. Schoof-van Pelt (1973) considered eutrophication to be the main cause for the impoverishment of Littorellion communities in moorland pools between 1935 and 1970. Strijbosch (1976) demonstrated that eutrophication by agricultural drainage water initially promoted the extension of *Juncus effusus* and Bidention communities, while the onset of eutrophication by guanotrophy (bird excrements) causes an increase of Parvocaricetea communities. The final stages in both cases are the Phragmitetea communities.

The impact of acidification was studied more recently. Van Dam & Kooyman-van Blokland (1978), Roelofs (1983) and Arts *et al.* (1990b) found a decrease of Littorellion taxa and an increase of *Juncus bulbosus* and *Sphagnum* in moorland pools all over the country at the expense of Littorellion taxa, particularly since 1950. *Lobelia dortmanna* was recorded at 183 sites in The Netherlands between 1834 and 1960 and only at 4 sites after 1980. At 72% of the sites which were not reclaimed and where the cause of the decline was known, the species disappeared due to acidification and at the remaining 28% of the sites *L. dortmanna* disappeared due to eutrophication.

The species holds only at those sites where specific conservation measures (regular removal of the organic top layer) are practiced. Formerly many Littorellion species were found in hydrologically isolated moorland pools, but in the present situation they are mainly found in water bodies where buffered ground- or surface water is introduced and where active management measures are taken (Arts *et al.*, 1988).

Primary production and litter breakdown

The gross primary production and respiration in Gerritsfles during the years 1988 were 190 and 205 g C m^{-2} yr^{-1} respectively, resulting in a net primary production of -15 g C m^{-2} yr^{-1}, as was calculated from daily variations of the oxygen concentrations at various depths and the intensity of incoming light by K. Kersting (pers. comm.), according to the method of Kersting & Kouwenhoven (1989). The negative production is surprising, because most of the production is due to the assimilation of the bottom covering *Sphagnum*, which is flourishing (plankton is nearly absent). Apparently more research is needed to get reliable results of primary production measurements in moorland pools.

The biomass of aquatic macrophytes in moorland pools is generally below 100 g m^{-2} dry weight (Roelofs & Bloemendaal, 1988), but in originally oligotrophic pools with phosphate enriched sediments far higher values can be found: in Beuven the total biomass of *Littorella uniflora* before the restoration was 469 g m^{-2} ash-free dry weight (Brock, 1988).

Acidification usually results in changes in carbon availability for primary production and may cause shifts in species composition (Roelofs *et al.*, 1984; Schuurkes *et al.*, 1986; Paffen & Roelofs, 1991). In a mathematical model, describing the growth of *L. uniflora* and *Juncus bulbosus*, the availability of carbon dioxide largely explains the shift in dominance of the two species in acidifying moorland pools (Wortelboer, 1990).

As has been mentioned before, many moorland pools, including Gerritsfles and Beuven, had a sandy bottom, at least in the nearshore areas. The decomposition kept pace with the primary production. In eutrophied pools, like Beuven, the bottom became partially covered with particulate organic material because the primary production was much more accelerated than the decomposition.

In acidified moorland pools, including Gerritsfles and Goorven, the sandy bottom was buried under organic material due to decreased decomposition by acidification (Schuurkes & Kok, 1988; Arts *et al.*, 1990b). The low breakdown rates are due to inhibition of microbial activity at $pH < 4$, where fungi take over the role of bacteria. The decay rate is stimulated when the alkalinity is above ca. 0.5 meq l^{-1} (Leuven & Wolfs, 1988; Kok *et al.*, 1990)

The low decay rate in acid waters will lead to a slow recycling of nutrients from the decomposing plant material. A relatively large part of the N and P in acid systems is located in organic matter and not available for primary production. Addition of acid neutralizing substances can lead to a sharp increase in decay rate and a rapid release of N and P from the detritus. Removal of the organic layer from the sediment is therefore necessary to keep acidified pools in an oligotrophic condition after restoration (Kok *et al.*, 1990). However, by raising the pH to circumneutral levels and keeping the alkalinity at a low level it might be possible to prevent undue decomposition and subsequent eutrophication (Kok & Van de Laar, 1991).

Microphytes

In Table 6 an impression is given of the few surveys of all groups of microphytes in a large number of pools. The rank order of the numbers of taxa in each of the classes of algae is strikingly similar. The similarity between the data sets from Drenthe and Noord-Brabant still increases when the 50% reduction of the number of desmid (Desmidiaceae) taxa in the province of Drenthe between 1926 and 1973 is taken into account (Coesel & Smit, 1977). The desmids as the most

Table 6. Numbers of taxa in main taxonomic groups of microphytes in the surveys of Beijerinck (1926) and J.E.G.M. de Bie & M.M.J. Maenen (unpubl.).

Period	1923–1926	1983
Province	Drenthe[a]	Noord-Brabant[b]
Number of pools	63	35
Chlorophyceae	312	166
(Desmidiaceae)	(211)	(79)
(Chlorococcales)	(30)	(54)
Bacillariophyceae	39	63
Cyanobacteria	27	27
Pyrrhophyceae	24	10
Euglenophyceae	18	20
Chrysophyceae	18	17
Xanthophyceae	13	3
Cryptophyceae	2	3
Chloromonadophyceae	2	1
Rhodophyceae	1	1
Sum	456	311

[a] The province where Kliplo (Fig. 1) is located; [b] The province where Goorven and Beuven (Fig. 1) are located (also some pools in other provinces are included).

important and the diatoms (Bacillariophyceae) as the second most important group will be discussed in separate sections.

The most frequent and abundant genera in both data sets are *Cryptomonas* (Cryptophyceae), *Synura, Dinobryon, Mallomonas* (Chrysophyceae) and *Peridinium* (Pyrrhophyceae) in the plankton, and the filamentous *Binuclearia, Microspora* and *Oedogonium* (Chlorophyceae) in the benthos. Also many desmid genera, particularly *Closterium, Cosmarium* and *Staurastrum* are present in the plankton and between macrophytes. *Frustulia, Tabellaria, Eunotia* and *Pinnularia* are the most important diatom genera. *Asterococcus, Oocystis*, and *Botryococcus* (Chlorophyceae) are very common in Drenthe, but much rarer in Noord-Brabant. Some of the genera, *e.g.* the chrysophyte *Rhipidodendron*, which is very common in Drenthe, are in The Netherlands virtually confined to moorland pools (Suykerbuyk & Roijackers, 1988).

The most important factor for the distribution of microphytes in moorland pools is pH. In the pools in Noord-Brabant the number of taxa of diatoms, Chlorococcales and desmids was positively correlated with pH, in the range 3.3–10.4. Most desmid species were found between pH 6 and 7. *Binuclearia, Microspora* and *Mougeotia* were particularly abundant at pH$<$5. The blue-green *Oscillatoria* and the Chlorococcales *Ankistrodesmus* and *Pediastrum* were found at neutral to alkaline pH, which reflects the eutrophic character of the unacidified pools (J.E.G.M. de Bie & M.M.J. Maenen, unpubl.).

The median summer ash-free dry weight and chlorophyll-*a* concentration of the phytoplankton in Noord-Brabant were 6.7 g m^{-3} and 7.3 mg m^{-3}, respectively. For Gerritsfles these data were 7.1 and 0.3, reflecting the oligotrophic character, and for Beuven (before the restoration) 10.2 and 13.3 (J.E.G.M. de Bie & M.M.J. Maenen, unpubl.). Mean annual phytoplankton biomasses of 1.8 g m^{-3} in an oligotrophic and of 14.4 g m^{-3} in a eutrophied moorland pool were measured. Summer values of respectively 3.3 and 22.7 g m^{-3} were found (Geelen, 1969). The higher summer values are particularly due to the development of Chrysophyta and Chlorophyta, including desmids (see also Beijerinck, 1926).

Desmids

Desmid species are numerically the most important group of algae in moorland pools (Table 6). Moorland pools are also the most important habitat for desmids in The Netherlands, as desmids are successfull competitors with other groups of algae in oligo- and mesotrophic water bodies, due to their comparatively low growth rate and economical life strategy. The richest assemblages are encountered at sites with moderately high conductivity, especially in situations with some spatial or temporal variation (Coesel, 1983).

The richest site of the four pools considered is the pool Goorven (Table 7). The richness is probably due to the trophic gradient, which was present between 1916 and 1955 (Van Dijk *et al.*, 1960), but impoverished drastically between 1955 and 1975 due to acidification (Coesel *et al.*, 1978).

Table 7. Selected desmid species, arranged in habitat groups according to Coesel (1975). Presence is marked with ×. Status: T = threatened species, V = very threatened species, E = presumably extinct species in The Netherlands (P.F.M. Coesel, unpubl.). Data from Beijerinck (1926), Dresscher *et al.* (1952), Van Dijk *et al.* (1960), Coesel *et al.* (1978), Buskens (1989), N. Bouton (unpubl.), F. Goessens (unpubl.), J.E.G.M. de Bie & M.M.J. Maenen (unpubl.), P.J. Schroevers (unpubl.), J.A. Sinkeldam (unpubl.), H.D.W. Smit (unpubl.), A.J. van Tooren (pers. comm.) and A.J. Wartena (unpubl.).

Habitat	Status	Taxon	Goorven			Gerritsfles			Kliplo			Beuven	
		From: 19..	19	50	75	18	50	77	23	54	72	68	87
		To: 19..	25	55	89	18	65	83	26	54	86	69	87
		Number of samples[a]	46[b]	9	47	10	26	37	15[c]	12	41	11	17
Moderately eutrophic, alkaline to mesotrophic, weakly acid water bodies	–	*Closterium moniliferum* (Bory) Ehr. ex Ralfs										×	×
	–	*Cosmarium humile* (Gay) Nordst.			×								×
	–	*C. laeve* Rab.											×
	–	*C. regnellii* Wille			×						×	×	
	–	*C. subprotumidum* Nordst.										×	
	–	*Staurastrum avicula* Bréb.										×	
	–	*S. polymorphum* Bréb.	×	×									
	–	*S. cuspidatum* Bréb.	×						×	×			
	T	*Xanthidium antilopaeum* (Bréb.) Kütz.	×		×					×	×		
Mesotrophic, slightly acid, (un)disturbed water bodies	T	*Closterium lunula* (Müll.) Nitzsch ex Ralfs	×			×			×	×	×		
	–	*Cosmarium quadratum* Ralfs	×		×								
	–	*Euastrum ansatum* Ehr. ex Ralfs	×	×	×	×		×			×		
	T	*Micrasterias denticulata* Bréb.	×	×	×				×				
	–	*Tetmemorus granulatus* (Bréb.) Ralfs ex Ralfs	×	×	×	×	×	×	×	×	×		
Mesotrophic, slightly acid, undisturbed water bodies	–	*Arthrodesmus convergens* Ehr.	×						×				
	T	*Closterium gracile* Bréb.		×	×			×			×		
	T	*Cosmarium margaritiferum* (Turp.) Ralfs	×	×	×			×			×		
	–	*Euastrum bidentatum* Näg.	×										
	T	*E. elegans* (Bréb.) Kütz.	×		×								
Mesotrophic, Slightly acid pools and bogs with high diversity of habitats	T	*Closterium turgidum* Ehr. ex Ralfs	×	×									
	T	*Cosmarium connatum* (Bréb.) Ralfs	×		×								
	T	*Micrasterias papillifera* Bréb.								×			
Extremely oligotrophic, slightly disturbed and undisturbed pools with *Sphagnum*	–	*Bambusina brebissonii* Kütz.						×	×	×	×		
	–	*Closterium striolatum* Ehr.	×	×	×	×	×	×	×		×		
	–	*Cylindrocystis brebissonii* (Menegh. ex Ralfs) De Bary		×	×	×		×	×	×			×
	–	*Micrasterias truncata* (Corda) Bréb.	×	×	×	×	×	×	×	×	×		
	–	*Staurastrum margaritaceum* (Ehr.) Menegh.			×						×		
	–	*S. punctulatum* Bréb.			×		×	×	×	×	×		×
Oligotrophic acid pools and depressions in peat bogs	–	*Closterium ulna* Focke	×	×	×	×	×						
	–	*Cosmarium cucurbita* Bréb.	×	×	×	×	×	×	×	×	×		
	–	*C. pyramidatum* Bréb.	×	×	×				×	×	×		
	–	*Spondylosium pulchellum* Arch.			×			×			×		
	T	*Tetmemorus brebissonii* (Menegh.) Ralfs	×		×	×	×		×		×		
Oligotrophic, acid pools in *Sphagnum* quaking bogs	V	*Cosmarium cucurbitinum* (Biss.) Lütkem.	×	×						×			
	V	*Euastrum ampullaceum* Ralfs	×	×	×	×	×		×		×		
	T	*Pleurotaenium minutum* (Ralfs) Delp.	×	×	×			×	×	×	×		
	T	*Staurastrum furcatum* (Ehr.) Bréb.	×										
	V	*S. hystrix* Ralfs	×	×	×	×	×		×				
Acid pools in weakly minerotrophic quaking bogs	V	*Cosmarium ralfsii* Bréb.							×				
	V	*Docidium baculum* Bréb.	×	×	×			×					
	E	*D. undulatum* Bréb.	×										
	V	*Euastrum crassum* (Bréb.) Kütz.	×	×	×				×	×	×		
Number of taxa per period			114	79	115	31	26	59	60	56	67	24	17
Number of taxa per pool				161			73			137		35	

[a] A sample may consist of several subsamples (plankton, bottom material, moss squeeze etc.), [b] maximum, [c] estimated.

According to the more recent data in Table 7 the desmids seem to recover from acidification. However, it does not appear from this table that both the species from acidified and eutrophic water were particularly abundant between 1982 and 1989, in contrast with earlier periods. Apparently the increased pH and alkalinity after 1982 (Fig. 8) have contributed to an increased mineralization of the accumulated organic material and an improved availability of nutrients for the algae. These changes may explain why some very sensitive and probably extinct species (*e.g. Docidium undulatum*) have not been found again since 1975, despite intensive sampling.

The isolated pool Kliplo also has many taxa which are characteristic of oligo- and mesotrophic water bodies. However, as a spatial gradient could not develop in this pool, taxa from undisturbed and mesotrophic pools and bogs with a high diversity of habitats are rare. Although the number of taxa is still relatively high, the flora has impoverished over the years. Probably a high number of species is still present due to the relatively constant chemistry of the pool, as the major part of the pool bottom is never exposed to the atmosphere (Figs 8 and 3). A gradual eutrophication of this pool is very probable, because of the intensity of the mineralization processes, which is indicated by the high sulphate reduction rate (Table 2). Also the changes in macrophyte vegetation and the continuous bloom of an unidentified coccoid green algae since the spring of 1989 indicate gradual eutrophication of Kliplo.

In Gerritsfles, taxa typical of undisturbed mesotrophic pools and bogs with high diversity of habitats never have been found. Heimans (in Dresscher *et al.*, 1952) recorded an increase, between 1918 and 1950, in taxa usually found in extremely oligotrophic, acid and disturbed pools, and was the first to assess the effects of acidification on surface waters in The Netherlands.

No data on Beuven in the oligotrophic situation exist, but pools with stands of *Isoetes-Lobelia* vegetation should be poor in desmid taxa (Coesel, 1983). In 1968 and 1969 species from eutrophic habitats were found (Table 7). After the restoration, both, species from the most eutrophic and oligotrophic habitats were present, but the species from mesotrophic habitats were lacking. Such a situation has been denoted as metatrophic by Leentvaar (1958) and is often observed in originally oligotrophic moorland pools which are heavily influenced by agricultural drainage water.

Our results are consistent with those of other moorland pools and lakes in The Netherlands, Belgium, Denmark and Germany (Coesel & Smit, 1977; Coesel *et al.*, 1978; Weddigen & Geissler, 1980; N. Bouton, unpubl.).

Diatoms

The number of diatom taxa in moorland pools is often rather low in comparison with other types of wetlands, but highly characteristic and otherwise rare taxa do occur, particularly in the mesotrophic and slightly acid pools. Diatoms are excellent indicators for lake acidification and methods for inferring pH from diatom assemblages in moorland pools have been developed (Ter Braak & Van Dam, 1989; Charles *et al.*, 1989).

The numerically most important diatom taxa from the four pools considered are listed in Table 8. Six ecological groups are distinguished: (1) *Eunotia exigua*: very resistant to acidification and the associated high concentrations of heavy metals and aluminium; (2) acid-water taxa: widespread from (extremely) acid sites; (3) soft-water taxa: rarer taxa from acid sites and threatened by acidification, (4) eurytopic taxa: in a wide variety of water types; (5) eutraphentic taxa: typical for eutrophic, often alkaline, water bodies; and (6) saprophilous taxa: in moderately acid to alkaline water, loaded with biodegradable organic material. The temporal changes in ecological groups of diatoms as well as the diatom-inferred pH in Goorven, Gerritsfles and Kliplo are given in Fig. 10.

Both, acid-water taxa and soft-water taxa were well represented in the old samples (1916–1929) from these pools. *Eunotia exigua* increased enormously after the extremely dry year 1959 in Gerritsfles, and after the extremely dry year 1976 in

Table 8. Optimal pH-values and average percent abundance (weighted for the number of old and recent samples) of the most common diatom taxa, number of taxa and number of samples from each moorland pool. Taxa are classified according to ecological groups. Data from Van Dam (1988), Buskens (1989), Ter Braak & Van Dam (1989), and Van Dam & Mertens (1990).

pH optimum	**Ecological group**/taxon	Goorven	Gerritsfles	Kliplo	Beuven	
	From: 19 . .	19	16	24	38	88
	To: 19 . .	90	90	90	42	88
	Number of samples	32	38	25	3	3
	Eunotia exigua					
4.1	*Eunotia exigua* (Bréb.) Rab.	40	34	0	0	1
	Acid-water taxa					
4.2	*Frustulia rhomboides* var. *saxonica* (Rab.) De Toni	14	13	12	2	0
4.3	*Eunotia bilunaris* (Ehr.) Mills	6	5	7	1	6
4.6	*Tabellaria quadriseptata* Knudson	12	8	2	0	0
4.9	*Eunotia rhomboidea* Hust.	1	17	8	3	3
	Soft-water taxa					
4.0	*Navicula parasubtilissima* Kobayasi & Nagumo	4	3	9	0	0
–	*N. subtilissima* Cleve	1	0	2	0	0
5.0	*Eunotia incisa* Ehr.	10	11	10	3	0
5.0	*Tabellaria flocculosa* (Roth) Kütz.	0	3	14	5	0
5.6	*Fragilaria virescens* Ralfs	1	0	2	5	6
5.9	*Navicula heimansii* Van Dam & Kooyman	6	0	13	0	0
5.9	*Brachysira vitrea* fo. *lanceolata* (Mayer) Van Dam	2	2	14	5	4
–	*Brachysira vitrea* (Grun.) Ross	0	0	0	19	24
	Eurytopic taxa					
6.8	*Achnanthes minutissima* Kützing	0	0	1	29	4
	Eutraphentic taxa					
–	*Fragilaria capucina* Desmazières	0	0	0	10	2
5.6	*F. construens* var. *venter* (Ehr. Grun)	0	0	0	1	16
	Saprophilous taxa					
–	*Navicula minima* Grun.	0	0	0	0	7
–	*Nitzschia palea* (Kütz.) W. Smith	0	0	0	2	4
Number of taxa		56	73	53	44	53

Goorven. The pH declined severely as a result of sulphate release from the sediments in subsequent years (Fig. 8). The diatom assemblages of Gerritsfles recovered more rapidly from the perturbation of 1976, induced by acid precipitation and an extremely dry summer, than those of Goorven, because of the longer retention time of water and solutes in the latter pool (Van Dam & Mertens, 1990). *E. exigua* did not increase in Goorven after the extreme drought of 1921, because few sulphides had accumulated in the sediment by that time (Van Dam, 1988). In Kliplo, where only a small fraction of the sediments is exposed to the atmosphere in extremely dry years (Fig. 3), no *E. exigua* and still many soft-water taxa are present, including very characteric ones, *e.g. Navicula heimansii* (Table 8 and Fig. 10).

From Beuven diatoms on the leaves and adhering bottom sand of herbarium material and recently collected isoetid macrophytes have been studied. Soft-water taxa are well represented in the recent and old materials, but the percentage abundance of saprophilous taxa has increased over time (Table 8).

Diatoms have also been used to infer past changes of pH in Goorven, Gerritsfles and Kliplo

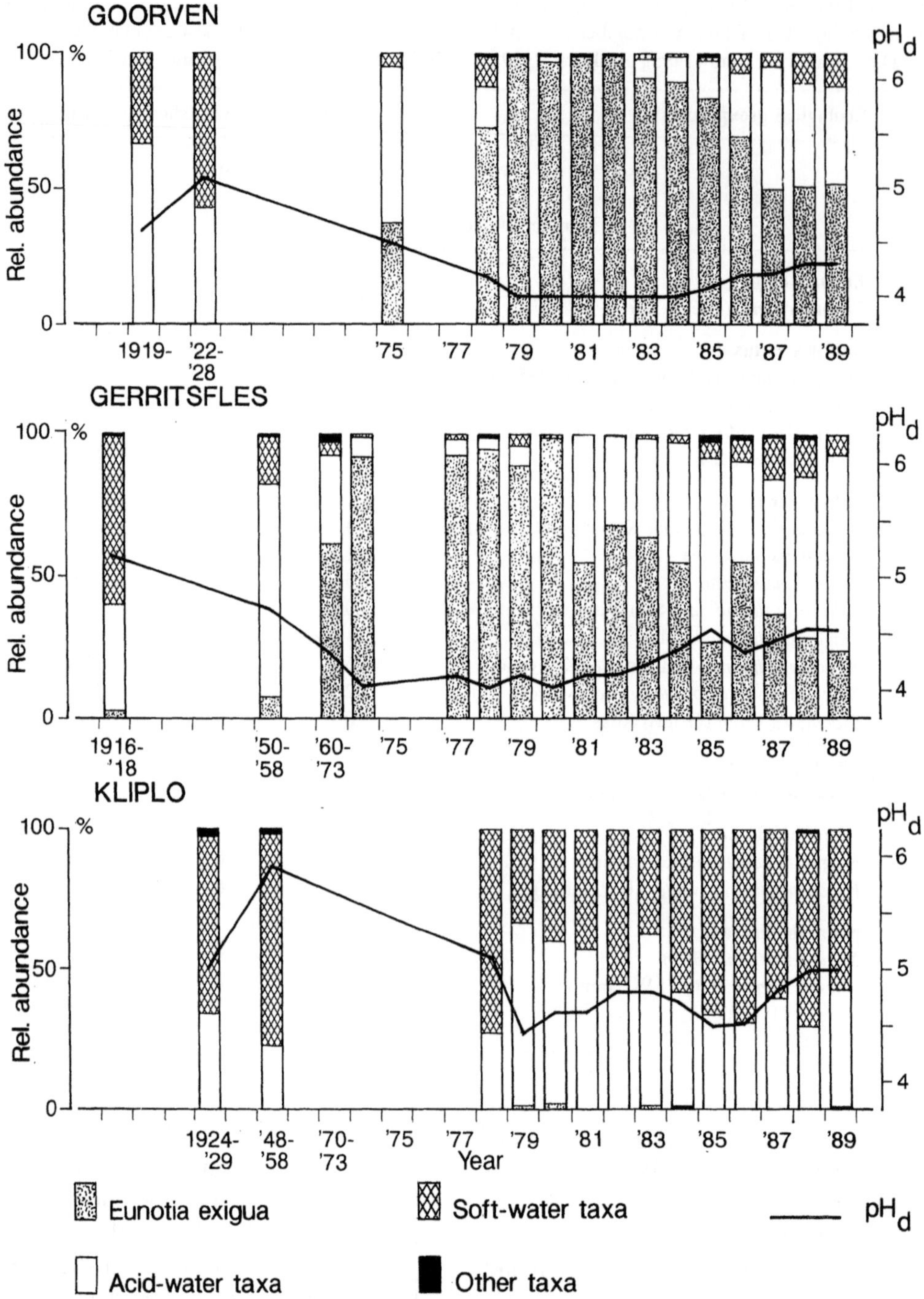

Fig. 10. Changes in percent abundance of ecological groups of diatoms (see text) in three moorland pools (bars) and diatom-inferred pH (pH_d) according to equation (5) with extra regression in Ter Braak & Van Dam (1989). Data from Van Dam & Mertens (1990).

from sediment core samples (Fig. 11). The weakly acid conditions (pH 5–6) around 1900 are anthropogenic. The subsequent decrease of pH is due both to termination of direct nutrient enrichment of the pools and of increased acid deposition. Consequently it is hard to determine

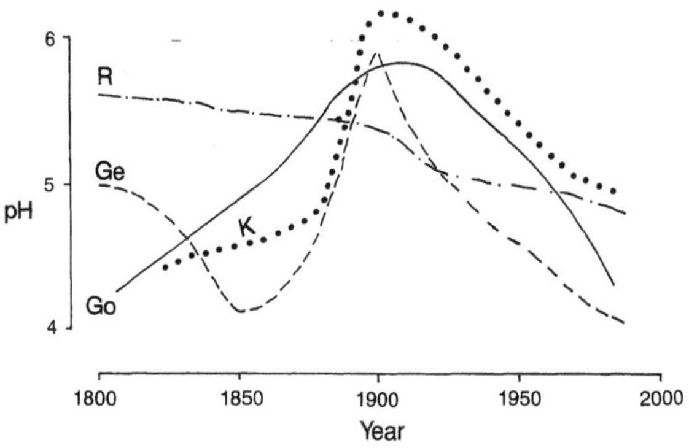

Fig. 11. Changes of pH over time, inferred from diatoms in sediment cores from the pools Goorven (Go), Gerritsfles (Ge) and Kliplo (K) and the Round Loch of Glenhead (R). Data from Flower & Battarbee (1983), Dickman *et al.* (1987), Van Dam *et al.* (1988b).

when acidification of the pools by atmospheric deposition started. Moorland pools are more dynamic systems than acidifying lakes in Europe and North America, for which the Round Loch of Glenhead (Scotland) is a typical example (Fig. 11).

A decrease of the diversity of diatom assemblages, caused by the increase of *Eunotia exigua*, has been found in all isolated moorland pools which are not naturally stained by humic acids in The Netherlands, Belgium and Germany. These pools belong to the most severely acidified aquatic ecosystems in the world (Van Dam & Kooyman-van Blokland, 1978; Van Dam *et al.*, 1981; Charles *et al.*, 1989). In non-isolated pools the diatom assemblages have been altered by eutrophication, mainly due to inflow of agricultural drainage water (Van Dam & Kooyman-van Blokland, 1978; De Vries, 1984).

Invertebrates

Microfauna

Rhizopoda from peat-mosses in moorland pools and bogs were studied by Hoogenraad (1935). Characteristic species belonging to *Arcella, Assulina, Difflugia* and other genera were found in Gerritsfles and other moorland pools. Unfortunately this study has not been repeated. Occasionally other groups of microfauna were studied. Although several characteristic taxa have been found, including the rotifers *Lecane ligona* (Dunlop), *Keratella serrulata* (Ehr.), the copepods *Diacyclops languidus* (Lilljeborg) and *Elaphiodella gracilis* Sars it is impossible to present a complete survey of these groups (*e.g.* Redeke & De Vos, 1932; Redeke, 1948; Dresscher *et al.*, 1952).

Only Cladocera were inventoried in all four pools (Table 9). Gerritsfles has been most intensively investigated over the years and most of the species characteristic for moorland pools were found (Notenboom-Ram, 1981). These species are detritivorous. Most species live on the bottom or between the aquatic macrophytes, including *Sphagnum*. Species feeding on phytoplankton and other suspended particles in water have only been found in Beuven, particularly before the restoration, when high biomass of plankton still occurred in that pool. The reduction of the number of species from eutrophic water bodies in Beuven after the restoration indicates the success of the measures taken.

Although Gerritsfles has been inventoried several times, it is difficult to draw conclusions about changes in species composition in relation to environmental changes, due to the use of different methods by different authors and also the diffences in the time of the year of the collections. Therefore Table 9 gives no indications for a shift in trophic state or acidity of this pool (Van Dam, 1988). One would expect such a shift, as acidification has a profound influence on the species composition of the assemblages of cladocerans (Geelen & Leuven, 1986; Brett, 1989).

The studies on cores from Gerritsfles and Goorven by R. van der Heijden (unpubl.) were not hampered by methodological discrepancies. The acidobiontic species *Bosmina longispina, Eurycercus glacialis* and *Acantholeberis curvirostris* had a mean relative abundance of 12.2, 0.3 and 0.2% respectively in these cores. Their relative abundance did not change significantly over time (Fig. 12). However, the taxonomic composition of the cladoceran assemblages has been changed over time, particularly by the increase of *Alonella*

Table 9. Presence of Cladocera in four moorland pools, arranged in ecological groups. Acidobiontic species are marked by * (Notenboom-Ram, 1981). Data from Beijerinck (1924), Dresscher *et al.* (1952), Van Dijk *et al.* (1960), Van Dam *et al.* (1988), Buskens (1989), J.E.G.M. de Bie & M.M.J. Maenen (unpubl.), J.F.M. Geelen (unpubl.), J.A. Sinkeldam (unpubl.) and A.P.C. de Vos (unpubl.).

Ecological group	Species	Goorven	Gerritsfles		Kliplo	Beuven		
	From: 19..	52	30	63	24	69	83	87
	To: 19..	53	58	83	29	69	83	87
	Number of samples	8	33	31	?	11	2	17
Rare in benthos of oligotrophic, acid moorland pools, detritivores	**Eurycercus glacialis* Lilljeb.		×					
	**Streblocerus serricaudatus* (Fischer)		×					
	Rhynchotalona falcata (Sars)		×	×				
	**Drepanothrix dentata* (Eurèn)		×		×			
	Acroperus elongatus (Sars)		×	×	×			
Common in benthos of oligotrophic, acid moorland pools, detritivores	**Acantholeberis curvirostris* Lilljeb.	×	×	×	×			
Plankton of oligotrophic, acid moorland pools, detritivores	*Alona rustica* Scott			×				
	**Bosmina longispina* Leydig		×	×	×		×	×
Common in benthos or on macrophytes of oligo- to eutrophic water bodies, substrate filtrators	*Alona quadrangularis* (O.F. Müller)	×	×					
	Acroperus harpae (Baird)	×	×	×				
	Alonella exigua (Lilljeb.)		×	×				
	Iliocryptus sordidus (Liénin)	×		×				
	Alona affinis (Leydig)	×		×				
	Ceriodaphnia quadrangula (O.F. Müller)	×		×		×		
	Chydorus sphaericus (O.F. Müller)	×	×	×	×	×	×	×
	Diaphanosoma brachyurum (Liévin)	×	×	×				×
	Scapholeberis mucronata (O.F. Müller)		×	×	×		×	×
	Alona rectangula Sars		×	×			×	
	Alonella excisa (Fischer)		×	×				×
	Alona guttata (Sars)		×	×				×
	Alonella nana (Baird)		×	×	×			
	Bosmina longirostris (O.F. Müller)			×	×		×	
	Pleuroxus truncatus (O.F. Müller)						×	
	Iliocryptus agilis Kurz						×	
Common in oligo- to eutrophic water bodies, carnivores	*Polyphemus pediculus* (L.)	×			×	×		×
Common in eutrophic water bodies, substrate filtrators	*Disparalona rostrata* (Koch)		×	×				
	Leydigia quadrangularis (Leydig)		×	×				
	Camptocercus rectirostris Schoedler			×				
	Pleuroxus aduncus (Jurine)			×	×			
	Eurycercus lamellatus (O.F. Müller)	×				×		×
	Daphnia pulex Leydig						×	×
	Pleuroxus uncinatus Baird						×	
Rare in eutrophic water bodies, substrate filtrator	*Ceriodaphnia megops* Sars						×	
Common in eutrophic water bodies, filtrators	*Simocephalus vetulus* (O.F. Müller)	×				×		
	Daphnia longispina O.F. Müller					×		
	Ceriodaphnia pulchella Sars					×		
	Daphnia hyalina Leydig						×	
	Daphnia cucullata Sars						×	×
Number of species		11	19	22	10	7	12	10

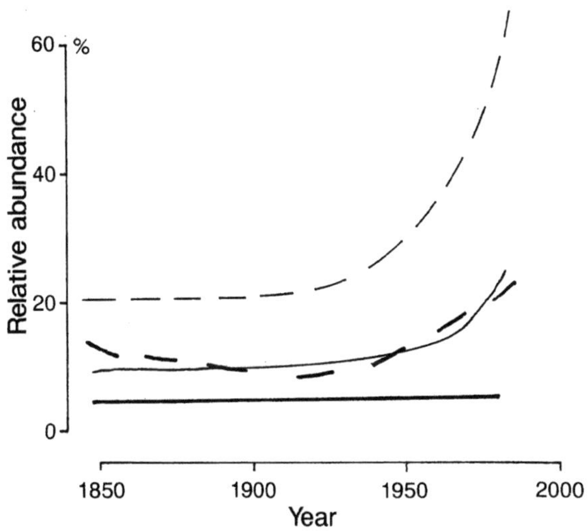

Fig. 12. Changes in percent abundance of Cladocera over time, inferred from sediment cores from Goorven (bold lines) and Gerritsfles (thin lines). Solid lines: acidobiontic species (*Bosmina longispina, Alonella excisa* and *Acantholeberis curvirostris*), broken lines: acidobiontic species plus *Alonella excisa*. Data from R. van der Heijden (unpubl.).

excisa since 1950. This species, which had a mean relative abundance of 25.5% in the cores, is considered to be acidobiontic by Krause-Dellin & Steinberg (1986), in accordance with its distribution in acidified German lakes. Apparently the cladocerans are less sensitive to pH-changes than the diatoms, as the acidobiontic cladocera increase much later than the acidobiontic diatoms, and also the pH-maximum around 1900 is not reflected by a decrease of acidobiontic cladocerans in Goorven and Gerritsfles (Fig. 11). Other factors than pH, *e.g.* indirect (biotic) influences are often as important if not more so to the zooplankton (Brett, 1989).

Macrofauna

The macrofauna of freshwater habitats in the province of Overijssel, including moorland pools was surveyed by Verdonschot (1990) (Table 10). The results are probably rather representative for those pools in The Netherlands where only selected groups of macrofauna have been studied (*e.g.* Redeke & De Vos, 1932; Leuven *et al.*, 1986b).

In comparison with average surface water bodies (last column of Table 10), Coleoptera, Diptera, Heteroptera and Odonata are overrepresented in moorland pools. Malacostraca, Hirudinea and Mollusca are clearly underrepresented. The low abundance of the latter group is due to the low calcium concentrations (Table 10). The Chironomidae make up the most abundant group, as in the average surface water body.

The high relative abundance (ca. 50%) of carnivores and the low relative abundance of detritivores and herbivores is striking from the functional point of view (Table 10). The proportion of carnivores is probably estimated too high, because of the uncertainty of the place of many species and life stages in the food web, but in comparison with other water types the carnivores are more important. Also in the litter bag experiments, mentioned earlier in this paper, a similar assemblage of functional groups was found. The high proportion of carnivores seems to be a quite general phenomenon in water bodies with low pH-values (*e.g.* Sutcliffe & Carrick, 1973). The dominance of Heteroptera and other invertebrate predators in acid water bodies is attributed to the absence of fish predation (Økland & Økland, 1986). However, the way in which a water body can sustain such a high number of carnivores is clearly a topic for further investigation. Particularly in the intermittent oligotrophic pools insects which can invade water bodies rapidly, *e.g.* Coleoptera and Diptera, are abundant (Table 10).

The number of taxa, particularly in the samples from the most acid pools, is fairly low, but a relatively high number of individuals belongs to taxa which are rare in other habitats and characteristic for moorland pools (Table 10).

From incomplete studies on the macrofauna of Goorven, Gerritsfles, Kliplo and Beuven (*e.g.* Dresscher *et al.*, 1952; Van Dijk *et al.*, 1960) it is obvious that Gerritsfles is an exellent example of Verdonschot's (1990) type P2. Kliplo has mainly species from P2, but also some from P3. Goorven has a P3 assemblage, but is not a very good example for this type. Prior to the restoration, Beuven had species from P4, but also common ones

Table 10. Average values of selected environmental variables, relative abundance (percent values) of main taxonomical groups and position in food web of the macrofauna in surface waters in the province of Overijssel. 0 = abundance <0.5%, – = not present (Verdonschot, 1990).

	Moorland pools				Springs, streams, canals, pools, lakes
	Oligotropic		Meso-tro-phic	Eu-tro-phied	
	Intermittent	Permanent			
Type	P1	P2	P3	P4	–
Number of samples	9	15	11	21	651
Environmental variables					
pH	4.2	4.3	5.1	7.3	7.3
Alkalinity (mmol m^{-3})	0	33	131	1409	1585
Ortho-PO_3 (mmol m^{-3})	0.1	0.3	0.5	3.8	2.8
SO_4^{2-} (mmol m^{-3})	177	125	146	73	546
Cl^- (mmol m^{-3})	254	226	282	592	1497
Ca^{2+} (mmol m^{-3})	25	75	125	574	1271
Taxonomical groups					
Chironomidae	18	36	34	19	23
Coleoptera	27	10	18	17	10
Diptera s.s.*	29	13	9	9	6
Heteroptera	2	9	13	9	4
Oligochaeta	10	5	5	10	16
Odonatae	2	12	7	6	2
Acarina	9	7	4	5	7
Trichoptera	1	6	4	1	5
Ephemeroptera	0	3	3	5	3
Mollusca	–	0	–	10	13
Hirudinea	0	0	0	5	4
Turbellaria	–	1	1	1	1
Lepidoptera	–	0	0	2	0
Malacostraca	–	–	0	1	5
Megaloptera and Neuroptera	–	0	0	0	0
Plecoptera	0	–	–	–	1
Position in food web					
Carnivores	53	49	50	46	30
Detritivores	9	11	13	19	26
Herbivores	5	7	12	21	21
Omnivores	1	9	5	2	5
Detriti-herbivores	31	17	17	12	18
Carni-detritivores	0	8	2	0	1
Number of taxa					
Number within group	87	129	213	278	842
Mean number per sample	23	31	51	61	55
Rare taxa (perc. abund.)	24	22	16	10	8

* Exclusive of Chironomidae.

from eutrophic water bodies. After the restoration Beuven harbored a fragmentary P3 assemblage. Some rare animals, *e.g.* the northern beetle *Dytiscus lapponicus* (Gyllenhål) were recorded in Gerritsfles and Kliplo, and differentiate these pools from those in Overijssel (Weber, 1919; Beijerinck, 1931). This species, and also the cladoceran *Eurycercus glacialis*, have been considered as glacial relict species and were a 'proof' for the glacial age of Gerritsfles (Dresscher *et al.*, 1952). The incorrectness of this speculation has been discussed in the section on origin.

The chironomids, the most common group of macro-invertebrates, have been investigated most thoroughly (Table 11).

Only a few species are acidobiontic. *Acamptocladius submontanus* and *Zalutschia humphriesiae* are only known from a few sites in Western Europe, particularly bogs and associated environments; *Psectrocladius platypus* is common in acid waters (Moller Pillot & Buskens, 1990).

The absence of circumneutral taxa in Goorven,

Table 11. Selected taxa of chironomid larvae in different periods, arranged in ecological groups; rare taxa, occurring in less than 5% of the 25 km^2-grid squares in The Netherlands are marked with * (Moller Pillot & Buskens, 1990). Size of symbol is proportional to estimated abundance (smallest dot: rare species, largest dot: massive abundance. Data prior to 1980 from paleolimnological analysis by Dickman *et al.* (1987), Klink (1986) and R.F.M. Buskens (unpubl.), other data from inventories by Leuven *et al.* (1987), Buskens (1989) and F.F. Repko (unpubl.).

	Goorven					Gerritsfles			Kliplo	Beuven					
From:	1800	1850	1900	1950	1983	1850	1900	1983	1980	1850	1900	1955	1971	1981	1986
To:	1850	1900	1950	1980	1983	1900	1950	1983	1982	1900	1945	1966	1980	1985	1989
Acidobiontic taxa															
**Acamptocladius submontanus* (Edwards)		·													
**Zalutschia humphriesiae* (Dowling & Murray)	·	·	·	·											
Psectrocladius platypus (Edwards)	·	·	·	·	•	·	·	·		·	·	·	·		·
Acidophilous taxa															
**Heterotanytarsus apicalis* Kieffer	·	·													
**Labrundinia longipalpis* (Goethgebuer)		·	·	·											
Dicrotendipes gr. *tritomus* sensu Moller Pillot		·	·	·		·	●	·	●						
Psectrocladius psilopterus Kieffer	●	·	•	·		·	•		●	·	·		·		·
Endochironomus gr. *dispar* sensu Moller Pillot						·		·		·	·	·	·	·	
Circumneutral taxa															
Parachironomus gr. *vitiosis* sensu Moller Pillot	·	·													
**Omisis? caledonicus* (Edwards)	·	·	·	·											
**Pagastiella orophila* (Edwards)	·	·									·	·			
Parakiefferiella bathophila (Kieffer)	·	·	·	·		·				·	·	·			
Microtendipes chloris agg. sensu Moller Pillot	·	·	·	·		●	•			•	•	•	•	·	●
Einfeldia dissidens agg. sensu Moller Pillot	·	·	·								·		·	·	
Parachironomus gr. *arcuatus* sensu Moller Pillot	·	·								·	·	·	·	·	·
Dicrotendipes sp.	•	•	•	•						·	·	·	·		
Cladotanytarsus gr. *mancus*	•	•	·	·						●	●	•	•	•	
Acricotopus lucens (Zetterstedt)		·	·							·	·	·	·		
Endochironomus albipennis (Meigen)		·	·			·				·	·	·	·	·	
Cricotopus (*Isocladius*)	·	·				·				·	•	·	·	·	
Nanocladius sp.										·	·		·		
Microchironomus tener Kieffer													·		
Anatopynia plumipes (Fries)														·	
pH-indifferent taxa															
Dicrotendipes lobiger (Kieffer)		·	·						·						
Stenochironomus sp.	·	·					·	·							
Cricotopus brevipalpis Kieffer	·			·		·	·			·	·	·	·		
Demicryptochironomus vulneratus (Zetterstedt)		·	·	·						·	·		·	·	
Tribelos intextus (Walker)	·	·	·			•	·		•	·	·				
Endochironomus tendens Fabricius				·	·	·	·	•	·					·	·
Polypedilum cf. *uncinatum* sensu Moller Pillot	·	·	·	·		·	·	·						·	·
Pseudochironomus sp.	·	•	·	•		·	•	·		·	·	·	·	·	·
Tanytarsini	•	●	●	•	·	·	•	·	•	•	●	●	●	●	•
Procladius s.a. sensu Moller Pillot	·	·	·	·	•	•	•	•	•	•	•	•	•	●	•
Cricotopus sp.	·	·	·	·					·	·	·	·	·		
Corynoneura scutellata agg. sensu Moller Pillot	·	·	·	·			·			·	·		·		·
Cryptochironomus sp.						·	·	·		·	·	·	·	·	·
Psectrotanypus varius (Fabricius)														·	
Phaenopsectra sp.															·
Number of taxa															
Acidobiontic	2	3	2	2	1	1	1	1	0	1	1	1	1	0	1
Acidophilous	3	5	4	4	0	4	3	2	2	3	3	1	2	1	1
Circumneutral	14	17	11	9	0	5	1	0	0	11	13	11	12	9	2
Indifferent	13	16	14	13	6	13	13	12	10	14	13	9	13	13	11

Gerritsfles and Kliplo (and also in other moorland pools in The Netherlands with pH < 5) after 1980 is striking. The presence in the uppermost part of the core in Goorven is probably due to bioturbation (Dickman *et al.*, 1987).

In Goorven, acidification has resulted in a decline of circumneutral taxa, including rare species (*e.g. Pagastiella orophila*). The presence of acidobiontic and circumneutral taxa in this pool in the 19th century is probably associated with the humic environment left after excavation of peat on one side of the pool and the development of a rich assemblage of aquatic macrophytes after inlet of water at the other side of the pool (Dickman *et al.*, 1987).

The higher number of circumneutral species (including *Parakiefferiella bathophila*) in Gerritsfles between 1850 and 1900 than in more recent times indicates acidification of this pool. A number of taxa, including *Cricotopus* (*Isocladius*) indicate the presence of an abundant vegetation of aquatic macrophytes, which corroborate the paleobotanical data.

In Beuven some acidophilous and circumneutral species decreased (*e.g. Parakiefferiella bathophila*) and other circumneutral (*e.g. Anatopinia plumipes*) and indifferent species (*e.g. Psectrotanypus varius*) invaded after 1950 due to eutrophication. After the restoration species from eutrophic water bodies (*e.g. Cladotanytarsus* gr. *mancus*) disappeared (Buskens, 1989).

Most of the chironomid taxa from the moorland pools in The Netherlands have been found in Goorven, Gerritsfles, Kliplo or Beuven. Characteristic for moorland pools are taxa from bogs (acidobiontic taxa in Table 11), from organic and mineral substrates (*e.g. Demicryptochironomus vulneratus*), oxyphilous taxa (*Parakiefferiella bathophila*), temporary environments and shores from water bodies with fluctuating water level (*Polypedilum* cf. *uncinatum*) (Moller Pillot & Buskens, 1990).

The assemblages of chironomid larvae in moorland pools differ from those in Holarctic, holomictic lakes, as examined by *e.g.* Brundin (1949) and Saether (1979), because of the shallowness or even the temporary character of the pools. The process of accumulation of organic matter and the differentiation of the substrate by wind action in saucer-shaped moorland pools are master factors for the distribution of chironomids (Buskens, 1987).

Vertebrates

Fishes

Leuven & Oyen (1987) investigated the species richness and composition of fish assemblages in 54 oligo- to eutrophic soft water bodies, of which 47 were moorland pools, with pH-values between 3.4 and 9.5. In total, 24 fish species were encountered. The composition of the assemblages was strongly related to pH, alkalinity, trophic state and ionic composition of the water.

In slightly acid and alkaline waters (pH ≥ 5) Roach, Rudd, Perch, Brown Bullhead and Carp (respectively *Rutilus rutilus* (L.), *Scardinius erytrophtalmus* (L.), *Perca fluviatilis* L., *Ictalurus nebulosus* (Le Sueur) and *Cyprinus carpio* L.) accounted for about 90% of the total number of specimens in the catches. The percentage of water bodies which harboured fish as well as the average number of fish species per pool decreased steeply between pH 6 and 4.

Strongly acid waters (pH < 5) generally were fishless, because of hampered reproduction of most fish species at such acid sites. Only in the south-eastern part of the country the exotic Striped Mudminnow (*Umbra pygmaea* (De Kay)) was common in acid moorland pools. Other species were seen in exceptional cases only. The lower pH-limit for the occurrence of the striped mudminnow was 3.1 and the maximum standing stock was 21.3 kg ha^{-1}, which was attained by the consumption of mainly macro-invertebrates (of which 84% are chironomids) by adults and cladocerans by juveniles. Striped Mudminnows are consumed by waterfowl and frogs (Leuven *et al.*, 1984; Dederen *et al.*, 1986).

At least 67% of the nowadays extremely acid waters formerly harboured fish populations. Leuven & Oyen (1987) attribute the impoverishment of fish communities or the total loss of fish primarily to acidification by acid deposition. However, about 50% of the 23 acidified moorland pools which were investigated in the southern part of the country by P.J.J. van den Munckhof (unpubl.) had been managed as fish ponds in the past. The originally acid, oligotrophic water, had been enriched by water inlet from eutrophic areas to increase the fish stock. Many of these pools were isolated again in this century, and the fish decreased.

Measures to increase the fish stock of moorland pools by inlet of buffered water and introduction of Carps in the moorland pool district of Goorven are already known from the 17th century. When the inlet water is oligo- to mesotrophic the pool water remains clear and can have a well developed fish stock, with Pike on top of the food chain. In this way the management of moorland pools as fish ponds can promote the biodiversity in the pools (Van Dam *et al.*, 1988a). However, in modern times many moorland pools are used by sport-fishermen and are heavily manured by foddering and inlet of agricultural drainage water. The water becomes turbid and Pike disappears. The Bream (*Abramis brama* (L.)) is one of the few surviving species, which feeds on the chironomids in the bottom and attributes to the release of nutrients from the bottom to the open water by stirring the sediments (Lammens, 1987).

Fish have never been reported from Goorven, but the introduction of relatively eutrophic water and the subsequent opulent development of desmids was associated with the management of adjacent pools as fish ponds in the past, as was discussed earlier in this paper.

There are accounts of a rich assemblage of Eel (*Anguilla anguilla* (L.), Pike ((*Esox lucius* L.) and Perch in Beuven in the 18th century. Around 1950 the Striped Mudminnow and Pike were abundant in this pool. Between 1970 and 1985 the Striped Mudminnow disappeared from Beuven, due to eutrophication, but returned after the restoration in 1986. Pike and Perch have been abundant since 1970. Between 1970 and 1986 Tench (*Tinca tinca* (L.)), Eel and Roach were also common (Buskens, 1989).

Fish were caught from Gerritsfles after the eutrophication during a few decades in the late 19th century (Van Dam *et al.*, 1988b). No records of fish from Kliplo are known.

Amphibians

During an inventory of the amphibians in 96 soft water systems in The Netherlands (alkalinity ≤ 1 meq l^{-1}), mainly moorland pools, two species of toads, three species of frogs and four species of newts were found (Table 12). The number of amphibians is lowest in extremely acid waters (pH < 4). Egg mortality increases sharply when pH drops below 4.0–4.5, because of infection with saprolegnious fungi (Leuven *et al.*, 1986a). The culture experiments conducted by these authors demonstrated that apart from low affected the reproductive success. The absence of submersed macrophytes in acidified moorland pools may negatively affect the reproductive success of amphibians, because of the excretion of phenolics (Saber & Dunson, 1978).

Acidification affects the reproduction success and the local populations of amphibians. Populations of the Moorfrog (*Rana arvalis*) in particular are strongly reduced. This species has its optimum in moderately acid waters (4 ≤ pH < 5) of moorland pools and bogs of The Netherlands, Belgium and Germany (Table 12). In Europe, it is threatened by drainage, eutrophication and acidification (Strijbosch, 1979; Corbett, 1989).

Table 12. The percentage of 96 soft water systems in The Netherlands harbouring amphibian populations (Leuven *et al.* 1986a).

pH	< 4	4–5	≥ 5
Number of sites	28	31	37
Bufo bufo L.	12	58	84
B. calamita Laurenti	0	3	13
Rana arvalis Nilsson	35	63	22
R. esculenta-group	63	77	77
R. temporaria L.	18	55	69
Triturus alpestris Laurenti	0	10	9
T. cristatus Laurenti	0	7	8
T. helveticus Razoumowsky	11	32	22
T. vulgaris L.	0	7	19

Birds

The omnivorous Moorhen (*Gallinula chloropus* (L.)), Coot (*Fulica atra* L.), Mallard (*Anas platyrhynchos* L.), Teal (*Anas crecca* L.), Tufted Duck (*Aythya fuligula* (L.)) and Pochard (*A. ferina* (L.)) are regular breeding birds of moorland pools. The insectivorous Black-necked Grebe (*Podiceps nigricollis* (Brehm)), Little Grebe (*Tachybaptus ruficollis* (Pallas)) and Black Tern (*Chlidonias niger* (L.)) occur more rarely (Bekhuis *et al.*, 1987).

Schuurkes & Starmans (1987) investigated the relationship between the presence of breeding birds in moorland pools and several environmental parameters. Surface correlated most closely with species richness: pools with an area less than 1 ha had only 6 species, pools with an area of more than 2.5 ha had 11–13 species. In pools with a zone of littoral vegetation less than 10 m in width, the number of species was lower than in pools with a wider belt of shore vegetation. Species numbers were correlated negatively with the disturbance by recreation and positively with the distance to human settlements, particularly in the smallest pools. Also, the land-use in the surrounding of the pool (cultivated land, moorland, forest) played a role. Acidification did not severely affect the number of species per pool, but the species composition was affected by pH. Fish-consuming species, *e.g.* the Great Crested Grebe (*Podiceps cristatus* (L.)), were absent in pools with pH below 4, except when the acidified pools had a population of Striped Mudminnows. Although Schuurkes & Starmans (1987) present no data about the toxic effects of the increased levels of heavy metals by acidification, they expect that these affect the occurrence of waterfowl of acidified pools (Diamond, 1989).

Moorland pools are used as roosts by Crane (*Grus grus* (L.)), Snipe (*Gallinago gallinago* (L.)), Curlew (*Numenius arquata* (L.)), Whimbrel (*N. phaeopus* (L.)), Black-tailed Godwit (*Limosa limosa* (L.)) and other waders (Rijksinstituut voor Natuurbeheer, 1979; Bekhuis *et al.*, 1987). Some wintering species, *e.g.* Bean Goose (*Anser fabalis* (Latham)) use moorland pools as roosts (Van den Bergh, 1985). Some moorland pool districts have been included in the list of wetlands with important ornithological values, primarily because of their international significance for birds in winter (Rooth, 1989).

Goorven, Gerritsfles and Kliplo have the basic assemblage of Little Grebe, Mallard and Teal. In Gerritsfles and Beuven the Hobby (*Falco subbuteo* L.) hunts on Dragonflies (Odonata). In Kliplo the Blue Heron (*Ardea cinerea* L.) consumes frogs. In Beuven and other large, eutrophic, moorland pools other Herons, including Bittern (*Botaurus stellatus* (L.)) and raptors, *e.g.* Osprey (*Pandion haliaetus* (L.)) and Black Kite (*Milvus migrans* (Boddaert)) have also been reported by local birdwatchers.

During the present century moorland pools have been increasingly used as breeding sites by the Black-headed Gull (*Larus ridibundus* L.). This development was stimulated by the reclamation of moorlands and bogs. The new cultivated land is a rich food source for the gulls. In winter they take advantage of the abundance of food (human garbage) in nearby urban areas and waste deposits (Bekhuis *et al.*, 1987). As one gull excretes about 16 g P per year, and often more than thousand gulls breed in one pool, the breeding sites are heavily eutrophied. The average concentrations of ortho-phosphate in guanotrophic moorland pools are between 3 and 74 mmol m^{-3}, with maximum concentrations of 116 mmol m^{-3}. As these values are far above those in unaffected pools (≤ 0.5 mmol m^{-3}) the vegetation of macro- and microphytes, characteristic for the oligotrophic environment, is changed by guanotrophy. Often *Juncus effusus* forms a broad belt along the shore of pools where gulls are breeding, while the biomass of the plankton increases by the development of Chlorococcales and/or flagellates, like *Chlamydomonas* (Leentvaar, 1967; Geelen, 1969; Strijbosch, 1976).

Threats and measures necessary for conservation and re-creation

The moorland pools in The Netherlands harboured a rich variety of biota of oligo- to mesotrophic habitats up to the first decades of the present century. The diversity was due to the impact of traditional small-scale activities, which emphasized the natural differences between the pools and added new environmental gradients. Major changes in the agricultural system and the landscape of The Netherlands took place after the introduction of artificial fertilizer in the second half of the nineteenth century and these affected the moorland pool systems severely (Higler, 1993).

Reclamation

As the moorlands and heathlands were no longer necessary for grazing of the sheep, which until then supplied the manure for the growth of crops, large areas were reclaimed or afforested. The surface area of heath- and moorlands decreased from 8,000 km^2 in 1900 to 420 km^2 in 1980 (Weinreich & Musters, 1989). In the province of Noord-Brabant the number of moorland pools decreased by 60% between 1900 and 1980 (A.W.M Mol, unpubl.). Over the same period 42% of 161 moorland pools with Littorellion-species in The Netherlands disappeared by reclamation (Table 13).

Moorland pools do not disappear by reclamation any longer. However, it is useful to consider the possibilities for the creation of new moorland pools, *e.g.* by digging shallow pits for extraction of sand instead of deep ones in areas were suitable buffered surface- or groundwater is available for use as a buffer for atmospheric (potential) acid deposition. Another opportunity is re-creation of pools in areas which were reclaimed less than a century ago, and are to be abandoned in the near future because of restriction of agricultural production.

Table 13. Environmental changes in 161 moorland pools in The Netherlands from c. 1900 to c. 1980, inferred from changes in the occurrence of macrophytes belonging to the Littorellion alliance (Arts & Buskens, 1989)

Inferred change	Percentage of pools
Acidified	43
Reclaimed	42
Eutrophied	10
Not changed	5
Alkalinized	1

Afforestation

The oldest moorland pools were formed in an open tundra landscape, which was invaded by trees later on. Since the Atlantic time (6,000 y BP) the forest was cleared and changed into heath- and moorlands. Since the late 19th century large areas of heathlands have been afforested, and the majority of the moorland pools which were not reclaimed were enclosed in forests, with Scots pine as the principal tree species (Table 14).

The following negative effects of afforestation on the present biological condition of moorland pools are mentioned by Westhoff *et al.* (1973), Coesel *et al.* (1978), Rijksinstituut voor Natuurbeheer (1979), Van Dam *et al.* (1988a), Arts (1990) and Vahle (1990): (1) the presence of the forests increased interception of wet and dry (potential) acidifying atmospheric deposition, which can enhance acidification of the pools; (2) the Pine pollen accumulating in the pools is a source of nutrients, which can contribute to eutrophication; (3) Pine needles falling in the pool contribute both to acidification and eutrophication; (4) afforestation causes evapotranspiration to increase and the water table falls by the interception

Table 14. Changes of land-use around 860 moorland pools in the province of Noord-Brabant. Entries are percentages of the pools that have the particular type of land-use in their immediate environment. As combinations of land-use occur, the sum of entries in each column may be greater than 100% (A.W.M. Mol, unpubl.).

Land-use type	Year		
	1850	1900	1980
Heathland	83	83	51
Forest	27	47	74
Agricultural land	1	8	20

of precipitation; (5) prior to afforestation the ground- and surface water level was lowered; (6) the interception of light by the trees hampers the development of macrophytes in the littoral zone and the associated assemblages of algae, including characteristic desmids; (7) the reduction of wind erosion stimulates the formation and accumulation of organic sediments, which are unfavorable to isoetids, other soft-water macrophytes and macrofauna from mineral soils, and hampers the formation of new moorland pools.

As positive effects of afforestation on moorland pools are not known, the forest around the moorland pools should be cleared as much as possible, but slowly (over a time-course of some decades), in order to prevent shock-effects by sudden release of nutrients. To keep the wind at full speed, a strip of heathland or unfertilized grassland with a width of twenty times the height of the adjacent forest should be situated on the lee-side of the pool. However, a woodstrip of at least 15 m wide should be maintained between the agricultural area and the heathland or unfertilized grassland to reduce the input of fertilizers in the form of aerosols or dust (Vahle, 1990).

Cessation of human activities

When the moorland pools were still a part of the agricultural system the sediment was irregularly disturbed by cutting peat for use as fuel, removal of aquatic macrophytes for use as fertiler, fishing, hunting, sheep washing, swimming, etc. Like wind erosion these activities stimulated the development of isoetids, and macrofauna from mineral bottoms. Such activities are rarely practized in moorland pools in modern nature reserves (Westhoff, 1979). Therefore, sods should be cut nearshore, or the shores should be planed (smoothed with a big iron scraper) at irregular intervals (Schaminée *et al.*, 1990). Also strictly regulated tourist trips with glass-bottom boats, or diving excursions (in relatively deep pools) could be a source of the necessary small-scale dynamics for the growth of some rare aquatic macrophytes (Vahle, 1990).

Drainage

In addition to afforestation of heathland the drainage of agricultural land and collection of drinking water has also contributed to a lowering of the water table of moorland pools. Isolated pools with a perched water table may be affected by lowering of the groundwater level if the depth of the unsaturated zone below the pool bottom is <1 m (Bannink *et al.*, 1989). The size of 52% of the still-existing moorland pools in the province of Noord-Brabant decreased between 1850 and 1980 (A.W.M. Mol, unpubl.). All except one of 41 moorland pools surveyed by P.J.J. van den Munckhof (unpubl.) suffered from drainage.

The lowered water table increases the frequency of desiccation, which stimulates the growth of *Littorella uniflora*, but exacerbates the negative effects of acidification on water chemistry and diatoms. The increase of *Juncus effusus*, also a negative sign for the conservationist, is often observed in areas where both temporal desiccation and eutrophication occur (Van Dam, 1988; Arts, 1990). In future, desiccation will occur more often, as the frequency of extremely dry years will increase 5–10 times, due to global warming (Wigley, 1989).

Measures to increase the groundwater level should be implemented in regional integrated water management plans. Sometimes it is possible to increase the surface water level by retention of the local precipitation water, *e.g.* by the construction of dams. These measures can counteract the negative effects of the lowering of the groundwater table on a local scale (Cals & Roelofs, 1990; Cals *et al.*, 1991). Inlet of surface water from adjacent agricultural areas is rarely possible without measures to improve the quality of the inlet water, in order to prevent eutrophication, as will be discussed below.

Eutrophication

After 1850 the percentage of moorland pools which receive drainage via ditches and trenches from agricultural land or are bordering directly by

Table 15. Changes of hydrological isolation of 860 moorland pools in the province of Noord-Brabant. Entries are percentages of the pools that have the particular type of hydrology (A.W.M. Mol, unpubl.).

Isolation type	Year		
	1850	1900	1980
Isolated	86	78	72
Connected with ditches or trenches	7	16	26
Part of stream system	6	4	1
Other	1	2	2

agricultural land greatly increased (Tables 14 and 15). The input of nutrients by agricultural drainage water, often associated with the use of the pools as fish ponds, stimulated the primary production. Until recently eutrophication was considered as the main cause for the decline of the soft water macrophytes in The Netherlands (Schoof-van Pelt, 1973). Arts *et al.* (1989) investigated the changes in aquatic macrophytes by comparing the presence of herbarium specimens and other documentation with the recent distribution and discovered that isoetids had declined from 1900 to 1950, due to eutrophication (Fig. 13).

When mitigation measures are planned, the first step should be a search for existing data on chemistry, biology and management in the literature, archives and biological collections. If it appears that macrophytes from eutrophic waters have replaced those from strongly acid water (Table 5) and that species from weakly acid and weakly acid to neutral water (Table 5) never have been present, the vegetation can be restored by removal of the sediment, without measures to compensate for subsequent acidification (Cals & Roelofs, 1990).

In the past, isolation from the unwanted input of nutrients, has been used as an effective measure against eutrophication in those pools where species from weakly acid and weakly acid to neutral water were important, like some eutrophied moorland pools near Goorven. After eutrophication by agricultural drainage water and other

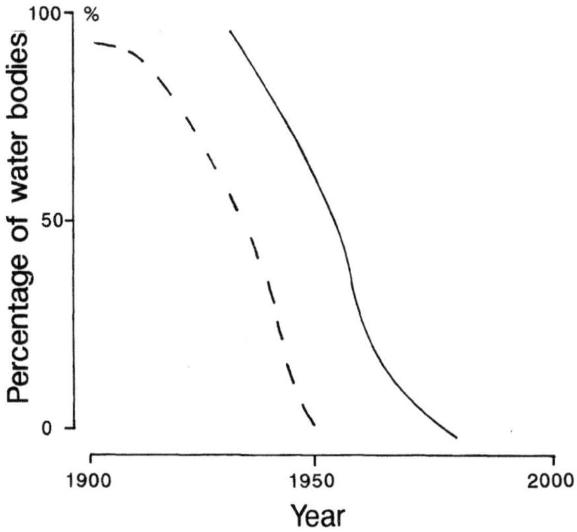

Fig. 13. Last record of a well developed soft-water vegetation in nowadays acidified (solid line, n = 44) or eutrophied (broken line, n = 16) moorland pools and other soft-water bodies (alkalinity ≤ 2 meq.l^{-1}) in The Netherlands. After Arts *et al.* (1989).

sources these pools were dredged in 1950 and the inflow of eutrophic water ceased. Initially the soft-water flora recovered. Later, however, the macrophytes from strongly acid water (Table 5) became dominant, because no measures had been taken to add buffering material to the pools to compensate for acidification by atmospheric deposition (Van Dijk *et al.*, 1960; Van Dam & Kooyman-van Blokland, 1978; Cals & Roelofs, 1990).

The restoration of Beuven was performed by removal of the mud layer and most of the reeds and the diversion of the rivulet which transported agricultural drainage water to the pool. A part of the well-buffered agricultural drainage water enters into a sedimentation basin (i.e. the separated southern part of the original pool) where the nutrients are removed by chemical and microbial processes as well as uptake by algae and macrophytes. After a period of three weeks the concentration of nutrients in the water layer has decreased sufficiently, so that the water from the sedimentation basin can be used as a buffer to counteract the acidification of Beuven by atmospheric deposition (Buskens, 1989).

Eutrophication by Black-headed Gulls can be

prevented effectively by removing eggs and nests a few times each season, with the first raid at the end of the egg-laying period, and the next one three weeks later (Rijksinstituut voor Natuurbeheer, 1979; A.L. Spaans, pers. comm.). Such actions are most effective when alternative breeding sites are present within 20–30 km of the pool which is to be protected from the settlement of gulls. The destruction of eggs in the nests is time-saving, but is discouraged because of the possibility of introducing avian botulism on the spot (A.L. Spaans, pers. comm.).

Acidification

Although the acidification of moorland pools is at least as or even more important than eutrophication for the deterioration of the biological quality of moorland pools, the negative consequences became manifest much later (Table 13 and Fig. 13).

Through comparison of historical and recent data on deposition and aquatic macrophytes it has been determined that the maximal load of (potential) acid deposition, including dry deposition, for Littorellion communities is 80 mmol m^{-2} yr and the maximal load for the decline of acid neutralizing capacity is 40 mmol m^{-2} yr. For NH_x these values are 50 and 30 mmol m^{-2} yr repectively (Arts, 1987). From experiments with small-scale ecosystems in a greenhouse, Schuurkes *et al.* (1987) concluded that the critical loads for (potential) acid deposition and nitrogen are 25 and 140 mmol m^{-2} yr, respectively. According to Erisman *et al.* (1989) the present depositions of (potential) acid and NH_x are 500 and 150 mmol m^{-2} yr respectively; thus large reductions are necessary. As such reductions are not feasible within the next decade, measures have to be taken in or near the moorland pools to mitigate the most serious effects of acidification.

If macrophytes from strongly acid water have been dominant, and macrophytes from weakly acid to neutral water (Table 5) have been absent in historical times, measures are not useful. After dredging the same species would return (Cals & Roelofs, 1990). In the past strong acidity in isolated moorland pools was often due to the presence of humic acids, nowadays strong acidity is more often caused by mineral acids. The microphyte assemblages of these two types of acid waters may be very different (Van Dam *et al.*, 1981; 1988). Those of humic acid waters are probably difficult to get back by restoration measures.

Only in special cases it is useful to manage pools with an original vegetation of macrophytes of strongly acid water bodies (Table 5), *e.g.* when habitats of the Moorfrog should be preserved. It is possible to reduce the fungal infection rate by the addition of powdered limestone to small moorland pools (Table 16). However, particularly in permanent pools, liming caused undesirable eutrophication by mineralization of organic material, even if the bulk of it (*Sphagnum*) was removed. The eutrophication was indicated by the increase of the relative abundance of eurytopic, eutraphentic and saprophilous diatoms, which are allochtonous to naturally acid moorland pools. An increase of nutrients, *e.g.* phosphate, could not be demonstrated (Table 16). The eutrophication was relatively low in intermittent pools.

In pools where well-buffered, relatively oligo- to mesotrophic surface water was introduced in the past, macrophytes from weakly acid water and weakly acid to neutral water (Table 5), desmids from mesotrophic habitats (Table 7) and soft-water diatoms (Table 8) were present. After

Table 16. Summary of liming experiments in small moorland pools by Bellemakers & Van Dam (in press). Treatment: C = control, L = limed, S = *Sphagnum* removed, i = intermittent, p = permanent. FIR = fungal infection rate of eggs of *Rana arvalis*, AD = relative abundance of allochtonous diatoms. Values are means over 1988 and 1989.

Treatment	Area (10^3 m^2)	Depth (cm)	pH	$H_2PO_4^-$ (mmol m^{-3})	FIR (%)	AD (%)
Ci	1.0	25	4.1	0.40	65	0
Cp	3.1	43	4.1	0.34	85	0
Si	0.2	13	3.7	0.31	85	0
Sp	1.5	20	4.0	0.43	85	0
Li	0.8	16	4.5	0.29	5	1
Lp	0.6	40	5.5	0.23	10	60
SLi	0.5	19	4.6	0.38	13	4
SLp	1.5	32	5.3	0.41	10	26

termination of the inflow the characteristic organisms disappeared, due to acidification, like in the moorland pools near Goorven. Inlet of buffered surface water, after nutrient removal, like in Beuven, could compensate for acidification. The introduction of relatively lime-rich groundwater might be an alternative, but this method still has to be tested in The Netherlands (Cals & Roelofs, 1990; Garrison & Rose, 1990).

Monitoring

Compared to other types of wetlands, biological and chemical changes in moorland pools are well documented. Moorland pools have attracted the attention of botanists and hydrobiologists since the beginning of this century. The preserved materials and the reports of the early investigators can potentially be used both for the detection of change in the pools and their environment and for the formulation of standards for the protection of their biological quality.

Few data are available on the moorland pool ecosystems and their management before 1900 and retrospective monitoring by paleolimnological and archive studies is necessary, particularly in humic-acid pools.

Regional and national surveys of selected chemical and biological parameters, particularly the species composition of macrophyte and algal assemblages, with time intervals of one to two decades in a comparatively large number of pools, are necessary for the detection of changes by human impact on moorland pools.

Intensive sampling, several times each year, of selected chemical and biological parameters, of a small number of carefully selected moorland pools is necessary to detect the effects of meteorological variations and climatic changes and to follow the effects of processes like acidification more closely.

Restoration projects should include evaluation studies. Historical biological information should be made available and the chemical and biological situation should be monitored before and after execution of the measures to assess their effect.

Acknowledgements

We are indebted to Dr G.H.P. Arts, Dr J.P. Bakker, Dr P.H. Best, Dr D.F. Charles, Dr P.F.M. Coesel, Dr L.W.G. Higler, Ir H. Houweling, Ir E.C.L. Marnette, Mrs E. Notenboom-Ram, Dr A.L. Spaans and Dr P.F.M. Verdonschot for commenting on the manuscript and to Dr K. Kersting and Mr A.J. van Tooren for making available their unpublished data.

References

Arts, G.H.P., 1987. Geschiedenis van de verzuring van zwak gebufferde wateren in Nederland onder invloed van atmosferische depositie. Laboratorium voor Aquatische Oecologie, Katholieke Universiteit, Nijmegen, 51 pp.

Arts, G.H.P., 1990. Deterioration of Atlantic soft-water systems and their flora, a historical account. Ph.D. Thesis, Katholieke Universiteit Nijmegen, 197 pp.

Arts, G.H.P. & R.F.M. Buskens, 1989. Aanvoer van venvreemd water: een noodzaak? In J.G.M. Roelofs (ed.), Aanvoer van gebiedsvreemd water. Vakgroep Aquatische Oecologie en Biogeologie, Katholieke Universiteit Nijmegen, Nijmegen: 100–110.

Arts, G.H.P. & C. den Hartog, 1990. Phytogeographical aspects of the West-European soft-water macrophyte flora. Acta bot. Neerl. 39: 369–378.

Arts, G.H.P., J.G.M. Roelofs & M.J.H. de Lyon, 1990a. Differential tolerances among soft-water macrophyte species to acidification. Can. J. Bot. 68: 2127–2134.

Arts, G.H.P., J.H.H. Schaminée & P.J.J. van den Munckhof, 1988. Human impact on origin, deterioration and maintenance of Littorelletalia-communities. In M. Zaliberová (ed.), Proceedings of the 5th Symposium on Synanthropic Flora and Vegetation, Martin, Czechoslovakia, 22–27 August, 1988: 11–18.

Arts, G.H.P., A.J. de Haan, M.B. Siebum & G.M. Verheggen, 1989. Extent and historical development of the decline of Dutch soft waters. Proc. K. Ned. Akad. Wet., Ser. C 92: 281–295.

Arts, G.H.P., G. van der Velde, J.G.M. Roelofs & C.A.M. van Swaay, 1990b. Successional changes in the soft-water macrophyte vegetation of (sub)atlantic, sandy, lowland regions during this century. Freshwat. Biol. 24: 287–294.

Asman, W.A.H., B. Drukker & A.J. Janssen, 1988. Modelled historical concentrations and depositions of ammonia and ammonium in Europe. Atmosph. Envir. 22: 725–735.

Bakker, T.W.M., I.I.Y. Castel, F.H. Everts & N.P.J. de Vries, 1986. Het Dwingelderveld, een Drents heidelandschap. Pudoc, Wageningen, 198 pp.

Bannink, M.H., L.W. Dekker, J.M.H. Hendrickx & H.C. van Ommen, 1989. Wegzijging van water uit hooggelegen vennen: een gevoeligheidsanalyse. H_2O 22: 456–459, 464.

Beijerinck, W., 1924. Het Lheebroeker Zand. De Levende Natuur 29: 97–102.

Beijerinck, W., 1926. Over verspreiding en periodiciteit van de zoetwaterwieren in Drenthsche heideplassen. Verh. K. Akad. Wet. 25(2): 1–211.

Beijerinck, W., 1931. Over Eurycercus glacialis Lillj. en andere arctische levensvormen in de plassen der 'Dwingeloo'sche Heide". De Levende Natuur 36: 238–248.

Bekhuis, J., R. Bijlsma, A. van Dijk, F. Hustings, R. Lensink & F. Saris (eds), 1987. Atlas van de Nederlandse vogels. SOVON, Arnhem, 595 pp.

Bellemakers, M.J.S. & H. van Dam. Improvement of breeding success of the moor frog (*Rana arvalis*) by liming of acid moorland pools and the consequences of liming for water chemistry and diatoms. Envir. Poll. (in press).

Brett, M.T., 1989. Zooplankton communities and acidification processes (a review). Water, Air, and Soil Pollution 44: 387–414.

Brock, T.C.M., 1988. De invloed van waterplanten op hun omgeving. In F.J.H.L. Bloemendaal & J.G.M. Roelofs (eds), Waterplanten en waterkwaliteit. Koninklijke Nederlandse Natuurhistorische Vereniging, Utrecht: 27–41.

Broertjes, J.P., 1977. Het ontstaan van de Brabantse vennen. Bijdr. Stud. Brab. Heem 16: 19–25.

Brundin, L., 1949. Chironomiden und andere Bodentiere der südschwedischen Urgebirgsseen. Freshwater Institute, Drottningholm, 914 pp.

Buskens, R.F.M., 1987. The chironomid assemblages in shallow lentic waters differing in acidity, buffering capacity and trophic level in The Netherlands. Entomologica Scandinavica, Suppl. 29: 217–224.

Buskens, R.F.M., 1989. Beuven: herstel van een ecosysteem. Vakgroep Aquatische Oecologie en Biogeologie, Katholieke Universiteit, Nijmegen, 154 pp.

Cals, M.J.R. & J.G.M. Roelofs, 1990. Prae-advies effectgerichte maatregelen tegen verzuring en eutrofiëring in oppervlaktewateren. Vakgroep Aquatische Oecologie en Biogeologie, Katholieke Universiteit Nijmegen, Nijmegen, 30 pp.

Cals, M.J.R., J.G.M. Roelofs, M.J.S. Bellemakers, M.M. Maessen, M.C.C. de Graaf & P.J.M. Verbeek, 1990. Monitoring van effectgerichte maatregelen tegen verzuring en eutrofiëring in oppervlaktewateren en heide-en schraallanden. Vakgroep Aquatische Oecologie en Biogeologie, Katholieke Universiteit Nijmegen, Nijmegen, 57 pp.

Charles, D.F., R.W. Battarbee, I. Renberg, H. van Dam & J.P. Smol, 1989. Paleoecological analysis of lake acidification trends in North America and Europe using diatoms and chrysophytes. In S.A. Norton, S.E. Lindberg & A.L. Page (eds), Soils, aquatic processes, and lake acidification. Springer, New York: 207–276.

Clymo, R.S. & P.M. Hayward, 1982. The ecology of *Sphagnum*. In A.J.E. Smith (ed.), Bryophyte ecology. Chapman & Hall, London: 229–289.

Coesel, P.F.M., 1975. The relevance of desmids in the biological typology and evaluation of fresh waters. Hydrobiol. Bull. 9: 93–101.

Coesel, P.F.M., 1983. The significance of desmids as indicators of the trophic status of freshwaters. Schweiz. Z. Hydrol. 45: 388–393.

Coesel, P.F.M. & H.D.W. Smit, 1977. Jukwieren in Drente, vroeger en nu. Veranderingen in de desmidiaceeënflora van enige Drentse vennen gedurende de laatste 50 jaar. De Levende Natuur 80: 34–44.

Coesel, P.F.M., R. Kwakkestein & A. Verschoor, 1978. Oligotrophication and eutrophication tendencies in some Dutch moorland pools, as reflected in their desmid flora. Hydrobiologia 61: 21–31.

Corbett, K. (ed.), 1989. The conservation of European reptiles and amphibians. Helm, London, 274 pp.

Dederen, L.H.T., R.S.E.W. Leuven, S.E. Wendelaar Bonga & F.G.F. Oyen, 1986. Biology of the acid-tolerant fish species *Umbra pygmaea*. J. Fish Biol. 28: 307–326.

De Gans, W., 1982. Location, age and origin of pingo remnants in the Drentsche Aa valley area (The Netherlands). Geol. Mijnb. 61: 147–158.

De Lyon, M.J.H. & J.G.M. Roelofs, 1986. Waterplanten in relatie tot waterkwaliteit en bodemgesteldheid. II. Laboratorium voor Aquatische Oecologie, Katholieke Universiteit, Nijmegen, 126 pp.

De Vries, B.J., 1984. Diatom assemblies in some moorland pools in the Drenthian district (The Netherlands). Hydrobiol. Bull. 18: 3–10.

Diamond, A.W., 1989. Impacts of acid rain on aquatic birds. Envir. Monit. Assessm. 12: 245–254.

Dickman, M.D., H. van Dam, B. van Geel, A.G. Klink & A. van der Wijk, 1987. Acidification of a Dutch moorland pool – a palaeolimnological study. Arch. Hydrobiol. 109: 377–408.

Dresscher, T.G.N., F. de Graaf, A.A. de Groot, J. Heimans, G.P.H. van Heusden, J.T. Koster, W. Meyer, M.F. Mörzer Bruyns, H.J.W. Schimmel, A.P.C. de Vos, H.F. de Vries & A. van der Werff, 1952. De Gerritsflesch bij Kootwijk. Hydrobiologische Vereniging, Amsterdam, 22 pp.

Erisman, J.W., F.A.A.M. de Leeuw & R.M. van Aalst, 1989. Deposition of the most acidifying components in The Netherlands during the period 1980–1986. Atmosph. Envir. 23: 1051–1062.

Feijtel, T.C., Y. Salingar, C.A. Hordijk, J.P.R.A. Sweerts, N. van Breemen & T.E. Cappenberg, 1989. Sulfur cycling in a Dutch moorland pool under elevated atmospheric S-deposition. Wat. Air Soil Poll. 44: 215–234.

Flower, R.J. & R.W. Battarbee, 1983. Diatom evidence for recent acidification of two Scottish lochs. Nature 305: 130–133.

Garrison, P.J. & W.J. Rose, 1990. Increasing the pH of acidic lakes by augmenting the ground-water input: an alternative to liming. Conference Abstracts. International Conference on Acidic Deposition, its Nature and Impacts, Glasgow, 16–21 September, 1990. Royal Society of Edinburgh, Edinburgh: 605.

Geelen, J.F.M., 1969. Vergelijkend planktononderzoek in twee Hatertse Vennen. Ph.D. Thesis, Katholieke Universiteit, Nijmegen, 112 pp.

Geelen, J.F.M. & R.S.E.W. Leuven, 1986. Impact of acidification on phytoplankton and zooplankton communities. Experientia 42: 486–494.

Hanssen, J.E., U. Pedersen, J. Schaug, H. Dovland, J.M. Pacyna, A. Semb & J.E. Skjelmoen., 1990. Summary report from the Chemical Co-ordinating Centre for the fourth phase of EMEP. Norwegian Institute for Air Research, Lillestrøm, 149 pp.

Higler, L.W.G., 1993. Nutrient-poor freshwater wetlands. Hydrobiologia 265: 195–201.

Hoogenraad, H.R., 1935. Studien über die sphagnicolen Rhizopoden der niederländischen Fauna. Arch. Protistenk. 84: 1–100.

Kersting, K. & P. Kouwenhoven, 1989. Annual and diel oxygen regime in two polder ditches. Hydrobiol. Bull. 23: 111–123.

Klink, A., 1986. De geschiedenis van de verzuring in Nederland: een palaeolimnologische studie naar de invloed van verzuring op levensgemeenschappen in enige zwak gebufferde wateren. Hydrobiologisch Adviesbureau Klink, Wageningen, 43 pp.

Kok, C.J., H.W.G. Meesters & A.J. Kempers, 1990. Decomposition rate, chemical composition and nutrient recycling of *Nymphaea alba* L. floating leaf blade detritus as influenced by pH, alkalinity and aluminium in laboratory experiments. Aquat. Bot. 37: 215–227.

Kok, C.J. & B.J. van de Laar, 1991. Influence of pH and buffering capacity on the decomposition of *Nymphaea alba* L. detritus in laboratory experiments: a possible explanation for the inhibition at low alkalinity. Verh. int. Ver. Limnol. 24: 2689–2692.

Krause-Dellin, D. & C. Steinberg, 1986. Cladoceran remains as indicators of lake acidification. Hydrobiologia 129–134: 129–134.

Lammens, E.H.R.R., 1987. De rol van de brasem in het Nederlandse oppervlaktewater. De Levende Natuur 88: 238–242.

Leeflang, K.W.H., 1938. De chemische samenstelling van den neerslag in Nederland. Chem. Weekbl. 35: 658–664.

Leentvaar, P., 1958. Hydrobiologie van de Bergvennen. Kruipnieuws (Twente-nr.): 21–28.

Leentvaar, P., 1967. Observations in guanotrophic environments. Hydrobiologia 29: 441–489.

Leuven, R.S.E.W., F.G.F. Oyen & J.F.M. Geelen, 1984. De Oost-Amerikaanse hondvis. Natura 81: 271–275.

Leuven, R.S.E.W., 1988. Impact of acidification on aquatic ecosystems in The Netherlands with emphasis on structural and functional changes. Ph.D. Thesis, Katholieke Universiteit Nijmegen, 181 pp.

Leuven, R.S.E.W. & F.G.F. Oyen, 1987. Impact of acidification and eutrophication on the distribution of fish species in shallow and lentic soft waters of The Netherlands: an historical perspective. J. Fish Biol. 31: 753–774.

Leuven, R.S.E.W. & W. Wolfs, 1988. Effects of water acidification on the decomposition of *Juncus bulbosus* L. Aquat. Bot. 31: 57–81.

Leuven, R.S.E.W., C. den Hartog, M.M.C. Christiaans & W.H.C. Heijligers, 1986a. Effects of water acidification on the distribution pattern and the reproductive success of amphibians. Experientia 42: 495–503.

Leuven, R.S.E.W., G. van der Velde, J.A.M. Vanhemelrijk & R.L.E. Eeken, 1986b. Impact of acidification on the distribution of aquatic insects in lentic soft waters. Proceedings 3rd European Congress of Entomology. Dutch Entomological Society, Amsterdam: 103–106.

Leuven, R.S.E.W., J.A. van der Velden, J.A.M. Vanhemelrijk & G. van der Velde, 1987. Impact of acidification on chironomid communities in poorly buffered waters in The Netherlands. Entomol. Scand., Suppl. 29: 269–280.

Leuven, R.S.E.W., H.L.M. Kersten, J.A.A.R. Schuurkes, J.G.M. Roelofs & G.H.P. Arts, 1986c. Evidence for recent acidification of lentic soft waters in The Netherlands. Wat. Air Soil Poll. 30: 387–392.

Lijklema, L., 1980. Interaction of orthophosphate with iron (III) and aluminum hydroxides. Environ. Sci. Technol. 14: 537–541.

Moller Pillot, H.K.M. & R.F.M. Buskens, 1990. De larven der Nederlandse Chironomidae (Diptera). Deel C. Autoekologie en verspreiding. Nederl. Faunist. Meded. 1C: 1–87.

Nature Policy Plan, 1989. SDU, 's-Gravenhage, 179 pp.

Notenboom-Ram, E., 1981. Verspreiding en ecologie van de Branchiopoda in Nederland. Rijksinstituut voor Natuurbeheer, Leersum, 95 pp.

Økland, J. & K.A. Økland, 1986. The effects of acid deposition on benthic animals in lakes and streams. Experientia 42: 471–486.

Paffen, B.G.P. & J.G.M. Roelofs, 1991. Impact of carbon dioxide and ammonium on the growth of submerged *Sphagnum cuspidatum*. Aquat. Bot. 40: 61–71.

Posthumus, N.W., 1911. Keurboek van Oosterwijk. Versl. Med. Ver. Uitg. Bronnen Oud-Vaderl. Recht 6(2): 174–202.

Redeke, H.C., 1948. Hydrobiologie van Nederland: de zoete wateren. De Boer, Amsterdam, 580 pp.

Redeke, H.C. & A.P.C. de Vos, 1932. Beiträge zur Kenntnis der Fauna niederländischer oligotropher Gewässer. Int. Revue ges. Hydrobiol. Hydrogr. 28: 1–45.

Reuss, J.O., D.W. Johnson, 1986. Acid deposition and the acidification of soils and waters. Springer, New York, 119 pp.

Rijksinstituut voor Natuurbeheer, 1979. Natuurbeheer in Nederland: levensgemeenschappen. Pudoc, Wageningen, 392 pp.

Roelofs, J.G.M., 1983. Impact of acidification and eutrophication on macrophyte communities in soft waters in The Netherlands. I. Field observations. Aquat. Bot. 17: 139–155.

Roelofs, J.G.M. & F.J.H.L. Bloemendaal, 1988. Trofie. In F.J.H.L. Bloemendaal & J.G.M. Roelofs (eds), Waterplanten en waterkwaliteit. Koninklijke Nederlandse Natuurhistorische vereniging, Utrecht: 113–126.

Roelofs, J.G.M., J.A.A.R. Schuurkes & A.J.M. Smits, 1984. Impact of acidification and eutrophication on macrophyte communities in soft waters in The Netherlands. II. Experimental studies. Aquat. Bot. 18: 389–411.

Rooth, J., 1989. Lijst van internationaal belangrijke wetlands in Nederland. In A.L. Spaans (ed.), Wetlands en watervogels. Pudoc, Wageningen: 95–106.

Saber, P.A. & W.A. Dunson, 1978. Toxicity of bog water to embryonic and larval anuran amphibians. J. exp. Zool. 204: 33–42.

Saether, O.A., 1979. Chironomid communities as water quality indicators. Holarct. Ecol. 2: 65–74.

Sawhney, B.L., 1972. Selective sorption and fixation of cations by clay minerals: a review. Clays Clay Miner. 20: 93–100.

Schaminée, J.H.J., G.H.P. Arts & V. Westhoff, 1990. Plantengemeenschappen van Nederland. 6. Littorelletea. Rijksinstituut voor Natuurbeheer, Leersum, 44 pp.

Schimmel, H.J.W. & J. ter Hoeve, 1952. Bodemgesteldheid en waterhuishouding van de 'Gerritsflesch' bij Kootwijk. Tijdschr. K. Ned. Aardrijksk. Gen. 69: 16–28.

Schindler, D.W., 1987. Detecting ecosystem responses to anthropogenic stress. Can. J. Fish. aquat. Sci. 44 (Suppl. 1): 6–25.

Schindler, D.W., 1988. Confusion over the origin of alkalinity in lakes. Limnol. Oceanogr. 33: 1637–1640.

Schoof-van Pelt, M.M., 1973. Littorelletea: a study of the vegetation of some amphiphytic communities of Western Europe. Ph.D. Thesis, Katholieke Universiteit Nijmegen, 216 pp.

Schuurkes, J.A.A.R. & C.J. Kok, 1988. In vitro studies on sulphate reduction and acidification in sediments of shallow soft water lakes. Freshwat. Biol. 19: 417–426.

Schuurkes, J.A.A.R. & P. Starmans, 1987. Effecten van verzuring op het voorkomen van watervogels in kalkarme oppervlaktewateren. Vogeljaar 35: 57–64.

Schuurkes, J.A.A.R., J. Jansen & M. Maessen, 1988a. Water acidification by addition of ammonium sulphate in sediment-water columns and in natural waters. Arch. Hydrobiol. 112: 495–516.

Schuurkes, J.A.A.R., A.J. Kempers & C.J. Kok, 1988b. Aspects of biochemical sulphur conversions in sediments of a shallow soft water lake. J. freshwat. Ecol. 4: 369–381.

Schuurkes, J.A.A.R., C.J. Kok & C. den Hartog, 1986. Ammonium and nitrate uptake by aquatic plants from poorly buffered and acidified waters. Aquat. Bot. 24: 131–146.

Schuurkes, J.A.A.R., M.M.J. Maenen & J.G.M. Roelofs, 1988c. Chemical characteristics of precipitation in NH_3-affected areas. Atmosph. Envir. 22: 1689–1698.

Schuurkes, J.A.A.R., M.A. Elbers, J.J.F. Gudden & J.G.M. Roelofs, 1987. Effects of simulated ammonium sulphate and sulphuric acid rain on acidification, water quality and flora of small-scale soft water systems. Aquat. Bot. 28: 199–226.

Strijbosch, H., 1976. Een vergelijkend syntaxonomische en synoecologische studie in de Overasseltse en Hatertse Vennen bij Nijmegen. Ph.D. Thesis, Katholieke Universiteit Nijmegen, 335 pp.

Strijbosch, H., 1979. Habitat selection of amphibians during their aquatic phase. Oikos 33: 363–372.

Sutcliffe, D.W. & T.R. Carrick, 1973. Studies on mountain streams in the English Lake District. I. pH, calcium and the distribution of invertebrates in the River Duddon. Freshwat. Biol. 3: 437–462.

Suykerbuyk, R.E.M. & R.M.M. Roijackers, 1988. Typologie van Nederlandse zoete en brakke oppervlaktewateren op basis van het fytoplanktonbestand. In R.M.M. Roijackers (ed.), Hydrobiologisch onderzoek in Nederland: fundamentele en toepassingsgerichte aspekten. Hydrobiologische Vereniging, Amsterdam: 105–116.

Sykora, K.V., 1979. The effects of the severe drought of 1976 on the vegetation of some moorland pools in The Netherlands. Biol. Conserv. 16: 145–162.

Ter Braak, C.J.F. & H. van Dam, 1989. Inferring pH from diatoms: a comparison of old and new calibration methods. Hydrobiologia 178: 209–223.

Tutin, T.G., V.H. Heywood, N.A. Burges, D.M. Moore, D.H. Valentine, S.M. Walters & D.A. Webb, 1964–1980. Flora Europaea, Vols 1–5. Cambridge University Press, Cambridge.

Vahle, H.C., 1990. Grundlagen zum Schutz der Vegetation oligotropher Stillgewässer in Nordwestdeutschland. Natursch. Landschaftspf. Nieders. 22: 1–159.

Van Dam, H., 1987. Monitoring of chemistry, macrophytes, and diatoms in acidifying moorland pools. Research Institute for Nature Management, Leersum, 91 pp.

Van Dam, H., 1988. Acidification of three moorland pools 18 in The Netherlands by acid precipitation and extreme drought periods over seven decades. Freshwat. Biol. 20: 157–176.

Van Dam, H. & H. Kooyman-van Blokland, 1978. Man-made changes in some Dutch moorland pools, as reflected by historical and recent data about diatoms and macrophytes. Int. Revue ges. Hydrobiol. 63: 587–607.

Van Dam, H. & A. Mertens, 1990. Lange-termijnonderzoek naar verzuring van vennen. De Levende Natuur 91: 184–192.

Van Dam, H., G. Suurmond & C.J.F. ter Braak, 1981. Impact of acidification on diatoms and chemistry of Dutch moorland pools. Hydrobiologia 83: 425–459.

Van Dam, H., B. van Geel, A. van der Wijk & M.D. Dickman, 1988a. Beheer van vennen in historisch perspectief. De Levende Natuur 89: 66–73.

Van Dam, H., B. van Geel, A. van der Wijk, J.F.M. Geelen, R. van der Heijden & M.D. Dickman, 1988b. Palaeolimnological and documented evidence for alkalization and acidification of two moorland pools (The Netherlands). Review Palaeobot. Palynol. 55: 273–316.

Van den Bergh, L.M.J., 1985. Het voorkomen van de Taiga-rietgans *Anser fabalis fabalis* in Nederland. Limosa 58: 17–22.

Van der Voo, E.E. & P. Leentvaar, 1959. Het Teeselinkven. De Levende Natuur 62: 128–135.

Van Dijk, J., F. de Graaf, W. Graafland, A.A. de Groot, J. Heimans, J.T. Koster, A.P.C. de Vos, H.F. de Vries, A. van der Werff & V. Westhoff, 1960. Hydrobiologie van de Oisterwijkse Vennen. Hydrobiologische Vereniging, Amsterdam, 90 pp.

Van Dobben, H.F., J. Mulder, H. van Dam & H. Houweling, 1993. The impact of atmospheric deposition on the biogeochemistry of moorland pools and the surrounding terrestrial environment. Pudoc, Wageningen, 232 pp.

Vangenechten, J.H.D., F. Bosmans & H. Deckers, 1981. Effects of short-term changes in rain water supply on the ionic composition of acid moorland pools in the Campine of Antwerp (Belgium). Hydrobiologia 76: 149–159.

Verdonschot, P.F.M., 1990. Ecologische karakterisering van oppervlaktewateren in Overijssel. Rijksinstituut voor Natuurbeheer, Leersum/Provincie Overijssel, Zwolle, 301 pp.

Weber, M., 1919. Beschouwingen over de fauna van Nederland. Bijdr. Dierk. 21: 175–193.

Weddigen, U. & U. Geissler, 1980. Vergleichende Untersuchungen zur Algenflora zweier Berliner Naturschutzgebiete (Pech- und Barssee) unter besonderer Berücksichtigung der Desmidiaceen. Nova Hedwigia 33: 95–144.

Weinreich, J.A. & C.J.M. Musters, 1989. Toestand van de natuur: veranderingen in de Nederlandse natuur. SDU, 's-Gravenhage, 241 pp.

Westhoff, V., 1979. Bedrohung und Erhaltung seltener Pflanzengesellschaften in den Niederlanden. In O. Willmans & R. Tüxen, Werden und Vergehen von Pflanzengesellschaften. Cramer, Vaduz: 285–313.

Westhoff, V. & A.J. den Held, 1969. Plantengemeenschappen in Nederland. Thieme, Zutphen, 324 pp.

Westhoff, V., P.A. Bakker, C.G. van Leeuwen, E.E. van der Voo & I.S. Zonneveld, 1973. Wilde planten, flora en vegetatie in onze natuurgebieden. III. De hogere gronden. Vereniging tot Behoud van Natuurmonumenten in Nederland, Amsterdam, 359 pp.

Wigley, T.M.L., 1989. Effects of changing climate on the frequency of extreme events. Climate Monitor 17: 44–55.

Wortelboer, F.G., 1990. A model on the competition between two macrophyte species in acidifying shallow soft-water lakes in The Netherlands. Hydrobiol. Bull. 24: 91–107.

Hydrobiologia **265**: 265–279, 1993.
E. P. H. Best & J. P. Bakker (eds), Netherlands-Wetlands.

Stream valleys as wetlands

L.W.G. Higler & P.F.M. Verdonschot
Institute for Forestry and Nature Research, P.O. Box 46, 3956 ZR Leersum, The Netherlands

Key words: lowland streams, nutrient status

Abstract

Wetlands as a component of discharge areas have been reduced to mere stream channels in the larger part of The Netherlands. The original shape and condition are described with help of data from the past and from some small remnants. The typical Dutch type of running water is the lowland stream, fed predominantly by rain and groundwater. Its nature is nutrient-poor in the upper course and eutrophying gradually downstream. The chemical composition of streamwater, the vegetation in the floodplain and the flora and fauna in the stream reflect this gradient. Some attention is paid to the amount and cycling of minerals and nutrients. It is shown that the input of nutrients is not a limiting factor for primary production, but under natural conditions light is limiting. Threats to stream wetlands are the composition of the precipitation, fertilization of the discharge area and water management measures such as draining and canalization. As a result perspectives for a qualitative recovery are unfavourable. Restoration of the shape of streams, however, is more prospective and several restoration programs are in execution.

Introduction

Stream valleys as wetlands may seem a strange conception in The Netherlands nowadays. The overall picture of streams in The Netherlands is that of neat canals with dry banks or even low dikes in cultivated land or at best slightly meandering water courses in nature reserves. Sporadically the shores are too wet to reach the proper stream and in these places one can brood about the wetland that stream valleys ought to be. This contribution will be focussed on general characteristics of lowland streams and small lowland rivers as wetlands, some ecological concepts on running water with emphasis on hydrological, sediment and nutrient cycles in the Dutch situation and the consequences for flora and fauna, human influence and management in stream valleys and perspectives for the future.

Types of running water in The Netherlands

Redeke (1948) discerned three types of running water in The Netherlands:

1. Springs/sources.
2. Streams:
   a. Mountain streams (hill streams would be a better term).
   b. Lowland streams.
3. Rivers:
   a. Small rivers.
   b. Large rivers (Rhine and Meuse).

The functioning of running waters is primarily defined by hydraulic factors. Figure 1 shows a proposal for a typology based on the slope of terrain, the hydraulic radius, the roughness and the mean current velocity (Higler & Mol, 1984). By using this diagram the types can be defined

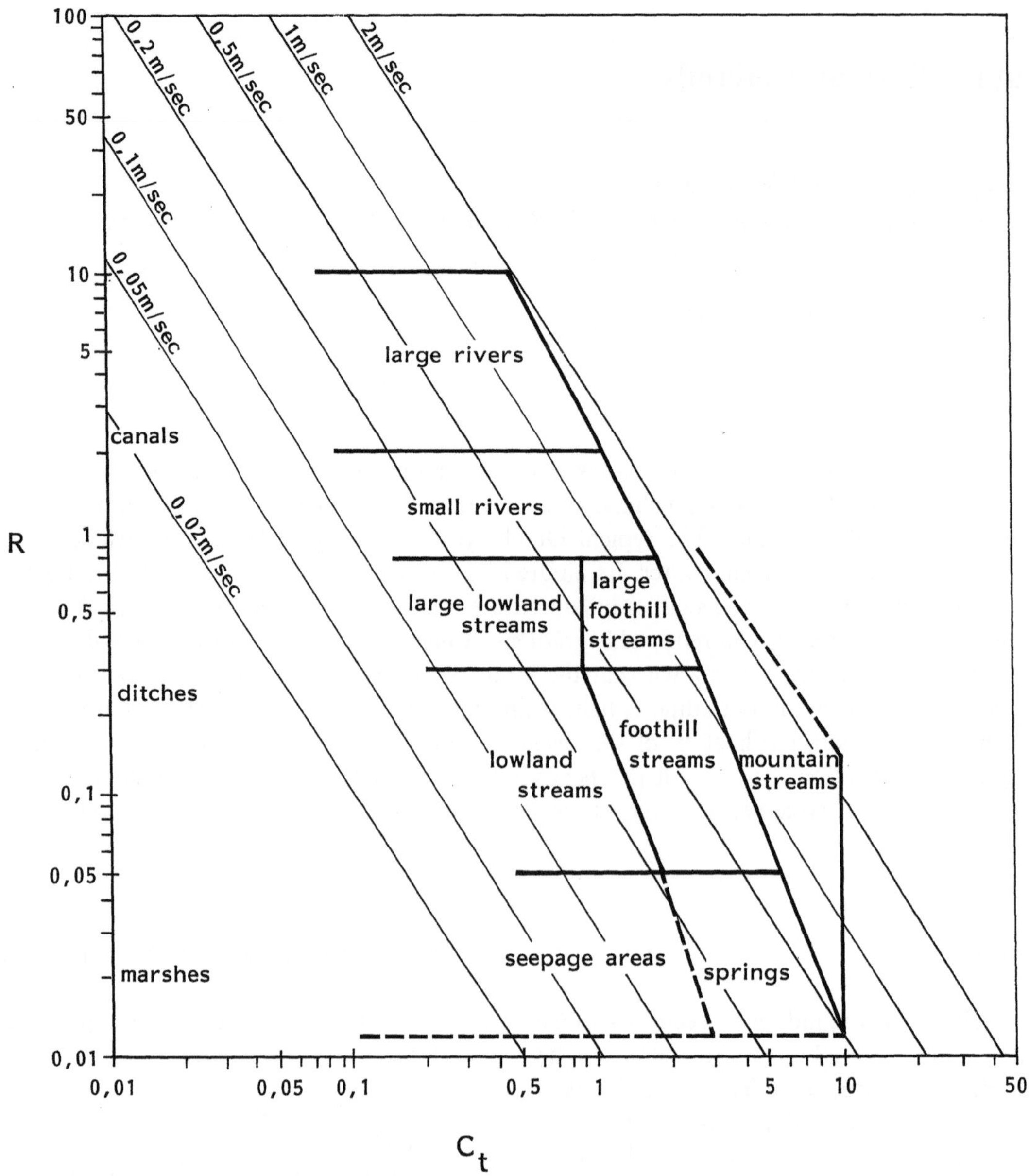

Fig. 1. Typology of running waters based on hydraulic factors (Higler & Mol, 1984; slightly modified).R = hydraulic radius (m); C_t = terrain factor (1 n $J^{1/2}$); n = roughness; J = ground slope.

acccording to easily measurable hydraulic variables.

For the purpose of this study the sequence from seepage area or helocrene source via lowland stream to small river (Fig. 1) is by far the most important one and the following will be restricted to this type of running water.

Diffuse seepage areas (helocrene sources) or a number of marshy places often form the beginning of especially lowland streams.

Springs as the distinct beginning of streams are only found in hilly regions, mostly ice-pushed ridges, in The Netherlands.

In other cases, lowland streams originate from

one or more small trenches, which become marshy and further downstream fill up with water. These are true rainwater-fed lowland streams.

Lowland streams and rivers are defined as streams with a slight slope and a low mean current velocity ($v < 50$ cm s^{-1}). In Fig. 2 the slope line of the river Regge represents a typical small lowland river. Natural lowland streams are free meandering and have an asymmetric cross profile with one beetling bank. The bottom is sandy and substrates are fine (silt and sand, with occasionally coarse gravel). The input of leaves and branches from the banks forms different organic substrates which can cause local changes in the discharge pattern. The lowland stream can be subdivided into the upper course (streamorder 1–3; width 0.5–5 m), the middle course (streamorder 3–4; width 5–10 m) and the lower course (streamorder >4; width >10 m).

Hill streams and a part of the rivers have a steep slope and consequently a high current velocity ($v > 50$ cm s^{-1}), a firm bottom and coarse substrates. In Fig. 2 the slope line of the river Geul is an example of a small hill river.

Lowland streams as wetland

The drainage area of lowland streams is determined by climatological and geomorphological processes. In the larger part of The Netherlands, where streams are present, the soils consist of fine and coarse cover sands over a deeper boulder-clay layer (Pleistocene district). The precipitation surplus forms the main source of the streams. Heavy rainfall is followed by large discharges, either immediate or somewhat delayed. In general the discharge pattern follows, though delayed, the precipitation pattern. Consequently there is a relationship between the season of maximal precipitation, from July up to and including January, and the water supply to the stream system (Fig. 3). More water will be supplied by seepage and capillary rise, as soon as the storage capacity in the underground is filled up. This water is then gradually released into the stream through spring areas and marshes or discharged immediately by the stream channel, dependent on the structure of the stream system. Springs and upper courses of streams are fed by local groundwater masses or

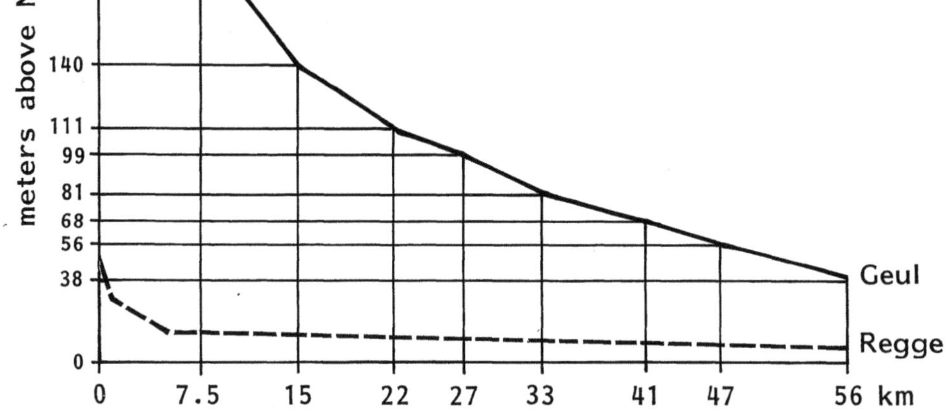

Fig. 2. Slope lines of the river Geul (Meerman, 1975) and the river Regge (Redeke, 1948; adapted).

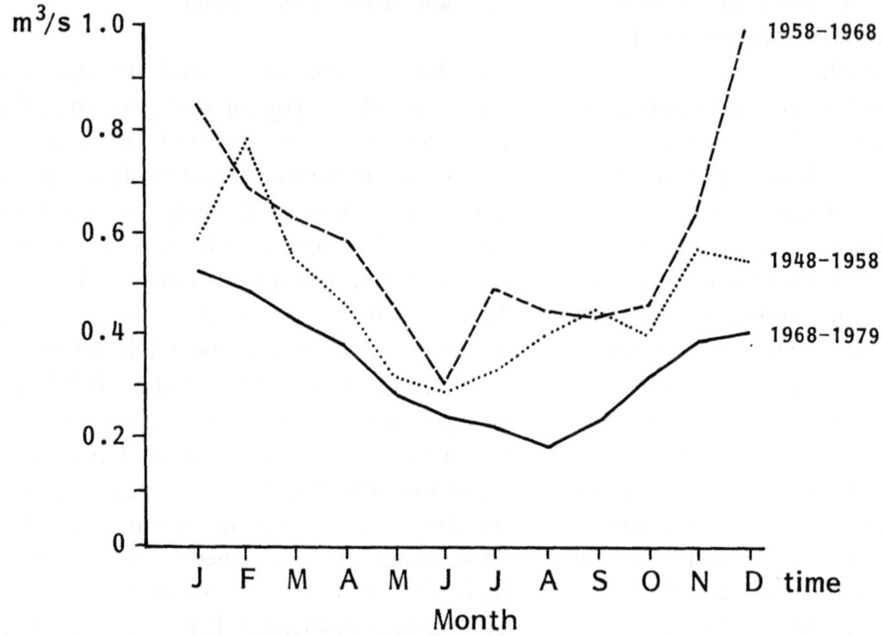

Fig. 3. Mean discharge in the Hierden stream per decade (Higler, 1980).

directly by rainwater input (mineral-poor). Small peat layers on clay, formed in the mini floodplains gather rainwater which is released to the upper courses only slowly. Middle courses receive deeper groundwater (mineral-rich) or water leaching from thick (up to 4 m) peat layers. The lower courses also receive deep groundwater (mineral-rich) or can merely function as drainage canals.

The presence of marshes is considered to be the natural situation alongside many lowland streams in the past. Their former presence is only indicated nowadays by peaty soils (Fig. 4). The increase in discharge causes higher current velocities and, consequently, transportation of sediments and organic material. A decrease in discharge leads to sedimentation. A naturally shaped lowland stream, that meanders through the landscape, regularly overflows and gives then rise to the formation of bank levees and shallow marshes outside these levees. Stagnant waters come into existence, when meanders are cut off. These will be overgrown in time and again form marshes, nowadays still recognized as mucky brook pools (Fig. 4). The mosaic pattern of sandy bank levees, peaty soils, oxbows and lower overflown loamy sands form mini floodplains. Lowland streams with their adjoining mini floodplains form long and narrow wetlands that used to be found all over the Pleistocene district.

The lowland stream habitat

The substratum of the streambed forms the main physical component of the system. Both rooting macrophytes and benthic organisms are dependent of the structure of the substratum. Lowland streams lack a hyporheic region, the area below the bed of a stream where interstitial water moves by percolation. The velocity of running water influences the size and distribution of substrate particles. Substrates in lowland streams range from gravel to silt and form a mosaic pattern of microhabitats (Tolkamp, 1980). The substrates in hill streams are more varied.

There is a linear relationship between the light availability at the stream surface and the submerged plant standing crop or the ratio of the biomass (Dawson & Kern-Hansen, 1979). The upper courses of natural streams are shaded by

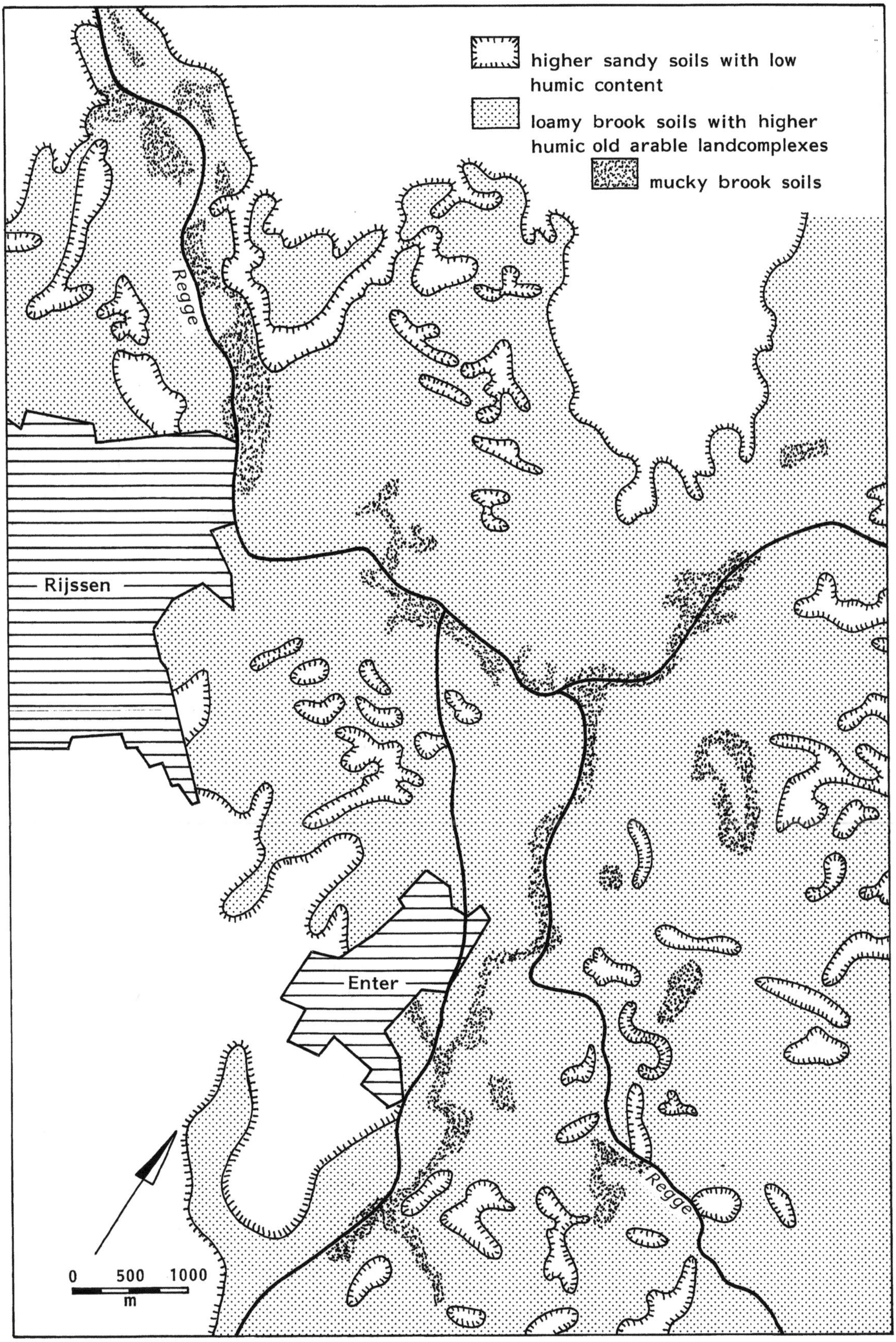

Fig. 4. Soil map indicating the present and former course of the river Regge system (province of Overijssel) and its influence on soil composition (Stiboka, 1983; adapted).

their riparian vegetation and consequently lack submerged plants. The role of macrophytes in the foodchain of streams is limited, especially in the upper course. In the natural lowland stream here and there patches of *Ranunculus fluitans* and *Potamogeton densus* occurred. More downstream a number of macrophytes could be found, all typical for slowly flowing and stagnant water like *Nuphar lutea, Sagittaria sagittifolia, Potamogeton natans, Sparganium emersum, Potamogeton alpinus, Ranunculus aquatilus, Elodea canadensis, E. nuttalii*, and *Callitriche spp.*

Lowland streams have a relatively low current velocity which is reflected by the absence of typical rheophilic species with a flattened or streamlined body. The distribution of aquatic invertebrate habits or modes of existence is illustrated in Table 1. Most taxa are burrowers and sprawlers in and on the sandy and silty substrates, especially dominant in the helocrene springs, and clingers and climbers in leaf packets and in the aquatic vegetation. Divers and swimmers become more important in the lower reaches.

Most aquatic insects are omnivorous with respect to their feeding mechanisms (Cummins & Klug, 1979). The type of food may change considerably, however, depending on the instar. Therefore, the aquatic organisms are not classified according to trophic level but to morpho-behavioral mechanisms of food acquisition (or 'functional groups'). Table 2 shows the distribution of functional feeding groups along a lowland stream.

Table 1. Frequency of different types of habit of aquatic invertebrates in lowland streams in the province of Overijssel (94 sampling stations) (Verdonschot, 1990).

Lowland stream part Habit	Helocrene spring	Spring Stream	Upper reach	Middle reach	Lower reach
Skaters	0.0	0.1	0.4	0.4	0.3
Divers	0.2	0.1	0.1	0.1	0.5
Swimmers	3.1	3.0	6.8	6.2	11.4
Clingers	18.8	24.4	18.3	13.5	18.2
Sprawlers	43.1	25.4	17.1	25.4	15.3
Climbers	11.9	21.8	22.8	15.0	25.7
Burrowers	23.0	25.2	35.5	39.4	28.5

Table 2. Frequency of different types of functional feeding groups in lowland streams in the province of Overijssel (94 sampling stations) (Verdonschot, 1990).

Lowland stream part Functional feeding group	Helocrene spring	Spring stream	Upper reach	Middle reach	Lower reach
Shredders	25.7	31.5	20.1	13.0	12.4
Filtering collectors	5.1	4.2	2.8	6.1	8.9
Gathering collectors	52.7	35.6	45.8	45.4	38.9
Scrapers	7.9	6.4	4.4	6.3	12.5
Predators	28.4	12.2	26.8	29.0	27.0

We attempted to apply the River Continuum Concept (RCC; Vannote *et al.*, 1980) to lowland streams in The Netherlands. The RCC is the conceptualization of the stream as a continuum of communities. The entire changes in a stream are thought of as a gradual overall shift in the downstream direction or a continuum of community replacements. Theoretically, the upper course of a stream is situated in a forested area and the stream is characterized by organisms which utilize the dominating leaf input (shredders and collectors). In the middle course the importance of leaf input is reduced in favour of primary production and the stream is characterized by collectors and grazers. If we apply this concept to the Dutch lowland stream, the role of collectors and predators proves to be constant throughout the stream (Table 2), the shredders decrease while the scrapers increase in the lower reach. Despite this general tendency the numbers fluctuate due to difficulties in classifying each taxon in a specific functional feeding group. In general the numbers do not agree with the RCC.

Vegetation pattern in mini floodplains

The natural lowland stream is accompanied by different types of bank vegetation, each related to the local conditions in hydrology, soil composition, groundwater tables and chemical composition of the groundwater. Furthermore, the vegetation types were mosaic-like scattered over the mini floodplain. Most of the woody vegetation is nowadays removed. Still a general pattern from source area to lower course can be seen (Fig. 5). In this figure only short vegetations are indicated for a stream in the province of Drenthe (Drentse Aa). In the past, most source areas and streams were overgrown by woodland.

The source areas are predominantly influenced by directly percolating rainwater. The short vegetation types are shown in Fig. 5a. The natural woodland vegetation would belong to the Sphagneto-Alnetum (under acid, nutrient- poor conditions) or the Carici elongatae-Alnetum (under weakly acid, nutrient-rich conditions) communities. These associations are not restricted to the sources, but may follow upper and middle courses whith stagnating groundwater. Typical associations for sources with moving groundwater are the Chrysosplenio oppositifolii-Alnetum (lime-poor, high watertable) and the Carici remotae-Fraxinetum (lime-rich, generally lower watertable).

In the upper course the vegetation is influenced by superficial ground-water seepage. Figure 5b shows some of the pertinent short vegetation types. The original vegetation type would be woodland belonging to the Alnion glutinosae communities, as indicated above, or as a succession on soils with moving groundwater, a Pruno-Fraxinetum or Cirsio-Alnetum.

The middle course is strongly influenced by deep groundwater seepage and sometimes inundation of the mini floodplain occurs. Figure 5c shows some of the related grassy vegetation types. The natural vegetation type would be woodland belonging to the Pruno-Fraxinetum or Cirsio-Alnetum and in the floodplain with stagnating water the Carici elongatae-Alnetum.

The lower course receives less groundwater seepage and sometimes only has a drainage function. The mini floodplain is often inundated whereby rich clay and silt are deposited. Figure 5d shows some of the related short vegetation types. The natural woodland vegetation would belong to the Fraxino-Ulmetum and in the floodplains again the Carici elongatae-Alnetum.

The natural chemical composition of precipitation and groundwater

The original chemical composition of the streamwater is determined by those of the precipitation and the groundwater, and by natural processes in the total discharge area. Meinardi (1988) lists the rainwater composition in different periods and different places in The Netherlands (Table 3).

Sulphate and nitrogen have obviously increased this century, the other components not. The chemical composition of unpolluted groundwater in the Veluwe region (central part of the country; Pleistocene district) is similar to that of rainwater, multiplied by an evapotranspiration factor (Meinardi, 1988). This latter factor depends on vegetation cover and ranges between 2 and 5.4. The chemical composition of deep groundwater beneath the entire Veluwe region and in upper parts of the aquifer in heath and forest areas resembles that of 'concentrated' rainwater if not polluted, because the water accumulates calcium and HCO_3 ions only from the poor sandy soils on its way down (Vasak, Krajenbrink & Appelo, 1981). These authors cite from Meinardi (1974) the numbers for concentrations in the annual precipitation excess (Table 4), in which the evapotranspiration factor has been used. They comment that sulphate concentrations may have changed recently, which also holds for NO_3 and NH_4. These numbers are similar to those of the nitrogenous components as given by Steenvoorden (1976) for the upper meter of the groundwater of some nature areas (Table 5).

The phosphate concentrations in rainwater and groundwater must have been negligible up to the Sixties. This statement may, however, be biased by the rather high detection limit in the past. The general conception was that phosphates were adsorbed to organic particles in the upper layers of the bottom and, therefore, never reach the groundwater. Steenvoorden (1976) measured phosphate concentrations on the same places as nitrogen (Table 5).

The precipitation that penetrated into the sandy soil reaches the stream after a shorter (weeks) or longer (months to years) period and was changed

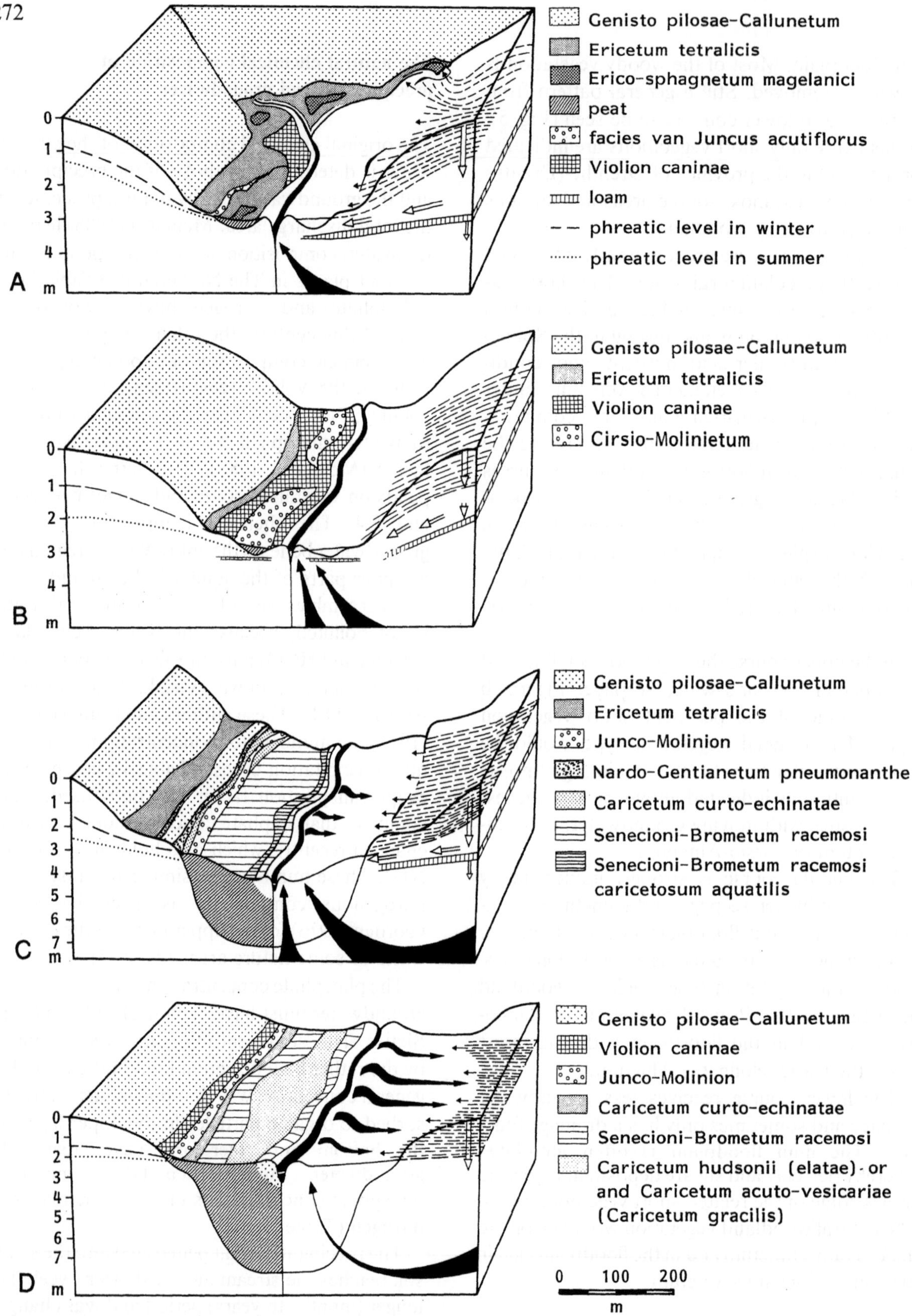

Fig. 5. The pattern of short vegetations in the floodplain of a lowland stream from source (A) to lower course (D) around 1900 (Grootjans, 1980). Further explanation in the text.

Table 3. Composition of precipitation in The Netherlands (Meinardi, 1988). All locations are situated in the central part of the Netherlands (province of Gelderland) with the exception of Hilversum (south-east of the province of Noord-Holland) and De Bilt (province of Utrecht).

Location Component mg l^{-1}	Veluwe 1900	Veluwe 1900–1950	Veluwe 1932–1937	Hilversum 1932–1937	De Bilt 1957–1962	Epe 1978–1982	Deelen 1978–1983
Cl^-	3.0	3.0	3.2	4.0	2.8	3.0	3.0
SO_4^{2-}	3.0	4.0	4.4	4.0	7.4	6.0	6.5
N_{tot}	0.3	0.5	–	0.5	1.1	2.0	2.6
Ca^{2+}	1.0	1.0	1.7	1.7	1.0	0.6	0.5
Mg^{2+}	0.5	0.5	0.8	1.0	0.4	0.3	0.2
Na^+	1.8	1.8	1.7	1.7	2.1	1.6	1.7
K^+	0.15	0.2	0.2	0.2	0.2	0.2	0.16

Table 4. Composition of precipitation multiplied by the evapotranspiration factor (Vasak *et al.*, 1981).

Component	Cl^-	SO_4^{2-}	HCO_3^-	NO_3^-	Na^+	Mg^{2+}	Ca^{2+}	NH_4^+
(mg l^{-1})	9	10	6	1	7	2.5	3.5	1

Table 5. Nitrogenous components and phosphates in the upper meter of groundwater in some nature reserves (Steenvoorden, 1976).

Soil type Component (mg l^{-1})	Poor sand with conifers	Poor sand with deciduous trees
NO_3^-	<0.2	<0.2
NO_2^-	<0.01	<0.01
NH_4^+	0.5	0.4
Org. N	0.5	0.5
Total N	1.0	0.9
Ortho-P	0.02	0.04
Total-P	0.06	0.08

in composition according to the retention time and the soil composition. The main component added was calcium and differences in calcium-content, therefore, constituted the most important differences between streams. These conditions were probably still present in the beginning of this century. Minerals leached to the groundwater due to mineralization of organic material in the discharge area outside the floodplain. Nitrates sometimes have reached the stream or its tributaries, but phosphates were retained in the upper soil layers and never reached the stream. In the mini floodplain more direct contact with the stream existed and especially organic material from trees and shrubs falling in the stream enriched the water with minerals. The ammonium from allochthonous organic material was converted to nitrate. Phosphate was the limiting factor and the streams were calcium-poor or calcium-rich depending on the subsoil. Streams that drained nutrient-poor raised bogs were acid. In general, the trophic status of streams gradually increased downstream from nutrient-poor sources to moderately nutrient-rich middle and lower courses.

Few records of the chemical composition of pristine streams exist. One of the best studied streams in the Veluwe area, the Hierden stream, was analysed twice in the early Sixties near the spring and the mouth (20 km distance; Higler, 1964) (Table 6).

The reliability of some of the measurements is perhaps not optimal, but the overall impression of the 1960 situation is that of an unpolluted stream with a remarkable high Cl-content and a very low Ca-content. Analyses of 22 springs in the southern part of the Veluwe area in 1954–1956 indicated wide ranges, however, for several components (Maas, 1959) (Table 7). The high values of *e.g.* NO_3 were explained by enrichment during

Table 6. Records of the chemical composition of the Hierden stream in the early Sixties (Higler, 1964). All measurements in mg l^{-1}, except pH.

Lowland stream part	Spring area		Mouth	
	1960	1964	1960	1964
Component				
pH	5.4	6.1	6.0	7.1
Cl^-	20	37	22	25
Ca^{2+}	0.1	1	0.7	1.9
Alkalinity	0.1	0.4	0.6	1.9
Mg^{2+}	0	0.4	0.2	1.1
NO_3^-	Trace	8	2	12
PO_4^{3-}	0	0.1	Trace	0.4

Table 7. Analysis of some components in Veluwe springs (central Pleistocene part of The Netherlands) (Maas, 1959).

Component		
pH	4–7.5	
E.C.	88–388	$\mu S\ s^{-1}$
Hardness	1.2–8.2	D °
Ca^{2+}	4.6–44	mg l^{-1}
Mg^{2+}	1.2–8.8	mg l^{-1}
NO_3^-	0–67.6	mg l^{-1}
NO_2^-	0–0.4	mg l^{-1}

the passage of groundwater through nitrogen-rich soils. Meinardi (1988) calculated the 'natural' composition of the water of springs from the eastern and southern Veluwe areas by using the composition of the precipitation (first column of Table 3), the retention time in the soil, the vegetation cover and the evapotranspiration factor. Recent samples taken from these springs were compared with the calculations and in this way he could indicate changes in the composition, which were apparent for pH and NO_3.

In the middle course the floods brought silt to the floodplain, which mixed with sand and formed loamy sands, and locally mesotrophic peat was formed (Fig. 4).

Nutrient cycles in a lowland stream

Cycling of nutrients hardly occurs in the stream and the mini floodplain. The groundwater in the catchment area moves into the direction of the stream and sooner or later all water and its solutes leave the catchment area. This also holds for the floodplain, but here nutrients which may be transported from the stream by flooding, are retained for one growing season, or longer where trees are concerned. The uptake, release, new uptake, new release etc. of nutrients by organisms from upstream to downstream is called nutrient spiralling (Webster, 1975). As an example we consider the Hierden stream in the northern Veluwe.

Only a small part of the total 20 km of stream length is exposed to direct sunlight and on these places aquatic vegetation develops, consisting of *Callitriche sp., Glyceria fluitans, Myriophyllum sp., Myosotis sp.* and *Ranunculus aquatilis.* Dense aquatic vegetation is only found over a total length of about 1 km. The plants are cut in May, float downstream and leave the system. Large stretches of the stream are under a tree canopy, either from alder trees on the banks or beech trees in the wood. Large amounts of leaves from the bordering trees fall into the stream. They are partially used as allochthonous energy source, but largely float downstream and form banks in the Veluwemeer. Beech leaves especially, leave the system in this way. Alder leaves are being utilized more easily than beech leaves by microbes and invertebrates and therefore constitute the majority of allochthonous energy.

The flows of phosphorus and nitrogen through the system are illustrated by a flow diagram in Fig. 6. The initial nutrient source of a pristine stream is groundwater which is supplemented by dissolved nutrients and allochthonous material, in this case predominantly leaves from alder and beech. The nutrients in the groundwater have been estimated to be 1,100 kg $P_{tot}\ yr^{-1}$ and 19,560 kg $N_{tot}\ yr^{-1}$. These figures are based on concentrations of 0.07 mg $P_{tot}\ l^{-1}$ and 1.25 mg $N_{tot}\ l^{-1}$ as an estimation of the natural situation.* The esti-

* The figure of 0.07 is the mean of Steenvoorden's (1976) figures. Appelo (1988) gives the figure of 1.25 for total N as a value around 1940 in the Hierden stream area.

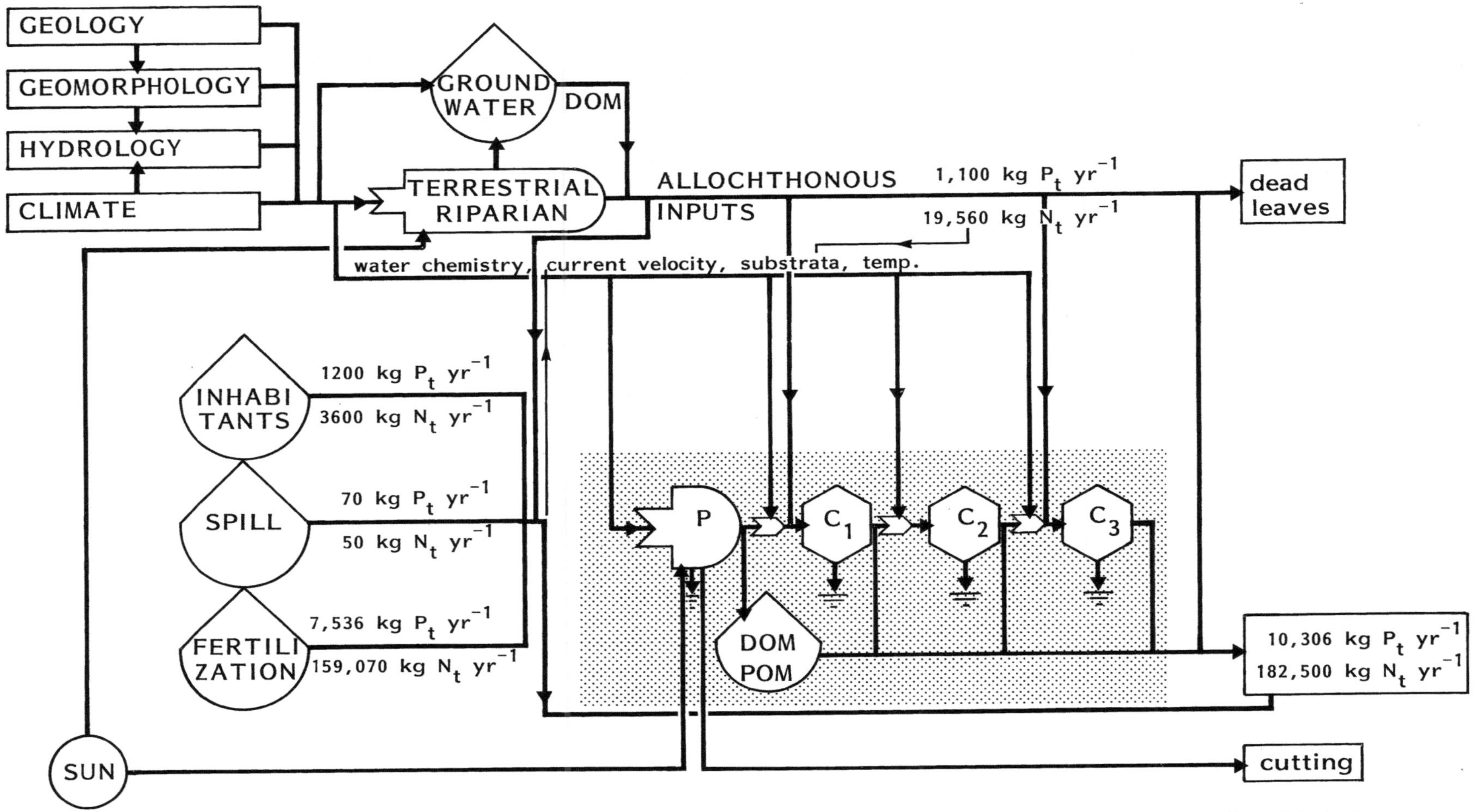

Fig. 6. Flow diagram of nutrients in the Hierden stream wetland. Adaptation of a diagram in Benke *et al.* (1988) with data from Water Pollution Control Authority Veluwe from 1984–1987. DOM = Dissolved Organic Material; POM = Particulate Organic Material; P = Producers; C = Consumers.

mated quantities are certainly not limiting for the growth of aquatic vegetation, observed in the early Sixties in the Hierden stream. The latter conclusion is based on work of Kouwe (1982) on aquatic vegetation in similar streams. Here, less than 1% of the supplied quantities of 700 to 3000 kg P_{tot} and 16,800 to 43,000 kg N_{tot} (which are far lower than in the Hierden situation) was taken up by the aquatic plants in the summer half-year; P-uptake ranged from 4 to maximally 21 kg, which was probably partly absorbed from the bottom. Similar data hold for a stream in the province of Gelderland (Oosterloo, pers. comm.). Therefore, light is probably the limiting factor for the growth of aquatic vegetation in lowland streams partly shaded by trees.

As a matter of fact, the few tens of kilograms P leave the stream after weed cutting, and the biocommunity depends further completely on dead leaves, detritus and epilithic and epipsammic algae. Furthermore, some surface-dwelling insects prey on small terrestrial invertebrates. In The Netherlands no study has been undertaken to calculate the biomass or bio-energetics of the stream organisms. Studies from other countries show the importance of macro-invertebrates for leaf-processing (i.e. Petersen & Cummins, 1974), but also their minute role in a general nutrient budget of the whole stream. Only 10 and 9.5%, resp. of the total annual input of allochthonous and autochthonous organic material, was used by the stream communities of two German streams, calculated on energy basis (kcal; Statzner, 1979). Allochthonous material only represented 25% of the total input. The data in Fig. 6 are obviously not biased severely by only quantifying the flows and not measuring the compartments of P, C and DOM/POM, because most nutrients are accounted for. Nutrients are in no way limiting primary production and the total amounts of input equal the total amounts of output. Budget studies in lowland streams may, therefore, not be very useful, especially because the measurements of the numerous species and stages of invertebrates are extremely time-consuming (Benke *et al.*, 1988).

Threats to the stream wetland

The threats posed to stream wetlands may be divided into two categories: those related to water management and those related to changes in water chemistry (Table 8).

Many threats are obvious and well-known. A few examples will be given of changes in water chemistry as a result of agricultural activities.

In the Hierden stream and its discharge area agricultural use has been intensified during the last decades. Moderately nutrient-rich hay-fields in the valley and pastures higher up in the discharge area occurred still in the late Fifties. Water discharge was accelerated by drainage and more overflows occurred. Since the early Sixties many fattening farms for calves were built and problems with superfluous liquid manure were solved by over-fertilization of pastures and drainage into the stream directly or after storage indirectly via trenches. Corn fields were laid out and more liquid manure was applied. Direct drainage caused the extinction of the whole population of brown trout and probably more animals in the stream. Over-fertilization caused pollution of the groundwater, as became clear by the continuous increase in N_{tot} in the stream (Fig. 7) and by the high NO_3 content measured in groundwater (asterisk in Fig. 7). Recent measurements of the water authority, including permanent registration over the period 1984–1987 have been used for the construction of the flow diagram in Fig. 6. It is clear that the influence of run-off and polluted groundwater form the lion's share of the total P and N discharge. The diffuse nature of the drainage makes handling of this problem very difficult. Appelo (1988) calculated, for instance, that the main aquifer from which the Hierden stream draws its water can only be cleaned in a period of 8 to 12 years, provided no new nutrients reach the groundwater.

A practical problem inherent to the data used for Fig. 6 is the calculation of total discharges with occasional analyses. It became evident during a period of permanent registration in the Hierden stream that generally the calculations are too low, because extra nutrient quantities are added

Table 8. Threats to lowland stream wetlands by water management, changes in water quality and mechanical disturbances.

Part of the stream	Springs and upper courses	Middle courses Streams	Lower courses Small rivers
Management measure			
Lowering of groundwater table	Drying temporarily or permanently: spring vegetation disappears	Drying temporarily current velocity decreases	
External water supply	Generally no effects: but rise of water table is possible with higher discharge	Eutrophication; possible toxicity	Eutrophication; possible toxicity
Internal water supply		Change in ion-composition; eutrophication; saprobity; toxicity	Eutrophication; saprobity; toxicity
Changes in morphology of banks	Removal of canopy > spring vegetation disappears; drainage; profile larger	Profile larger; no meanders; drainage weirs	Large-scale canalization; weirs; enforcement of banks
Acid precipitation	Spring flora affected by acidification and higher amounts of nitrogen	Influence on bank- and marsh vegetation by acidification, higher NO_3 content or ammonium-effects	
Fertilization of land	Eutrophication of groundwater disappearance of typical spring flora; eutrophication of spring	Eutrophication of bank- and marsh vegetation in floodplain; eutrophication of stream water > changes in aquatic vegetation and consequently of fauna	
Pollution	Temporal or permanent disappearance of typical spring- and flood plain vegetation; changes in species composition of aquatic vegetation; disappearance of fauna elements by low oxygen contents or toxicity		
Mechanical disturbance	Spring and surrounding flora is vulnerable For tread and rubbish-dump	The same holds for floodplain and bank vegetation, especially oligotrophic marsh vegetation	

to the stream during floods. Phosphates show a sharp rise in concentration in the beginning of a heavy rain period and a drop to the original values after some time. Total nitrogen, however, shows an initial sharp drop and only after a serious flood a rise above the original level. This is caused by the dilution of especially nitrate, which is easily drained under normal circumstances. W.P.C.M. Bots (pers. comm.) found similar results for the Roelinksbeek (in the eastern part of The Netherlands). 73% of the total phosphate discharge and 63% of the total Kjeldahl-nitrogen discharge was drained during 12% of the investigation time of one year at the days with larger discharge. Thus underestimation in case of monthly registration is obvious.

Changes in the vegetation of source and floodplain areas, aquatic vegetation and fauna in the Hierden stream in investigations ranging from the early Sixties until recently can be described as disappearance of oligotrophic vegetation and organisms sensitive to oxygen deficits or toxic sub-

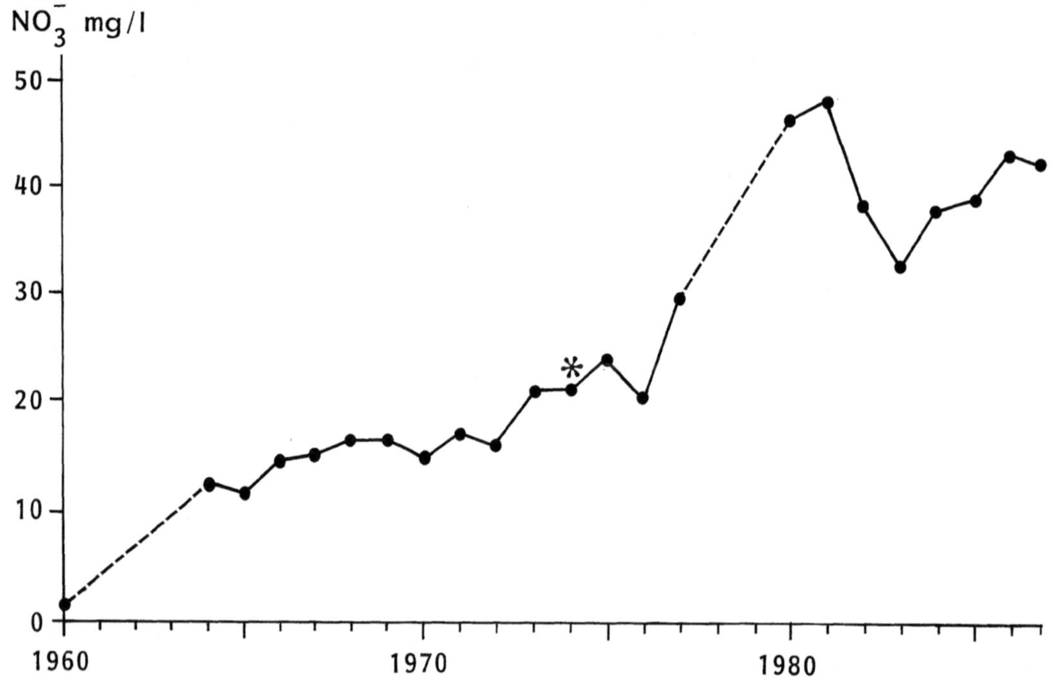

Fig. 7. Mean N_{tot} content in the Hierden stream and in a groundwater sample (Higler & Repko, 1981; with additional data from Water Pollution Control Authority Veluwe for 1980–1987). * = groundwater.

stances such as ammonia. Flora and fauna are generally impoverished and those sensible species that still are present occur in low numbers.

Perspectives

Basically there are two views in perspective. The pessimistic view is based upon the realization that both hydrological conditions and nutrient cycles in most present watersheds lead to a further deterioration of these ecosystems. The threats have been outlined above. On the other hand, the degradation of wetlands is a recognized problem nowadays and this inspires the optimistic view. There are initiatives both from governmental and private organizations to restore (parts of) streams. In many cases simple morphological adaptations have been planned. Canalized stretches will meander again, trees are planted alongside streams and salmonid fishes are being restocked.

It will be obvious in the light of the foregoing that there is a close relationship between hydrology of the watershed, land use and quality of the stream ecosystem. The morphology of a stream results from geomorphological structures, like terrain slope and soil composition and from the water supply. This means that restoration of stream valleys is not only a matter of creating meanders in a haphazard way, but it asks for thorough study of the actual possibilities. The restoration of a former shape may cause difficulties because of changed conditions in the watershed (*e.g.* drainage of agricultural land). The best way of restoration is offering free possibilities to the hydraulic forces of a stream. This process can be speeded up by careful planting of trees (the right species and ecotypes). Verweij (1984) discovered that only a certain ecotype of alder tree (*Alnus glutinosa*) must be used alongside streams and that most alder trees in Dutch marshes are not fit for this purpose.

Qualitative aspects are more difficult to deal with. Because of intensive land use in the drainage area the groundwater quality may be deteriorated in such a way that recovery of the stream water quality lasts many years. As a result, the ecosystem in the stream and its flood plain may develop more disappointingly than expected. This ought to be considered when restocking of fish is planned.

In conclusion it must be stressed that positive developments as mentioned better be accompanied by multidisciplinary research or as stated by Wetmore *et al.* (1990): "Where streams have been straightened and smoothly channelized, for example, benthic productivity is reduced and many fish habitats virtually eliminated. The design of stream rehabilitation works, with diverse habitats for fish and invertebrates, is a future challenge for all stream researchers."

References

Appelo, C.A.J., 1988. Beïnvloeding van de waterkwaliteit in het Hierdensche beek gebied (N.W. Veluwe). Inst. Earth Sciences, Free University Amsterdam, 100 pp.

Benke, A.C., C.A.S. Hall, C.P. Hawkins, R.H. Lowe-McConnell, J.A. Stanford, K. Suberkropp & J.V. Ward, 1988. Bioenergetic considerations in the analysis of stream ecosystems. J. N. Am. Benthol. Soc. 7: 480–502.

Cummins, K.W. & M.J. Klug, 1979. Feeding ecology of stream invertebrates. Ann. Rev. Ecol. System. 10: 147–172.

Grootjans, A.P., 1980. Distribution of plant communities along rivulets in relation to hydrology and management. In O. Wilmanns & R. Tüxen (eds), Epharmonie (Rinteln, 9.4.–11.4. 1979). Berichte der Internationalen Symposien der Internationalen Vereinigung für Vegetationskunde, Cramer, Vaduz: 143–169.

Higler, L.W.G., 1964. De Hierdense beek in gevaar. De Levende Natuur 67: 279–282.

Higler, L.W.G., 1980. Hydrologische, fysische en chemische gegevens van de Hierdense beek. Research Institute for Nature Management, Leersum. Report 80/3, 19 pp.

Higler, L.W.G. & A.W.M. Mol, 1984. Ecological types of running water based on stream hydraulics in The Netherlands. Hydrobiol. Bull. 18: 51–57.

Higler, L.W.G. & F.F. Repko, 1981. The effects of pollution in the drainage area of a Dutch lowland stream on fish and macro-invertebrates. Verh. Internat. Verein. Limnol. 21: 1077–1082.

Kouwe, F.A., 1982. Fosfaat en eutrofiëring in een laaglandbeek; een chemisch en hydrobiologisch onderzoek naar het effekt van defosfatering op fosfaatgehalte en waterplanten in de Beerze. Gemeenschappelijke Technologische Dienst Oost-Brabant, 86 pp.

Maas, F.M., 1959. Bronnen, bronbeken en bronbossen van Nederland, in het bijzonder die van de Veluwezoom. Veenman, Wageningen, 166 pp.

Meerman, M., 1975. De Geul, zijrivier van de Maas. Bijdrage tot de hydrografie van een uniek riviertje. Meerman, Kerkrade, 192 pp.

Meinardi, C.R., 1974. De chemische samenstelling van het grondwater van de Veluwe. RID Med. 74-4. Governm. Inst. for Water Supply, Voorburg, 48 pp.

Meinardi, C.R., 1988. Veranderingen in de samenstelling van het water in de Veluwse sprengen. H_2O 21: 52–57.

Petersen, R.C. & K.W. Cummins, 1974. Leaf processing in a woodland stream. Freshwat. Biol. 4: 343–368.

Redeke, H.C. 1948. Hydrobiologie van Nederland. De Boer, Amsterdam, 580 pp.

Statzner, B., 1979. Der Obere und Untere Schierenseebach (Schleswig-Holstein). Strukturen und Funktionen in zwei norddeutschen See-Ausfluss-Systemen, unter besonderer Berücksichtigung der Makroinvertebraten. Ph.D. Thesis, Kiel, 551 pp.

Steenvoorden, J.H.A.M., 1976. Nitrogen, phosphate and biocides in groundwater as influenced by soil factors and agriculture. CHO-TNO, Proceedings and Informations Nr. 21: 52–69.

Stiboka, 1983. Bodemkaart van Nederland, pag. 28 west (Almelo). Netherlands Soil Survey Institute (Stiboka; now part of The Winand Staring Centre), Wageningen.

Tolkamp, H.H., 1980. Organism-substrate relationships in lowland streams. Ph.D. Thesis, Agricultural University, Wageningen, 211 pp.

Vannote, R.L., G.W. Minshall, K.W. Cummins, J.R. Sedell & C.E. Cushing, 1980. The river continuum concept. Can. J. Fish. Aquat. Sci. 37: 130–137.

Vasak, L., G.J.W. Krajenbrink & C.A.J. Appelo, 1981. The spatial distribution of polluted groundwater from rural centres in a recharge area in The Netherlands – the Veluwe. The Science of the Total Environment 21: 105–112.

Verdonschot, P.F.M., 1990. Ecological characterization of surface waters in the province of Overijssel (The Netherlands). Ph.D. Thesis, Agricultural University, Wageningen, 255 pp.

Verweij, J.A., 1984. De zwarte els ..... een bosboom? Bosbouwvoorlichting 23: 84–87.

Webster, J.R., 1975. Analysis of potassium and calcium dynamics in stream ecosystems on three southern Apalachian watersheds of contrasting vegetation. Ph.D. Thesis, Univ. of Georgia, Athens. In A.C. Benke, C.A.S. Hall, C.P. Hawkins, R.H. Lowe-McConnell, J.A. Stanford, K. Suberkropp & J.V. Ward, 1988. Bioenergetic considerations in the analysis of stream ecosystems. J. N. Am. Benthol. Soc. 7: 480–502.

Wetmore, S.H., R.J. Mackay & R.W. Newbury, 1990. Characterization of the hydraulic habitat of Brachycentrus occidentalis, a filterfeeding caddisfly. J. N. Am. Benthol. Soc. 9: 157–169.

Hydrobiologia **265**: 281–304, 1993.
E. P. H. Best & J. P. Bakker (eds), Netherlands-Wetlands.

Wet dune slacks: decline and new opportunities

H. W. J. van Dijk[1] & A. P. Grootjans[2]
[1]*Provincial Board of Overijssel, Dept. of Water Management, P.O. Box 10078, 8000 GB Zwolle, The Netherlands;* [2]*State University Groningen, Department of Plant Ecology, P.O. 14, 9750 AA Haren, The Netherlands*

Key words: dune slacks, eutrophication, hydrology, infiltration, nitrate, phosphate, public water supply, vegetation, water table

Abstract

For a number of infiltrated coastal dune areas it is discussed to what extent artificial infiltration for the public water supply affects the quality of soil, groundwater and vegetation around pools and ponds, and what its effect is on the vegetation. Further, the results of investigations into the quality of vegetation, soil and water of a number of non-infiltrated, less affected dune areas are presented. The emphasis is on changes in groundwater flow pattern and on changes in the chemical composition of groundwater on the vegetation of wet dune slacks. Finally, recommendations for the management of wet dune slacks are presented.

It can be concluded that the introduction of nutrients through infiltration causes an abundance of nitrophilous herbaceous vegetation along the banks of all infiltration ponds and most dune pools. Of the three investigated macro-nutrients, nitrate, potassium and phosphate, the latter shows the most significant correlation with the composition, cover and biomass of the vegetation.

The moist biotopes of non-infiltrated dunes have largely disappeared because of desiccation, mainly as a consequence of water withdrawal, afforestation and coastal erosion. Relatively unaffected dune slacks can be found in the dunes on the Dutch Wadden Sea islands and a small number of dune areas on the mainland. In most areas, however, a serious decline in many rare species has been observed during the past twenty years because of eutrophic and acid precipitation, often in combination with disturbances of the groundwater regime.

Introduction

The coastal dunes are among the relatively least damaged and unaffected ecosystems in The Netherlands (Westhoff, 1989). The dry ecosystems in particular are unspoilt, although they have undergone changes due to such influences as dune reclamation, acidification and eutrophication through precipitation, and locally intensive recreation (Van Dorp *et al.*, 1985; Van der Meulen, 1982; Van der Zande, 1989). As for the present vegetation of wet dune slacks, only the dunes of the Wadden Sea Islands, the Zwanenwater dunes near Callantsoog and Voorne's dune still bear comparison with the natural abundance of the vegetation as found by Holkema (1870), Van Eeden (1886) and Vuyck (1898). Ninety years ago Vuyck described the first signs of deterioration of the wet dune slacks. Nowadays the character of almost all dune slacks has changed dramatically, especially on the mainland.

The origin and natural development of the dune

Fig. 1. Coastal dunes (hatched) in The Netherlands and location of areas mentioned in this paper. Wadden islands: S = Schiermonnikoog; A = Ameland; T = Terschelling; V = Vlieland; Tx = Texel. Mainland dunes: Z = Zwanenwater; Sl = Schoorl dunes; N = North-Holland Dune Reserve; A = Amsterdam Municipal Waterworks Dunes (Luchterdunes); B = Berkheide dunes; M = Meijendel dunes; W = Westland dunes; Ve = Voorne's dunes; OM = Oost dunes and Midden dunes; Sn = Schouwen dunes; Wa = Walcheren dunes.

area are characterized by natural and dynamic processes, more so than most other landscapes in the Netherlands. Whereas elsewhere, both before and during the Middle Ages, there was embankment, reclamation, development, canalization, etc., in the dune areas wind, water, sand, and vegetation were untampered with until a few centuries ago (Klijn, 1990). Human interference was then essentially limited to the gathering of wood, hunting and pasturing. The first major changes concerned the safeguarding of the inner dune belt from sand drift through afforestation and prohibition of pasturing. This introduced an era of landscape rigidity. Many parts of the inner dunes were afforested and/or levelled. During the past century, many of the middle dunes were planted with vast pine forests to be used in coal mines. Halfway through the 19th century in many dune areas rather extensive agriculture was attempted locally, as well as management for drinking water supply. The farming had to end soon but the use of dune areas for water catchment has been successful until now. Nowadays, the major functions of the dune areas are sea defence, water catchment, outdoor recreation and nature conservation. In most cases the above-mentioned sequence was the order of priority with the dune planning policy.

Apart from origin and use, the wet dune systems are different from other Dutch wetlands in other ways as well. For example, the substrate consists of fine sand, which is chemically rather inert and highly permeable. This feature allows for relatively high groundwater velocities. Hydrologically, the dunes do not just differ from other systems by the good permeability of the dune sand: the water in the dunes is quite isolated from surrounding water systems (the sea, polder water in the hinterlands, possible brackish water systems). This isolation can be explained as follows: over many centuries a large freshwater lens has developed in the entire dune area. This freshwater supply has a dynamic equilibrium with the underlying saline water and can be compared to an iceberg floating in the sea (Bakker, 1990). The development of the freshwater lens is closely related to the precipitation residue (about 300 mm year^{-1}) and the elevation of the dunes, up to tens of metres above the flat polder landscape behind it. The (potential) volume of freshwater stock is determined by the extent of the dune areas and the average height of the dunes. The shape of the freshwater lens is determined by the shape of the dunes as well as by the underlying, less permeable (semipervious) layers. The flow of precipitation residue occurs both towards the sea and towards the hinterland. However, despite the hydrological isolation and the rather inert dune sand the distinct origins of the groundwater in the Dutch dune areas can cause extremely complex hydrochemical situations in many places (Stuyfzand, 1990).

In coastal dunes of average size the freshwater lens reaches up to six or seven metres above the average sea level (in the central dune area) and it may reach down to a hundred metres below sea level. Theoretically, based on specific gravity and dynamic hydrological equilibrium only, the maximum distance from the depth of the freshwater lens to sea level would be forty times the distance measured from the top to sea level. Because of impervious peat and clay layers in the subsoil, in practice this factor is less, about fifteen to twenty-five (Bakker, 1981, 1990). From the top of the freshwater lens, i.e. the natural infiltration area, groundwater movements occur towards the sides. Where the freshwater lens intersects with low-lying areas, wet dune slacks are present. These slacks may contain marshes and in some cases also small pools.

Wet dune slacks originate from very dynamic processes in which the impact of wind and water on the sand movement is predominant. Two types of dune slacks can be distinguished based on their origin:

- Primary dune slacks; originated by separation from beach plains. In augmented dune areas this slack type is mainly restricted to the Wadden Islands and may develop in the near future at the southwest coast of The Netherlands.
- Secondary dune slacks; originated by blowing-out in the older dune areas.

The life expectancy of many of the *primary slacks* is not high: situated in the dynamic outer dune environment they are subject to strong blowing in of sand, which causes them to dry out. Apart from that, local abrasion of the dune area may also cause desiccation.

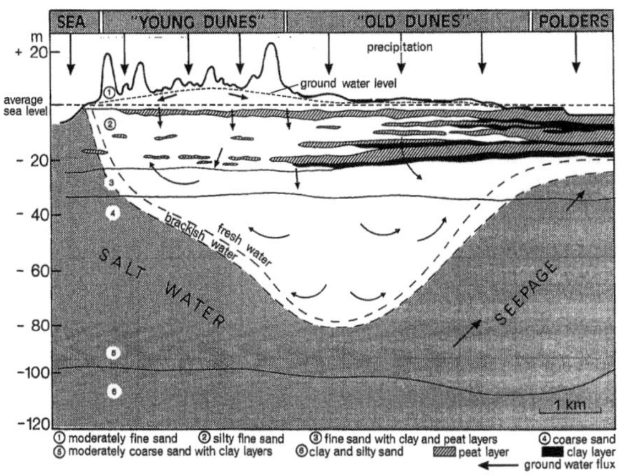

Fig. 2. Cross-section of a natural dune area (Berkheide dunes before artificial infiltration) showing the original water table and the freshwater front.

Essentially, recently formed primary slacks contain brackish water. High chloride concentrations may remain long after separation due to the salt spray as the location is close to the sea (Van Dijck & Meltzer, 1981). This sea-spray is greater when the surf is closer to the foot of the dune. Apart from an increased salt concentration, pools near the sea tend to have higher concentrations of nutrients, because the wind carries much organic material from the shore, which mineralizes rapidly in an alkaline environment. In addition sea-bird colonies near these pools may provide an extra input of nutrients. In the vicinity of the beach, the water and the soil contains so much calcium that the system will be sufficiently buffered against acidification for several decades.

If local circumstances allow complete blowing out of *secondary slacks*, the erosion will carry on till the capillary zone of the (summer) water table has been reached. In unaffected dune areas the local climate introduces seasonal fluctuations of the water table as a result of the alternation of high evaporation during summer and little during winter, which results in negative effective precipitation during summer and a large precipitation surplus in winter. The water level is lowest in August or September and rises in winter to a maximum in January or February. The annual fluctuation of the water table is 0.3–0.7 (exceptionally even 1.1) m; the pattern of fluctuation is mainly determined by the extent of open water and by the size of the surrounding water system. The result of the seasonal fluctuation pattern is that in summer there are vast, recently-drained sites, whereas in winter, pools with a depth of half a metre appear. Should there be accretion of the coastal area then locally the water level can increase strongly and these pools can achieve a more permanent lake-like character.

Secondary slacks may contain precipitation- and groundwater-type pools. The first type is characterized by a relatively low calcium concentration, and is generally found in the central dune areas, near the top of the freshwater lens, where infiltration predominates. Precipitation water-type pools are sometimes found near the borders of the dune area. In this case superficial impervious layers occur, which cause an apparent water table and stop the groundwater from entering the pool. Most water of the second (groundwater-) type has been in the soil for a long time (Van Dijk & Meltzer, 1981). In general precipitation water-type pools are characterized by lower nutrient concentrations than the groundwater pools and they are usually more sensitive to acidification and desiccation.

Originally, the moist to wet dune slacks in the Dutch coastal dunes took up one-third of the total surface area (Londo, 1971). In these slacks the surface water used to be oligotrophic to mesotrophic because of low nutrient loads from the substrate, the sparseness of forest and brushwood, and hydrological isolation (Leentvaar, 1963, 1967, 1981). In its natural state the phreatic groundwater has low nutrient concentrations as well (Bakker, 1981,1990). Depending on the content of calcium in the dune sand the grounwater may contain much or little calcium. The dune district south of Bergen is characterized by high lime contents of the dune sand, and therefore by higher concentrations of calcium in groundwater and surface water than the dunes in the Wadden Sea area, which are either moderately rich in lime or even completely leached and acid in the top layers.

In the moist to wet dune slacks, the original vegetation is very diverse. Many environmental gradients are reflected in it (Vuyck, 1898; Bijhouwer, 1926; Londo, 1971; Van Zadelhoff, 1981; Westhoff, 1989). The constancy of the annual water table fluctuation, the relative shortage of nutrients in the soil and in the groundwater, the gradients of wet and dry in the slacks, and the scant development of brushwood and forest under the influence of seawind and grazing, contributed to this diversity of the 'Young Dunes'.

The next sections provide a description of the changes which have occurred in the Dutch dune area during the past century as well as the ecological processes involved. The water cycle and the nutrient cycle in the past and at present are treated.

Water cycle: natural and affected geohydrology

Original situation (until ca. 1850)

After the formation of the present Young Dunes during the Middle Ages, several centuries passed before the freshwater lens reached its maximum size. Roughly speaking, the width of the dune area is directly proportional to the period required to build up the freshwater table. This process takes approximately 50 years for a dune area with a width of 0.5 km, approximately 200 years for a width of 2 km, and 300 to 400 years for the maximum width of 4 km (Bakker, 1981, 1990). From that moment on, a dynamic balance sets in with the total surplus precipitation (300 to 350 mm $year^{-1}$) draining off sideways. About half the surplus precipitation goes to the adjacent polders while the remainder flows seawards because of this sideways drainage. The groundwater flows horizontally at a speed of 0.05 to 0.15 m day^{-1} in the first few meters below the surface.

Water extraction and/or drainage result in a new balance. Accelerated drainage or the extraction of the total surplus precipitation will eventually cause the total salination of the groundwater. The extraction of smaller quantities causes the freshwater lens to shrink and the groundwater level will subside till a new dynamic balance sets in.

Interference with the hydrology

As for the disturbances of the natural water cycle in the dune area in The Netherlands we see a mosaic of effects: great differences between local situations may occur. In some parts, water extraction may have a strong impact, while in others, the planting of woods or the shifting of the coastline may be the dominant factor. We will therefore only indicate the intensity of the effects on the basis of a number of Dutch case studies.

Taking 1850 as their reference year, Bakker *et al.* (1979) trace the disturbances of the natural hydrological system to the following main factors: the afforestation and the formation of brushwood and reduced grazing, the shifting of the coastline, sand excavations in the inner dune edge, the lowering of the polder water level, surface drainage of dune water through excavated in the dune area and, finally, water extraction. For each disturbance factor Bakker (1981, 1990) indicates a mathematical approach to analyse the effect quantitatively. This section is based on the practical values which Bakker (1981) gives for the reduction in groundwater levels and for the disturbance of the groundwater flow in the dune area.

Changes in the vegetation

With the changes in the vegetation a distinction has to be made between the afforestation (mostly pine woods, occasionally larch or oak woods) and the more or less spontaneous formation of brushwood and natural forests.

A good example of the effects of planted pine woods are the Schoorl dunes. Here water levels in the central dune area, near the seabord dunes and in the north have dropped by approximately 1 m, and consequently all formerly humid-to-wet dune slacks have completely dried up. Bakker (1981) regards the planted pine woods as the main cause of the drop in the groundwater level in the central area. Near the seaboard dunes, however, coastal abrasion has also played an important role. Pine woods were planted here from the end of the 19th century onwards. As it can take approximately a century for a new hydrological balance to set in, this means that the total effects of desiccation due to the increase of the evaporation of these woods have only been realized for about 50 percent. The sharpest reductions in the water level are recorded at the actual sites of the woods. Some examples of hydrological data on the effects of planted woods are presented here:

- In the Bergen dunes the planted pine woods are causing a drop in water levels ranging from 0.1–0.2 m (inner dune edge) to 0.2–0.5 m (central, seaboard) (Bakker, 1981).
- For the dunes south of Zandvoort Bakker (1981) estimates that the groundwater levels have fallen by 0.1–0.2 m as a result of the newly planted woods.
- On the Wadden Sea island of Terschelling the pine woods planted between West-aan-Zee and Hoorn have led to a decrease in groundwater levels averaging 0.4–0.7 m in the central dune area (Bakker, 1981).
- The dunes at Schouwen have seen groundwater levels fall by 1 to 2 m at actual (pine) wood sites (Van Dijk, 1984b).

The spontaneous development of a vegetation of shrubs and deciduous trees may in due course result in the complete disappearance of the open character of the dune vegetation, particularly in the (moist) dune slacks. This development is not entirely part of a natural transition towards the final stage of the vegetation succession. Often human influences have stimulated the process. Such influences include the planting of beach grass and sea buckthorn to promote local binding, excavations that stimulated the buckthorn and the depletion of grazing levels due to hunting, and bans against cattle grazing. It is plausible that after the rabbit disease myxomatosis – which was introduced by man and decimated the rabbit population (the most important grazers in the dunes) in the Dutch dune area during the 1950s – the formation of dune scrubs accelerated. In the case of Voorne's dune, another contributory factor is the coastal works that reduced the amount of salt blown in from the sea (Van Dorp *et al.*, 1985). In addition, the planted pine woods (for instance on the Frisian islands) reduce the sea spray and subsequently encourage the growth of shrubs and trees.

Deciduous trees and shrubs have a lower evaporation rate than the pines, which remain green through the winter (coniferous wood: additional annual evaporation of approximately 25 m yr^{-1}). However, due to the more or less spontaneous encroachment of high and dense brushwood, water levels may still drop by 0.1–0.2 m (example: Terschelling to the east of Hoorn (Bakker, 1981).

Coastal abrasion or coastal formation

Local coastal formation and abrasion have caused proportional increases or reductions in groundwater levels. Some well known cases (Bakker, 1981) are presented:

- Since 1850 the coastal abrasion near Bergen aan Zee has varied between 50 to 100 metres. As a result, the water level behind the seaboard dunes has dropped by 0.5–1 m, in the central area by 0.2–0.5 m, and in the inner dune zone by 0.1–0.2 m.
- In West Terschelling coastal formation has occurred since 1859 with an average of 1 km (maximum 1.5 km). As a result, water levels have risen by 3 m at the original foot of the dune by approximately 1 m in the centre of the original area, and by 0.2 to 0.5 m at the inner dune zone.
- On the same island approximately 150 m of the originally 1.9 km wide dune corridor between Midsland and Oosterend has been eroded since 1850. In this context, Bakker (1981) mentions reduced water levels of 0.5–1 m (seaboard dunes), 0.25–0.5 m (central dune area) and 0.1–0.2 m (inner dune edge).

Excavation of the inner dune edge

Excavations have taken place in the inner dune edge to extract sand (e.g. for the sand-lime industry and for roads, as well as for residential and industrial estates). These do not have any noticeable effect on the hydrological system unless the excavation cuts into the freshwater lens. Near the Schoorl dunes, only the Hagergat quarrying has had a small and local effect on the groundwater level (Bakker, 1981). The reduction in groundwater levels caused by larger quarries (e.g. the Oosterduin lake near the dunes south of Zandvoort) has not been quantified so far.

Lowering of the polder level

Another factor mentioned by Bakker (1981) is the lowering of the water levels in polders located directly behind the dunes and the reclamation of land further away from the dunes. Examples of reduced polder water levels near the dune area include:

- The island of Terschelling: a reduction in the polder water level of 0.25–0.5 m caused a reduction in groundwater level in the dune area by a value decreasing almost linearly from 0.25–0.5 m at the inner dune edge to 0 at the seaboard dunes.
- Schoorl dunes: a reduction in the polder water level of 0.2 m (winter level) caused a reduction in groundwater levels in the dune area decreasing almost linearly from 0.1 m at the inner dune edge to 0 at the seaboard dunes.
- The effect of land reclamation further away from the inner dune edge can be seen in the Haarlemmermeer (reclaimed ca. 1850); the resulting reduction in water levels was 0.3 m at maximum in the central part of the Kennemer dunes while the effect at the inner dune edge and seaboard dunes was zero.
- In the dunes of Zandvoort which are further from the Haarlemmermeer and have more impervious and semipervious layers in the subsoil, reclamation caused the water level in the central area to drop by about 0.1 m.

Drainage ditches in the dune slacks

In the first half of the 19th century, extensive systems of drainage ditches were constructed with the intention of cultivating dune slacks in North and South Holland. At the turn of the century, similar systems were excavated in the dune slacks of the present North-Holland Dune Reserve, Terschelling, Vlieland, Texel and Schouwen, sometimes in combination with extensive forestry schemes. In some dune areas these ditches still exist causing a significant drop (in the order of decimetres) in water tables over extensive areas (Bakker, 1981). Bakker (1981) mentions that ditches excavated in West Terschelling around 1910 reduce water levels by an average of 0.3–0.5 m. In most dune areas (e.g. Central and East Terschelling, Luchter dunes), however, the old drainage systems have fallen into disuse and are of negligible importance.

Extraction of dune water and dune infiltration

Since the mid-19th century, the larger cities in the west of The Netherlands (excluding Rotterdam) have collected dune water on behalf of the public water supply. Today, virtually all mainland dunes are being used as water extraction areas. Parts of the dunes in Zeeland and the islands of South Holland and Friesland also serve as water extraction areas. In total, approximately 40 million m^3 of natural dune water is pumped up annually for the production of drinking water (Udo de Haes, 1982). The dunes in The Netherlands (40 000 hectares) have an overall annual precipitation water surplus of approximately 120 million m^3 per year. Water collection, therefore, accounts for a third of the surplus. At the actual extraction sites, a much larger proportion of the surplus precipitation is extracted, but it is subsequently recharged from other parts of the dune area. In many cases the extraction of dune water has caused the slacks to dry up entirely. The highest (winter) water levels dropped more than 1.5 m below the surface of the slacks. A clear example is the reduction in the water level in the dunes north of Noordwijk (1 to 2 m between 1850 and 1978) which is almost completely attributable to water extraction (Bakker, 1981).

Water extraction not only reduces the average water level, but it also affects the seasonal fluctuations of water tables. In most cases these fluctuations have increased because more water is extracted in periods of drought than in the winter season, and because a larger portion of the water remains underground (the so-called open water effect). The increased dynamics impoverish the richness and diversity of the vegetation, partly because desiccation combined with increased and accelerated fluctuations speed up the mineralization process, thereby increasing the availability of nutrients (Van der Laan, 1979; Grootjans, 1985).

The extraction of natural dune water from deep and surface layers is one of the major causes of the strong decline of Dutch slacks over the past century (Bakker *et al.*, 1979). It is worth mentioning that not all of the extracted dune water is used as drinking water; over half a million m^3 of dune water is pumped up annually to sprinkle the golf courses located in the dunes – and most of this water evaporates in the process.

Since the mid-50s, many dune areas in the west of The Netherlands have been infiltrated with river water or reservoir water from adjacent polders. Hydrologists saw the replenishment of the dune water with surface water from elsewhere as a logical and welcome step after the continually increasing dune water extraction had made great inroads into the freshwater lens (Carrière, 1929; Huisman & Oltshoorn, 1983). The fresh-saline water boundary had, for instance, risen tens of metres in many places while the groundwater level had dropped sharply. The depletion of the supply of pure groundwater in the dune area posed a direct threat to the future extraction of dune water and also threatened to increase salt seepage in the hinterland.

In the past nature conservationists, too, applauded the advent of dune infiltration. The reason for that is that between the 1930s and 1950s it was generally assumed that if, following the introduction of dune infiltration for water supply purposes, the dune valleys were made wet again, all phreatic plant species (e.g. *Parnassia palustris*, *Schoenus nigricans* and diverse species of *Orchideae*), which had almost disappeared because of desiccation, would regenerate themselves. At the same time, the fauna characteristic of moist places and surface waters was expected to recover.

Most infiltration ponds and accompanying extraction installations were constructed between 1955 and 1960. The technical background of this method of producing drinking water – virtually unique in the world – is described by Huisman & Oltshoorn (1983). Broadly speaking, two methods of dune infiltration can be distinguished. The first method is extensive infiltration whereby natural slacks in hilly dune territory are flooded; the natural contours of the dunes are only adapted to water extraction purposes on the side where the extraction installations – consisting of drain

Table 1. Quantification of the dune infiltration in The Netherlands (ca. 1985).

Dune area	Infiltrated water	
	Million m^3 $year^{-1}$	% in relation to natural surplus precipitation
North-Holland Dune Reserve (Castricum/Egmond)	48	460
Kennemer dunes (Bloemendaal)	10	210
Amsterdam Waterworks Dunes (south of Zandvoort)	70	640
Berkheide (Katwijk)	17	610
Meijendel (north of The Hague)	50	860
Solleveld (south of The Hague)	3.2	490
Middle and East dunes (Goeree)	2.2	100
Westerban dunes (Schouwen)	1.9	70

pipes or rows of wells – are located. The second method – intensive infiltration – is found in dune areas that have been levelled by natural causes or by large-scale excavations; the infiltration ponds are parallel, more or less straight channels and the extraction installations are situated between the channels.

With both methods of infiltration, the distances between infiltration ponds and extraction installations have been chosen to ensure that the infiltrated water takes two to three months to flow through the dune from the pond to the well or drain. However, particularly in extensively infiltrated areas, the water may remain in the dune soil much longer if it is free to flow from the infiltration ponds in the direction of the sea, e.g. due to the absence of extraction installations. In this case, water may resurface locally in the slacks resulting in seepage on the side of the infiltration pond and infiltration on the other side. If this process gives rise to permanent open water in these naturally depressed areas, we speak of seepage pools; these can be up to 2 metres deep. Many of the presented research results relate to these seepage pools. There are two reasons for the focusing of research on seepage pools. First, seepage pools were expected to have comparatively high natural values; and second, they appeared very convenient for performing hydrological and hydrochemical measurements.

In the past 10 years, the supply of infiltration water amounted to over 200 million m^3 per year. Table 1 indicates the maximum amounts of water which is filtered into various dune areas via the constructed infrastructure.

It appears that in the larger dune areas the quantity of infiltrated water is 5 to 9 times greater than the amount reaching the dune naturally, i.e. via surplus precipitation.

Infiltration has caused a marked change in the

Table 2. Maximum local change of the groundwater table by various factors.

Factor	Maximum change (m) in the total dune area	Maximum change (m) in the central dune area
1. Vegetation		
– Planted pine woods	– 0.7 (– 2)	– 0.7 (– 2)
– Natural forests of deciduous trees	– 0.2	– 0.2
2. Coastal abrasion/formation	– 3/ + 1	– 0.5/ + 0.5
3. Excavation of inner dune edge	– 0.1	> – 0.1
4. Lowering of polder level	– 0.5	– 0.25
5. Surface drainage	– 0.5	– 0.3 (?)
6. Water works		
– Extraction of dune water	– 3	– 3
– Infiltration	– 4/ + 3	– 4/ + 3

Table 3. Quantification of landscape and hydrological changes in dune areas between 1850 and 1980.

Type of change	Cause	Surface-area (% of current dune area)	
Disappearance	Coastal erosion	5	
Complete loss of original character	Industrial development	3	
	Closed construction	4	
	Quarrying of inner dune zone	1	
Loss of total dune area (% of current total)		**13**	
Severe negative impact on original character	Infiltration area	6	
	Open construction	4	
Severe negative impact, total		**10**	**10**
Moderate negative impact on original character	Afforestation	15	
	Total dessication	33	
Moderate negative impact, total		**48**	**58**
Disappearance of dune flows	Fall in groundwater	p.m.	
Small negative impact on original character*		34	
Small negative impact, total		**34**	**92**
New formation	Dry	1	
	With wet slacks	7	
Total new formation		**8**	
Total current dune area, approximately 40,000 ha			**100**

* 'Small' means there may well have been a moderate fall in the water table (as much as 3 dm), considerable scrub growth in the valleys or thick wooding.

original, gradually bulging shape of the groundwater level between the beach and the polders in the hinterland. With infiltration, the groundwater level is in some places many metres higher than its original level – i.e. near infiltration ponds on high ground – while in others – i.e. near wells or drains – the actual water level is metres lower than the situation prior to the water extraction activities. As a result, the gradient of the groundwater level is much steeper than it has ever been causing the groundwater to flow much faster than in the original situation. in the natural situation groundwater never flows than 0.1 to, at maximum, 0.2 m day^{-1}, while through infiltration the velocity can run up to 1.5 m day^{-1} (Van Dijk 1984a, 1985a; Van Dijk & De Groot, 1987).

Summary of the disturbances of the water cycle

The maximum local impact of the factors mentioned in the preceding sections is presented in Table 2.

The consequences since 1850 of the factors discussed above and of their cumulative local effects have been estimated and are presented in Table 3 (from: Bakker *et al.*, 1979):

Only one third of the Dutch dune area can be characterized by a 'small negative impact' which still may include a drop in the water table of as much as 0.3 m as well as the effects of acid rain. During the last century over 10% of the total dune area was lost because of coastal erosion, industrial development, closed construction and excavation of the inner dune zone.

Cycle of dissolved substances: abiotic effects of interferences

Original situation (until ca. 1950)

In the original situation, the availability of nutrients in the dune area was the result of dissolved minerals from the dune sand, from precipitation water and from what was blown in from the sea. In the past, the latter two sources were minor suppliers of macro-nutrients (nitrogen and phosphate), while the mineral yield from the dune sand (particularly calcium) varied widely from one dune area to another being strongly dependent on the local age of the dunes within a particular dune area. The older the substratum in other words the farther from the sea and the lower the calcium content, the more organic material and accompanying nitrogen and accumulated in the upper soil layers (Ranwell, 1972; Gerlach, 1990).

Interference by air pollution

As in the rest of The Netherlands, the deposition of acid components has risen considerably in the dunes due to the effects of industry, domestic heating and traffic. In 1983, the highest average concentrations in the air of hydrogen fluoride, sulphur dioxide and nitrogen dioxide were measured in, respectively, southwest Zeeland, near Rotterdam, and near Leiden. To the north, total acid deposits are decreasing from more than 2,200 acidity equivalents to less than 1,800 ha^{-1} $year^{-1}$. In the Wadden Sea area, the sulphur dioxide concentration increases to the east, due to the effects of industrial plants in Delfzijl and in Germany (Van Dijk, 1985c).

The effects on the dune pools and lakes which are fed directly by precipitation are evident: particularly in the leached sand soil (Wadden Sea district and inner dune area of the calcium-rich mainland dunes) ecological processes and the changes recorded are rather similar to those observed in oligo- to mesotrophic fens (Arts, 1990; Van Dam, 1987). In naturally calcium-deficient dune areas even pools fed by groundwater are affected severely. The reason is that the surrounding soil loses its buffering capacity to a considerable depth within a few decades (Manual *et al.*, 1984), particularly when (semi)pervious layers are lacking and the groundwater contains oxygen. The pH of the groundwater then decreases dramatically and the mineral composition changes radically. Significant changes in the quality of the groundwater due to the effects of precipitation have been observed even in calcium-rich dune soils, for example a doubling of the nitrate content of the drainage water from the top 2.5 m of dune soil in the course of 20 years (in the same period local NOx emissions also doubled) (Stuyfzand & Moberts, 1987b). In addition, the surface water in the (non-infiltrated) dunes not only becomes more acidic but also richer in nutrients. Dune lakes that were still oligotrophic to mesotrophic around 1960 have become eutrophic almost without exception (Van Dijk & Meltzer, 1981). This could very well be attributed to polluted precipitation.

Interference by scrub growth and afforestation

Apart from changes in the quality of precipitation, the more rapid growth of scrub and woodland in dry dune areas possibly cause the eutrophication of originally mesotrophic dune lakes (Van Dijk & Meltzer, 1981). It has been observed that a dune area that was open fifty years ago, such as Voorne's dune, is now transformed completely into scrub and woodland (Van Dorp *et al.*, 1985). The tendency is the same in most other dune areas. In the event of abundant scrub growth, a large accumulation of organic substances occurs in the top soil layers; mineralization and leaching increase the nutrient content of the shallow groundwater and neighbouring surface water. Abundance of the nitrogen-fixating sea buckhorn, in particular, may result in considerably increased concentrations of nitrates in the ground and surface water (Van Dijk, 1985a; Van Dijk & De Groot, 1987).

Eutrophication by avifauna

Between the 1930s and the 1970s many (herring) gull colonies have increased rapidly in size since they are no longer controlled (Van Ommering & Verstrael, 1987; Van Ommering & Van der Salm, 1990; Texeira, 1979; Van Dijk & Meltzer, 1981). Colonies of cormorants have also established themselves in some places. Van Dijk & Meltzer (1981) consider these developments to have contributed to the eutrophication of larger dune lakes.

Dune infiltration

Artificially infiltrated water has been pre-purified to a greater extent in recent years but due to its scale and use, dune infiltration affects the cycle of dissolved substances more than any of the factors referred to above.

From the end of the 1960s until far into the 1970s, infiltration water contained large quantities of macro-nutrients. This caused problems in the production of drinking water. Algal blooms blocked the bottoms and banks of infiltration ponds increasingly so that the penetration of the water into the soil stagnated. Another problem was the seepage of substances into the dune soil and into the drinking water. For these reasons, most dune water companies switched to more stringent pre-purification of inflow water. This strongly reduced the ammonia and phosphate contents. Thanks mainly to phosphate stripping the original clarity of the water and a natural richness of water plants was restored in many infiltration ponds (Van Dijk, 1989a).

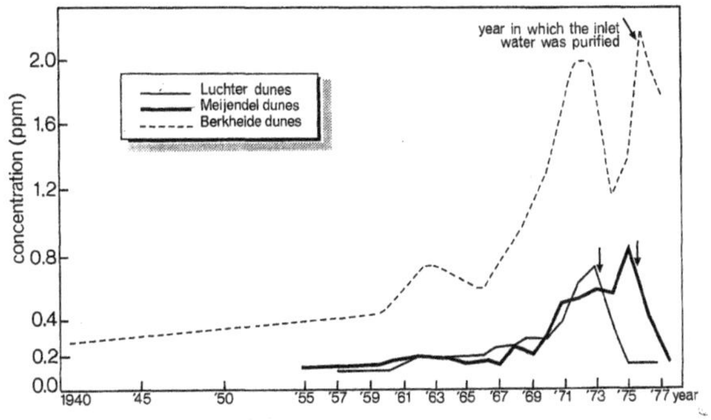

Fig. 3. Changes in the orthophosphate concentration in the inlet water of infiltrated dune areas during 37 years (after: Van Dijk, 1989).

The investigations presented here largely relate to situations encountered before or immediately after these phosphate strippings. Figure 3 shows the concentrations of orthophosphate in the infiltration water, which, between 1973 and 1978, increased to 0.9 (river water) and even to over 2 mg PO_4 l^{-1} (source of inlet water: reservoir water of polder). For reference: in natural dune pools the average annual orthophosphate content is between 0.01 and 0.1, with incidental peaks <0.5 mg PO^4 l^{-1} (Van Dijk & Meltzer, 1981; Van Dijk & Bakker, 1984; Van Dijk & De Groot, 1987). The nitrogen and potassium contents of the infiltration water were, and still are, much higher than the natural concentrations (Van Dijk, 1982, 1984a).

The water quality in the infiltration ponds strongly reflects the quality of the inlet water because it only stays there for a short time (a few days to weeks). Local improvements in the water quality can be observed at points situated far from the inlet. This is due to the longer residence time and therefore the extraction of nutrients by water plants and dying and sinking algae (Hoekstra, 1974; Van Dijk, 1989a).

As for the increase of the actual nutrient load caused by infiltration in dune areas Van Dijk & Bakker (1984) record some estimations for a number of larger infiltrated dune areas. For example the potassium loads have increased by a factor 12 to 200 and the orthophosphate loads up to a factor of 200. These estimated values depend strongly on the calculation method used.

Before infiltration started it was expected that the ecosystems of wet dune slacks which had disappeared by desiccation would return in general. This recovery would occur in the seepage pools and marshes which would develop in the infiltrated dunes. The original assumption about seepage pools was that valleys situated several hunderd metres away from the infiltration ponds would be safeguarded from infiltration water and related increases of nutrient load. This was based on the hypothesis that there should be a precipitation water lens on top of the flowing infiltration water. If, for example, the underlying infiltration water had travelled for three years since penetrating the dune soil, the assumption was that three years' effective precipitation would be present above this infiltration water. After having made corrections for the pore volume of dune sand, this would mean a 3 m thick precipitation water lens. Figure 4 shows this hypothesis diagrammatically. In practice, however, it was observed (Van der Werf, 1974; Londo, 1975) that seepage pools also displayed vegetation development – albeit later and less extreme than observed on the borders of the infiltration ponds – towards nitrophilous phreatophytes. This indicates that the infiltration water forces its way to the upper groundwater even in the case of seepage pools located far away from the infiltration ponds. A three-year investigation of flow lines in seepage pools in Meijendel gives the definite answer (Van Dijk, 1984a; Stuyfzand & Moberts, 1987a, 1987b).

In the Meijendel research project the origin of the groundwater and seepage pool water was determined with the aid of tracer substances (chloride, potassium, fluoride), which are, in principle, conservative substances with higher concentrations in the infiltration water than in the precipitation water and the original dune water. The concentrations of nitrogen and phosphate were also measured. Investigation of the top 2.5 m of groundwater near seepage pools revealed that, in the case of rapid flow (around 3 m day^{-1}), 50% of this groundwater still consisted of original infiltration water even 200 m from the infiltration pools (Fig. 5).

The conclusions of the investigations are:

- residence time is a main factor in the proportion of infiltration water in the upper groundwater and seepage pool water (Fig. 6); therefore the impact of the nearby infiltration pond decreases faster at an increasing distance from the infiltration pond in the case of slower groundwater flow;
- the size and depth of the pools are a second important factor.

The second conclusion indicates that the explanation for the occurrence of infiltration water in seepage pools should not be sought in hydrodynamic (macro) dispersion (De Groot, 1984), but in characteristics unique to the seepage pool (Stuyfzand & Moberts, 1987; Van Dijk & De Groot, 1987). The somewhat deeper seepage pools attracted groundwater from several metres down, while this water – after being mixed in the pool by the wind – infiltrated again across a broad front in the bottom and the banks. As a result of this hydrological mechanism less than 20 of some

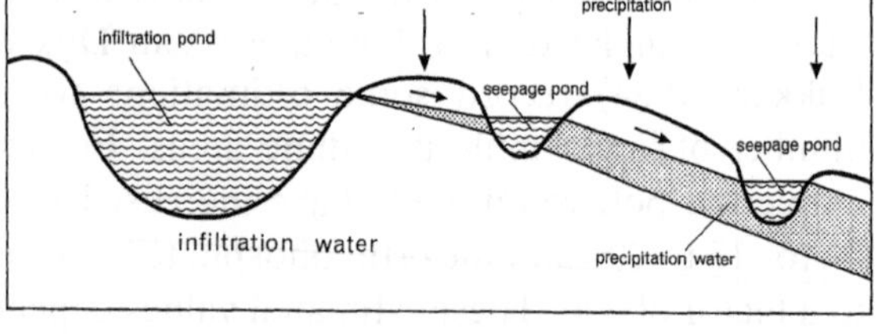

Fig. 4. Model of the 'precipitation water lens' hypothesis.

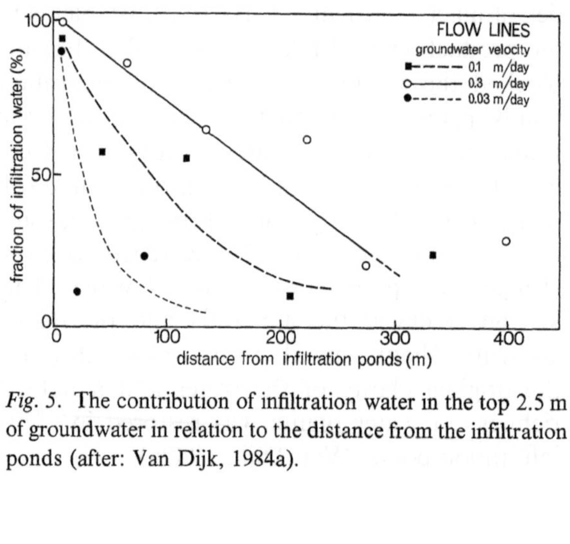

Fig. 5. The contribution of infiltration water in the top 2.5 m of groundwater in relation to the distance from the infiltration ponds (after: Van Dijk, 1984a).

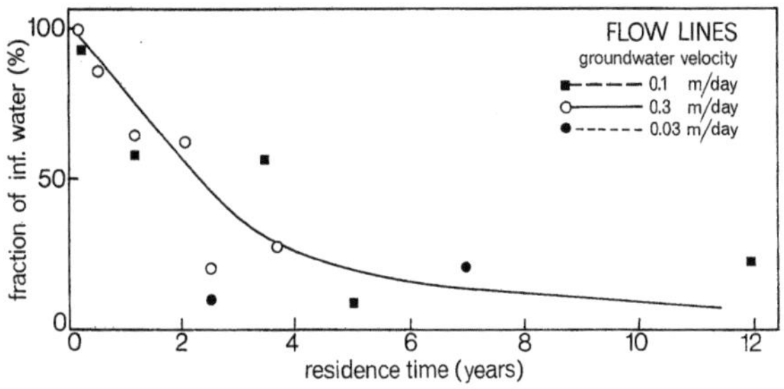

Fig. 6. The contribution of infiltration water in the top 2.5 m of groundwater in relation to the residence time (after: Van Dijk, 1984a).

300 seepage pools in the Meijendel and Berkheide infiltration areas appear to be safeguarded from mixture with infiltration water (Van Dijk, 1986, 1989b).

The relation between the hydrological mechanism described above and the nutrient contents in groundwater and seepage pool water was studied. In the case of the flow lines presented in Fig. 6, hypothetical concentrations of nitrate and phosphate were calculated on the basis of the concentration of the substances used as tracers. The projected value was based on the assumption that only dilution with precipitation water determines the concentration. The results of these calculations, together with the concentrations measured, are presented in Figs 7 and 8. Nitrate, representing 90% of the nitrogen in the groundwater, proved to have natural sources in the dune which overruled the effect of infiltration (Fig. 7). Obvious sources are eutrophication from gull colonies and nitrogen-fixing sea buckthorn. In addition, the effect of atmospheric deposits cannot be excluded.

Generally speaking phosphate (Fig. 8), unlike nitrate, behaves as a conservative substance in groundwater; in other words, the groundwater concentration is in equilibrium with the dune soil, i.e. the dune sand is more or less saturated with phosphate. The results are, however, obscured by the dephosphatization of the infiltration water carried out by some waterworks companies shortly before the observations presented here.

Far higher phosphate concentrations than in Meijendel were revealed in the seepage pool water

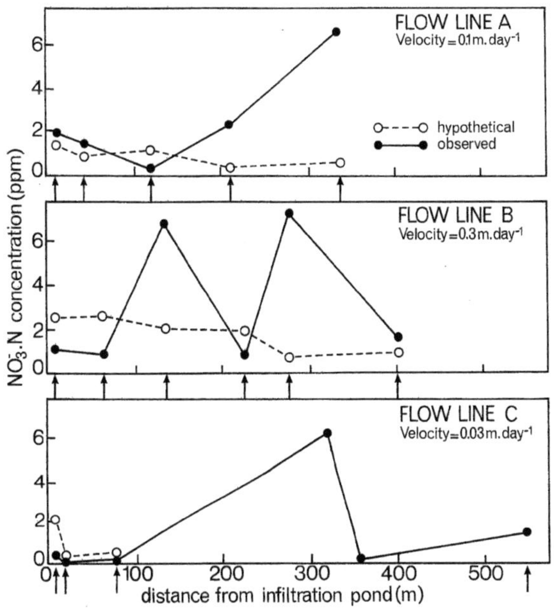

Fig. 7. The nitrate concentration in the top 2.5 m of groundwater in relation to the distance from infiltration ponds (West Meijendel) (Van Dijk, 1984a; Van Dijk & De Groot, 1987).

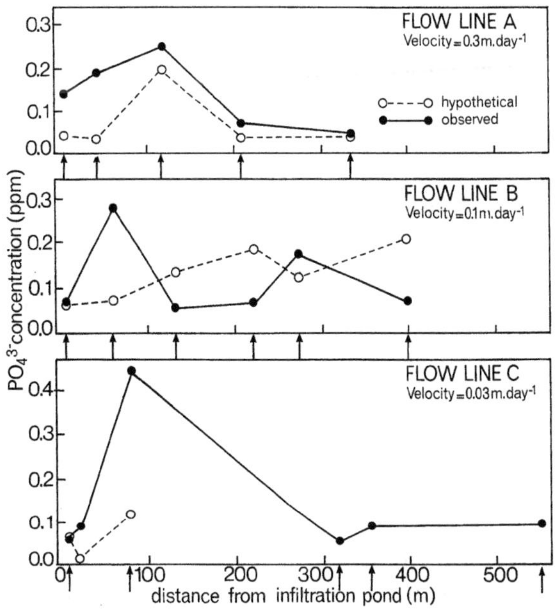

Fig. 8. The orthophosphate concentration in the top 2.5 m of groundwater in relation to the distance from infiltration ponds (West Meijendel) (Van Dijk, 1984a; Van Dijk & De Groot, 1987).

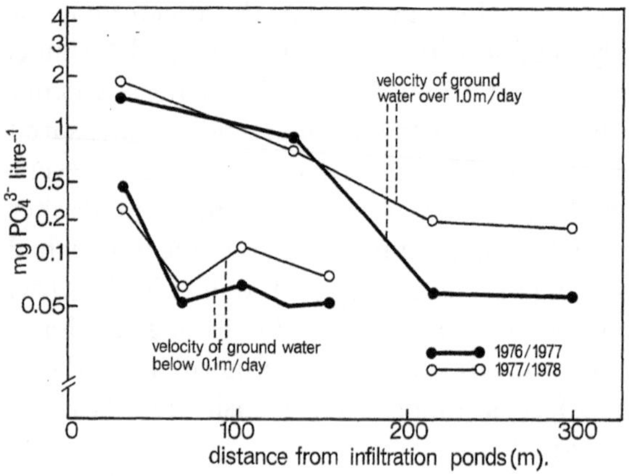

Fig. 9. The orthophosphate concentration in seepage pool water in relation to the distance from infiltration ponds (Berkheide) (Van Dijk, 1984a and 1985).

in the Berkheide dune area, were phosphate concentrations in the infiltration water were some three times higher than in Meijendel (Fig. 9). An originally clear gradient of concentration decreasing with distance to infiltration levelled out markedly within one year, which points to strong signs of saturation of the subsoil. In addition, it was found that the highest phosphate concentrations occurred where the groundwater velocities were highest.

The nutrient contents in the dune soil appear to reflect the observed gradations of the nutrient concentrations of the groundwater between infiltrated and non-infiltrated areas as well as within the infiltrated areas.

Obvious differences occurred in the soil nutrient contents of banks in infiltrated and non-infiltrated dune areas (Fig. 10). In Fig. 10 Terschelling (T), Zwanenwater (Z), Voorne (V) are non-infiltrated areas. Kennemer dunes (K), North-Holland Dune reserve/southern infiltration area (N) and Luchterdunes (L) represent infiltrated dune areas where the infiltration started very recently (K, N) or where infiltration occurred with strongly dephosphatized inlet water. Infiltration with very phosphate-rich inlet water has been going on for 20 years in Meijendel (M, s = seepage pools; i = infiltration ponds) and in Berkheide (B, only seepage pools). In Fig. 10 the infiltration influence increases from left to right.

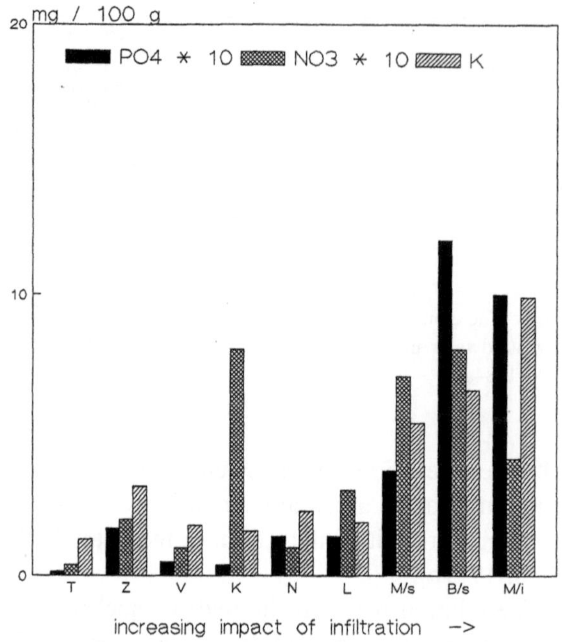

Fig. 10. Nutrient amounts (grams m^{-2}) of the upper 30 cm subsoil of the banks of pools and ponds (average values; sampling August–September/after: Van Dijk, 1985b). T = Terschelling; Z = Zwanenwater; V = Voorne's dunes; K = Kennemer dunes; N = North-Holland Dune Reserve; L = Amsterdam Municipal Waterworks Dunes (Luchter dunes); B = Berkheide dunes; M = Meijendel dunes;/ s = seepage pools; /i = infiltration ponds. (*left (T,Z,V)*: non-infiltrated areas; *middle (K,N,L)*: areas very recently infiltrated or infiltrated with good prepurification of inlet water; *right (M,B)*: areas infiltrated during about twenty years).

As can be seen, large differences between the infiltrated and non-infiltrated areas occurred in the potassium and orthophosphate contents of the soil. As far as phosphates are concerned, the seepage pools in Berkheide were much more polluted than the infiltration ponds in Meijendel.

The gradient in the concentration of phosphate in the Berkheide pool water (Fig. 9) recurs in the phosphate concentrations in the soil of the pool banks (Fig. 11). The phosphate concentrations in the soil, however, showed much more variance compared to the concentrations recorded for the pool water. The highest soil concentrations per pool or pond were found in the places with the strongest seepage and infiltration – in other words, in the places with the highest temporary and cumulative phosphate load.

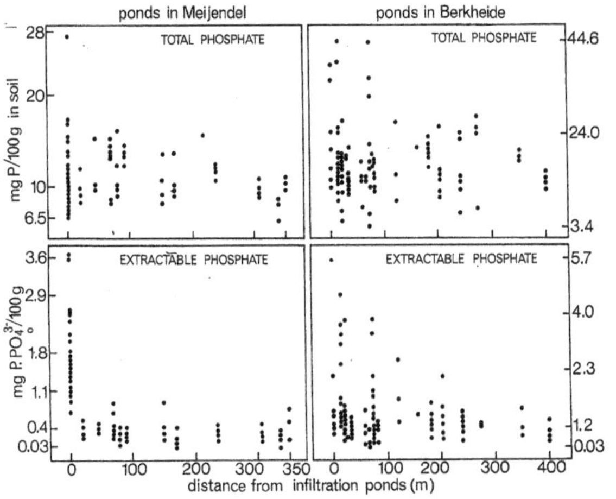

Fig. 11. Phosphate levels of the top 3 dm in the banks of seepage pools in relation to the distance from infiltration ponds (after: Van Dijk, 1985b).

Ecological consequences of interfering with water, calcium and nutrient cycles

A great deal is known about the consequences of affecting natural cycles for higher plants. Causal ecological chains are less known for the bird world, and even lesser for mammals, amphibians and reptiles (Croin Michielsen, 1974). Ecological knowledge about the lower animals in the dune area is hardly adequate. The ensuing text is restricted to the consequences of interfering with the water and nutrient cycle for the phreatophytic dune vegetation.

Decline in the average water level and changes in water table fluctuation

Even before the turn of the century, Vuyck (1898) pointed to a strong deterioration in phreatophytic plant species which he attributed to extracting water from the dunes. Local research such as done in the Verbrande Pan near Bergen (N.H.) by Bijhouwer (1926) and Hoffman & Westhoff (1951) confirmed that many wet plant species and vegetation types shifted to lower places or ultimately disappeared.

Examples of phreatophytic plant species that have substantially declined since 1850, especially between The Hague and Den Helder, include *Epipactis palustris*, *Parnassia palustris*, *Equisetum variegatum*, *Orchis morio*, *Carex pulicaris*, *Gentianella amarella* and *Echinodorus ranunculoides* (Londo, 1971; Van Zadelhoff, 1981). Eight phreatophytes, including two *Rhynchospora* species, *Eleocharis acicularis* and *Narthecium ossifragum*, completely vanished from the dune area during this period (Van Zadelhoff, 1981). Strangely enough, these eight species mainly occurred to the north of Petten and on the Wadden Sea Islands which are the hydrologically least affected dune areas (Van Zadelhoff, 1981). The largest decline in phreatophyte species was assessed among the moisture-loving species tied to the less dynamic, relatively nutrient poor and vulnerable biotopes. The aquatic macrophytes have also dwindled and often disappeared completely (Van Dijk, 1989a).

Changes in the water table fluctuation pattern can lead to eutrophication due to accelerated mineralization, especially in lime- and humus-rich situations. The vegetation, therefore, acquires a bushier and more nitrophilous character. The very large water fluctuations that sometimes occur in infiltration areas severely affect the vegetation. Only a few plant species such as *Agrostis stolonifera* can survive whereas a few pioneering species such as the *Chenopodium rubrum*, *Limosella aquatica* and *Cirsium arvense* can germinate in such situations.

Changes in groundwater flows, without artificial infiltration

Intensive extraction of groundwater sometimes leads to serious disturbances in the direction and volume of groundwater flows. Bakker (1981) reports that near the extraction installations on Texel, the groundwater velocity has increased fivefold from 0.1–0.2 to 0.5–1 m day^{-1}. According to Bakker *et al.* (1979), the concurrent increase in the supply of (nutritional) substances per time unit is less clear than the drop in water tables. By contrast, other authors (Grootjans *et al.*, 1988, 1990) claim that the consequences of the disturbance of the groundwater flow – with the ground-

Table 4. Decline of nine plant species measured as a percentage of the number present in 1964 in permanent plots (10 m × 10 m) in the Kapenglop (Grootjans *et al.*, 1990).

Species	Occurrence in number of plots (1964)	Decline (% of permanent plots)	
		1964–77	1964–87
Antennaria dioica	171	75	>90
Liparis loeselii	8	100	100
Pedicularis palustris	151	68	97
Epipactis palustris	143	29	95
Dactylorhiza incarnata	39	7	97
Schoenus nigricans	64	?	>80
Parnàssia palustris	68	55	91
Pyrola rotundifolia	14	7	100
Gentianella amarella	13	?	100

water level remaining largely unaffected – can be extremely serious.

The detailed studies cited here were carried out in three dune valleys on the Wadden Sea island Schiermonnikoog: the Arnica marsh, the Grïenglop and the Kapenglop. The Grïenglop has for years consisted exclusively of heavily acidified and impoverished vegetation, while the lower parts of the Arnica marsh still have many rare plants, such as *Pedicularis palustris*, *Carex pulicaris* and various species of orchids, that more or less point to calcium-rich groundwater. The Kapenglop, which comprised some of the best developed dune valley vegetation in the 1950s with lots of *Schoenus nigricans*, *Pedicularis palustris* and *Litorella uniflora*, had been monitored in 1964, 1977, 1983 and 1987; now it is strongly acidified and tall bushy hemicrytophytes and shrubs dominate locally. More than 90% of the following rare plant species of vulnerable gradient situations have disappeared within a detailed and frequently investigated 0.6 ha area (Table 4).

Grootjans *et al.* (1988, 1990) show that differences in the development of the valleys are not primarily attributable to desiccation. The water level is such that all three examples can still be categorized as moist to wet valleys. The influence of water extraction is not so large that drying out occurs in the strict sense, but groundwater flows have been shifted. The groundwater begins to flow in new directions because the groundwater level elsewhere has experienced a significant drop, which diminishes hydrostatic pressure. The seepage and run-off are weakened, while the direction in which the groundwater flows may change. The low-lying wet-to-dry gradient situations in the Arnica marsh with extremely high numbers of species, appear to be coupled to a stable hydrological situation in which calcium-rich groundwater can reach the surface of the lower lying seepage areas throughout the entire growing season and thereby pushes the acidifying precipitation aside (Fig. 12).

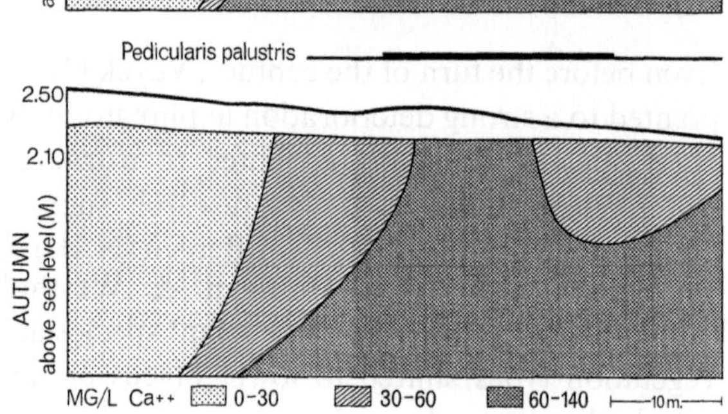

Fig. 12. Cross sections of the Arnica marsh, showing the calcium content of the groundwater in relation to the occurrence of *Pedicularis palustris*.

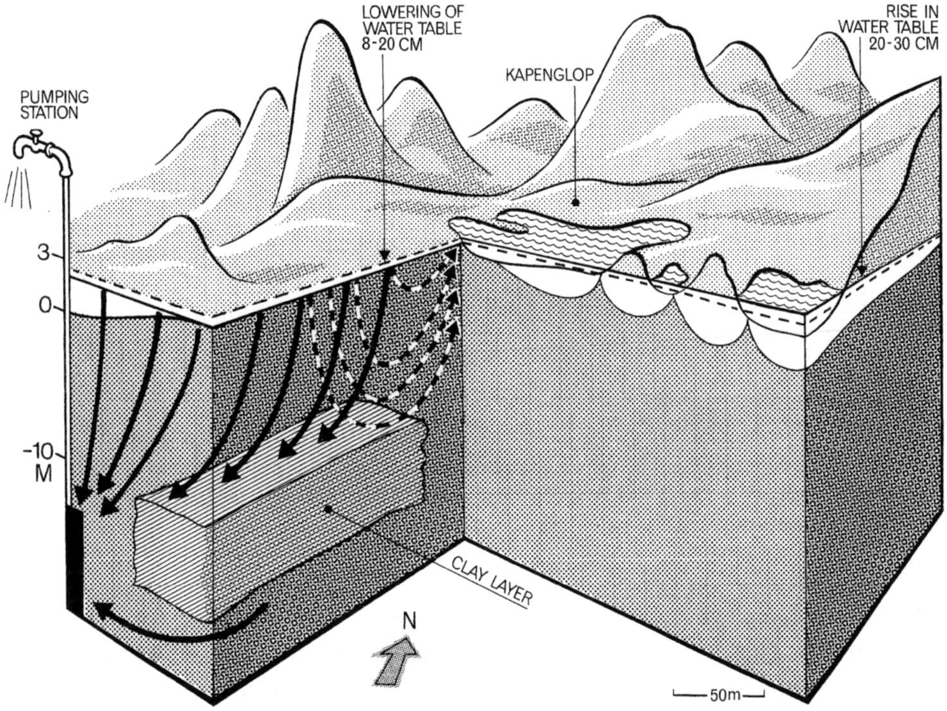

Fig. 13. Hypothesis if the hydrological situation around the Kapenglop valley; solid lines indicate the present flow lines and dotted lines indicate former flow lines.

Small decalcified sand hills function as local infiltration areas and the presence of less impervious layers of peat enables the calcium-deficient infiltration water to run off to more calcium-rich seepage situations. The zone where calcium-poor groundwater meets the calcium-rich seepage water is characterized, among other things, by the frequent occurrence of *Carex pulicaris*. This species is still found in the same zone as it was over 25 years ago, illustrating the stability of the hydrological gradient.

This situation contrasts with the nearby Grïenglop, where the whole structure of local hydrological systems has collapsed and the vegetation has acidified. This has been caused by hydrological measures in the polder and probably also by the excavation of a sand extraction hole in the Grïenglop itself.

The acidification in the Kapenglop was attributed by Grootjans *et al.* (1990) to changes in the local hydrological system that provided the Kapenglop with calcium-rich groundwater (see Figs 13 and 14). The local acidification is assumed *NOT* to originate from a natural de-calcification process because the top soil layers had already been totally decalcified for many decades. The underlying causes here are the increasing influence of water extraction in the nearby Hertenbos valley from about 1960 and the growing evaporation through the dune vegetation.

Some measures taken in the dry years of the 1970s, to save the valley from drying out, have had an adverse effect on the natural values during the wet 1980s. Specifically, the measures taken to retain precipitation water in the valley seem in retrospect to have put an *EXTRA* brake on the inflow of calcium-rich groundwater, which has led to reinforcement of the acidification phenomenon.

The data described show that the dune valleys in the centre of the dune system are very vulnerable systems for several reasons: infiltration is very prevalent in these valleys; they are situated in the midst of leached-out dunes; they rely on

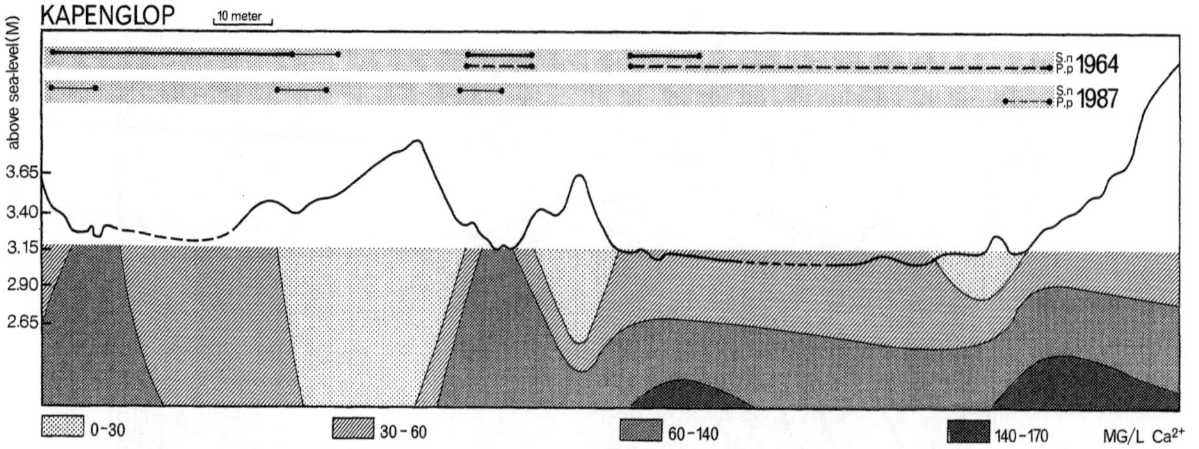

Fig. 14. Cross section of the Kapenglop valley, showing the calcium content of the groundwater in relation to the occurrence of two plant species (*Schoenus nigricans* (S.n.) and *Pedicularis palustris* (P.p.) in 1964 and 1987. The tickness of the lines indicates the relative abundance of the species.

local hydrological systems which provide calcium-rich groundwater. These small hydrological systems can thus be adversely affected by very minor influences, such as a slight increase in water extraction, the digging of run-off ditches in source areas, or increasing evaporation from woods. All these influences lead to shifts in groundwater flows, after which the acidified precipitation can begin to dominate the local surface water (Grootjans *et al.*, 1990).

Outside Schiermonnikoog the occurrence of endangered calciphilous plant species (among others, those mentioned in Table 4) has declined sharply (Table 5). This decline does not necessarily always point to a disturbance in the hydrological systems, due to which calcium-rich seepage has declined or disappeared. A species such as *Pedicularis palustris*, for example, has declined mainly because currently fewer primary dune valleys are being formed than was previously the case. The optimal environment for this species is the boundary zone of this kind of valley, where the influence of the sea, rather than seepage, takes care of hydrological 'buffering'.

From Table 5 it also becomes clear that the decline of endangered dune slack species in the Wadden Sea Islands is much less pronounced than elsewhere in The Netherlands.

Table 5. Percentage decline since 1950 in the occurrence of calcium and moisture-loving plant species in squares of 5 × 5 km (Grootjans *et al.*, 1988).

Species	Wadden Sea dune area	Dune district	Elsewhere in The Netherlands
Pedicularis palustris	17	53	58
Dactylorhiza incarnata	0	11	70
Liparis loeseli	20	71	61
Parnassia palustris	33	46	74
Anagallis tenella	38	71	83

Dune infiltration

Occurrence of nitrophilous tall forbs in the bank vegetation

To get a rough impression of the effects of artificial infiltration, the species composition of four infiltrated and thirteen non-infiltrated dune areas were compared (Van Dijk, 1989a). Fifty plants species, indicative of oligotrophic and mesotrophic habitats, have declined as a result of infiltration. Nineteen species have increased, all of which demand a high nutrient availability. The development of the vegetation of the banks of infiltration ponds and seepage pools since the start of the infiltration to the present situation has been described by Van der Werf (1974) and Londo (1975). In a period of ever increasing nutrient concentration in the infiltration water they observe a shift in vegetation towards dominance of only a few species, which succeed each other in the following sequence: *Calamagrostis epigejos*, *Mentha aquatica*, *Lycopus europaeus*, *Epilobium hirsutum*, *Eupatorium cannabinum*, and finally *Urtica dioica*.

At the present the herbaceous bank vegetation along most of the infiltration ponds is strongly dominated by these tall forbs, which are fast-growing, have large leaves and are very productive. When fully grown they intercept much of the light from the small, slowly growing, indigenous dune valley plants. Figure 15 shows the average cover of two ecological groups of herbaceous plant species occurring along the pond banks of various dune areas. The first group is formed by the tall forbs *Epilobium hirsutum*, *Eupatorium canabinum*, *Cirsium arvense* (grey in Fig. 15) and *Urtica dioica* (dark grey), which are extremely

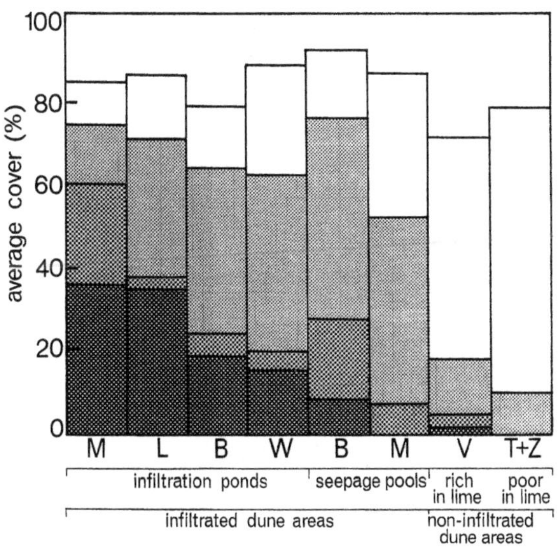

Fig. 15. The average of nitrophilous tall forbs in the bank vegetation of pools and ponds in heavily to non-infiltrated dune areas (after: Van Dijk, 1989). The most extremely 'nitrophilous' species are marked grey (*Epilobium hirsutum*, *Eupatorium canabinum*, *Cirsium arvense*) and dark grey (*Urtica dioica*). The three less nitrophilous species *Calamagrostis epigejos*, *Mentha aquatica* and *Lycopus europeus* are marked light grey. (M = Meijendel dunes; L = Amsterdam Municipal Waterworks Dunes (Luchter dunes); B = Berkheide dunes; W = Westland dunes; V = Voorne's dunes; T = Terschelling; Z = Zwanenwater).

'nitrophilous' according to Ellenberg (1979). The second group consists of three species (light grey): *Calamagrostis epigejos*, *Mentha aquatica* and *Lycopus europeus*. The species of the second group are less nitrophilous and much smaller, but may dominate the vegetation too when the nutrient availability is increased by infiltration.

Figure 15 clearly shows that species of both ecological groups hardly occur among the bank vegetation in non-infiltrated dune areas. In the areas with recent infiltration or with infiltration with pre-purified inlet water the cover of the seven 'nitrophilous' species is also relatively low, but in areas that have been infiltrated with very eutrophic water for about 20 years (Berkheide and Meijendel dunes), the four most nitrophilous tall forbs have left little room for the original vegetation. Not only the banks of the infiltration ponds are covered with tall forbs, but also the banks of seepage pools situated up to 100 metres further down.

To investigate the ecological mechanisms involved in this eutrophication process, the vegetation of the pond banks was related to nutrient concentrations in soil and groundwater, and finally to nutrient supply in the rooting zone, where nutrient concentrations and flow velocities are combined.

Biomass and nutrient concentration in soil and groundwater

The relationship between vegetation characteristics and nutrient concentration in the soil and incoming groundwater was studied in eight infiltrated and eight non-infiltrated areas. The dry biomass values of the bank vegetation were averaged per area and related to the annual average nutrient concentrations of surface water, soil and vegetation (Fig. 16).

The phosphate concentrations – of surface water and soil as well as above-ground dry weight – clearly show the most positive relationship with regard to biomass. The potassium and phosphate concentrations have a strong mutual correlation (Meltzer & Van Dijk, 1986), but potassium only

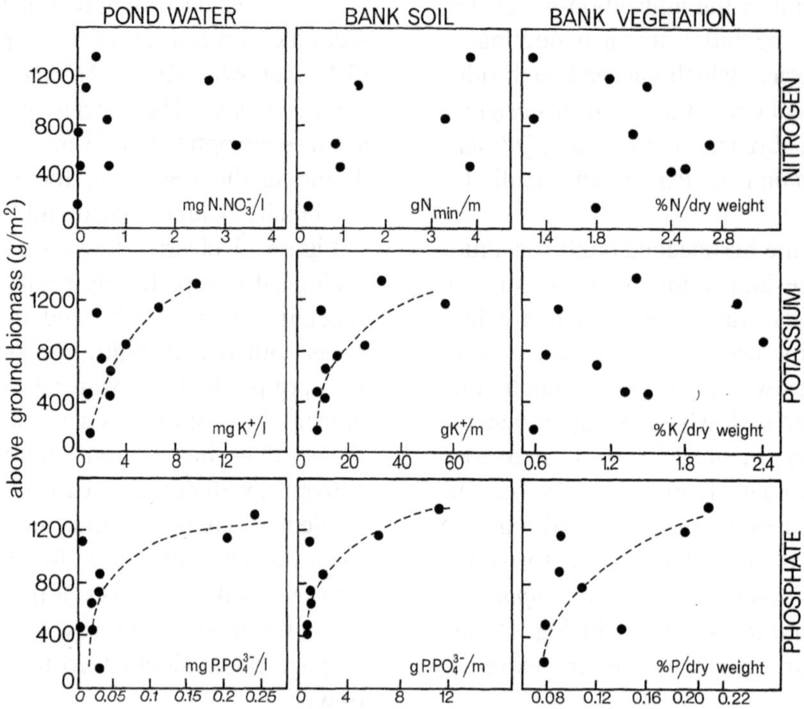

Fig. 16. Relationship between dry biomass of herbaceous bank vegetation and nutrient contents in surface water, soil and surface vegetation sections (average values of eight areas / after: Van Dijk, 1985b).

showed a clear relation with regard to surface water and bank soil. The nitrogen concentrations demonstrated no clear relations with the biomass. The right-hand graphs of Fig. 16 (bank vegetation) show strong indications that changes of the phosphate supply have the greatest influence on the occurrence of nitrophilous macrophytes along the banks of dune ponds and pools. The other graphs of Fig. 16 (surface water and bank soil) appear to indicate that the nitrogen level has a smaller influence than the potassium level.

Nutrient loading via groundwater flow

Great differences were often encountered in the cover of tall nitrophilous plant species along the banks of one seepage or infiltration pond. In general the water quality was the same here. Since large differences in flow velocity may occur in such situations, the nutrient load via the groundwater might play a key role here; extremely nitrophilous species reached the highest cover on those parts of the bank where the seepage or infiltration was greatest (Van Dijk, 1984a).

The nutrient loading via flowing groundwater was estimated from the product of groundwater flow rate and nutrient concentration. Annual average nutrient concentrations of the surface water were determined per pond or pool from monthly concentration measurements. In addition the velocity of the groundwater flow was measured (method: see Van Dijk, 1984a).

When the maximum covers of the seven 'nitrophilous' species of ecological species group R7 were related to the momentary phosphate load, a sequence in response curves was obtained (Fig. 17). *Mentha aquatica*, for instance, increases at a relatively low loading, while *Urtica dioica* increases significantly at a high loading. It is striking that the same succession in species was described by Londo (1975) and Van der Werf (1974) in a sequence of species gaining dominance during the time that the nutrient concentrations in the infiltration water increased strongly.

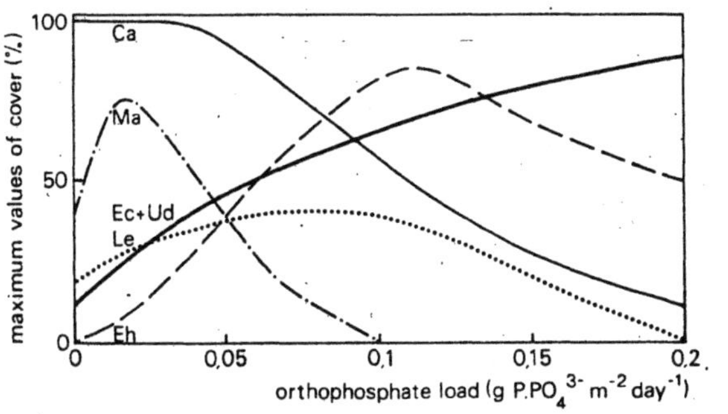

Fig. 17. The relationship between the maximum covers of six nitrophilous species and the annual average orthophosphate loading via infiltration pond water (after: Van Dijk, 1984a) (Ca = Calamagrostis epigejos; Ma = Mentha aquatica; Le = Lycopus europaeus; Eh = Epilobium hirsutum; Ec = Eupatorium cannabinum; Ud = Urtica dioica).

Conclusions

As for artificial dune infiltration, it must be concluded that the strongly increased velocity of the groundwater results in unnaturally high nutrient loads, even after thorough pre-treatment of the infiltration water. In addition, the small-scale hydrological gradients that determine the quality of the groundwater in wet dune slacks have become greatly affected.

Further it must be concluded that all ecological observations in infiltrated and non-infiltrated dune areas indicate that the supply of phosphate via the groundwater plays a key role in the occurrence of 'nitrophilous' tall herbaceous plant species in dune valleys under the influence of artificial dune infiltration. This is in accordance with earlier eco-physiological work on *Urtica dioica* (Pigott & Taylor, 1964) which showed that this 'nitrophilous' species requires a high local concentration of phosphate.

The last conclusion is supported by the observation that in the Meijendel area the cover of most 'nitrophilous' species has substantially decreased since 1976 when the intake water was de-phosphated. This decrease in cover was not followed, however, by an increase in the number of species nor by the re-establishment of original dune valley species.

Both conclusions are supported by data on nutrient loading via mineralization in both non-infiltrated and infiltrated dune areas (Van Dijk, Noordervliet & De Groot, 1985). It appears that within the non-infiltrated areas the cover of the nitrophilous species of the bank vegetation primarily corresponds with nitrogen mineralization; but the infiltration areas, however, phosphate loading via the groundwater flow dominates over mineralization and over potassium and nitrogen loading via flowing groundwater.

Recommendations

Revised starting points for drinking water extraction.

Surface infiltration must ultimately be stopped. Thorough pre-treatment of the infiltration waters offers no guarantee whatsoever for the return of the complete ecosystem of the wet dune slacks with this type of infiltration. Under certain circumstances, depth infiltration with intensively pre-treated water may offer a better alternative from an environmental point of view (Van Dijk, 1989a). The freshwater lens above the impervious soil layers can restore itself, and the dune area still retains the function of a safe repository for drinking water. Other alternatives for surface infiltration are also at hand (Udo de Haes *et al.*, 1980; Anonymous, 1982).

Both the regional authorities of North and South Holland and the central Dutch government have made political statements to the effects that surface infiltration is to be reduced in due course. Nevertheless, the infiltration and extraction installations at the Berkheide infiltration area have recently been greatly expanded and there are also plans to expand the infiltration capacity of the Amsterdam Waterboard Dunes.

For the Wadden Sea islands alternatives for dune-water catchment must be more intensively initiated. The still increasing influx of recreational visitors and the greater disturbance of the hydrology related to this, is taking an ever-increasing toll on the relatively unaffected ecosystem – at least in comparison with the mainland dunes – of the wet dune slacks.

Renewed out-blowing of dune slacks

Many factors such as coastal erosion, afforestation, water extraction, lowering of the water level in polders and drainage systems cause strong reductions of the water table, with the result that more than half of the dune area suffers severely from desiccation. At the same time, the dune area is strongly immobilized by the measures against sand-drifts which have been carried out intensively for many centuries. If the causes of the desiccation can only partially be dealt with, then stimulation of blow outs in the valleys will often allow the groundwater to reach the surface again.

While choosing locations for sand-drift experiments, very close account must be taken of the chemical and hydrological factors of the soil that control the external conditions for a return of the rich variety of the natural vegetation of the dune slacks.

Revision of forestry management

It has been shown that in many places the non-indigenous planted woodlands may exert a dominant local influence on the hydrology. Possibilities to reduce the negative effects of the woodlands are the reduction of the woodland area or – a less good solution – wettening of the present woodland by means of removing the drainage ditches that were excavated when the woodland was planted. In any case it is not recommanded to consider new pine-tree planting for the dune area and in more cases, and more quickly, the existing pinewood should be cut and contingently replaced with deciduous trees. In addition, drainage systems in or near the woods should be eliminated.

Choice for a withdrawing or an accreting coast

The renewed interest in scenarios for coastline management, partly prompted by discussions over a possible rise in the sea level, has important implications for wet dune slacks. The choice for extensive sand supplementation is the choice for wettening of and decreasing sea influence in existing dune slacks and possibly for beaches and primary dune formation. The choice for the local withdrawal of the primary sea barrier can mean the new creation of sea inlets ('slufters'), an almost extinct type of natural landscape. In the first case the development of marsh vegetation in the secondary dune slacks will be stimulated; in the second case the secondary dune slacks will be further dried up, but new salt and brackish vegetation may start the succession anew.

Acknowledgements

The authors are indebted to Dr T. W. M. Bakker and Drs W. Hurkmans for critically reviewing the manuscript, and J. Seebus and Drs A. B. Kaal for taking care of the translation. E. Leeuwinga and D. Visser drew the figures.

References

Anonymus, 1982. Integraal Onderzoek Drinkwatervoorziening Zuid-Holland. Final report. Provincial Board of South-Holland/State Institute for Drinking Water Supply. The Hague.

Arts, G. H. P., 1990. Detoriation of atlantic soft-water systems and their flora; A historical account. Ph.D. Thesis, Nijmegen.

Bakker, T. W. M., 1981. Nederlandse Kunstduinen; Geohydrologie. Ph.D. Thesis, Wageningen.

Bakker, T. W. M., 1990. The geohydrology of coastal dunes. Th. W. Bakker, Jungerius, P. D. & Klijn, J. A (eds), Dunes of the European coasts; Geomorphology – Hydrology – Soils. Catena Suppl. 18. Cremlingen-Destedt: P. 109–121.

Bakker, T. W. M., J. A. Klijn, F. J. Van Zadelhoff, 1979. Duinen en duinvalleien. PUDOC, Wageningen.

Bijhouwer, J. T. P., 1926. Geobotanische studies van de Berger duinen. PH.D. Thesis, Wageningen.

Carrière, J. E., 1929. De waterleidingen in vele gevallen het behoud van onze duinen. De Levende Natuur 34, 126–135: 161–173.

Croin Michielsen, N., (ed.) 1974. Meijendel, Duin-Water-Leven. W.v.Hoeve, The Hague/Baarn.

Gerlach, A., 1990. Biochemistry of nitrogen in coastal dune ecosystems. Landscape Ecology (in press).

Groot, W. T. de, 1984. Hydrodynamic macro-dispersion as the causal back-ground of the phreatic water type mixing in infiltrated dunes. P. 208–212 in: Van Dijk, H. W. J., 1984.

Invloeden van Oppervlakte-Infiltratie ten behoeve van Duinwaterwinning op Kruidachtige Oevervegetaties. Ph.D. Thesis, Wageningen.

Grootjans, A. P., 1985. Changes of groundwater regime in wet meadows. Ph.D. Thesis, Groningen.

Grootjans, A. P., P. Hendriksma, M. Engelmoer & V. Westhoff, 1988. Vegetation dynamics in a wet dune slack I: rare species decline on the Wadden island of Schiermonnikoog in The Netherlands. Acta Bot. Neerl. 37: 265–278.

Grootjans, A. P., 1990. Hydrologische randvoorwaarden voor natuurherstel op de Waddeneilanden. P. 33–45 in: J. Abrahamse & A. Woudstra (eds), Duinen in beweging. LVBW, Harlingen.

Hoekstra, A. C., 1974. De infiltratievijver en zijn levende have. P. 216–222 in: N. Croin Michielsen (ed.), Meijendel; duinwater-leven. Baarn/Den Haag.

Hoffman, M. E. & V. Westhoff, 1951. Flora en vegetatie van de Verbrande Pan bij Bergen. De Levende Natuur, 54.

Holkema, F., 1870. De plantengroei der Nederlandse Noordzee-eilanden; een bijdrage tot de Flora van Nederland. Ph.D. Thesis, Groningen.

Huisman, L. & T. N. Oltshoorn, 1983. Artificial groundwater recharge. Pitman, Boston/London/Melbourne.

Klijn, J. A., 1990. The younger dunes in The Netherlands: chronology and causation in: Bakker, Th. W., P. D. Jungerius & J. A. Klijn, (eds). Dunes of the European coasts; Geomorphology – Hydrology – Soils. Catena Suppl. 18. Cremlingen-Destedt: p. 89–101.

Leentvaar, P., 1963. Dunewaters in The Netherlands. Quakjeswater, BreedeWater and Vogelmeer. Act. Bot. Neerl. 12: 489–520.

Leentvaar, P., 1967. Duinmeren II. Zwanenwater, Muy, Oerd en Hunenplak. Biol. Jaarboek Dodonaea 35: 228–266.

Leentvaar, P., 1981. Hydrobiologie van natuurlijke duinmeren (with English summary). H2O 14, 9, 188–191.

Londo, G., 1966a. Veranderingen in flora en vegetatie van het Lekwater-infiltratiegebied in de duinen bij Zandvoort. De Levende Natuur 69, 6: 121–129.

Londo, G., 1966b. De huidige flora van het infiltratiegebied bij Zandvoort in vergelijking met andere natte duinvalleien in heden en verleden. De Levende Natuur 69: 145–151.

Londo, G., 1971. Patroon en proces in duinvalleivegetaties. Ph.D. Thesis, Nijmegen.

Londo, G., 1975. Infiltreren is nivelleren. De Levende Natuur 78: 74–79.

Manuel, A. R. *et al.*, 1984. Acidification by atmospheric deposition. Evaluation report. Ministries VROM/L&V, The Hague.

Meltzer, J. A. & H. W. J. Van Dijk, 1986. Effects of dissolved macro-nutrients on herbaceous vegetations around dune pools. Vegetatio 65: 53–61.

Nygaard, G., 1949. Hydrobiological studies on some Danish ponds and lakes. II. Kongl. Danske Vid. Selsk. Biol. Skr. 7, 293 pp.

Pigott, C. D. & K. Taylor, 1964. The distribution of some woodland herbs in relation to the supply of nitrogen and phosphorus in the soil. J. Ecol. 52: 175–185.

Ranwell, D. S., 1972. Ecology of salt marshes and sand dunes. London.

Stuyfzand, P. J., 1990. Hydrochemical facies analysis of coastal dunes and adjacent low lands: The Netherlands as an example. P. 121–132 in: Bakker, Th. W., P. D. Jungerius & Klijn, J. A., (eds). Dunes of the European coasts; Geomorphology – Hydrology – Soils. Catena Suppl. 18. Cremlingen-Destedt.

Stuyfzand, P. J. & R. J. Stuurman, 1985. Experimenteel bewijs en modellering van een zoetwaterlens op kunstmatig geïnfiltreerd water (with English summary). H2O 18: 408–416.

Stuyfzand, P. J. & F. M. L. Moberts, 1987a. De bijzondere hydrologie van kwelplassen in duinen met kunstmatige infiltratie (with English summary). H2O, 20, 3, 52–57/62.

Stuyfzand, P. J. & F. M. L. Moberts, 1987b. Hydrochemie en hydrologie van 3 soorten (ver)nat duinterrein langs Hollands kust. Report SWE 86.006, KIWA, Nieuwegein.

Texeira, R. M., 1979. Atlas van de Nederlandse Broedvogels. Vereniging tot Behoud van Natuurmonumenten in Nederland/SOVON, 's Graveland.

Udo de Haes, H. A., 1982. De gevolgen van waterwinning in de duinen. P. 133–157 in: Ten Houten, J. G.(ed.), Waddenduinen-delta. PUDOC, Wageningen.

Udo de Haes, H. A., G. J. Baeijens & H. W. J. van Dijk, 1980. Drinkwaterwinning en natuurbehoud: conflicten en mogelijkheden tot integratie (with English summary). H2O 19: 441–449.

Van Dam, H., 1987. Acidification of moorland pools: a process in time. Ph.D. Thesis, Wageningen.

Van der Laan, D., 1979. Spatial and temporal variation in the vegetation of dune slacks in relation to the groundwater régime. Vegetatio 39:43–51.

Van der Meulen, F., 1982. Vegetation changes and water catchment in a Dutch coastal dune area. Biol. Cons. 24: 305–316.

Van der Zande, A. N., 1989. Outdoor recreation and dune conservation in The Netherlands. p. 207–217 in: Van der Meulen, F., P. D. Jungerius & J. H. Visser (eds), Perspectives in coastal dune management. SBP Academic Publishing, The Hague.

Van Dijk, H. W. J., 1984a. Invloeden van Oppervlakte-Infiltratie ten behoeve van Duinwaterwinning op Kruidachtige Oevervegetaties (with English summary). Ph.D. Thesis, Wageningen.

Van Dijk, H. W. J., 1984b. Duininfiltratie op Schouwen: invloed op het natuurlijke milieu. Duin 7, 2, 19–22.

Van Dijk, H. W. J., 1985a. The impact of artificial dune infiltration on the nutrient content of ground- and surface water. Biol. Cons. 34: 149–167.

Van Dijk, H. W. J., 1985b. Bodemchemische processen bij herstel van bodem en vegetatie. p. 9–16 in Report of symposium 'Return of the wet dune slacks'. Foundation for Dune Conservation, Leiden.

Van Dijk, H. W. J., 1985c. Luchtverontreiniging: effekten op duinmilieu. Duin 1985, 2, 3–7.

Van Dijk, H. W. J., 1986. Duininfiltratie en zoetwaterlenzen in de Nederlandse kustduinen (with English summary). H2O 19, 12, 265/276–279.

Van Dijk, H. W. J., 1989a. Ecological impact of drinking-water production in Dutch coastal dunes. Pag. 163-183 in: Meulen, F. van der, P. D. Jungerius & Visser, J. (eds), Perspectives in coastal dune management. SPB Academic Publishing, The Hague.

Van Dijk, H. W. J., 1989b. Effekten van infiltratie op de vegetatie en bodem in duingebieden. Pag. 8–32 in: Roelofs, J. G. M. (ed.), Aanvoer van gebiedsvreemd water: omvang en effecten op oecosystemen. Dept. Aquatic Ecology, Nijmegen University.

Van Dijk, H. W. J. & T. W. M. Bakker, 1984. Duininfiltratie: invloed op balans en concentraties van voedingsstoffen (with English summary). H2O 17: 597–600.

Van Dijk, H. W. J. & W. T. De Groot, 1987. Eutrophication of coastal dunes by artificial infiltration. Wat. Res. 21: 11–18.

Van Dijk, H. W. J. & Meltzer, J. A., 1981. Hydrobiologie van natuurlijke duinmeren: een commentaar (with English summary). H2O 14: 564–567.

Van Dijk, H. W. J., M. A. M. Noordervliet & W. T. De Groot, 1985. Nutrient supply of herbaceous bank vegetations in Dutch coastal dunes: the importance of nutrient mobilization in relation to artificial infiltration. Acta Bot. Neerl. 34: 301–321.

Van Dijk, J. (ed.), 1981. Randstad en broedvogels. Vogelwerkgroep Avifauna West-Nederland. Tilburg.

Van Ommering, G. & J. N. C. Van der Salm, 1990. Ontwikkelingen in de broedvogelbevolking van Meijendel. Meijendel Comité, New Series no. 115. Duin + Kust, Leiden.

Van Ommering, G. & T. Verstrael, 1987. Vogels van Berkheide; 10 jaar vogelpopulatie-onderzoek in het duingebied Berkheide. Working-group Berkheide/Stichting Publikatiefonds Duinen. Leiden.

Van Dorp, D., R. Boot & E. Van der Maarel, 1985. Vegetation succession on the dunes of Oost-Voorne, The Netherlands, since 1934 interpreted from air photographs and vegetation maps. Vegetatio 58: 123–136.

Van Eeden, F. W., 1886. Onkruid. Haarlem.

Van Oosterhoud, E., G. C. Janze, W. T. De Groot & H. W. Van Dijk, 1982. Fosfaat en duininfiltratie: een experimentele benadering (with English summary). H2O 15: 497–502.

Van Zadelhoff, F. J., 1981. Nederlandse Kustduinen; Geobotanie. Ph.D. Thesis, Wageningen.

Vuyck, L., 1898. De plantengroei der duinen. Ph.D. Thesis, Leiden.

Van der Werf, S., 1974. Infiltratie: met water meer plant. P. 226–231 in: N. Croin Michielsen (ed.). Meijendel, Duin-Water-Leven. W.v.Hoeve, The Hague/Baarn.

Westhoff, V., 1989. Dunes and dune management along the North Sea coasts. p. 41–51. In: van der Meulen *et al.* (eds), Perspectives in coastal dune management. SBP Academic Publishing, The Hague.

Hydrobiologia **265**: 305–320, 1993.
E. P. H. Best & J. P. Bakker (eds), Netherlands-Wetlands.

The ecology of The Netherlands wetlands: characteristics, threats, prospects and perspectives for ecological research

E. P. H. Best[1], J. T. A. Verhoeven[2] & W. J. Wolff[3]
[1] *DLO Centre for Agrobiological Research, P.O.Box 14, 6700 AA Wageningen, The Netherlands;*
[2] *Department of Plant Ecology and Evolutionary Biology, University of Utrecht, Sorbonnelaan 16, 3584 CA Utrecht, The Netherlands;* [3] *DLO Institute for Forestry and Nature Research, P.O. Box 46, 3956 ZR Leersum and Dept. of Nature Conservation, Agricultural University, Wageningen, The Netherlands*

Key words: wetland characteristics, threats, conservation, rehabilitation, management, ecological research, The Netherlands

Abstract

Wetlands are among the worlds' most important, but also most threatened, environmental resources. Wetland losses have been in progress particularly from the industrial revolution onwards, because wetland functions could not successfully compete for space with other land uses. Wetlands became recently foci of conservation efforts because of the increased awareness of their importance in water management and wildlife conservation, and because of the diversity of their habitats.

The Netherlands are relatively rich in wetlands: 16% of its' territory is regarded as internationally important wetland and 7% has been registered as such. The major Dutch wetland types are: coastal ecosystems, large riverine systems, base-rich freshwater systems, and nutrient-poor freshwater systems.

Most threats to the Dutch wetlands are of man-made origin. They comprise: (1) Changes in hydrology leading to changed discharges, currents and desiccation; (2) Acidification; (3) Eutrophication; and (4) Toxification. Long-term threats are largely climate-change related, and concern temperature rise and the UV-B increase in irradiation.

General conservation goals also apply to wetlands but Ramsar-registered wetlands have a special status. Conservation of the Dutch wetlands is difficult, because of the high population density of the country and its inherent threats. However, ecological targets and standards are increasingly set in national Policy Plans and international agreements. Rehabilitation and creation of wetlands is presently widely advocated, and sometimes realised.

For ecological research, the sustainability of wetlands should get top priority. Such a research programme would focus on understanding the underlying ecological processes in natural and man-dominated wetland systems to prescribe conservation, rehabilitation and management strategies that would enhance the sustainability of these systems. Within this framework special attention should be directed to studies (1) At the ecosystem level of ecosystem parameters, of which natural oscillations and trends in time, and on which the impact of disturbances are quantified. Particularly these studies, in which often simulation models are used as tools for interpretation, can provide the basis for extrapolations in space and time; (2) On adaptation capacity and mechanisms of (groups of) species to extreme environmental conditions; (3) On (mutual) relationships between plants, animals and microorganisms (e.g. competition, grazing and mineralization); (4) On dispersion between small wetlands.

For the contemporary quantitative assessment of the long-term effects of climate changes, the effects of temperature rise and increase in UV-B irradiation on individual species, communities and ecosystems should also be studied.

Introduction

Wetlands are among the worlds' most important, but also most threatened, environmental resources. Wetland losses have been in progress particularly from the industrial revolution onwards, because wetland functions could not successfully compete for space with other land uses. In the twentieth century, wetlands have become the foci of conservation efforts because of the increased awareness of their importance in water management and wildlife conservation, and because of the diversity of their habitats. Unfortunately, many of the benefits of wetland conservation have been neither clearly perceived nor wisely used. Understanding wetland ecosystems is a major goal of contemporary ecological science, and is a prerequisite to land use management philosophies that will permit them to be a harmonious interface between human populations and nature.

Wetlands generally are very dynamic systems: they are subject to strong forces, largely linked to the interaction between wind and water. The wetland ecosystem type, and its' structure and functioning vary according to genesis, geographical location, water regime, water and soil chemistry, as well as dominant plant and animal species (Maltby, 1988). This is also true for the wetlands in The Netherlands, but here the situation of the inland wetlands in a largely cultural landscape where watertable and water quality are continuously managed by man is also a paramount feature.

In this paper the characteristics of the major Dutch wetland types are summarized and their threats discussed. Contemporary conservation and management issues are indicated, and actual and potential rehabilitation and creation plans described. Finally, perspectives for ecological research are indicated.

Characteristics of the major wetland types of The Netherlands

Wetlands presently (1991) cover a considerable part of The Netherlands. About 16% of this countries' territory is regarded as internationally important wetland and 7% has been registered as such (Wolff, 1993). This is a high share, because on a global scale, wetlands cover an estimated 6% of the world's land surface (Maltby & Turner, 1983). The situation of the major wetland types in The Netherlands is indicated in Fig. 1. A short list of characteristics concerning hydrology, nutrient sources, area and conservation status is given in Table 1.

The Dutch coastal wetlands consist of estuaries (water, tidal flats and salt marshes) of the Rivers Rhine, Meuse, Scheldt and Ems, and of the extensive tidal wetland of the Wadden Sea. They are heavily influenced by the discharge of the R. Rhine, which causes a local decrease in salinity, and large differences in sedimentation rates and nutrient cycling. Eutrophication is evident, but varies in intensity and substance with locality. Their main conservation value lies in serving as forage and breeding area for water birds and in their nursery function for marine fishes (Nienhuis, 1993; De Jonge *et al.*, 1993; Wolff, 1993).

The large rivers Rhine and Meuse play a vital role in the hydrology of The Netherlands. The Rhine receives meltwater from the Alps and rainwater from its catchment, and has a large, relatively stable discharge. The Meuse receives rainwater only, and has a smaller discharge with a distinct summer minimum. The water systems of both rivers comprise both fast-flowing riverine reaches and (partly) tidal estuaries. They encompass a great variety of wetlands in their floodplains. Both rivers are eutrophic and carry high loads of toxicants (Admiraal *et al.*, 1993).

Classification of the other Dutch freshwater

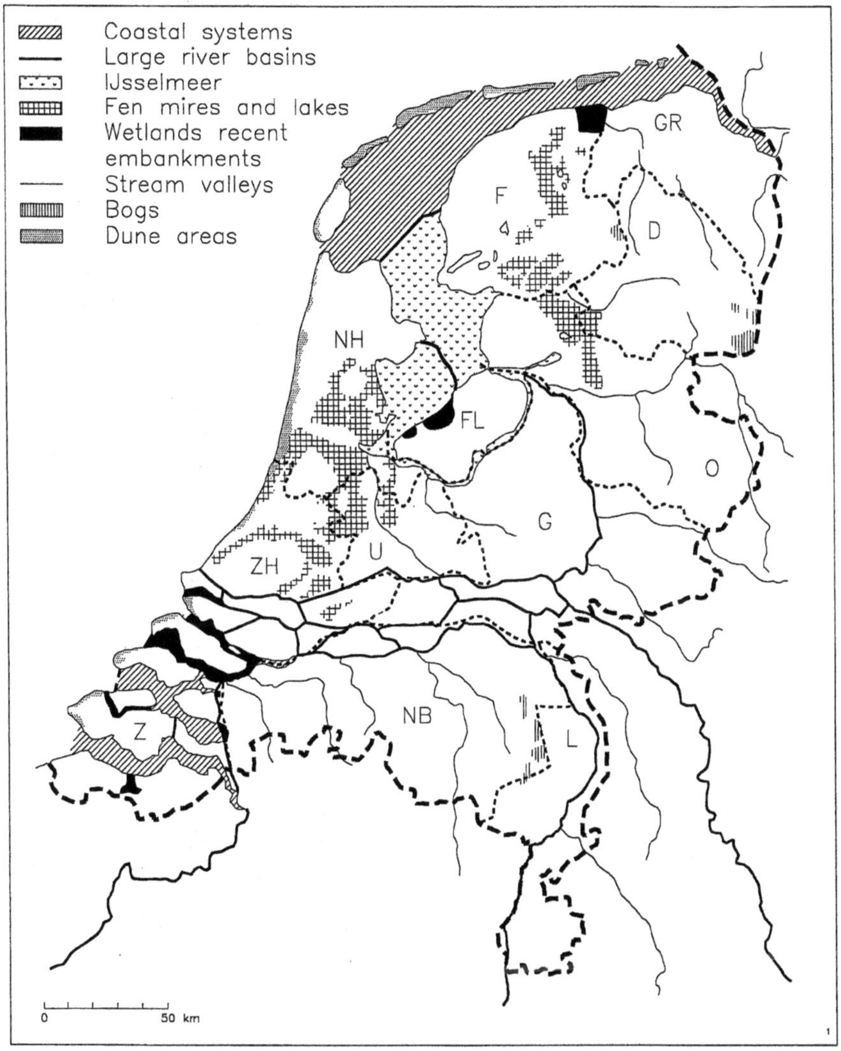

Fig. 1. Major wetland types in The Netherlands. After Bosatlas (1976) and De Bakker & Locher (1991).
Initials indicate the provinces. NH, Noord-Holland; ZH, Zuid-Holland; Z, Zeeland; NB, Noord-Brabant; L, Limburg; U, Utrecht; FL, Flevoland; G, Gelderland; O, Overijssel; D, Drente; F, Friesland; GR, Groningen.

wetlands is difficult. According to their ontogenesis, they can be divided into two main groups, being base-rich or base-poor (Van Wirdum *et al.*, 1992).

The base-rich freshwater wetlands consist largely of primary and secondary mires, including marshes, swamps, lakes and fens. These systems formerly occurred under a range of trophic conditions with many oligotrophic-eutrophic gradients. Most of them are now eutrophic (Van Wirdum, 1993; Van Wirdum & Koerselman, 1993; De Haan *et al.*, 1993).

Young base-rich freshwater wetlands originate from more recent reclamation from the sea and from inland lakes (Joenje & Verhoeven, 1993). Several estuaries have been reclaimed from the Wadden Sea in the north and from the North Sea in the south-west. Over 400 inland lakes were reclaimed and turned largely into agricultural land, parts of which became wetland. The wetland IJsselmeer originates from 1932, when the brackish lagoon Zuiderzee was dammed and desalinised. This system serves as the countries' largest surface freshwater buffer, which is largely

Table 1. Characteristics of the major wetland types of The Netherlands, in 1991. After Bosatlas (1976), De Bakker & Locher (1991) and various contributions to this volume.

System type	Water		Nutrient source	Surface area (km^2)	Characteristic important for conservation
	Table	Movement			
Coastal systems					
Wadden Sea and Delta area	Fluct. (1.5–4 m)	Strong current	Atm. deposition, water transport	2400 600–700	Sand-, tidal flats, creeks and open water; forage and breeding area for water fowl
Salt marshes	(Ir)regular inundation	Depending on tide	Atm. deposition, sedimentation	128	
Large river systems					
Rhine and Meuse basins					
– water	Fluct.	Strong current	Atm. deposition, large catchment area	385 [a]	Open, fast flowing waters, relatively poor in plankton and fish
– river flood-plains	Spring and winter inundation	Small; strong current at inundation	Atm. deposition, river sediment, and -water	270	Natural levees and hollows, covered by grassland and scrubs between summer bedding and winter dikes
– former river branches	Small fluct., occasional to regular inundation	Small; strong current at inundation	Atm. deposition, groundwater sometimes, river water	[b]	Areas of extensive marshes and shallow open water developed in the standing water of the floodplains
– IJsselmeer	Small fluct.	Flow and wave action depend on local hydrol. and fetch	River water, precipitation	2020 [c]	Shallow open water, forage area for waterfowl
Base-rich freshwater systems					
Fen mires	Fluct. <1 m	Small or absent	Atm. deposition, river water, pollution by agriculture and sewage, soil mineralization after drainage	140 [d]	Species-rich marshy areas dominated by sedges, situated in agricultural landscape; often developed on former peat excavations
Fen lakes	Fluct. <1 m	Flow and wave action depend on local hydrol. and fetch	Atm. deposition, river water, pollution by agriculture and domestic wastewater	240 [d]	Shallow open water, forage and breeding area for waterfowl
Wetlands recent embankments	Small fluct.	Small or absent	Atm. deposition, surface water, groundwater	50 [e]	Large swamps dominated by reed; breeding and overwintering area for waterfowl

Table 1. (Continued)

System type	Water		Nutrient source	Surface area (km^2)	Characteristic important for conservation
	Table	Movement			
Nutrient-poor freshwater systems					
Bogs	Near surface	Slow seepage	Atm. deposition, soil mineralization after drainage	36	Hummocks and hollows in flat landscape; vegetation dominated by Sphagnum species
Moorland pools	Small fluct.	Small or absent	Atm. deposition, bird excrements, sometimes surface water	7	Shallow pools in formerly extensive moorlands; aquatic flora dominated by desmids and isoetids
Stream valleys	Large fluct.	Strong current	Atm. deposition, regional drainage, pollution by agriculture	[b]	Small streams accompanied by grasslands; most original forest lost
Wet dune slacks	Small fluct.	Small, increased by river water infiltration	Atm. deposition, infiltration water, ground water, salt spray	400[f]	Oblong valleys between high dunes; characteristic species-rich vegetation

[a] Rhine and Meuse: 135 km^2; Hollands Diep, Haringvliet and Biesbosch: 250 km^2
[b] Surface area unknown.
[c] Including Markermeer
[d] The total fen area in which the mires, lakes, arable land and pastures are situated measures 3773 km^2. The fens are important foraging and breeding areas for meadow birds.
[e] Oostvaardersplassen area only (centre of the country).
[f] Total dune area, of which the wet dune slacks are a small part.

Table 2. Threats of the major wetland types of The Netherlands.

System type	Desiccation	Acidification	Eutrophication	Toxification	Climate change
Coastal systems					
Wadden Sea and			+	+	+
Delta area			+	+	+
Salt marshes			+	+	+
Large river systems					
Rhine and Meuse basins			+	+	+
Base-rich freshwater systems					
Fen mires and lakes	+		+	+	+
Wetlands recent embankments			+	+	+
Nutrient-Poor Freshwater Systems					
Bogs	+	+	+		+
Moorland pools	+	+	+		+
Stream valleys	+	+	+	+	+
Wet dune slacks	+	+	+	+	+

fed, indirectly, by Rhine water. All reclaimed systems are eutrophic.

The base-poor freshwater wetlands comprise tertiary mires (bogs). All have once been oligotrophic. Some of them, however, are being enriched due to the increase in nutrients of precipitation and the inflow of nutrient-rich surface water.

The stream valley wetlands consist of mere stream channels (Higler & Verdonschot, 1993). They are both fed by rain and groundwater, and are oligotrophic in the upper courses but progressively eutrophic downstream.

The Dutch wet dune slacks originate from the impact of wind and water on the sand dunes along the coast (Van Dijk & Grootjans, 1993). Wetlands developed mostly in the secondary dune slacks in the inner, relatively old, dune zone which is chalk-rich south of the village of Bergen (halfway the province of Noord-Holland) and chalk-poor northward. The water in these slacks comes in the first place from a nutrient-poor aquifer, which is hydrologically isolated from other water bodies, and secondly from precipitation. The ratio between the inputs from both water sources determines the nutrient status of the water, which is, however, mostly nutrient-poor.

Threats

The threats to the various wetlands are related to the morphology and the forcing variables of the system. They can be divided into current and long-term ones (Table 2).

Current threats

Current threats are due largely to human activities: (1) Changes in hydrology leading to increased discharges, decreases in water level fluctuation, drawdown, desiccation and subsequent mineralization of the topsoil; (2) Air pollution leading to acidification, eutrophication and toxification of terrestrial and aquatic ecosystems; and (3) Direct eutrophication and toxification of surface waters.

Desiccation

Desiccation is widespread in The Netherlands. The causes of desiccation are drainage and increased use of surface water for agriculture, together with greater withdrawal of groundwater for drinking water purposes. About thirty percent of the land suffers severely (classes moderately-severely, severely desiccated; Intergroep Verdroging, 1989) from desiccation. The affected areas are: (1) Wetlands in the elevated Pleistocene grounds in the east and south-east, comprising largely the bogs, moorland pools and stream valleys; (2) The dunes all along the coast, comprising the wet dune slacks; and (3) The eutrophic fen mires in the center of the country (Fig.2).

In the dune areas most current threats are linked to hydrology, and involve changes in the mixing ratio between groundwater and precipitation. The man-made extraction of groundwater, and infiltration of surface water (and subsequent extraction) for drinking water purposes also contribute. Extraction of a part of the precipitation surplus causes shrinking of the subterranean freshwater lens and subsidence of the groundwater level, while total extraction may ultimately cause salinization of the aquifer. Desiccation occurs often as a side-effect, which can, however, also be caused by land use changes such as pine afforestation which increases evapotranspiration. Infiltration has locally similar effects as extraction.

In peaty areas, eutrophication occurs as a side effect because of increased mineralization in the drying peat (Grootjans, 1985). Allochthonous water supply has been used extensively to alleviate desiccation problems (Fig.3). However, this also usually causes deterioration of the original flora and fauna due to changes in habitat factors, since the qualities of allochthonous and autochthonous water are different (Barendregt *et al.*, 1985). Moreover, most surface water in The Netherlands comes from the heavily polluted river Rhine.

Acidification

Acid deposition affects the whole country severely. It is the highest of Western Europe and

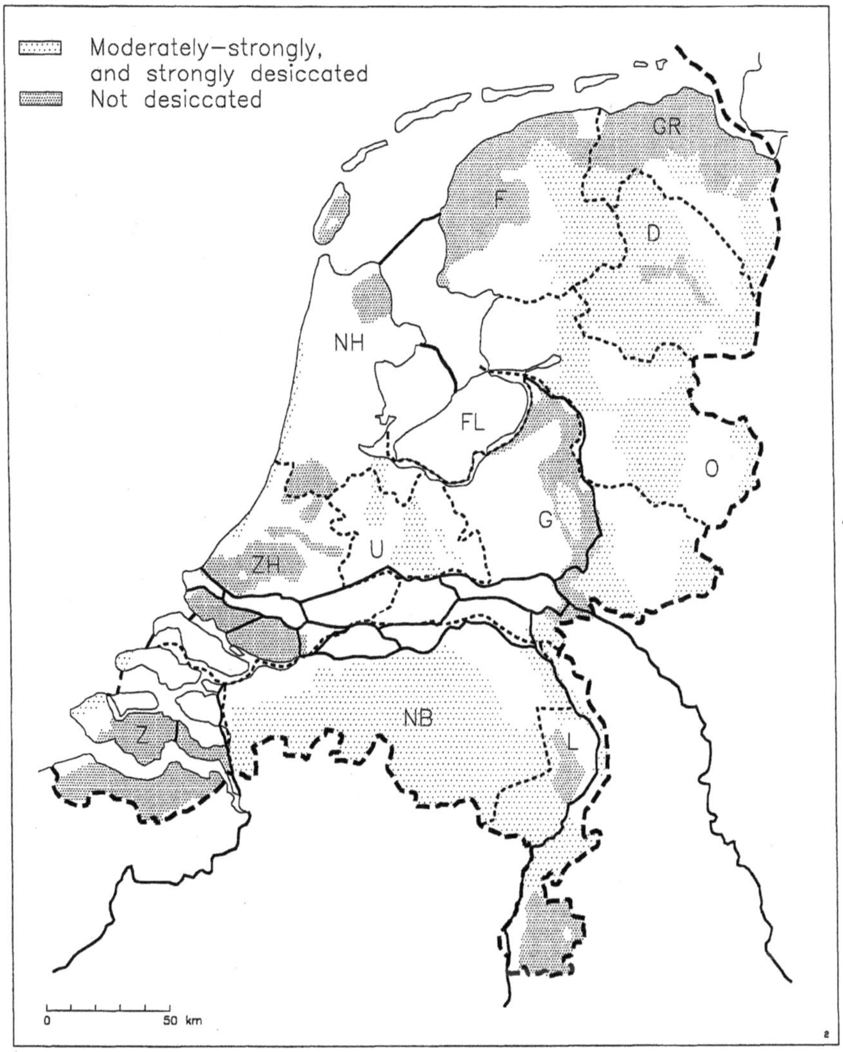

Fig. 2. Distribution of desiccated (moderately-strongly, strongly) and non-desiccated areas in The Netherlands. The blank areas suffer in various degrees from desiccation, between not and moderately-strongly. After Intergroep Verdroging (1989). Initials indicate the provinces.

originates largely (85%) from industrial activity outside the country (Care for Tomorrow, 1989). The critical value of 1400 mol.ha^{-1}.yr^{-1}, i.e. the value at which long-term detrimental effects on the biota are expected, has been surpassed everywhere (Table 3).

Effects on the abiotic environment are decreases in pH of the open and pore water, release of heavy metals and increases in nitrate and sulphate, which usually result in decreases in primary production and deterioration of flora and fauna. The effects on the base-rich freshwater and coastal wetlands are delayed because here the base-containing substances act first, as a buffer. The effects on the base-poor wetlands are particularly severe. Of the flora, especially the submerged aquatic macrophytes of such wetlands are affected because they can only obtain their carbon from the water at a suitable pH: many species, therefore, disappear. Of the fauna, many invertebrate animals are very sensitive to $pH < 6.0$, while fishes are not. However, the reduction of the biomass at the lower trophic levels will cause starvation of the higher trophic levels and thus affect fish populations indirectly.

The prognoses for the year 2010 are potential

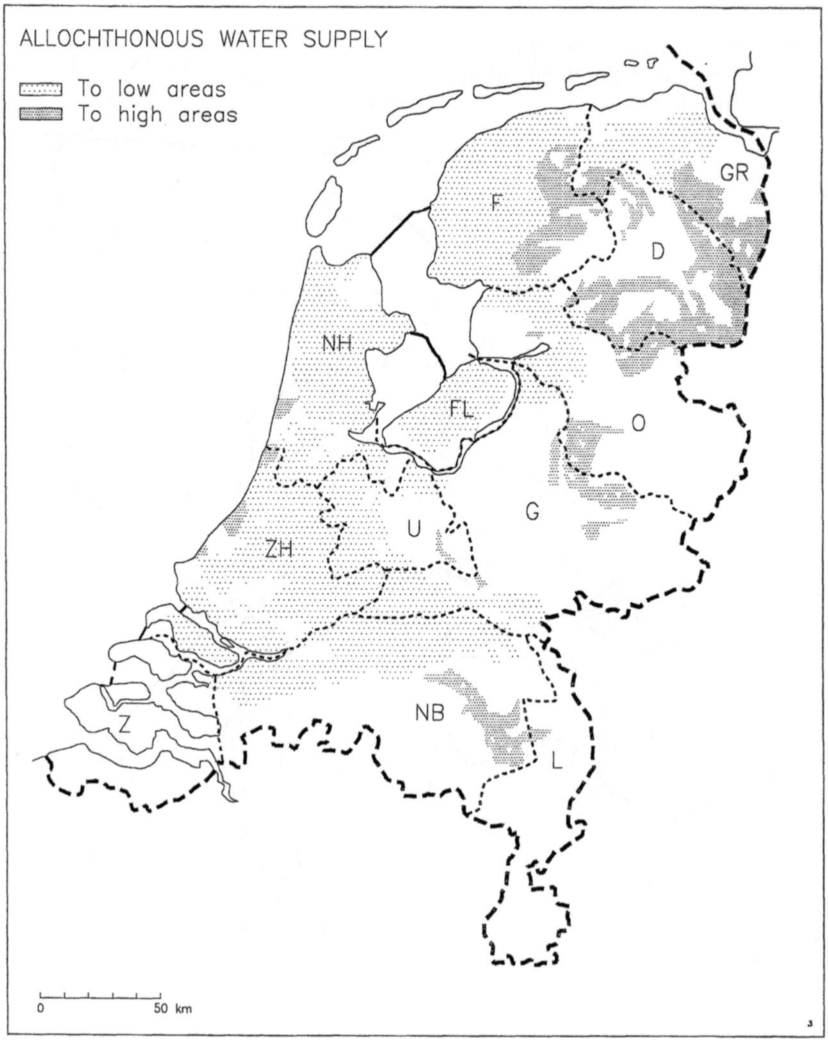

Fig. 3. Supply of allochthonous water in The Netherlands. After Unie van Waterschappen (1978) in Udo de Haes (1983). Initials indicate the provinces.

acid deposition values of >2000 mol.ha^{-1}.yr^{-1} under the present environmental regulations, and >1500 mol.ha^{-1}.yr^{-1}, when severe measures would be enforced by the Dutch government limiting emission by Dutch industry, bio-industry and car-users. Acid deposition may decrease to acceptable levels only by a coordinated European effort to stem the emission of acidifying substances (<1000 in the North of the provinces Noord-Holland, Friesland and Groningen, 1000–1500 in the rest of the country, except for the northern part of Limburg, where it will remain 1500–2000 mol.ha^{-1}.yr^{-1} in the year 2010; Care for Tomorrow, 1989).

Eutrophication

Eutrophication is caused by NOx, NHy, $K+$ and phosphate. In terrestrial and marine systems, eutrophication is mainly caused by nitrogenous substances and potassium. In freshwater aquatic systems phosphate concentration is more critical. Wetlands, which contain both terrestrial and aquatic landscape elements, often suffer, therefore, from all substances responsible for eutrophication.

Many nature reserves are mainly impacted by atmospheric deposition of nitrogenous substances, but also by polluted surface- and groundwater. General effects are loss of species diversity

Table 3. Potential acidification in The Netherlands. The reference value is 1400 mol $ha^{-1} yr^{-1}$. After Care for Tomorrow (1989).

Measures	Potential acid deposition in 2010 (mol $ha^{-1} yr^{-1}$)
The Netherlands	
Present governmental measures	> 2000; whole country
Extra technical measures	> 1500; whole country
Europe	
Control measures, aimed at	< 1000; northern part of the country
2400 mol $ha^{-1} yr^{-1}$ in 2000 and	1500–2000; northern part of Limburg
1400 mol $ha^{-1} yr^{-1}$ in 2010	1000–1500; remainder of the country

and increases in primary production. The no-effect level (critical load, i.e. level at which the basic cation content- mainly Ca and Mg- of the soil remain unaffected) of loading for vulnerable systems like fens, estimated at 400 mol $N.ha^{-1}.yr^{-1}$, is presently surpassed and it will remain so with any probable scenario. This implies that all sensitive systems are in danger both at present and in the near future. The nitrate concentration in groundwater is steadily increasing. High concentrations occur particularly in sandy areas, surpassing the drinking water standard of 50 $mg.l^{-1}$. It is expected that this situation will not have ameliorated significantly in 2010 (Third Water Policy Plan, 1989).

High concentrations of phosphate stimulate growth of highly productive algae and cyanobacteria in saline and fresh surface waters. This leads to a decrease in water transparency and consequent elimination of low-productive algae and submerged aquatic macrophytes. The plankton blooms are often succeeded by fish kills caused by decreases in oxygen concentration due to decomposing algal cells or toxic algae. The large rivers are the main source of phosphates and nitrogenous compounds for The Netherlands. The Rhine water causes severe eutrophication of the sea along the Dutch coast North of Hoek van Holland and in the western Wadden Sea. River water, largely from the River Rhine and to some extent from the River Meuse (Udo de Haes, 1983), is used to supply the country with water at times of shortage. Water shortage occurs relatively often because a large part of the country is drained intensively, mainly for agriculture. The Rhine water is supplied to high as well as low parts of the country (Fig.3), and poses a threat to all wetlands receiving this water. Presently the no-effect level for any natural system (ecological standard which varies with the water system type; CUWVO, 1988) has been surpassed and it will remain so, unless an extra reduction in phosphate loading within the Rhine catchment areas to about 75% occurs. Now and in the near future particularly the Westerschelde estuary, the Ems-Dollard estuary, the Wadden Sea and the large river basins (including all sensitive ecosystems dependent on the Rhine water) are threatened.

Toxification

Xenobiotics and heavy metals can have disruptive effects on the functioning of wetlands in The Netherlands. Most effect studies pertain to individual species, but studies at the ecosystem level are scarce and often biased because of the simultaneous occurrence of acidification, desiccation or eutrophication. It can be expected that systems which are suffering already from acidification are extra sensitive for heavy metal deposition (Hekstra & Van Linden, 1991).

Long-term threats

Long-term threats are expected to occur mainly because of climate changes which are also partly of man-made origin. Because the climate changes strongly affect the hydrological cycle, which is so critical for wetlands, wetlands are expected to be very sensitive to these changes.

Two climate change-related factors are considered important for wetlands: the rise in temperature and the increase in ultraviolet-B irradiation.

– The increase in temperature, in The Netherlands (53° N; Atlantic climate) expected to be about 0.3 °C.decade^{-1} up to the year 2000 and certainly not less later on, may change the direction of the Gulf Stream in the Atlantic Ocean, affecting the climate of NW-Europe (Bakker *et al.*, 1991; cf. IPCC workgroup I). However, also without a change of the Gulf Stream, climate change may result in shorter winters, warmer winters and summers, and a 10% increase in precipitation with peaks in winter and early spring leading to greatly enhanced discharges of rivers but desiccation due to increased evapotranspiration in other areas. The growing season may increase as may primary production and mineralization. A shift in species composition may occur due to an influx of southern species and an efflux of temperate species northwards, leading to either an increase or a decrease in species diversity.

The expected changes in climate will affect the coastal systems largely by their impact on the dynamics of wind-water interactions and hydrological cycle (Bakker *et al.*, 1991; cf. IPCC workgroup II). The sea level rise (mean relative sea level rise estimated at 0.10–0.16 cm.yr^{-1} over the last two decades, cf. Dijkema *et al.*, 1990) threatens mainly the coastal wetlands, tidal flats as well as salt marshes. The speed with which the coastal wetlands will disappear depends on the actual sea level rise, of which predictions are very difficult to make and which varies with location. The mainland salt marshes can withstand a rate of sea level rise similar to the present one because of increased accretion, but the marshes on the barrier islands probably can not. The low-lying wetlands inside the dykes (Fig.1, Fig.4) will be protected by increasing the dyke height. The climate changes may cause the disappearance of bogs from The Netherlands (Casparie, 1993).

– The UV-B irradiation reaching the earth surface is expected to increase in total dose and in exposure period. Organisms may protect themselves by changes in their behaviour (migration and avoidance by animals; e.g. zooplankton and fish), shielding (pigment migration in diatoms of the tidal flats) and pigmentation (lower in crab carapaces in temperate regions). Ultimately, however, photosynthesis of particular species may be inhibited, as may growth, development, and fecundity, and, therefore, also survival. This will lead to shifts in species composition. Relatively small UV-B effects are expected in the coastal waters because of their high turbidity. Wetlands with temporarily dry parts (salt and freshwater marshes, bogs), very shallow water systems (small streams) and clear deep water systems, however, will be greatly affected (Kramer, 1990).

Conservation and management of present wetlands

General

The Netherlands are rich in wetlands. Most of these wetlands are currently protected, others are not. A large part of the present area of nature reserves (63%) contains wetlands. It is to be expected that the area of wetlands will increase in the near future (national Nature Policy Plan, 1989). The locations of these future wetlands are not really known, although government policy is to create a wet axis from the south-west to the north-east of the country composed by ecological nuclei interconnected by corridors (national Nature Policy Plan, 1989).

Contemporary wetland conservation goals are: (1) The preservation of wetland ecosystem types registered on the international wetland list in their present state; and (2) The protection of higher organisms listed nationally and/or internationally as rare, vulnerable and/or endangered. However, relatively few wetlands are registered on the Ramsar list. Recently a large fen mire in Overijssel (Weerribben; 3,400 ha) has been registered, and a formerly tidal wetland (Biesbosch) and a raised bog system in Noord-Brabant (Grote Peel; na-

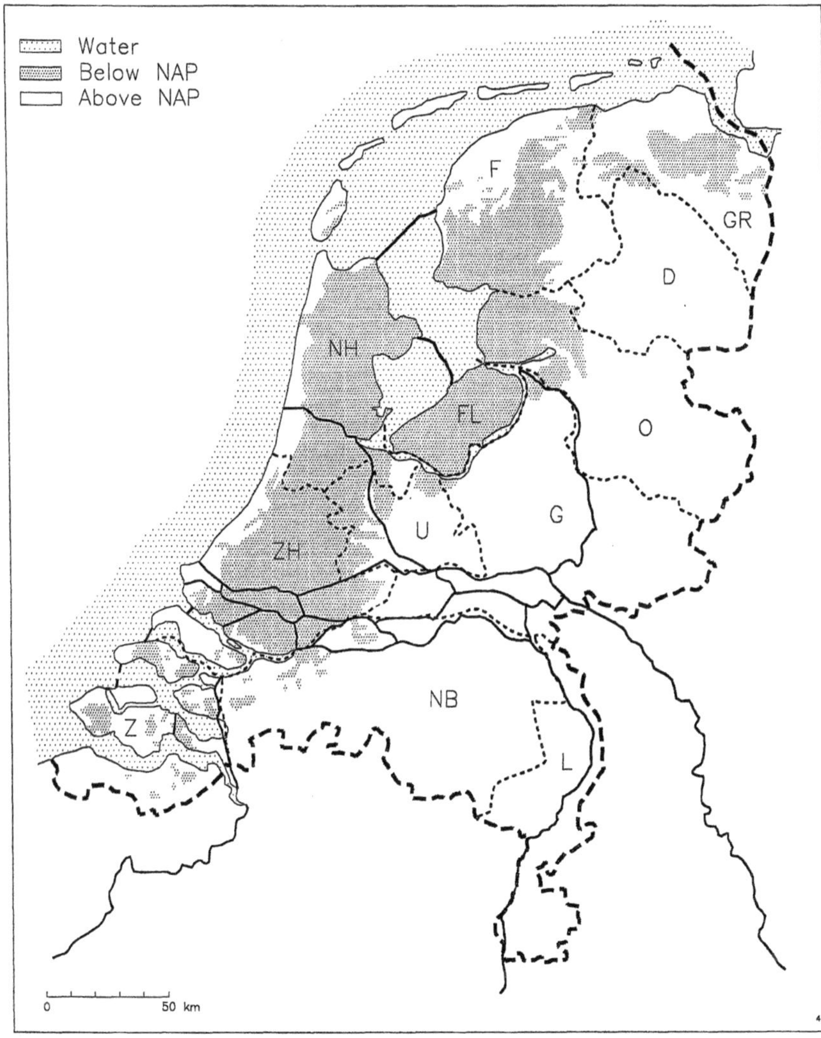

Fig. 4. Situation of The Netherlands relative to sea level. NAP: Nieuw Amsterdams Peil (Dutch ordnance level). After Bosatlas, 1976.

tional Nature Policy Plan, 1989) will be added in the future. It is planned by the government to expand the list by adding on average five wetlands per year from 1992 onwards (Tweede Kamer, 1990–1991).

Management for conservation

Conservation of wetlands depends largely on water management.

Integrated water management

The present, national Dutch approach of integrated water management aims at managing water systems (or land systems where water is an essential part) together with the associated lakes and riverbeds, banks and groundwater, as one complete unit in relation to the human interests (Third Water Policy Plan, 1989; Saeijs, 1991). It is carried out to meet preset standards with regard to quantity, and chemical and ecological quality.

Important premisses are, that: (1) Water systems function as an entirity, and that coherence in diversity must be preserved; (2) A large number of methods is available to control a system (level, salinity, residence time, collecting, transporting, extracting, infiltrating, constructing barrages etc.); many techniques used have their roots in ecology; however, application must be carefully considered; (3) Many parties are involved in management of water systems, but all may have different and sometimes conflicting demands on the system; interests and possibilities must be weighed; (4) Water, and everything related to it, is a dynamic continuum and subject to changing jurisdiction.

Desiccation

Desiccation is widespread and countermeasures are required urgently. The transport to and the inlet of allochthonous water, however, has proven to be detrimental to most wetland types. Therefore, the inlet of Rhine water in sensitive wetlands must be limited as far as possible by autochthonous water conservation and sensible water management. The perpetuation of extremely sensitive systems like bogs and fens, however, will only be possible by complete, artificial hydrological isolation on top of that (Schouwenaars, 1993; Van Dam & Buskens, 1993). These measures are expensive, especially the latter, and it has to be seen whether they will be successful because of other threats (e.g. acidification).

Acidification

Acidification is severe and should be limited, which can only be enforced by international agreements and inherent measures. If these measures take effect relatively rapidly, the wetland types in which the hydrological cycle is dominated by precipitation may be preserved. These systems are bogs, moorland pools and stream valleys. This means for example for moorland pools, a potential acid deposition of <400 $mol.ha^{-1}.yr^{-1}$ and ammonia deposition of <300 $mol.ha^{-1}.yr^{-1}$ (Van Dam & Buskens, 1993).

Eutrophication and toxification

International agreements and consequent measures to limit the nitrogen, phosphorus and toxicant loads of the river Rhine are already having their effect today. Standard water quality levels are generally attained, except for chloride. A Rhine Salt Agreement was signed by the Rhine-bordering countries in September 1991, which sets an upper limit of 200 mg $Cl.l^{-1}$. However, standard Rhine water quality is often still detrimental for other wetland systems, particularly the nutrient-poor ones. These systems should preferably receive water of an ionic composition similar to that of its own. No similar agreements exist for the river Meuse.

Specific management

Some wetland types require specific measures on top of those required in general management practices. The most important ones are salt marshes and moorland pools.

Salt marshes

Salt marshes depend for their very existence on the balance between erosion and sedimentation, which in turn is strongly influenced by water movements, mean high tide level and sediment discharge of the large rivers. Salt marshes used to be exploited by man, but this has strongly decreased today. Management included hay-making and grazing by domestic animals. The marshes have a history of agricultural use, which strongly determines the ecosystem structure and functioning, and led in some cases to high species richness. Few marshes have never been exploited which gives them their intrinsic value. International agreement was reached between the Wadden Sea-bordering countries on the following recommendations: erosion-sedimentation interactions and drainage should not be affected artificially, and grazing should be stimulated in such a way that grazed and ungrazed areas alternate producing a diverse vegetation pattern (see Bakker *et al.*, 1993).

Moorland pools

The Dutch moorland pools usually lie in a sandy area (former bog) and, consequently, the water level is linked to the watertable in the adjoining area. Thus, a suitable water level in the pool requires the planning of coinciding regional watertable management.

In the past the isolated hydrology of bogs was

favourable, but nowadays it often is not because it increases acidification and eutrophication. Presently, addition of buffering capacity (by liming or the inlet of nutrient-poor, hard water) is a prerequisite for the preservation of these pools (Arts & Buskens, 1989).

The moorland pools should be situated in an open landscape, because nearby afforestation affects these systems negatively (increased acidification and eutrophication, inhibition of the formation of shore vegetation; Van Dam & Buskens, 1993; Arts & Buskens, 1989).

Rehabilitation and creation

Rehabilitation and creation of any particular ecosystem requires first of all clearly formulated criteria. General criteria for the ecosystems described in this volume can be given obviously. Specific criteria, however, for the ranges within which nutrient cycling should take place, for the type of food chain, for typical species composition etc. are far more difficult to give and can often not be formulated because of lack of comprehensive information. Establishment of criteria for a network of ecosystems required to maintain migratory animals like birds is an even more daunting task.

The present Dutch approach to rehabilitation and creation is largely as follows (national Nature Policy Plan, 1989).

To plan the rehabilitation of an ecosystem it should be clear what the system was like in the past, what it is at present and what the potential rehabilitation measures are. Ecosystem types change with time and, therefore, we should be able to choose a moment in time when a particular system was functioning in the desired manner. Moreover, the system should have functioned in the presence of man. This reference point has been suggested to have occurred around 1930 for the Dutch coastal waters and before 1900 for the major rivers (together comprising 80–90% of the wetlands), i.e. when the climate was similar, and eutrophication and toxification had not yet taken place. The numbers of macroscopic animals and plants recorded at that time are taken as reference due to lack of information on functional system parameters. The system to be rehabilitated is compared with the reference system. Deviations are quantified and differences acceptable to the public formulated. This process usually is followed by taking measures which influence the forcing variables of the system, like changing the water table, nutrient inputs etc. This Dutch approach is called the AMOEBE approach (Algemene Methode OEcosysteem BEoordeling; General Method of Ecological Assessment; Ten Brink & Colijn, 1990; Colijn *et al.*, 1993).

To plan the creation of a particular system it should be clear what the reference system is like, what the present area is like and what the potential creation measures are. Natural ecosystems are self-regulatory, and have a sustainable production and species composition. Virtually all Dutch wetlands are presently situated in a cultural landscape and several of them, therefore, lack these characteristics. Consequently, in the latter cases reference systems from other European countries with a similar climate as The Netherlands could serve as example. Creation of new wetlands can be executed largely along lines similar to those for rehabilitation, but creation usually involves more drastic changes in morphology and hydrology.

Coastal areas

In the coastal areas new salt marshes may develop temporarily along the coast and at the landside of the barrier islands due to increased sedimentation, partly of anthropogenic origin (including coastal engineering and dredging activities).

Large rivers

The Rhine Action Programme (IKSR, 1987) provides a framework for the large-scale rehabilitation of the Rhine. It is the first pollution abatement programme to include ecological objectives for the Rhine (Smit & Van Urk, 1987). It is aimed at: (1) Creating conditions which would enable the return of higher organisms (rehabilitation of floodplains, restocking with the Atlantic salmon);

(2) Safeguarding the use of Rhine water for the supply of drinking water; and (3) Decreasing the contamination of sediments with toxic compounds. A similar programme could be developed for other large rivers, such as the Meuse and Scheldt.

However, the large changes in the ecosystem functioning of the large rivers are mainly due to physical changes and only partly because of eutrophication and toxification. The AMOEBE approach for the Rhine shows that there is a large discrepancy between the biotic components around 1930 and at present. Because the physical changes in the system are largely irreversible, it can not be expected that even if eutrophication and toxification are halted the original communities will return (Admiraal *et al.*, 1993). Therefore, it remains to be seen whether the AMOEBE approach results in effective restoration.

Stream valleys

Small-scale rehabilitation projects are presently in progress, initiated by governmental and private organizations for nature conservation. These involve relatively simple adaptations in morphology often combined with measures like afforestation of the stream sides and restocking of salmonids. These measures, however, would not ameliorate water quality (Higler & Verdonschot, 1993).

Large-scale rehabilitation is expected to be more effective. This would affect large part of, or the whole catchment area; involve restoration of former discharge patterns by reconstruction of stream beds, and changing land use and consequent draining and fertilization practices (e.g. Bakker & Grootjans, 1991). However, even if these measures are taken, it is still to be seen whether and when the surface and ground water quality will be good enough to support biotic communities similar to those which used to live in these ecosystems.

Moorland pools

Moorland pools can be suitable objects for rehabilitation and creation. Successful rehabilitation measures have been proven to be the removal of polluted sediment, followed by the inlet of chalk-rich, nutrient-poor water or liming (Van Dam & Buskens, 1993). Creation of new moorland pools is feasible in (former bog) areas with suitably buffered qualities of surface and groundwater. It involves the excavation of relatively shallow sand pits. The sand can then again be used for other, possibly economic, purposes.

Wet dune slacks

To rehabilitate the wet dune slacks first of all the viewpoints for drinking water extraction should be revised. Keeping the original mixing ratio between autochthonous groundwater and precipitation intact would probably be a good criterion. This necessitates cessation of the surface water infiltration, for which depth infiltration with pretreated water may be an alternative in some cases. Although the regional authorities and Dutch parliament largely agree to this viewpoint, so far practice has proven differently. Secondly, evapotranspiration by forests should be limited by either decreasing the planted area or afforestation with less transpiring species.

New wet dune slacks can be created in existing dune areas by allowing new slacks to be blown out by wind, thus omitting sand-dune reclamation. In the long term, the political choice for a dynamic but mainland-preserving coastline is of great importance for the dune areas. In this way the strong wave action in front of the primary sea barrier on the newly sedimented areas allows the initiation of a relatively rare, virgin landscape. Intensive sand supplementation at the seaside of the dykes for maintenance of the present coastline, causes a relative rise in groundwater level in existing dune areas and thus establishes better conditions for wet dune slacks, and the formation of new, primary dunes (Van Dijk & Grootjans, 1993).

Perspectives for ecological research in The Netherlands wetlands

The very existence of the Dutch wetlands is at stake. The sustainability of wetlands should therefore be the prime ecological research priority. Such a research programme would focus on understanding the underlying ecological processes in natural and man-dominated wetland systems to prescribe conservation, rehabilitation and management strategies that would enhance the sustainability of these systems.

Within this framework the following topics deserve special attention:

- Within the specific wetland types

1. Studies at the ecosystem level of ecosystem parameters, of which natural oscillations and trends in time, and on which the impact of disturbances are quantified (e.g. Carpenter *et al.*, 1989; Jassby & Powell, 1990). These studies would require the (long-term) monitoring of ecosystems, of which the biotic parameters would be the same but the prevalent conditioning (abiotic) factors may vary with wetland type (see: characteristics of Dutch wetland types). Particularly these studies, in which often simulation models are used as tools for interpretation, can provide the basis for extrapolations in space and time.
2. Studies on adaptation capacity and mechanisms of (groups of) species to extreme environmental conditions. These conditions should represent the current threats to the Dutch wetlands, being any change in water characteristics (quality, quantity, current velocity, level etc.) and acidification.
3. Studies on (mutual) relationships between plants, animals and microorganisms (e.g. competition, grazing and mineralization) in recently disturbed and natural wetland communities, to enable recognition and prediction of short- and long-term response patterns to disturbance.

- Between the specific wetland types

Dispersion studies to assess the potential for survival of species in relatively small wetlands, interconnected by ribbonlike landscape elements like ditches and small streams, as is typical in The Netherlands.

For the contemporary quantitative assessment of the long-term effects of climate changes, the effects of temperature rise and increase in UV-B irradiation on individual species, communities and ecosystems should also be studied. In this context studies on the impact of introduced species on local communities could provide suitable models to mimic the future community changes due to the temperature-rise-driven migration of species. Particularly in Netherlands wetlands, the invasion of new species can be rapid and considerable because of its open infrastructure maintained by waterways.

We are aware of the fact that to indicate what research would be feasible to perform does not guarantee its' implementation. It requires a concerted support and effort of ecologists and their institutions. Meeting the financial needs of a concerted wetland research programme requires significantly increased funding from both public and private sources. Public acceptance of a national wetland research programme would be greatly facilitated by convincing the general public of the uniqueness of The Netherlands as wetland, and the wetlands therein. Moreover, opportunities to obtain funding for wetland research will increase when research proposals fit into an accepted general (national or international) framework. Part of the recommendations for ecological research of wetlands fits into a larger, international framework proposed for global ecological research (Lubchenko *et al.* 1991), and links up with a proposal for land/inland water ecotone studies (Decamps *et al.*, 1990).

References

Admiraal, W., G. van der Velde, H. Smit & W.G. Cazemier, this volume. The rivers Rhine and Meuse in The Netherlands: present state and signs of recovery.

Arts, G.H.P & R.F.M. Buskens, 1989. Aanvoer van venvreemd water: een noodzaak?. In: J.G.M. Roelofs (Ed.). Aanvoer van gebiedsvreemd water: omvang en effecten op oecosystemen. Proc.Symp.21 december 1988, Fac. Natuurwetenschappen, Katholieke Univ.Nijmegen: 100–109 (In Dutch).

Bakker, H.J., J. Dronkers & P. Vellinga, 1991. Veranderend wereldklimaat. Nederlandse visies, reacties en commentaren. Delwel, 's Gravenhage. 120 pp. (In Dutch).

Bakker, J.P. & A.P. Grootjans, 1991. Potential for vegetation regeneration in the middle course of the Drentsche A brook valley (The Netherlands). Verh.Gesellschaft fur Okologie (Freising-Weihenstephan 1990) Band 20: 249–263.

Bakker, J.P., J. de Leeuw, K.S. Dijkema, P. Leendertse,

H.H.T. Prins & J. Rozema, 1993. Salt marshes along the coast of The Netherlands. Hydrobiologia 265: 73–95.

Barendregt, A., J.T. de Smidt & M.J. Wassen, 1985. Relaties tussen milieufaktoren en water- en moerasplanten in de Vechtstreek en de omgeving van Groet. Rapport Interfak. Vakgr.Milieukunde, Univ.Utrecht. 47 pp. (In Dutch).

Bosatlas, 1976. F. J. Ormeling (ed.). Wolters-Noordhoff, Groningen. 195 pp. (In Dutch).

Care for Tommorrow, 1989. Zorgen voor Morgen. Nationale milieuverkenning 1985–2010. Samsom H.D.Tjeenk Willink, Alphen aan de Rijn. 456 pp. (In Dutch).

Carpenter, S.C., T.M. Frost, D. Heisy & T.K. Kratz, 1989. Randomized intervention analysis and the interpretation of whole ecosystem experiments. Ecol. 70: 1142–1152.

Casparie, W.A., 1993. The Bourtanger Moor: endurance and vulnerability of a raised bog system. Hydrobiologia 265: 203–215.

Colijn, F., R. Uyterlinde & H.W. Kroes, this volume. Wetlands policy of the Dutch government: failure or success?.

CUWVO, 1988. Coordinatie Commissie Uitvoering Wet Verontreiniging oppervlaktewateren, Werkgroep V-1. Ecologische normdoelstellingen voor Nederlandse oppervlaktewateren. 154 pp. (In Dutch).

De Bakker, H. & W. P. Locher, 1991. Bodemkunde van Nederland. Deel 1. Algemene bodemkunde. Malmberg, Den Bosch (In Dutch).

Decamps, H., F. Fournier, R.J. Naiman & R.C. Petersen Jr., 1990. An international research effort on land/inland water ecotones in landscape management and restoration 1990–1996. Ambio 19: 175–176.

De Haan, H., L. van Liere, Sj.P. Klapwijk & E. van Donk, 1993. Structure and function of fen lakes in relation to water table management in The Netherlands. Hydrobiologia 265: 155–177.

De Jonge, V.N., K. Essink & R. Boddeke, 1993. The Dutch Wadden Sea: a changed ecosystem. Hydrobiologia 265: 45–71.

Dijkema, K. S., J.H. Bossinade, P. Bouwsema, 1990. Salt marshes in the Netherlands Wadden Sea: rising high-tide levels and accretion enhancement. In: J.J. Beukema, W.J. Wolff & J.J.W.M. Brouns (eds), Expected effects of climatic change on marine coastal ecosystems. Kluwer Academic Publishers, Dordrecht: 173–188.

Grootjans, A. P., 1985. Changes of groundwater regimes in wet meadows. PhD.Thesis, Univ.of Groningen.

Hekstra, G.P. & F.J.M. van Linden (eds), 1991. Flora en fauna chemisch onder druk. Verslag van een nationaal symposium georganiseerd door de Nederlandse Ecologenvereniging - Oecologische Kring, Arnhem, 9–10 December 1990. Pudoc, Wageningen. 376 pp. (In Dutch).

Higler, L.W.G. & P.F.M. Verdonschot, 1993. Stream valleys as wetlands. Hydrobiologia 265: 265–279.

IKSR, 1987. Aktionsprogramm 'Rhein'. Internationale Kommission zum Schutze des Rheins gegen Verunreinigung, Koblenz, F.R.G. 18 pp.

Intergroep Verdroging, 1989. Verdroging van natuur en landschap in Nederland. Beschrijving en analyse. In opdracht van het ministerie van Verkeer en Waterstaat, het ministerie van Landbouw en Visserij, en het ministerie van Volkshuisvesting, Ruimtelijke Ordening en Milieubeheer. 67 pp. (In Dutch).

Jassby, A.D. & T.M. Powell, 1990. Detecting changes in ecological time series. Ecol.71: 2044–2052.

Joenje, W. & B. Verhoeven, 1993. Wetlands of recent Dutch embankments. Hydrobiologia 265: 179–193.

Kramer, K.J.M., 1990. Effects of increased solar UV-B radiation on coastal marine ecosystems: an overview. In: J.J.Beukema, W.J.Wolff & J.J.M.Brouns (eds), Expected effects of climatic change on marine coastal ecosystems. Kluwer Academic Publishers, Dordrecht: 195–210.

Lubchenko, J., A.M. Olson, L.B. Brubaker, S.R. Carpenter, M.M. Holland, S.P. Hubbell, S.A. Levin, J.A. MacMahon, P.A. Matson, J.M. Melillo, H.A. Mooney, C.H. Peterson, H.R. Pulliam, L.A. Real, P.J. Regal, P.G. Risser ,1991. The sustainable biosphere initiative: an ecological research agenda. Ecol.72: 371–412.

Maltby, E., 1988. Global wetlands – history, current status and future. In: D.D. Hook, W.H. McKee, H.K. Smith *et al.* (eds), The Ecology and management of wetlands. Vol.1. Ecology of wetlands. Croom Helm, London & Sydney: 3–14.

Maltby, E. & R.E. Turner, 1983. Wetlands are not wastelands. Geogr. Mag. 55: 346–351.

Nature Policy Plan, 1989. Natuurbeleidsplan. SDU, 's Gravenhage. 179 pp. (In Dutch).

Nienhuis, P.H., 1993. Nutrient cycling and foodwebs in Dutch estuaries. Hydrobiologia 265: 15–44.

Saeijs, H.L.F., 1991. Integrated water management: a new concept. From treating of symptoms towards a controlled ecosystem management in the Dutch Delta. Landscape & Urban Planning 20: 245–255.

Schouwenaars, J.M., 1993. Hydrological differences between bogs and bog-relicts and consequences for bog restoration. Hydrobiologia 265: 217–224.

Smit, H. & Van Urk, G., 1987. Het herstel van de ecologische waarden van de Rijn: over de zalm en ecologische doelstellingen. H2O 20: 427–430.

Ten Brink, B.J.E. & F. Colijn, 1990. Ecologische ontwikkelingsrichtingen zoute wateren. RWS/DGW, Nota GW-90 (In Dutch).

Third Water Policy Plan, 1989. Derde Nota Waterhuishouding. SDU, 's Gravenhage. 297 pp. (In Dutch).

Udo de Haes, H.A., 1983. Infiltratie van Rijnwater in hogere zandgebieden. In: G.P.Hekstra & W.Joenje (eds). Rijnwater in Nederland. Oecologische consequenties in het stroomgebied van de vervuilde Rijn. Oecologische Kring: 81–101 (In Dutch).

Tweede Kamer, 1990–1991. Vergaderstukken 1990–1991, 21 8789, no 1–2: 54.

Van Dam, H. & R.F.M. Buskens, 1993. Ecology and management of moorland pools: balancing acidification and eutrophication. Hydrobiologia 265: 225–263.

Van Dijk, H.W.J. & A.P. Grootjans, 1993. Wet dune slacks: decline and new opportunities. Hydrobiologia 265: 281–304.

Van Wirdum, G., 1993. An ecosystems approach to base-rich freshwater wetlands, with special reference to fenlands. Hydrobiologia 265: 129–153.

Wolff, W.J., 1993. Netherlands-Wetlands. Hydrobiologia 265: 1–14.

Van Wirdum, G., A.J. den Held & M. Schmitz, 1992. Terrestrializing fen vegetation in former turbaries in The Netherlands. In: J.T.A. Verhoeven (ed.). Fens and bogs in The Netherlands: vegetation, history, nutrient dynamics and conservation. Kluwer Academic Publishers, Dordrecht: 331–370.

Hydrobiologia **265**: 321–328, 1993.
E. P. H. Best & J. P. Bakker (eds), Netherlands-Wetlands.

Wetlands policy of the Dutch government: failure or success?

F. Colijn [1], R. Uyterlinde [2] & H.W. Kroes [3]
[1] *Rijkswaterstaat/Tidal Waters Division, P.O. Box 20907, 2500 EX The Hague, The Netherlands;*
[2] *NBLF/Ministry of Agriculture, Nature Management and Fisheries, P.O. Box 20401, 2500 EK The Hague, The Netherlands;* [3] *DGM/VROM, P.O. Box 450, 2260 MB Leidschendam, The Netherlands*

Key words: wetlands policy, ecological objectives, The Netherlands

Abstract

An overview is presented of Dutch wetlands and the perspectives and threats to which they are exposed. The general environmental problems of The Netherlands are reflected in the problems with the preservation of wetlands: deterioriation of water quality is one of the main problems, directly followed by the desiccation of former wet areas. Also splitting up of larger coherent areas into smaller fragments due to infrastructural works does not favour long time existence of complete (eco)systems.

The policy of the Dutch government and its different ministries towards the protection of wetlands will be described.

The policy to protect the Wadden Sea and which is based on concepts of wise use and sustainability will be presented as an example of Dutch integral environmental policy for wetlands particularly. Our conclusion is that wetlands policy is tardy but successful.

Introduction

The wetlands policy of The Netherlands central government is primarily the responsibility of the Minister of Agriculture, Nature Management and Fisheries. Wetlands of international significance are designated in accordance with the 1985 list of the former Research Institute for Nature Management (RIN) using criteria developed within the framework of the Wetlands Convention drawn up at Ramsar in 1971, and revised at meetings at Heiligenhafen, Groningen and Regina.

Wetlands are defined as "marshes, bogs, lake areas, of natural or artificial origin, permanent or temporary, with stagnant or running water, fresh, brackish or salt water, including marine waters, with a depth at lowtide of less than six meters."

A recent overview of The Netherlands as a wetland and of wetlands in The Netherlands is given by Rooth (1989), Best *et al.* (1993) and Wolff (1993); a total of 103 areas varying in size from 225,000 hectares (the Wadden Sea) to no more than 10 hectares (Springendal in the province of Overijssel) are listed according to the revised criteria in Regina in 1987.

Located in the delta of Scheldt, Meuse, Rhine and Ems the pristine Netherlands can be conceived as one large wetland. However since ages the Dutch have been changing and protecting their country with dikes against flooding. In modern times large infrastructural works (*e.g.* the Delta plan) have caused a large reduction in former extensive areas of brackish waters as the famous Biesbosch area, due to damming of the Haringvliet and Grevelingen, and the IJsselmeer, a former part of the Wadden Sea. Other important wetland areas are the fens of Holland and NW Overijssel and the bogs in several provinces. These areas are treated in much more detail in other chapters. The wetlands with a current protected status based on the Wetlands Convention are listed in the next chapter. Potential wetlands with such a status can be found along the North Sea coast (the Voordelta area), some of the former estuaries in the southwestern part of The Netherlands (Oosterschelde, Grevelingen, Haringvliet) and along the main rivers and the IJsselmeer.

In principle, the Ministry of Agriculture, Nature Management and Fisheries (LNV) would be solely responsible for implementing wetlands management, were it not for the fact that 80–90% (in terms of surface area) of the areas meeting the criteria for wetlands of international importance are also designated as national waters. Since these waters come under the jurisdiction of the Department of Public Works (Rijkswaterstaat) it is not surprising that Rijkswaterstaat too makes a contribution to wetlands management. The Ministry of Housing, Physical Planning and the Environment (VROM) has no specific competence with regard to wetlands management. However, in view of this Ministry's mandate in the field of general and specific standards for emissions into the whole range of national and provincial waters and the reduction of atmospheric pollution, it has a clear involvement in maintaining conditions which will permit the existence of wetlands on the long term.

In conclusion, it is a collective responsibility of the Ministries to accomplish the maintenance of a wetlands policy, not frustrated by solistic actions of one of the partners.

Assignment of wetlands

In The Netherlands the nomination for registration of wetlands is closely linked to the protection of the areas concerned. This means that beforehand it must be clear that the ecological values of the area can be maintained or developed into the desired direction. The need for a sufficient scientific and political basis is evident. Hereafter the procedure for nomination is started.

The scientific criteria for identifying wetlands of international importance were defined in 1980 in Cagliary. These criteria are:

1. Quantitative criteria for identifying wetlands of importance to waterfowl. A wetland is considered internationally important if it:

(a) regularly supports either 10,000 ducks, geese and swans; or 10,000 coots; or 20,000 waders, or:
(b) regularly supports 1% of the individuals in a population of one species or subspecies of waterfowl, or:
(c) regularly supports 1% of the breeding pairs in a population of one species or subspecies of waterfowl.

2. General criteria for identifying wetlands of importance to plants or animals. A wetland is considered internationally important if it:

(a) supports an appreciable number of rare, vulnerable or endangered species or subspecies of plants or animals, or:
(b) is of social value for maintaining the genetic and ecological diversity of a region because of the quality and peculiarities of its flora and fauna, or:
(c) is of special value as the habitat of plants or animals at a critical stage in their biological cycles, or:
(d) is of special value for its endemic plant or animal species or communities.

3. Criteria for assessing the value of representative or unique wetlands. A wetland is considered to be internationally important if it is a particularly good example of a specific type of wetland characteristic of its region. A sufficient policy basis means protection by:

(a) political agreement upon the headlines of the designation and management of the area, or:
(b) safeguarding of the ecological importance in physical planning, or:
(c) designation under the Nature protection act and/or ownership by a (semi-) governmental institution.

The link of the nomination of wetlands to nature protection implies that the use of the areas is not only based upon the maintenance of the potential values of the area, but also upon the maintenance and development of the actual values. This is in fact a more specified interpretation of 'wise use' than defined by the parties of the Ramsar convention in Regina in 1987 (cf. Anonymous, 1985). After the scientific and policy basis has been considered to be sufficient, the Minister of Agriculture, Nature Management and Fisheries starts the procedure for nomination. A motivated proposal is being discussed on interministerial level, after the provinces and municipalities have been given the opportunity to react to this proposal. The same procedure is followed in changing the boarders of a registered wetland.

On a scientific basis, in 1985 103 areas were selected to be nominated for registration as a wetland of international importance. Until now 11 wetlands have in fact been nominated for registration. They differ in characteristics and size ranging from about 200 ha (Zwanewater) until 225,000 ha (Wadden Sea).

Threats to wetlands

The heavy burden on the environment in The Netherlands is caused by the high number of inhabitants and the high industrial activity in the country and in the neighbours (Germany and Belgium). Altogether these human activities cause water pollution, both from industry, households and agriculture and cattle-raising. Also direct human use as *e.g.* recreation causes disturbance of natural areas. Some of the main threats for wetlands will be illustrated in more detail.

Despite the large amount of rivers, canals, ditches and lakes and the low level of the country the freshwater supply is rather limited: in dry summers groundwater is used on a large scale. This has caused a drop in groundwater level (desiccation); in a large number of places this has dropped more than 50 cm as a result of water level management measures associated with land consolidation, groundwater extraction for the supply of water to households and industry and agricultural irrigation. The degree of desiccation in biotopes which derive their value from a humid environment is shown in the summary of the Third Water Policy Plan, a time for action (1987).

Another problem affecting the inland waters in The Netherlands is the overall influence of the water quality of the main rivers (Rhine, Meuse). During warm summers with high evaporation rates river water is used to keep a specific level in the Dutch water system. This can only be reached by transport of Rhine water to areas with a lack of water. This however often conflicts with the natural characteristic of these waters being less eutrophic and polluted, and with a distinct ionic composition from the original waters.

In a small country as The Netherlands splitting up of the country into smaller and smaller areas by canals, roads and railways and increasing urbanization, makes it more and more difficult to protect and conserve large intact areas. Loss of biotopes also is one of the main threats of wetlands like for other nature areas. The Dutch nature policy is aimed at a sustainable preservation, restoration and development of nature and scenic values. Since 1990 first priority has been given to what is called the 'ecological main structure' of The Netherlands. This network also covers the areas which have sufficient basis for wetland nomination. Besides the protected areas themselves special attention is given to the ecological relations between them.

This results in an ecological main structure consisting of core areas, buffer areas, corridors and stepping stones. Partly these areas already exist, partly they are to be restored or newly created on suitable locations, occasionally in compensation for lost biotopes.

Atmospheric deposition including acidification is still a major threat to all ecosystems, particularly wetlands in The Netherlands. High amounts of nitrogen containing deposition from cattle raising affects peat-bogs and fens, leading to acidification and eutrophication of former oligotrophic fens.

Disturbance by intensive human use as a consequence of the high populationnumber can be observed everywhere: economic activities such as fisheries in coastal areas and the Wadden Sea conflict with natural values (see last section), oil and gas exploitation may cause great hazards when calamities would occur, disturbance by military activities in the Wadden Sea are obvious but their ecological effects are difficult to quantify. Recreational use of the coastline and beaches, but also in other natural reserves causes disturbance of birds and mammals. Again their effects are difficult to quantify.

In conclusion, the exact nature of the adverse effects of these factors on the sustainability of the wetlands is not precisely known and needs further investigation.

Principles of wetlands policy

Several important developments for wetlands policy have been taken place in recent years. The Ministry of Agriculture, Nature Management and Fisheries published a memorandum devoted to wetlands stating the criteria for nomination (see above) (Anonymous, 1985).

Rijkswaterstaat published a memorandum called 'Living with water' (1985) in which it sets out its philosophy on an integrated water system policy. New policy lines include devoting attention to the relationship between quantity and quality in watersystems. The policy document on Nature Development produced by the Ministry of Agriculture, Nature Management and Fisheries (Anonymous, 1989) sets out a new approach to our natural heritage and the ways of developing it. It is inspired on a new philosophy not merely based on conservation of natural areas but much more on a natural development of large areas such as the Oostvaardersplassen, the river forelands and the Voordelta area.

Following the appearance of 'Care for tomorrow' (1988) published by the National Institute for Public Health and the Environment (RIVM), three new policy documents were published in rapid succession in 1989: the national Environmental Policy Plan (NMP), the national Nature Policy Plan (NBP) and the Third Water Policy Plan (NW3). The three policy documents all have ramifications for water and environmental policy and are therefore of significance for the prior conditions for wetlands policy.

These three documents have in common the point of departure that it is desirable to create a policy on sustainable development and use, along the lines set out in the Brundtland Commission's report and incorporated into Dutch governmental policy. An important aspect of this policy is the reduction of pollution on all fronts at both local and global level. Also within the policy of the three ministries the ecosystem approach plays a major role. According to this approach a wetland is apprehended specifically as an ecosystem, the emphasis being on the fact it is one system with intricate relationships between the different trophic levels, and it consists of open, running or stagnant waters, banks and beds. Besides sustainable use, wise use also is adopted as a policy concept. In the declaration of the Trilateral Wadden Sea Conference in Esbjerg (Anonymous, 1991) conservation and wise use are taken as the basic concepts for protection. The definition of wise use links sustainable use and natural properties of ecosystems: "The wise use of wetlands is their sustainable utilisation for the benefit of humankind in a way compatible with the maintenance of the natural properties of the ecosystem."

Problem analyses in the above mentioned documents have shown the main threats of wetlands, but also that reduction of contaminants alone does not give a final solution to pollution problems. A multitrack approach combining reduction of contaminants and organization is needed to improve natural values. This has given rise to integral projects in which *e.g.* reduction of nutrients and other additional measures have been successfully combined. (See also next section.)

Perspectives

In The Netherlands a policy has been adopted to tackle environmental problems in an integrated way: insertion on the list of wetlands of international importance will take place only when administrative consensus has been attained in the region in question and when the integral weighing of interests has resulted in a prime function 'nature' for the area.

This, however, due to the lengthy discussion, leads to a tardy policy in assigning wetlands with protective status. Because of conflicting economic and nature objectives (see also Best *et al.*, 1993), *e.g.* of the main rivers which have a prime function of shipping routes, it is impossible to list these rivers or parts of them as wetlands.

Nevertheless, concepts of sustainable and wise

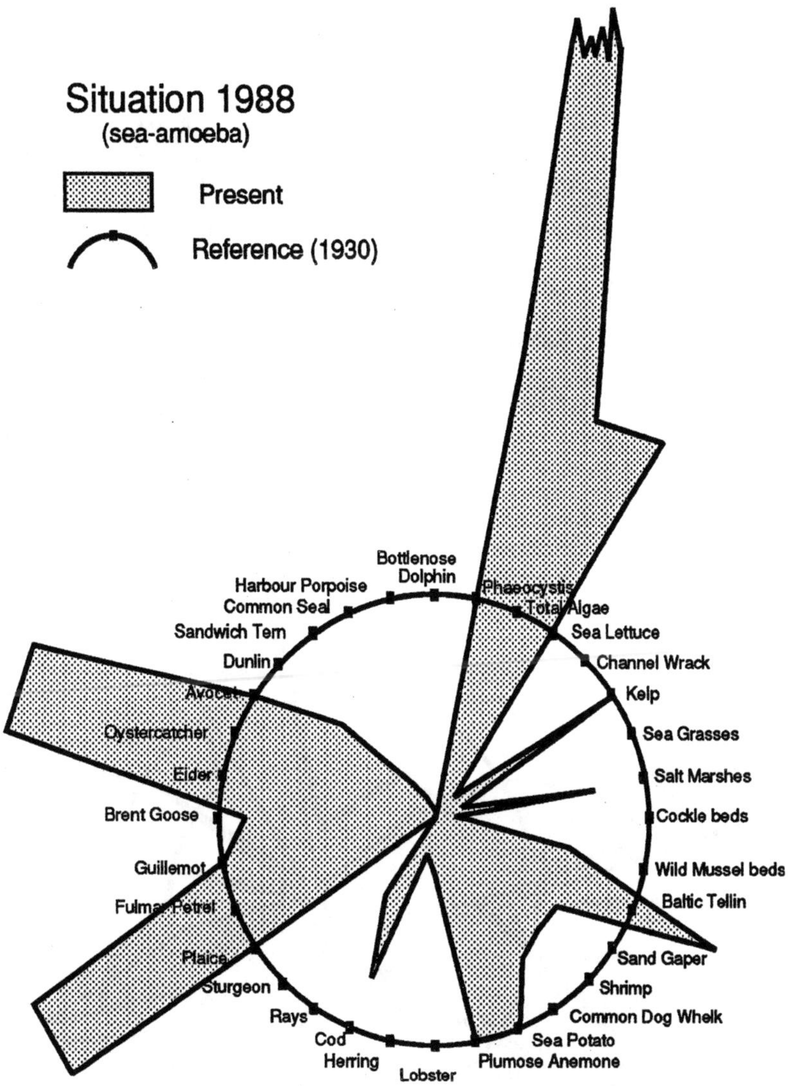

Fig. 1. AMOEBA for the sea in 1988. For all species or entities shown the reference values around 1930 are represented on the circle (= 100 %). The present values are given by connecting the values to the grey area, the amoeba-like figure. Examples are the common seal with a reference value of 6,000 individuals and a present value of 600 individuals, and sea grasses with a reference value of 16,350 ha and a present value of 2,970 ha. All values are given in ten Brink & Colijn (1990). It can be clearly seen that under the influence of eutrophication, phytoplankton has shown a strong increase, that sea grasses and certain brown algae have been substantially reduced, that some fish have disappeared entirely, that some bird populations have increased, and that marine mammals have largely disappeared owing to the presence of PCB's.

use have brought us much further in the right direction: these policies are incorporated into the forementioned documents and worked out in such a way that they enable a concrete implementation of management of wetlands. Although up to now these policies look top-down, we have some very good examples of bottom-up management in the so-called integral water management projects.

Of course the measures differ per area. In the smaller and isolated wetlands for instance considerable attention is paid to the water control, including the water quality. This sometimes results in projects to improve the situation drastically by an integrated package of measures customised for the regional circumstances. Good results are obtained from projects like Botshol and Nieuwkoopse Plassen in the reduction of eutrophication. This generally includes the reduc-

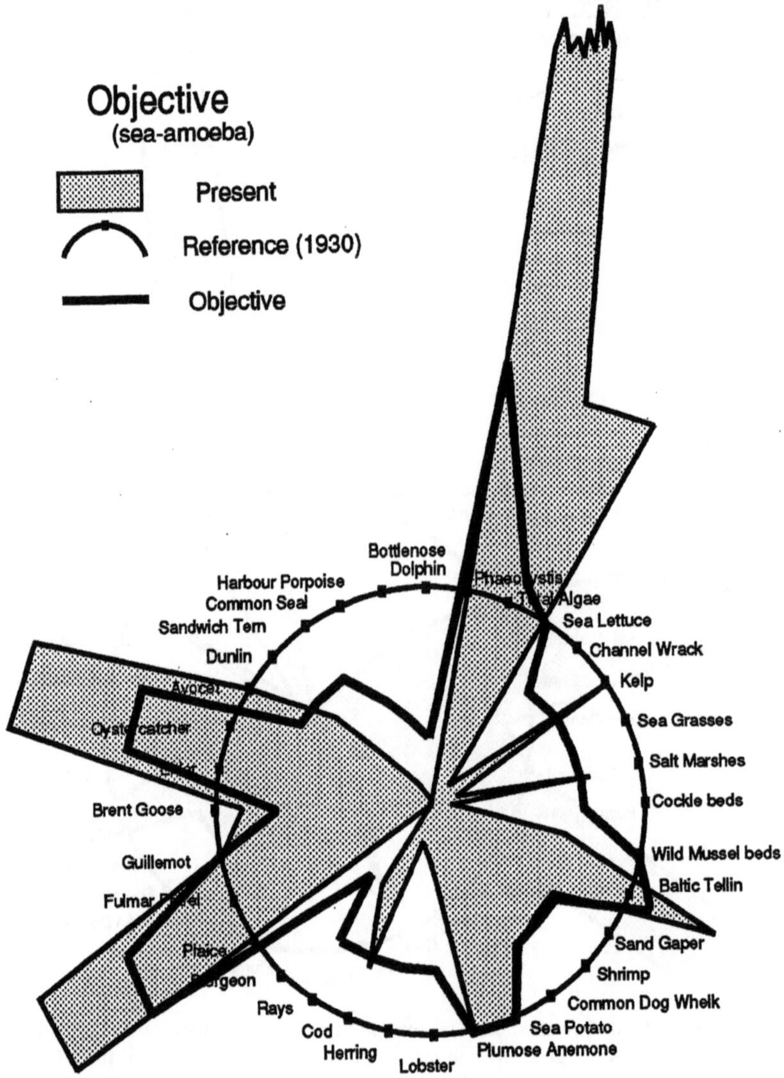

Fig. 2. AMOEBA for the sea including the objective (thick line). To set targets within the national Third Water Policy Plan ecological objectives were established. It was assumed that sustainable use was garanteed when these objectives were set between 75% and 200% of the reference values. To achieve these values a package of measures is needed with a combination of pollution reduction and structural measures.

tion of influx of nutrient-rich waters but also the fixation of phosphates in the bottom by injection with ironsulphate. Other promising results were obtained by dredging of the phosphate-rich sediments, or the biological manipulation of the fish populations.

Another example is the policy for the Wadden Sea, where international cooperation is needed to reach the common objectives.

For the coastal Wadden Sea extending eastward to Germany and Denmark a common trilateral approach on the wise-use concept is being developed on the basis of wetland-registrations in the three countries. This is the first attempt to define wise use for a transboundary wetland in the world.

In the Dutch coastal waters which are characterised by a high biological production, special attention is paid to the intensive fisheries. The aim is to combine interests of nature and fisheries in the whole area. In case when this is not possible activities will be size or area limited. This strategy will result in the closing of considerable areas for fisheries on mussels and cockles and may give characteristic biocoenoses, that have almost disappeared nowadays, chances to recover.

The efficacy of policy and management necessitates monitoring of the objectives and goals or targets. A good example is the Dutch Amoebe approach (Ten Brink *et al.*, 1991; Ten Brink & Colijn, 1990) which combines reference values, present values and predictions of numbers of a specific number of animal and plant species (Figs 1 and 2). According to their paper reference values guarantee fundamental values for mankind such as sustainable production and harvest, diversity of species and self-regulation. Comparable approaches have been developed within Anonymous (1991) for the protection of the Wadden Sea. A policy based on protection of areas by using species numbers and process parameters is most suitable to be evaluated. However the number of parameters should be kept to a minimum due to the enormous costs needed to describe and analyse samples statistically significant for trend analysis.

Conclusions

The central government has started out on the right track for the preservation of our wetlands. Although there is some division of responsibilities among the various authorities, we hope that 'united we stand' will be our motto. More and far-reaching agreements between ministries concerned and local authorities in such overlapping policy areas are needed.

More concrete ecological objectives are an important tool in the recovery of nature values. This, together with the Nature Policy Plan, restoring an ecological main structure in The Netherlands and connecting the areas with natural values, is an important step towards the preservation of wetlands in The Netherlands. However, this is not a simple process in a changing and intricate society. Therefore, the process is tardy, but we think successful.

References

Anonymous, 1985. Brief van de Minister van Landbouw en Visserij, Memorandum on Wetlands Convention, Tweede Kamer, 1985–1986, 17990, 24: 1–15.

Anonymous, 1989. Natuurontwikkeling: een verkennende studie. Ministerie Van Landbouw en Visserij, Directie Natuur, Milieu en Faunabeheer, SDU, The Hague, 102 pp.

Anonymous, 1991. Wise use and conservation of the Wadden Sea, Common Wadden Sea Secretariat, 50 p.

Best, E.P.H., J.T.A. Verhoeven & W.J.Wolff, 1993. The ecology of The Netherlands wetlands: characteristics threats, prospects and perspectives for ecological research. Hydrobiologia 265: 305–319.

Care for Tomorrow. Zorgen voor Morgen. Nationale Milieuverkenning 1985–2010. Samsom H.D. Tjeenk Willink, Alphen aan de Rijn, 456 pp.

Environmental Policy Plan. Nationaal Milieubeleidsplan. SDU, The Hague, 258 pp.

Living with Water, 1985. Omgaan met Water, Ministerie van Verkeer en Waterstaat, The Hague.

Nature Policy Plan, 1989. Natuurbeleidsplan. SDU, 's-Gravenhage, 179 pp.

Rooth, J., 1989. Lijst van internationaal belangrijke wetlands in Nederland. In A.L. Spaans (ed.), Wetlands en watervogels. PUDOC, Wageningen.

Ten Brink, B.J.E., S.H. Hosper & F. Colijn, 1991. A quantitative method for the description and assessment of ecosystems: the AMOEBA approach. Mar. Poll. Bull. 23: 265–270.

Ten Brink, B.J.E. & F. Colijn (eds), 1990. Ecologische ontwikkelingsrichtingen Zoute Wateren. Ministerie van Verkeer en Waterstaat, Rijkswaterstaat, Dienst Getijdewateren, GWWS-90.009, 289 pp.

Third Water Policy Plan, 1989. Derde Nota Waterhuishouding. SDU, The Hague, 297 pp.

Water in the Netherlands: a time for action, 1989. Summary of the Third Water Policy Plan. Ministry of Transport and Public Works, Information Department, The Hague.

Wolff, W.J., 1993. Netherlands-Wetlands. Hydrobiologia 265: 1–14.

FSC
www.fsc.org
MIX
Papier aus verantwortungsvollen Quellen
Paper from responsible sources
FSC® C105338

If you have any concerns about our products,
you can contact us on
ProductSafety@springernature.com

In case Publisher is established outside the EU,
the EU authorized representative is:
Springer Nature Customer Service Center GmbH
Europaplatz 3, 69115 Heidelberg, Germany

Printed by Libri Plureos GmbH
in Hamburg, Germany